Complex-Valued Matrix Derivatives

In this complete introduction to the theory of finding derivatives of scalar-, vector-, and matrix-valued functions in relation to complex matrix variables, Hjørungnes describes an essential set of mathematical tools for solving research problems where unknown parameters are contained in complex-valued matrices. Self-contained and easy to follow, this singular reference uses numerous practical examples from signal processing and communications to demonstrate how these tools can be used to analyze and optimize the performance of engineering systems. This is the first book on complex-valued matrix derivatives from an engineering perspective. It covers both unpatterned and patterned matrices, uses the latest research examples to illustrate concepts, and includes applications in a range of areas, such as wireless communications, control theory, adaptive filtering, resource management, and digital signal processing. The book includes eighty-one end-of-chapter exercises and a complete solutions manual (available on the Web).

Are Hjørungnes is a Professor in the Faculty of Mathematics and Natural Sciences at the University of Oslo, Norway. He is an Editor of the *IEEE Transactions on Wireless Communications*, and has served as a Guest Editor of the *IEEE Journal of Selected Topics in Signal Processing* and the *IEEE Journal on Selected Areas in Communications.*

This book addresses the problem of complex-valued derivatives in a wide range of contexts. The mathematical presentation is rigorous but its structured and comprehensive presentation makes the information easily accessible. Clearly, it is an invaluable reference to researchers, professionals and students dealing with functions of complex-valued matrices that arise frequently in many different areas. Throughout the book the examples and exercises help the reader learn how to apply the results presented in the propositions, lemmas and theorems. In conclusion, this book provides a well organized, easy to read, authoritative and unique presentation that everyone looking to exploit complex functions should have available in their own shelves and libraries.

Professor Paulo S. R. Diniz, Federal University of Rio de Janeiro

Complex vector and matrix optimization problems are often encountered by researchers in the electrical engineering fields and much beyond. Their solution, which can sometimes be reached from using existing standard algebra literature, may however be a time consuming and sometimes difficult process. This is particularly so when complicated cost function and constraint expressions arise. This book brings together several mathematical theories in a novel manner to offer a beautifully unified and systematic methodology for approaching such problems. It will no doubt be a great companion to many researchers and engineers alike.

Professor David Gesbert, EURECOM, Sophia-Antipolis, France

Complex-Valued Matrix Derivatives

With Applications in Signal Processing and Communications

ARE HJØRUNGNES
University of Oslo, Norway

CAMBRIDGE
UNIVERSITY PRESS

CAMBRIDGE
UNIVERSITY PRESS

University Printing House, Cambridge CB2 8BS, United Kingdom

One Liberty Plaza, 20th Floor, New York, NY 10006, USA

477 Williamstown Road, Port Melbourne, VIC 3207, Australia

314-321, 3rd Floor, Plot 3, Splendor Forum, Jasola District Centre, New Delhi - 110025, India

103 Penang Road, #05-06/07, Visioncrest Commercial, Singapore 238467

Cambridge University Press is part of the University of Cambridge.

It furthers the University's mission by disseminating knowledge in the pursuit of education, learning and research at the highest international levels of excellence.

www.cambridge.org
Information on this title: www.cambridge.org/9780521192644

First published 2011

A catalogue record for this publication is available from the British Library

Library of Congress Cataloging in Publication data
Hjørungnes, Are.
Complex-Valued Matrix Derivatives : With Applications in Signal Processing and Communications / Are Hjørungnes.
p. cm.
Includes bibliographical references and index.
ISBN 978-0-521-19264-4 (hardback)
1. Matrix derivatives. 2. Systems engineering. 3. Signal processing – Mathematical models. 4. Telecommunication – Mathematical models. I. Title.
TA347.D4H56 2011
621.382′2 – dc22 2010046598

ISBN 978-0-521-19264-4 Hardback

Additional resources for this publication at www.cambridge.org/hjorungnes

To my parents, Tove and Odd

Contents

Preface

This book is written as an engineering-oriented mathematics book. It introduces the field involved in finding derivatives of complex-valued functions with respect to complex-valued matrices, in which the output of the function may be a scalar, a vector, or a matrix. The theory of complex-valued matrix derivatives, collected in this book, will benefit researchers and engineers working in fields such as signal processing and communications. Theories for finding complex-valued derivatives with respect to both complex-valued matrices with independent components and matrices that have certain dependencies among the components are developed and illustrative examples that show how to find such derivatives are presented. Key results are summarized in tables. Through several research-related examples, it will be shown how complex-valued matrix derivatives can be used as a tool to solve research problems in the fields of signal processing and communications.

This book is suitable for M.S. and Ph.D. students, researchers, engineers, and professors working in signal processing, communications, and other fields in which the unknown variables of a problem can be expressed as complex-valued matrices. The goal of the book is to present the tools of complex-valued matrix derivatives such that the reader is able to use these theories to solve open research problems in his or her own field. Depending on the nature of the problem, the components inside the unknown matrix might be independent, or certain interrelations might exist among the components. Matrices with independent components are called *unpatterned* and, if functional dependencies exist among the elements, the matrix is called *patterned* or *structured*. Derivatives relating to complex matrices with independent components are called *complex-valued matrix derivatives*; derivatives relating to matrices that belong to sets that may contain certain structures are called *generalized complex-valued matrix derivatives*. Researchers and engineers can use the theories presented in this book to optimize systems that contain complex-valued matrices. The theories in this book can be used as tools for solving problems, with the aim of minimizing or maximizing real-valued objective functions with respect to complex-valued matrices. People who work in research and development for future signal processing and communication systems can benefit from this book because they can use the presented material to optimize their complex-valued design parameters.

Book Overview

This book contains seven chapters. Chapter 1 gives a short introduction to the book. Mathematical background material needed throughout the book is presented in Chapter 2. Complex differentials and the definition of complex-valued derivatives are provided in Chapter 3, and, in addition, several important theorems are proved. Chapter 4 uses many examples to show the reader how complex-valued derivatives can be found for nine types of functions, depending on function output (scalar, vector, or matrix) and input parameters (scalar, vector, or matrix). Second-order derivatives are presented in Chapter 5, which shows how to find the Hessian matrices of complex-valued scalar, vector, and matrix functions for unpatterned matrix input variables. Chapter 6 is devoted to the theory of generalized complex-valued matrix derivatives. This theory includes derivatives with respect to complex-valued matrices that belong to certain sets, such as Hermitian matrices. Chapter 7 presents several examples that show how the theory can be used as an important tool to solve research problems related to signal processing and communications. All chapters except Chapter 1 include at least 11 exercises with relevant problems taken from the chapters. A solution manual that provides complete solutions to problems in all exercises is available at www.cambridge.org/hjorungnes.

I will be very interested to hear from you, the reader, on any comments or suggestions regarding this book.

Acknowledgments

During my Ph.D. studies, I started to work in the field of complex-valued matrix derivatives. I am very grateful to my Ph.D. advisor Professor Tor A. Ramstad at the Norwegian University of Science and Technology for everything he has taught me and, in particular, for leading me onto the path to matrix derivatives. My work on matrix derivatives was intensified when I worked as a postdoctoral Research Fellow at Helsinki University of Technology and the University of Oslo. The idea of writing a book developed gradually, but actual work on it started at the beginning of 2008.

I would like to thank the people at Cambridge University Press for their help. I would especially like to thank Dr. Phil Meyler for the opportunity to publish this book with Cambridge and Sarah Finlay, Cambridge Publishing Assistant, for her help with its practical concerns during this preparation. Thanks also go to the reviewers of my book proposal for helping me improve my work.

I would like to acknowledge the financial support of the Research Council of Norway for its funding of the FRITEK project "Theoretical Foundations of Mobile Flexible Networks – THEFONE" (project number 197565/V30). The THEFONE project contains one work package on complex-valued matrix derivatives.

I am grateful to Professor Zhu Han of the University of Houston for discussing with me points on book writing and book proposals, especially during my visit to the University of Houston in December, 2008. I thank Professor Paulo S. R. Diniz of the Federal University of Rio de Janeiro for helping me with questions about book proposals and other matters relating to book writing. I am grateful to Professor David Gesbert of EURECOM and Professor Daniel P. Palomar of Hong Kong University of Science and Technology for their help in organizing some parts of this book and for their valuable feedback and suggestions during its early stages. Thanks also go to Professor Visa Koivunen of the Aalto University School of Science and Technology for encouraging me to collect material on complex-valued matrix derivatives and for their valuable comments on how to organize the material. I thank Professor Kenneth Kreutz-Delgado for interesting discussions during my visit to the University of California, San Diego, in December, 2009, and pointing out several relevant references. Dr. Per Christian Moan helped by discussing several topics in this book in an inspiring and friendly atmosphere. I am grateful to Professor Hans Brodersen and Professor John Rognes, both of the University of Oslo, for discussions related to the initial material on manifold. I also thank Professor Aleksandar Kavčić of the University of Hawai′i at Mānoa for helping arrange my sabbatical in Hawai′i from mid-July, 2010, to mid-July, 2011.

Thanks go to the postdoctoral research fellows and Ph.D. students in my research group, in addition to all the inspiring guests who visited with my group while I was writing this book. Several people have helped me find errors and improve the material. I would especially like to thank Dr. Ninoslav Marina, who has been of great help in finding typographical errors. I thank Professor Manav R. Bhatnagar of the Indian Institute of Technology Delhi; Professor Dusit Niyato of Nanyang Technological University; Dr. Xiangyun Zhou; and Dr. David K. Choi for their suggestions. In addition, thanks go to Martin Makundi, Dr. Marius Sîrbu, Dr. Timo Roman, and Dr. Traian Abrudan, who made corrections on early versions of this book.

Finally, I thank my friends and family for their support during the preparation and writing of this book.

Abbreviations

BER	bit error rate
CDMA	code division multiple access
CFO	carrier frequency offset
DFT	discrete Fourier transform
FIR	finite impulse response
i.i.d.	independent and identically distributed
LOS	line-of-sight
LTI	linear time-invariant
MIMO	multiple-input multiple-output
MLD	maximum likelihood decoding
MSE	mean square error
OFDM	orthogonal frequency-division multiplexing
OSTBC	orthogonal space-time block code
PAM	pulse amplitude modulation
PSK	phase shift keying
QAM	quadrature amplitude modulation
SER	symbol error rate
SISO	single-input single-output
SNR	signal-to-noise ratio
SVD	singular value decomposition
TDMA	time division multiple access
wrt.	with respect to

Nomenclature

$\otimes$	Kronecker product
$\odot$	Hadamard product
$\triangleq$	defined equal to
$\subseteq$	subset of
$\subset$	proper subset of
$\wedge$	logical conjunction
$\forall$	for all
$\sum$	summation
$\prod$	product
$\times$	Cartesian product
$\int$	integral
$\leq$	less than or equal to
$<$	strictly less than
$\geq$	greater than or equal to
$>$	strictly greater than
$\succeq$	$\boldsymbol{S} \succeq \mathbf{0}_{N\times N}$ means that $\boldsymbol{S}$ is positive semidefinite
∞	infinity
$\neq$	not equal to
$\mid$	such that
$\lvert\cdot\rvert$	(1) $\lvert z\rvert \geq 0$ returns the absolute value of the number $z \in \mathbb{C}$ (2) $\lvert\boldsymbol{z}\rvert \in (\mathbb{R}^+ \cup \{0\})^{N\times 1}$ returns the component-wise absolute values of the vector $\boldsymbol{z} \in \mathbb{C}^{N\times 1}$ (3) $\lvert\mathcal{A}\rvert$ returns the cardinality of the set $\mathcal{A}$
$\angle(\cdot)$	(1) $\angle z$ returns the principal value of the argument of the complex input variable z (2) $\angle\boldsymbol{z} \in (-\pi, \pi]^{N\times 1}$ returns the component-wise principal argument of the vector $\boldsymbol{z} \in \mathbb{C}^{N\times 1}$
$\sim$	is statistically distributed according to
$\mathbf{0}_{M\times N}$	$M \times N$ matrix containing only zeros
$\mathbf{1}_{M\times N}$	$M \times N$ matrix containing only ones
$(\cdot)^*$	$\boldsymbol{Z}^*$ means component-wise complex conjugation of the elements in the matrix $\boldsymbol{Z}$
$\emptyset$	empty set
$\setminus$	set difference

$(\cdot)^{-1}$	matrix inverse
$\lvert\cdot\rvert^{-1}$	if $z \in \{\mathbb{C} \setminus \{0\}\}^{N\times 1}$, then $\lvert z\rvert^{-1}$ returns a vector in $(\mathbb{R}^{+})^{N\times 1}$ with the inverse of the component-wise absolute values of z
$(\cdot)^{+}$	Moore-Penrose inverse
$(\cdot)^{\#}$	adjoint of a matrix
$\mathbb{C}$	set of complex numbers
$\mathcal{C}(\boldsymbol{A})$	column space of the matrix $\boldsymbol{A}$
$\mathcal{CN}$	complex normally distributed
$\mathcal{N}(\boldsymbol{A})$	null space of the matrix $\boldsymbol{A}$
$\mathcal{R}(\boldsymbol{A})$	row space of the matrix $\boldsymbol{A}$
$\delta_{i,j}$	Kronecker delta function with two input arguments
$\delta_{i,j,k}$	Kronecker delta function with three input arguments
$\lambda_{\max}(\cdot)$	maximum eigenvalue of the input matrix, which must be Hermitian
$\lambda_{\min}(\cdot)$	minimum eigenvalue of the input matrix, which must be Hermitian
μ	Lagrange multiplier
$\nabla_{\boldsymbol{Z}} f$	the gradient of f with respect to $\boldsymbol{Z}^*$ and $\nabla_{\boldsymbol{Z}} f \in \mathbb{C}^{N\times Q}$ when $\boldsymbol{Z} \in \mathbb{C}^{N\times Q}$
$\frac{\partial}{\partial z}$	formal derivative with respect to z given by $\frac{\partial}{\partial z} = \frac{1}{2}\left(\frac{\partial}{\partial x} - \jmath \frac{\partial}{\partial y}\right)$
$\frac{\partial}{\partial z^*}$	formal derivative with respect to z given by $\frac{\partial}{\partial z^*} = \frac{1}{2}\left(\frac{\partial}{\partial x} + \jmath \frac{\partial}{\partial y}\right)$
$\frac{\partial}{\partial \boldsymbol{Z}} f$	the gradient of f with respect to $\boldsymbol{Z} \in \mathbb{C}^{N\times Q}$ and $\frac{\partial}{\partial \boldsymbol{Z}} f \in \mathbb{C}^{N\times Q}$
$\frac{\partial}{\partial \boldsymbol{z}^T}\boldsymbol{f}(\boldsymbol{z}, \boldsymbol{z}^*)$	formal derivatives of the vector function $\boldsymbol{f} : \mathbb{C}^{N\times 1} \times \mathbb{C}^{N\times 1} \rightarrow \mathbb{C}^{M\times 1}$ with respect to the row vector $\boldsymbol{z}^T$, and $\frac{\partial}{\partial \boldsymbol{z}^T}\boldsymbol{f}(\boldsymbol{z}, \boldsymbol{z}^*) \in \mathbb{C}^{M\times N}$
$\frac{\partial}{\partial \boldsymbol{z}^H}\boldsymbol{f}(\boldsymbol{z}, \boldsymbol{z}^*)$	formal derivatives of the vector function $\boldsymbol{f} : \mathbb{C}^{N\times 1} \times \mathbb{C}^{N\times 1} \rightarrow \mathbb{C}^{M\times 1}$ with respect to the row vector $\boldsymbol{z}^H$, and $\frac{\partial}{\partial \boldsymbol{z}^H}\boldsymbol{f}(\boldsymbol{z}, \boldsymbol{z}^*) \in \mathbb{C}^{M\times N}$
π	mathematical constant, $\pi \approx 3.14159265358979323846$
a_i	i-th vector component of the vector $\boldsymbol{a}$
$a_{k,l}$	(k, l)-th element of the matrix $\boldsymbol{A}$
$\{a_0, a_1, \ldots, a_{N-1}\}$	set that contains the N elements $a_0, a_1, \ldots, a_{N-1}$
$[a_0, a_1, \ldots, a_{N-1}]$	row vector of size $1 \times N$, where the i-th elements is given by a_i
$a \cdot b,\ a \times b$	a multiplied by b
$\lVert\boldsymbol{a}\rVert$	the Euclidean norm of the vector $\boldsymbol{a} \in \mathbb{C}^{N\times 1}$, i.e., $\lVert\boldsymbol{a}\rVert = \sqrt{\boldsymbol{a}^H\boldsymbol{a}}$
$\boldsymbol{A}^{\odot k}$	the Hadamard product of $\boldsymbol{A}$ with itself k times
$\boldsymbol{A}^{-T}$	the transposed of the inverse of the invertible square matrix $\boldsymbol{A}$, i.e., $\boldsymbol{A}^{-T} = \left(\boldsymbol{A}^{-1}\right)^T$
$\boldsymbol{A}_{k,:}$	k-th row of the matrix $\boldsymbol{A}$
$\boldsymbol{A}_{:,k} = \boldsymbol{a}_k$	k-th column of the matrix $\boldsymbol{A}$

$(\boldsymbol{A})_{k,l}$	(k, l)-th component of the matrix $\boldsymbol{A}$, i.e., $(\boldsymbol{A})_{k,l} = a_{k,l}$
$\lVert\boldsymbol{A}\rVert_F$	the Frobenius norm of the matrix $\boldsymbol{A} \in \mathbb{C}^{N\times Q}$, i.e., $\lVert\boldsymbol{A}\rVert_F = \sqrt{\text{Tr}\left\{\boldsymbol{A}\boldsymbol{A}^H\right\}}$
$\mathcal{A} \times \mathcal{B}$	Cartesian product of the two sets $\mathcal{A}$ and $\mathcal{B}$, that is, $\mathcal{A} \times \mathcal{B} = \{(a, b) \mid a \in \mathcal{A}, b \in \mathcal{B}\}$
arctan	inverse tangent
argmin	minimizing argument
$c_{k,l}(\boldsymbol{Z})$	the (k, l)-th cofactor of the matrix $\boldsymbol{Z} \in \mathbb{C}^{N\times N}$
$\boldsymbol{C}(\boldsymbol{Z})$	if $\boldsymbol{Z} \in \mathbb{C}^{N\times N}$, then the matrix $\boldsymbol{C}(\boldsymbol{Z}) \in \mathbb{C}^{N\times N}$ contains the cofactors of $\boldsymbol{Z}$
d	differential operator
$\mathcal{D}_{\boldsymbol{Z}}\boldsymbol{F}$	complex-valued matrix derivative of the matrix function $\boldsymbol{F}$ with respect to the matrix variable $\boldsymbol{Z}$
$\boldsymbol{D}_N$	duplication matrix of size $N^2 \times \frac{N(N+1)}{2}$
$\det(\cdot)$	determinant of a matrix
$\dim_{\mathbb{C}}\{\cdot\}$	complex dimension of the space it is applied to
$\dim_{\mathbb{R}}\{\cdot\}$	real dimension of the space it is applied to
$\text{diag}(\cdot)$	diagonalization operator produces a diagonal matrix from a column vector
e	base of natural logarithm, $e \approx 2.71828182845904523536$
$\mathbb{E}[\cdot]$	expected value operator
$e^z = \exp(z)$	complex exponential function of the complex scalar z
$e^{\jmath\angle z}$	if $z \in \mathbb{C}^{N\times 1}$, then $e^{\jmath\angle z} \triangleq \left[e^{\jmath\angle z_0}, e^{\jmath\angle z_1}, \ldots, e^{\jmath\angle z_{N-1}}\right]^T$, where $\angle z_i \in (-\pi, \pi]$ denotes the principal value of the argument of z_i
$\exp(\boldsymbol{Z})$	complex exponential matrix function, which has a complex square matrix $\boldsymbol{Z}$ as input variable
$\boldsymbol{e}_i$	standard basis in $\mathbb{C}^{N\times 1}$
$\boldsymbol{E}_{i,j}$	$\boldsymbol{E}_{i,j} \in \mathbb{C}^{N\times N}$ is given by $\boldsymbol{E}_{i,j} = \boldsymbol{e}_i\boldsymbol{e}_j^T$
$\boldsymbol{E}_{-}$	$M_t \times (m+1)N$ row-expansion of the FIR MIMO filter $\{\boldsymbol{E}(k)\}_{k=0}^m$, where $\boldsymbol{E}(k) \in \mathbb{C}^{M_t\times N}$
$\boldsymbol{E}_{\mid}$	$(m+1)M_t \times N$ column-expansion of the FIR MIMO filter $\{\boldsymbol{E}(k)\}_{k=0}^m$, where $\boldsymbol{E}(k) \in \mathbb{C}^{M_t\times N}$
$\boldsymbol{E}_{\urcorner}^{(l)}$	$(l+1)M_t \times (m+l+1)N$ matrix, which expresses the row-diagonal expanded matrix of order l of the FIR MIMO filter $\{\boldsymbol{E}(k)\}_{k=0}^m$, where $\boldsymbol{E}(k) \in \mathbb{C}^{M_t\times N}$
$\boldsymbol{E}_{\llcorner}^{(l)}$	$(m+l+1)M_t \times (l+1)N$ matrix, which expresses the column-diagonal expanded matrix of order l of the FIR MIMO filter $\{\boldsymbol{E}(k)\}_{k=0}^m$, where $\boldsymbol{E}(k) \in \mathbb{C}^{M_t\times N}$
f	complex-valued scalar function
$\boldsymbol{f}$	complex-valued vector function
$\boldsymbol{F}$	complex-valued matrix function
$\boldsymbol{F}_N$	$N \times N$ inverse DFT matrix
$f : X \to Y$	f is a function with domain X and range Y

$(\cdot)^H$	$\boldsymbol{A}^H$ is the conjugate transpose of the matrix $\boldsymbol{A}$
$H(\boldsymbol{x})$	differential entropy of $\boldsymbol{x}$
$H(\boldsymbol{x} \mid \boldsymbol{y})$	conditional differential entropy of $\boldsymbol{x}$ when $\boldsymbol{y}$ is given
$I(\boldsymbol{x}; \boldsymbol{y})$	mutual information between $\boldsymbol{x}$ and $\boldsymbol{y}$
$\boldsymbol{I}$	identity matrix
$\boldsymbol{I}_p$	$p \times p$ identity matrix
$\boldsymbol{I}_N^{(k)}$	$N \times N$ matrix containing zeros everywhere and ones on the k-th diagonal where the lower diagonal is numbered as $N-1$, the main diagonal is numbered with 0, and the upper diagonal is numbered with $-(N-1)$
$\operatorname{Im}\{\cdot\}$	returns imaginary part of the input
$\jmath$	imaginary unit
$\boldsymbol{J}$	$MN \times MN$ matrix with $N \times N$ identity matrices on the main reverse block diagonal and zeros elsewhere, i.e., $\boldsymbol{J} = \boldsymbol{J}_M \otimes \boldsymbol{I}_N$
$\boldsymbol{J}_N$	$N \times N$ reverse identity matrix with zeros everywhere except $+1$ on the main reverse diagonal
$\boldsymbol{J}_N^{(k)}$	$N \times N$ matrix containing zeros everywhere and ones on the k-th reverse diagonal where the upper reverse is numbered by $N-1$, the main reverse diagonal is numbered with 0, and the lower reverse diagonal is numbered with $-(N-1)$
$\mathbb{K}^{N \times Q}$	$N \times Q$ dimensional vector space over the field $\mathbb{K}$ and possible values of $\mathbb{K}$ are, for example, $\mathbb{R}$ or $\mathbb{C}$
$\boldsymbol{K}_{Q,N}$	commutation matrix of size $QN \times QN$
$\mathcal{L}$	Lagrange function
$\boldsymbol{L}_d$	$N^2 \times N$ matrix used to place the diagonal elements of $\boldsymbol{A} \in \mathbb{C}^{N \times N}$ on $\operatorname{vec}(\boldsymbol{A})$
$\boldsymbol{L}_l$	$N^2 \times \frac{N(N-1)}{2}$ matrix used to place the elements strictly below the main diagonal of $\boldsymbol{A} \in \mathbb{C}^{N \times N}$ on $\operatorname{vec}(\boldsymbol{A})$
$\boldsymbol{L}_u$	$N^2 \times \frac{N(N-1)}{2}$ matrix used to place the elements strictly above the main diagonal of $\boldsymbol{A} \in \mathbb{C}^{N \times N}$ on $\operatorname{vec}(\boldsymbol{A})$
$\lim_{z \to a} f(z)$	limit of $f(z)$ when z approaches a
$\ln(z)$	principal value of natural logarithm of z, where $z \in \mathbb{C}$
$m_{k,l}(\boldsymbol{Z})$	the (k, l)-th minor of the matrix $\boldsymbol{Z} \in \mathbb{C}^{N \times N}$
$\boldsymbol{M}(\boldsymbol{Z})$	if $\boldsymbol{Z} \in \mathbb{C}^{N \times N}$, then the matrix $\boldsymbol{M}(\boldsymbol{Z}) \in \mathbb{C}^{N \times N}$ contains the minors of $\boldsymbol{Z}$
max	maximum value of
min	minimum value of
$\mathbb{N}$	natural numbers $\{1, 2, 3, \ldots\}$
$n!$	factorial of n given by $n! = \prod_{i=1}^{n} i = 1 \cdot 2 \cdot 3 \cdot \ldots \cdot n$
$\operatorname{perm}(\cdot)$	permanent of a matrix
$\boldsymbol{P}_N$	primary circular matrix of size $N \times N$
$\mathbb{R}$	the set of real numbers

$\mathbb{R}^+$	the set $(0, \infty)$
$\text{rank}(\cdot)$	rank of a matrix
$\text{Re}\{\cdot\}$	returns real part of the input
$(\cdot)^T$	$\boldsymbol{A}^T$ is the transpose of the matrix $\boldsymbol{A}$
$\mathcal{T}^{(k)}\{\cdot\}$	linear reshaping operator used in connection with transmitter FIR MIMO optimization
$\text{Tr}\{\cdot\}$	trace of a square matrix
$v(\cdot)$	return all the elements on and below main diagonal taken in the same column-wise order as the ordinary vec-operator
$\text{vec}(\cdot)$	vectorization operator stacks the columns into a long column vector
$\text{vec}_d(\cdot)$	extracts the diagonal elements of a square matrix and returns them in a column vector
$\text{vec}_l(\cdot)$	extracts the elements strictly below the main diagonal of a square matrix in a column-wise manner and returns them into a column vector
$\text{vec}_u(\cdot)$	extracts the elements strictly above the main diagonal of a square matrix in a row-wise manner and returns them into a column vector
$\text{vecb}(\cdot)$	block vectorization operator stacks square block matrices of the input into a long block column matrix
$\boldsymbol{V}$	permutation matrix of size $\frac{N(N+1)}{2} \times \frac{N(N+1)}{2}$ given by $\boldsymbol{V} = [\boldsymbol{V}_d, \ \boldsymbol{V}_l]$
$\boldsymbol{V}_d$	matrix of size $\frac{N(N+1)}{2} \times N$ used to place the elements of $\text{vec}_d(\boldsymbol{A})$ on $v(\boldsymbol{A})$, where $\boldsymbol{A} \in \mathbb{C}^{N\times N}$ is symmetric
$\boldsymbol{V}_l$	matrix of size $\frac{N(N+1)}{2} \times \frac{N(N-1)}{2}$ used to place the elements of $\text{vec}_l(\boldsymbol{A})$ on $v(\boldsymbol{A})$, where $\boldsymbol{A} \in \mathbb{C}^{N\times N}$ is symmetric
$\mathcal{W}$	set containing matrices in a manifold
$\mathcal{W}^*$	set containing all the complex conjugate elements of the elements in $\mathcal{W}$, that is, when $\mathcal{W}$ is given, $\mathcal{W}^* \triangleq \{\boldsymbol{W}^* \mid \boldsymbol{W} \in \mathcal{W}\}$
$\boldsymbol{W}$	symbol often used to represent a matrix in a manifold, that is, $\boldsymbol{W} \in \mathcal{W}$, where $\mathcal{W}$ represents a manifold
$\tilde{\boldsymbol{W}}$	matrix used to represent a matrix of the same size as the matrix $\boldsymbol{W}$; however, the matrix $\tilde{\boldsymbol{W}}$ is unpatterned
$[x_0, x_1]$	closed interval given by the set $\{x \mid x_0 \leq x \leq x_1\}$
$(x_0, x_1]$	semi-open interval given by the set $\{x \mid x_0 < x \leq x_1\}$
(x_0, x_1)	open interval given by the set $\{x \mid x_0 < x < x_1\}$
$\boldsymbol{x}(n)_{\mathsf{I}}^{(\nu)}$	column-expansion of vector time-series of size $(\nu + 1)N \times 1$, where $\boldsymbol{x}(n) \in \mathbb{C}^{N\times 1}$
$\mathbb{Z}$	the set of integers, that is, $\mathbb{Z} = \{\ldots, -2, -1, 0, 1, 2, \ldots\}$
$\mathbb{Z}_N$	the set $\{0, 1, \ldots, N-1\}$
z	complex-valued scalar variable
$\boldsymbol{z}$	complex-valued vector variable
$\boldsymbol{Z}$	complex-valued matrix variable

1 Introduction

1.1 Introduction to the Book

To solve increasingly complicated open research problems, it is crucial to develop useful mathematical tools. Often, the task of a researcher or an engineer is to find the optimal values of unknown parameters that can be represented by complex-valued matrices. One powerful tool for finding the optimal values of complex-valued matrices is to calculate the derivatives with respect to these matrices. In this book, the main focus is on complex-valued matrix calculus because the theory of *real-valued* matrix derivatives has been thoroughly covered already in an excellent manner in Magnus and Neudecker (1988). The purpose of this book is to provide an introduction to the area of complex-valued matrix derivatives and to show how they can be applied as a tool for solving problems in signal processing and communications.

The framework of complex-valued matrix derivatives can be used in the optimization of systems that depend on complex design parameters in areas where the unknown parameters are complex-valued matrices with independent components, or where they belong to sets of matrices with certain structures. Many of the results discussed in this book are summarized in tabular form, so that they are easily accessible. Several examples taken from recently published material show how signal processing and communication systems can be optimized using complex-valued matrix derivatives. Note that the differentiation procedure is usually not sufficient to solve such problems completely; however, it is often an essential step toward finding the solution to the problem.

In many engineering problems, the unknown parameters are complex-valued matrices, and often, the task of the system designer is to find the values of these complex parameters, which optimize a certain scalar real-valued objective function. For solving these kinds of optimization problems, one approach is to find necessary conditions for optimality. Chapter 3 shows that when a scalar real-valued function depends on a complex-valued matrix variable, the necessary conditions for optimality can be found by setting the derivative of the function with respect to the complex-valued matrix variable or its complex conjugate to zero. It will also be shown that the direction of the maximum rate of change of a real-valued scalar function, with respect to the complex-valued matrix variable, is given by the derivative of the function with respect to the *complex conjugate* of the complex-valued input matrix variable. This result has important applications in, for example, complex-valued adaptive filters.

This book presents a comprehensive theory on how to obtain the derivatives of scalar-, vector-, and matrix-valued functions with respect to complex matrix variables. The theory of finding complex-valued matrix derivatives with respect to unpatterned matrices is based on the *complex differential* of the function of interest. The method of using differentials is substantially different from the component-wise approach.[1] A key idea when using complex differentials is to treat the *differential* of the complex and the complex conjugate variables as *independent*. This theory will be applied to derive useful matrix derivatives that can be used, for example, in signal processing and communications.

The complex Hessian matrix will be defined for complex scalar, vector, and matrix functions, and how this matrix can be obtained from the second-order differential of these functions is shown. Hessians are useful, for example, to check whether a stationary point is a saddle point, a local minimum, or a local maximum; Hessians can also be used to speed up the convergence of iterative algorithms.

A systematic theory on how to find *generalized complex-valued matrix derivatives* is presented. These are derivatives of complex-valued matrix functions with respect to matrices that belong to a set of complex-valued matrices, which might contain certain dependencies among the matrix elements. Such matrices include Hermitian, symmetric, diagonal, skew-symmetric, and skew-Hermitian. The theory of manifolds is used to find generalized complex-valued matrix derivatives. One key point of this theory is the requirement that the function, which spans all matrices within the set under consideration, is diffeomorphic; this function will be called the *parameterization function*. Several examples show how to find generalized complex-valued matrix derivatives with respect to matrices belonging to sets of matrices that are relevant for signal processing and communications.

Various applications from signal processing and communications are presented throughout the book. The last chapter is dedicated to various applications of complex-valued matrix derivatives.

1.2 Motivation for the Book

Complex signals appear in many parts of signal processing and communications. Good introductions to complex-valued signal processing can be found in Mandic and Goh (2009) and Schreier and Scharf (2010). One area where optimization problems with complex-valued matrices appear is digital communications, in which digital filters may contain complex-valued coefficients (Paulraj, Nabar, & Gore 2003). Other areas include analysis of power networks and electric circuits (González-Vázquez 1988); control theory (Alexander 1984); adaptive filters (Hanna & Mandic 2003; Diniz 2008); resource management (Han & Liu 2008); sensitivity analysis (Fränken 1997; Tsipouridou & Liavas 2008); and acoustics, optics, mechanical vibrating systems, heat con-

[1] In the author's opinion, the current approach of complex-valued matrix derivatives is preferred because it often leads to shorter and simpler calculations.

duction, fluid flow, and electrostatics (Kreyszig 1988). Convex optimization, in which the unknown parameters might be complex-valued, is treated in Boyd and Vandenberghe (2004) and Palomar and Eldar (2010). Usually, using complex-valued matrices leads to fewer computations and more compact expressions compared with treating the real and imaginary parts as two independent real-valued matrices. The complex-valued approach is general and usually easier to handle than working with the real and imaginary parts separately, because the complex matrix variable and its complex conjugate should be treated as independent variables when complex-valued matrix derivatives are calculated.

One of the main reasons why *complex-valued matrix derivatives* are so important is that necessary conditions for optimality can be found through these derivatives. By setting the complex-valued matrix derivative of the objective function equal to zero, necessary conditions for optimality are found. The theory of complex-valued matrix derivatives and the generalized complex-valued matrix derivatives are useful tools for researchers and engineers interested in designing systems in which the parameters are complex-valued matrices. The theory of *generalized* complex-valued matrix derivatives is particularly suited for problems with some type of structure within the unknown matrix of the optimization problem under consideration. Examples of such structured matrices include complex-valued diagonal, symmetric, skew-symmetric, Hermitian, skew-Hermitian, orthogonal, unitary, and positive semidefinite matrices. Finding derivatives with respect to complex-valued structured matrices is related to the field of manifolds. The theory of manifolds is a part of mathematics involving generalized derivatives on special geometric constructions spanned by so-called diffeomorphic functions (i.e., smooth invertible functions with a smooth inverse), which map the geometric construction back to a space with independent components. Optimization over such complex-valued constrained matrix sets can be done by using the theory of generalized matrix derivatives.

Complex-valued matrix derivatives are often used as a tool for solving problems in signal processing and communications. In the next section, a short overview of some of the literature on matrix derivatives is presented.

1.3 Brief Literature Summary

An early contribution to real-valued symbolic matrix calculus is found in Dwyer and Macphail (1948), which presents a basic treatment of matrix derivatives. Matrix derivatives in multivariate analysis are presented in Dwyer (1967). Another contribution is given in Nel (1980), which emphasizes the statistical applications of matrix derivatives.

The original work (Wirtinger 1927) showed that the complex variable and its complex conjugate can be treated as independent variables when finding derivatives. An introduction on how to find the Wirtinger calculus with respect to complex-valued scalars and vectors can be found in Fischer (2002, Appendix A). In Brandwood (1983), a theory is developed for finding derivatives of complex-valued scalar functions with respect to complex-valued *vectors*. It is argued in Brandwood (1983) that it is better to use the complex-valued vector and its complex conjugate as input variables instead of the real and imaginary parts of the vector – the main reason being that the complex-valued approach often leads to a simpler approach that requires fewer calculations than the method that treats the real and imaginary parts explicitly. Mandic and Goh (2009, p. 20) mention that the complex-valued representation may not always have a real physical interpretation; however, the complex framework is general and more mathematically tractable than working on the real and imaginary parts done separately.

An introduction to matrix derivatives, which focuses on component-wise derivatives, and to the Kronecker product is found in Graham (1981). Moon and Stirling (2000, Appendix E) focused on component-wise treatment of both real-valued and complex-valued matrix derivatives. Several useful results on complex-valued matrices are collected into Trees (2002, Appendix A), which also contains a few results on matrix calculus for which a component-wise treatment was used.

Magnus and Neudecker (1988) give a very solid treatment of real-valued matrices with independent components. However, they do not consider the case of formal derivatives, where the differential of the complex-valued matrix and the differential of its complex conjugate should be treated as *independent*; moreover, they do not treat the case of finding derivatives with respect to complex-valued *patterned matrices* (i.e., matrices containing certain structures). The problem of finding derivatives with respect to *real-valued* matrices containing independent elements is well known and has been studied, for example, in Harville (1997) and Minka (December 28, 2000). A substantial collection of derivatives in relation to real-valued vectors and matrices can be found in Lütkepohl (1996, Chapter 10).

Various references give brief treatments of the case of finding derivatives of real-valued scalar functions that depend on complex-valued vectors (van den Bos 1994*a*; Hayes 1996, Section 2.3.10; Haykin 2002, Appendix B; Sayed 2008, Background Material, Chapter C). A systematic and simple way to find derivatives with respect to *unpatterned* complex-valued *matrices* is presented in Hjørungnes and Gesbert (2007*a*).

Two online publications (Kreutz-Delgado 2008) and (Kreutz-Delgado 2009) give an introduction to real- and complex-valued derivatives with respect to vectors. Both first- and second-order derivatives are studied in these references. Two Internet sites with useful material on matrix derivatives are *The Matrix Cookbook* (Petersen & Pedersen 2008) and *The Matrix Reference Manual* (Brookes, July 25, 2009).

Hessians (second-order derivatives) of scalar functions of complex vectors are studied in van den Bos (1994*a*). The theory for finding Hessian matrices of scalar complex-valued function with respect to *unpatterned* complex-valued matrices and its complex conjugate is developed in Hjørungnes and Gesbert (2007*b*).

The theory for finding derivatives of real-valued functions that depend on patterned real-valued matrices is developed in Tracy and Jinadasa (1988). In Hjørungnes and Palomar (2008*b*), the theory for finding derivatives of functions that depend on complex-valued *patterned* matrices is studied; this was extended in Hjørungnes and Palomar (2008*a*), where the connections to manifolds are exploited. In Palomar and Verdú (2006), derivatives of certain scalar functions with respect to complex-valued matrices are discussed, and some results for complex-valued scalar functions with respect to matrices that contain dependent elements are presented. Vaidyanathan et al. (2010, Chapter 20), presents a treatment of real- and complex-valued matrix derivatives; however, it is based on component-wise developments. Some results on derivatives with respect to patterned matrices are presented in Vaidyanathan et al. (2010, Chapter 20).

1.4 Brief Outline

Some of the important notations used in this book and various useful formulas are discussed in Chapter 2. These items provide background material for later chapters. A classification of complex variables and functions is also presented in Chapter 2, which includes a discussion of the differences between analytic functions – subject matter usually studied in mathematical courses for engineers, and non-analytic functions, which are encountered when dealing with practical engineering problems of complex variables.

In Chapter 3, the complex differential is introduced. Based on the complex differential, the definition of the derivatives of complex-valued matrix functions with respect to the unpatterned complex-valued matrix variable and its complex conjugate is introduced. In addition, a procedure showing how the derivatives can be found from the differential of a function when the complex matrix variable contains independent elements is presented in Chapter 3. This chapter also contains several important results stated in theorems, such as the chain rule and necessary conditions for optimality for real-valued scalar functions.

Chapter 4 states several results in tables and shows how most of these results can be derived for nine different types of functions. These nine function types result when the input and the output of the function take the form of a scalar, a vector, or a matrix.

The Hessian matrix of complex-valued scalar, vector, and matrix functions dependent on complex matrices is defined in Chapter 5, which shows how this Hessian matrix can be obtained from the second-order differential. Hessian matrices can, for example, be used to speed up convergence of iterative algorithms, to study the convexity and concavity of an objective function, and to perform stability analysis of iterative algorithms.

Often, in signal processing and communications, the challenge is to find a matrix that optimizes a problem when the matrix is constrained to belong to a certain set, such as Hermitian matrices or symmetric matrices. For solving such types of problems, derivatives associated with matrices belonging to these sets are useful. These types of derivatives are called *generalized complex-valued matrix derivatives*, and a theory for finding such derivatives is presented in Chapter 6.

In Chapter 7, various applications taken from signal processing and communications are presented to show how complex-valued matrix derivatives can be used as a tool to solve research problems in these two fields.

After the seven chapters, references and the index follow.

2 Background Material

2.1 Introduction

In this chapter, most of the notation used in this book will be introduced. It is *not* assumed that the reader is familiar with topics such as Kronecker product, Hadamard product, or vectorization operator. Therefore, this chapter defines these concepts and gives some of their properties. The current chapter also provides background material for matrix manipulations that will be used later in the book. However, it contains just the minimum of material that will be used later because many excellent books in linear algebra are available for the reader to consult (Gantmacher 1959*a*–1959*b*; Horn & Johnson 1985; Strang 1988; Magnus & Neudecker 1988; Golub & van Loan 1989; Horn & Johnson 1991; Lütkepohl 1996; Harville 1997; Bernstein 2005).

This chapter is organized as follows: Section 2.2 introduces the basic notation and classification used for complex-valued variables and functions. A discussion of the differences between analytic and non-analytic functions is presented in Section 2.3. Basic matrix-related definitions are provided in Section 2.4. Several results involving matrix manipulations used in later chapters are found in Section 2.5. Section 2.6 offers exercises related to the material included in this chapter. Theoretical derivations and computer programming in `MATLAB` are topics of these exercises.

2.2 Notation and Classification of Complex Variables and Functions

Denote $\mathbb{R}$ and $\mathbb{C}$ the sets of the real and complex numbers, respectively, and define $\mathbb{Z}_N \triangleq \{0, 1, \ldots, N-1\}$. The notation used for the two matrices consisting entirely of zeros and ones is $\mathbf{0}_{N\times Q}$ and $\mathbf{1}_{N\times Q}$, respectively, where the size of the matrices is indicated by the subindex to be $N \times Q$.

The following conventions are always used in this book:

- Scalar quantities are denoted by lowercase symbols.
- Vector quantities are denoted by lowercase boldface symbols.
- Matrix quantities are denoted by capital boldface symbols.

Table 2.1 Symbols and sizes of the most frequently used variables and functions.

Symbol	z	$\boldsymbol{z}$	$\boldsymbol{Z}$	f	$\boldsymbol{f}$	$\boldsymbol{F}$
Size	1×1	$N \times 1$	$N \times Q$	1×1	$M \times 1$	$M \times P$

2.2.1 Complex-Valued Variables

A function's complex input argument can be a scalar, denoted z, a vector, denoted $\boldsymbol{z}$, or a matrix, denoted $\boldsymbol{Z}$.

Let the symbol z denote a complex scalar variable, and let the real and imaginary part of z be denoted by x and y, respectively, then

$$z = x + \jmath y, \tag{2.1}$$

where $\jmath$ is the imaginary unit, and $\jmath^2 = -1$. The absolute value of the complex number z is denoted by $|z|$.

The real and imaginary operators return the real and imaginary parts of the input matrix, respectively. These operators are denoted by $\text{Re}\{\cdot\}$ and $\text{Im}\{\cdot\}$. If $\boldsymbol{Z} \in \mathbb{C}^{N\times Q}$ is a complex-valued matrix, then

$$\boldsymbol{Z} = \text{Re}\{\boldsymbol{Z}\} + \jmath\, \text{Im}\{\boldsymbol{Z}\}, \tag{2.2}$$

$$\boldsymbol{Z}^* = \text{Re}\{\boldsymbol{Z}\} - \jmath\, \text{Im}\{\boldsymbol{Z}\}, \tag{2.3}$$

where $\text{Re}\{\boldsymbol{Z}\} \in \mathbb{R}^{N\times Q}$, $\text{Im}\{\boldsymbol{Z}\} \in \mathbb{R}^{N\times Q}$, and the operator $(\cdot)^*$ denote the complex conjugate of the matrix it is applied to. The real and imaginary operators can be expressed as

$$\text{Re}\{\boldsymbol{Z}\} = \frac{1}{2}(\boldsymbol{Z} + \boldsymbol{Z}^*), \tag{2.4}$$

$$\text{Im}\{\boldsymbol{Z}\} = \frac{1}{2\jmath}(\boldsymbol{Z} - \boldsymbol{Z}^*). \tag{2.5}$$

2.2.2 Complex-Valued Functions

For complex-valued functions, the following conventions are always used in this book:

- If the function returns a scalar, then a lowercase symbol is used, for example, f.
- If the function returns a vector, then a lowercase boldface symbol is used, for example, $\boldsymbol{f}$.
- If the function returns a matrix, then a capital boldface symbol is used, for example, $\boldsymbol{F}$.

Table 2.1 shows the sizes and symbols of the variables and functions most frequently used in the part of the book that treats complex matrix derivatives with independent components. Note that $\boldsymbol{F}$ covers all situations because scalars f and vectors $\boldsymbol{f}$ are special cases of a matrix. In the sequel, however, the three types of functions are distinguished as scalar, vector, or matrix because, as we shall see in Chapter 4, different definitions of the derivatives, based on type of functions, are found in the literature.

2.3 Analytic versus Non-Analytic Functions

Let the symbol $\subseteq$ mean subset of, and $\subset$ proper subset of. Mathematical courses on complex functions for *engineers* often involve only the analysis of analytic functions (Kreyszig 1988, p. 738) defined as follows:

Definition 2.1 (Analytic Function) *Let $D \subseteq \mathbb{C}$ be the domain*[1] *of definition of the function $f : D \to \mathbb{C}$. The function f is an* analytic function in the domain *D if $\lim_{\Delta z \to 0} \frac{f(z + \Delta z) - f(z)}{\Delta z}$ exists for all $z \in D$.*

If $f(z)$ satisfies the Cauchy-Riemann equations (Kreyszig 1988, pp. 740–743), then it is analytic. A function that is analytic is also named complex differentiable, holomorphic, or regular. The Cauchy-Riemann equations for the scalar function f can be formulated as a single equation in the following way:

$$\frac{\partial}{\partial z^*} f = 0. \tag{2.6}$$

From (2.6), it is seen that any analytic function f is *not* dependent on the variable z^*. This can also be seen from Theorem 1 in Kreyszig (1988, p. 804), which states that any analytic function $f(z)$ can be written as a power series[2] with non-negative exponents of the complex variable z, and this power series is called the Taylor series. This series does *not* contain any terms that depend on z^*. The derivative of a complex-valued scalar function in mathematical courses of complex analysis for *engineers* is often defined only for analytic functions. However, in engineering problems, the functions of interest often are *not* analytic because they are often real-valued functions. If a function is dependent only on z, as are analytic functions, and is not implicitly or explicitly dependent on z^*, then this function *cannot* in general be real-valued; a function can be real-valued only if the imaginary part of f vanishes, and this is possible only if the function also depends on terms that depend on z^*. An alternative treatment for how to find the derivative of real functions dependent on complex variables other than the one used for analytic function is needed. In this book, a theory that solves this problem is provided for scalar, vector, or matrix functions and variables.

[1] If $f : \mathcal{A} \to \mathcal{B}$, then the set $\mathcal{A}$ is called the *domain* of f, the set $\mathcal{B}$ is called the *range* of f, and the set $\{f(x) \mid x \in \mathcal{A}\}$ is called the *image set* of f (Munkres 2000, p. 16).

[2] A power series in the variable $z \in \mathbb{C}$ is an infinite sum of the form $\sum_{n=0}^{\infty} a_n (z - z_0)^n$, where a_n, $z_0 \in \mathbb{C}$ (Kreyszig 1988, p. 812).

In engineering problems, the squared Euclidean distance is often used. Let $f : \mathbb{C} \rightarrow \mathbb{C}$ be defined as

$$f(z) = |z|^2 = zz^*. \tag{2.7}$$

If the traditional definition of the derivative given in Definition 2.1 is used, then the function f is not differentiable because

$$\begin{aligned}
\lim_{\Delta z \rightarrow 0} \frac{f(z_0 + \Delta z) - f(z_0)}{\Delta z} &= \lim_{\Delta z \rightarrow 0} \frac{|z_0 + \Delta z|^2 - |z_0|^2}{\Delta z} \\
&= \lim_{\Delta z \rightarrow 0} \frac{(z_0 + \Delta z)(z_0^* + (\Delta z)^*) - z_0 z_0^*}{\Delta z} \\
&= \lim_{\Delta z \rightarrow 0} \frac{(\Delta z) z_0^* + z_0 (\Delta z)^* + \Delta z (\Delta z)^*}{\Delta z},
\end{aligned} \tag{2.8}$$

and this limit does *not exist*, because different values are found depending on how Δz is approaching 0. Let $\Delta z = \Delta x + \jmath \Delta y$. First, let Δz approach 0 such that $\Delta x = 0$, then the last fraction in (2.8) is

$$\frac{\jmath (\Delta y) z_0^* - \jmath z_0 \Delta y + (\Delta y)^2}{\jmath \Delta y} = z_0^* - z_0 - \jmath \Delta y, \tag{2.9}$$

which approaches $z_0^* - z_0 = -2\jmath \operatorname{Im}\{z_0\}$, when $\Delta y \rightarrow 0$. Second, let Δz approach 0 such that $\Delta y = 0$, then the last fraction in (2.8) is

$$\frac{(\Delta x) z_0^* + z_0 \Delta x + (\Delta x)^2}{\Delta x} = z_0 + z_0^* + \Delta x, \tag{2.10}$$

which approaches $z_0 + z_0^* = 2 \operatorname{Re}\{z_0\}$ when $\Delta x \rightarrow 0$. For an arbitrary complex number z_0, in general, $2 \operatorname{Re}\{z_0\} \neq -2\jmath \operatorname{Im}\{z_0\}$. This means that the function $f(z) = |z|^2 = zz^*$ is *not* differentiable when the commonly encountered definition given in Definition 2.1 is used, and, hence, f is not analytic.

Two alternative ways (Hayes 1996, Subsection 2.3.10) are known for finding the derivative of a scalar real-valued function $f \in \mathbb{R}$ with respect to the unknown complex-valued matrix variable $\boldsymbol{Z} \in \mathbb{C}^{N \times Q}$. The first way is to rewrite f as a function of the real $\boldsymbol{X}$ and imaginary parts $\boldsymbol{Y}$ of the complex variable $\boldsymbol{Z}$, and then to find the derivatives of the rewritten function with respect to these two independent real variables, $\boldsymbol{X}$ and $\boldsymbol{Y}$, separately. Notice that NQ independent complex unknown variables in $\boldsymbol{Z}$ correspond to $2NQ$ independent real variables in $\boldsymbol{X}$ and $\boldsymbol{Y}$. The second way to deal with this problem, which is more elegant and is used in this book, is to treat the differentials of the variables $\boldsymbol{Z}$ and $\boldsymbol{Z}^*$ as independent, in the way that will be shown by Lemma 3.1. Chapter 3 shows that the derivative of f with respect to $\boldsymbol{Z}$ and $\boldsymbol{Z}^*$ can be identified by the differential of f.

Complex numbers cannot be ordered as real numbers can. Therefore, the objective functions of interest, when dealing with engineering problems, are usually real valued in such a way that it makes sense to minimize or maximize them. If a real-valued function depends on a complex matrix $\boldsymbol{Z}$, it must also be explicitly or implicitly dependent on $\boldsymbol{Z}^*$, such that the result is real (see also the discussion following (2.6)). A real-valued

Table 2.2 Classification of functions.

Function type	$z, z^* \in \mathbb{C}$	$\boldsymbol{z}, \boldsymbol{z}^* \in \mathbb{C}^{N\times 1}$	$\boldsymbol{Z}, \boldsymbol{Z}^* \in \mathbb{C}^{N\times Q}$
Scalar function $f \in \mathbb{C}$	$f(z, z^*)$ $f: \mathbb{C}\times\mathbb{C}\to\mathbb{C}$	$f(\boldsymbol{z}, \boldsymbol{z}^*)$ $f: \mathbb{C}^{N\times 1}\times\mathbb{C}^{N\times 1}\to\mathbb{C}$	$f(\boldsymbol{Z}, \boldsymbol{Z}^*)$ $f: \mathbb{C}^{N\times Q}\times\mathbb{C}^{N\times Q}\to\mathbb{C}$
Vector function $\boldsymbol{f} \in \mathbb{C}^{M\times 1}$	$\boldsymbol{f}(z, z^*)$ $\boldsymbol{f}: \mathbb{C}\times\mathbb{C}\to\mathbb{C}^{M\times 1}$	$\boldsymbol{f}(\boldsymbol{z}, \boldsymbol{z}^*)$ $\boldsymbol{f}: \mathbb{C}^{N\times 1}\times\mathbb{C}^{N\times 1}\to\mathbb{C}^{M\times 1}$	$\boldsymbol{f}(\boldsymbol{Z}, \boldsymbol{Z}^*)$ $\boldsymbol{f}: \mathbb{C}^{N\times Q}\times\mathbb{C}^{N\times Q}\to\mathbb{C}^{M\times 1}$
Matrix function $\boldsymbol{F} \in \mathbb{C}^{M\times P}$	$\boldsymbol{F}(z, z^*)$ $\boldsymbol{F}: \mathbb{C}\times\mathbb{C}\to\mathbb{C}^{M\times P}$	$\boldsymbol{F}(\boldsymbol{z}, \boldsymbol{z}^*)$ $\boldsymbol{F}: \mathbb{C}^{N\times 1}\times\mathbb{C}^{N\times 1}\to\mathbb{C}^{M\times P}$	$\boldsymbol{F}(\boldsymbol{Z}, \boldsymbol{Z}^*)$ $\boldsymbol{F}: \mathbb{C}^{N\times Q}\times\mathbb{C}^{N\times Q}\to\mathbb{C}^{M\times P}$

Adapted from Hjørungnes and Gesbert (2007*a*).

function can consist of several terms; it is possible that some of these terms are complex valued, even though their sum is real.

The main types of functions used throughout this book, when working with complex-valued matrix derivatives with independent components, can be classified as in Table 2.2. The table shows that all functions depend on a complex variable and the complex conjugate of the same variable, and the reason for this is that the complex *differential* of the variables $\boldsymbol{Z}$ and $\boldsymbol{Z}^*$ should be treated independently. When the function has two complex input variables of the same size (e.g., $\boldsymbol{F}: \mathbb{C}^{N\times Q}\times\mathbb{C}^{N\times Q}\to\mathbb{C}^{M\times P}$ for the general case), then two input variables should be the complex conjugate of each other. This means that they cannot be chosen independently of each other. However, in Lemmas 3.1 and 6.1, it will be shown that the *differentials* of the two input matrix variables $\boldsymbol{Z}$ and $\boldsymbol{Z}^*$ are independent. The convention of using both a complex variable and its complex conjugate explicitly in the function definition was used in Brandwood (1983). When evaluating, for example, the most general function in Table 2.2 (i.e., $\boldsymbol{F}: \mathbb{C}^{N\times Q}\times\mathbb{C}^{N\times Q}\to\mathbb{C}^{M\times P}$), the notation adapted is that the two complex-valued input variables should be the complex conjugates of each other. Hence, the two input arguments of $\boldsymbol{F}(\boldsymbol{Z}, \boldsymbol{Z}^*)$ are a function of each other, but as will be seen in Lemma 3.1, the *differentials* of the two input variables $\boldsymbol{Z}$ and $\boldsymbol{Z}^*$ are independent. When working with complex-valued matrix derivatives in later chapters, we will see that complex differentials are very important.

Definition 2.2 (Formal Derivatives) *Let $z = x + \jmath y$, where $x, y \in \mathbb{R}$, then the* formal derivatives, *with respect to z and z^* of $f(z_0)$ at $z_0 \in \mathbb{C}$ or* Wirtinger derivatives (Wirtinger 1927), *are defined as*

$$\frac{\partial}{\partial z} f(z_0) = \frac{1}{2}\left(\frac{\partial}{\partial x} f(z_0) - \jmath \frac{\partial}{\partial y} f(z_0)\right), \tag{2.11}$$

$$\frac{\partial}{\partial z^*} f(z_0) = \frac{1}{2}\left(\frac{\partial}{\partial x} f(z_0) + \jmath \frac{\partial}{\partial y} f(z_0)\right). \tag{2.12}$$

When finding $\frac{\partial}{\partial z} f(z_0)$ and $\frac{\partial}{\partial z^*} f(z_0)$, the variables z and z^* are treated as independent variables (Brandwood 1983, Theorem 1).

The formal derivatives above must be interpreted formally because z and z^* cannot be varied independently of each other (Kreutz-Delgado 2009, June 25th, Footnote 27, p. 15). In Kreutz-Delgado (2009, June 25th), the topic of Wirtinger calculations is also named $\mathbb{CR}$-calculus.

From Definition 2.2, it follows that the derivatives of the function f with respect to the real part x and the imaginary y part of z can be expressed as

$$\frac{\partial}{\partial x} f(z_0) = \frac{\partial}{\partial z} f(z_0) + \frac{\partial}{\partial z^*} f(z_0), \tag{2.13}$$

$$\frac{\partial}{\partial y} f(z_0) = \jmath \left(\frac{\partial}{\partial z} f(z_0) - \frac{\partial}{\partial z^*} f(z_0) \right), \tag{2.14}$$

respectively.

The results in (2.13) and (2.14) are found by considering (2.11) and (2.12) as two linear equations with the two unknowns $\frac{\partial}{\partial x} f(z_0)$ and $\frac{\partial}{\partial y} f(z_0)$.

If the function f is dependent on several variables, Definition 2.2 can be extended. In Chapters 3 and 4, it will be shown how the derivatives, with respect to a complex-valued matrix variable and its complex conjugate, of all function types given in Table 2.2 can be identified from the complex differentials of these functions.

Example 2.1 By using Definition 2.2, the following formal derivatives are found:

$$\frac{\partial z}{\partial z} = \frac{1}{2}\left(\frac{\partial}{\partial x} - \jmath \frac{\partial}{\partial y} \right)(x + \jmath y) = \frac{1}{2}(1 + 1) = 1, \tag{2.15}$$

$$\frac{\partial z^*}{\partial z^*} = \frac{1}{2}\left(\frac{\partial}{\partial x} + \jmath \frac{\partial}{\partial y} \right)(x - \jmath y) = \frac{1}{2}(1 + 1) = 1, \tag{2.16}$$

$$\frac{\partial z}{\partial z^*} = \frac{1}{2}\left(\frac{\partial}{\partial x} + \jmath \frac{\partial}{\partial y} \right)(x + \jmath y) = \frac{1}{2}(1 - 1) = 0, \tag{2.17}$$

$$\frac{\partial z^*}{\partial z} = \frac{1}{2}\left(\frac{\partial}{\partial x} - \jmath \frac{\partial}{\partial y} \right)(x - \jmath y) = \frac{1}{2}(1 - 1) = 0. \tag{2.18}$$

When working with derivatives of analytic functions (see Definition 2.1), only derivatives with respect to z are studied, and $\frac{dz}{dz} = 1$ but $\frac{dz^*}{dz}$ does not exist.

Example 2.2 Let the function $f : \mathbb{C} \times \mathbb{C} \to \mathbb{R}$ given by $f(z, z^*) = zz^*$. This function is differentiable with respect to both variables z and z^* (when using Definition 2.2), and the expressions for the formal derivatives are given by

$$\frac{\partial}{\partial z} f(z, z^*) = z^*, \tag{2.19}$$

$$\frac{\partial}{\partial z^*} f(z, z^*) = z. \tag{2.20}$$

When the complex variable z and its complex conjugate twin z^* are treated as independent variables (Brandwood 1983, Theorem 1), then the function f is differentiable in both of these variables. Remember that, as was shown earlier in this section, the same function is *not* differentiable in the ordinary way using the traditional expression for the derivative for analytic functions provided in Definition 2.1.

2.4 Matrix-Related Definitions

The matrix $\boldsymbol{Z}^T$ represents the transpose of the matrix $\boldsymbol{Z}$. The Hermitian operator, or the complex conjugate transpose of a matrix $\boldsymbol{Z}$, is given by $\boldsymbol{Z}^H$. The trace of a square matrix $\boldsymbol{Z}$ is denoted by $\operatorname{Tr}\{\boldsymbol{Z}\}$. The determinant of a square matrix $\boldsymbol{Z} \in \mathbb{C}^{N\times N}$ is denoted by $\det(\boldsymbol{Z})$. The inverse matrix of a square nonsingular[3] matrix $\boldsymbol{Z}$ is denoted by $\boldsymbol{Z}^{-1}$. The adjoint of a matrix $\boldsymbol{Z} \in \mathbb{C}^{N\times N}$ is denoted by $\boldsymbol{Z}^{\#}$ and is obtained by

$$\boldsymbol{Z}^{\#} = \det(\boldsymbol{Z})\boldsymbol{Z}^{-1}. \tag{2.21}$$

The rank of a matrix $\boldsymbol{A}$ is denoted by $\operatorname{rank}(\boldsymbol{A})$. The operators $\dim_{\mathbb{C}}(\cdot)$ and $\dim_{\mathbb{R}}(\cdot)$ return the complex and real dimension of the vector space they are applied to, respectively. $\mathcal{C}(\boldsymbol{A})$, $\mathcal{R}(\boldsymbol{A})$, and $\mathcal{N}(\boldsymbol{A})$ are the symbols used for the column, row, and null space of a matrix $\boldsymbol{A} \in \mathbb{C}^{N\times Q}$, respectively (i.e., $\mathcal{C}(\boldsymbol{A}) = \{\boldsymbol{w} \in \mathbb{C}^{N\times 1} | \boldsymbol{w} = \boldsymbol{A}\boldsymbol{z}$, for some $\boldsymbol{z} \in \mathbb{C}^{Q\times 1}\}$, $\mathcal{R}(\boldsymbol{A}) = \{\boldsymbol{w} \in \mathbb{C}^{1\times Q} | \boldsymbol{w} = \boldsymbol{z}\boldsymbol{A},$ for some $\boldsymbol{z} \in \mathbb{C}^{1\times N}\}$, and $\mathcal{N}(\boldsymbol{A}) = \{\boldsymbol{z} \in \mathbb{C}^{Q\times 1} | \boldsymbol{A}\boldsymbol{z} = \boldsymbol{0}_{N\times 1}\}$).

Definition 2.3 (Idempotent) *A matrix* $\boldsymbol{A} \in \mathbb{C}^{N\times N}$ *is* idempotent *if* $\boldsymbol{A}^2 = \boldsymbol{A}$.

Definition 2.4 (Moore-Penrose Inverse) *The* Moore-Penrose inverse *of* $\boldsymbol{Z} \in \mathbb{C}^{N\times Q}$ *is denoted* $\boldsymbol{Z}^+ \in \mathbb{C}^{Q\times N}$ *and is defined through the following four relations (Horn & Johnson 1985, p. 421):*

$$\left(\boldsymbol{Z}\boldsymbol{Z}^+\right)^H = \boldsymbol{Z}\boldsymbol{Z}^+, \tag{2.22}$$

$$\left(\boldsymbol{Z}^+\boldsymbol{Z}\right)^H = \boldsymbol{Z}^+\boldsymbol{Z}, \tag{2.23}$$

$$\boldsymbol{Z}\boldsymbol{Z}^+\boldsymbol{Z} = \boldsymbol{Z}, \tag{2.24}$$

$$\boldsymbol{Z}^+\boldsymbol{Z}\boldsymbol{Z}^+ = \boldsymbol{Z}^+. \tag{2.25}$$

The Moore-Penrose inverse is an extension of the traditional inverse matrix that exists only for square nonsingular matrices (i.e., matrices with a nonzero determinant). When designing equalizers for a memoryless MIMO system, the Moore-Penrose inverse can be used to find the zero-forcing equalizer (Paulraj et al. 2003, pp. 152–153). A zero-forcing equalizer tries to set the total signal error to zero, but this can lead to noise amplification in the receiver.

Remark *The indices in this book are mostly chosen to start with* 0.

Definition 2.5 (Exponential Matrix Function) *Let* $\boldsymbol{I}_N$ *denote the* $N \times N$ *identity matrix. If* $\boldsymbol{Z} \in \mathbb{C}^{N\times N}$*, then the* exponential matrix function $\exp : \mathbb{C}^{N\times N} \to \mathbb{C}^{N\times N}$ *is denoted*

[3] A nonsingular matrix is a square matrix with a nonzero determinant (i.e., an invertible matrix).

$\exp(\boldsymbol{Z})$ *and is defined as*

$$\exp(\boldsymbol{Z}) = \sum_{k=0}^{\infty} \frac{1}{k!} \boldsymbol{Z}^k, \tag{2.26}$$

where $\boldsymbol{Z}^0 \triangleq \boldsymbol{I}_N$, $\forall\, \boldsymbol{Z} \in \mathbb{C}^{N\times N}$.

Definition 2.6 (Kronecker Product) *Let* $\boldsymbol{A} \in \mathbb{C}^{M\times N}$ *and* $\boldsymbol{B} \in \mathbb{C}^{P\times Q}$. *Denote element number* (k, l) *of the matrix* $\boldsymbol{A}$ *by* $a_{k,l}$. *The* Kronecker product *(Horn & Johnson 1991), denoted* $\otimes$, *between the complex-valued matrices* $\boldsymbol{A}$ *and* $\boldsymbol{B}$ *is defined as the matrix* $\boldsymbol{A} \otimes \boldsymbol{B} \in \mathbb{C}^{MP\times NQ}$, *given by*

$$\boldsymbol{A} \otimes \boldsymbol{B} = \begin{bmatrix} a_{0,0}\boldsymbol{B} & \cdots & a_{0,N-1}\boldsymbol{B} \\ \vdots & & \vdots \\ a_{M-1,0}\boldsymbol{B} & \cdots & a_{M-1,N-1}\boldsymbol{B} \end{bmatrix}. \tag{2.27}$$

Equivalently, this can be expressed as follows:

$$[\boldsymbol{A} \otimes \boldsymbol{B}]_{i+jP,k+lQ} = a_{j,l} b_{i,k}, \tag{2.28}$$

where $i \in \{0, 1, \ldots, P-1\}$, $j \in \{0, 1, \ldots, M-1\}$, $k \in \{0, 1, \ldots, Q-1\}$, *and* $l \in \{0, 1, \ldots, N-1\}$.

Definition 2.7 (Hadamard Product[4]) *Let* $\boldsymbol{A} \in \mathbb{C}^{M\times N}$ *and* $\boldsymbol{B} \in \mathbb{C}^{M\times N}$. *Denote element number* (k, l) *of the matrices* $\boldsymbol{A}$ *and* $\boldsymbol{B}$ *by* $a_{k,l}$ *and* $b_{k,l}$, *respectively. The* Hadamard product *(Horn & Johnson 1991), denoted by* $\odot$, *between the complex-valued matrices* $\boldsymbol{A}$ *and* $\boldsymbol{B}$, *is defined as the matrix* $\boldsymbol{A} \odot \boldsymbol{B} \in \mathbb{C}^{M\times N}$, *given by*

$$\boldsymbol{A} \odot \boldsymbol{B} = \begin{bmatrix} a_{0,0}b_{0,0} & \cdots & a_{0,N-1}b_{0,N-1} \\ \vdots & & \vdots \\ a_{M-1,0}b_{M-1,0} & \cdots & a_{M-1,N-1}b_{M-1,N-1} \end{bmatrix}. \tag{2.29}$$

Definition 2.8 (Vectorization Operator) *Let* $\boldsymbol{A} \in \mathbb{C}^{M\times N}$ *and denote the* i*-th column of* $\boldsymbol{A}$ *by* $\boldsymbol{a}_i$, *where* $i \in \{0, 1, \ldots, N-1\}$. *Then the* $\operatorname{vec}(\cdot)$ *operator is defined as the* $MN \times 1$ *vector given by*

$$\operatorname{vec}(\boldsymbol{A}) = \begin{bmatrix} \boldsymbol{a}_0 \\ \boldsymbol{a}_1 \\ \vdots \\ \boldsymbol{a}_{N-1} \end{bmatrix}. \tag{2.30}$$

Let $\boldsymbol{A} \in \mathbb{C}^{N\times Q}$, then there exists a permutation matrix that connects the vectors $\operatorname{vec}(\boldsymbol{A})$ and $\operatorname{vec}(\boldsymbol{A}^T)$. The permutation matrix that gives the connection between $\operatorname{vec}(\boldsymbol{A})$ and $\operatorname{vec}(\boldsymbol{A}^T)$ is called the *commutation matrix* and is defined as follows:

Definition 2.9 (Commutation Matrix) *Let* $\boldsymbol{A} \in \mathbb{C}^{N\times Q}$. *The* commutation matrix $\boldsymbol{K}_{N,Q}$ *is a permutation matrix of size* $NQ \times NQ$, *and it gives the connection between* $\operatorname{vec}(\boldsymbol{A})$

[4] In Bernstein (2005, p. 252), this product is called the *Schur product*.

and $\text{vec}\left(\boldsymbol{A}^T\right)$ *in the following way:*

$$\boldsymbol{K}_{N,Q}\,\text{vec}(\boldsymbol{A}) = \text{vec}(\boldsymbol{A}^T). \tag{2.31}$$

Example 2.3 If $\boldsymbol{A} \in \mathbb{C}^{3\times 2}$, then by studying the connection between $\text{vec}(\boldsymbol{A})$ and $\text{vec}(\boldsymbol{A}^T)$, together with (2.31), it can be seen that $\boldsymbol{K}_{3,2}$ is given by

$$\boldsymbol{K}_{3,2} = \begin{bmatrix} 1 & 0 & 0 & 0 & 0 & 0 \\ 0 & 0 & 0 & 1 & 0 & 0 \\ 0 & 1 & 0 & 0 & 0 & 0 \\ 0 & 0 & 0 & 0 & 1 & 0 \\ 0 & 0 & 1 & 0 & 0 & 0 \\ 0 & 0 & 0 & 0 & 0 & 1 \end{bmatrix}. \tag{2.32}$$

Example 2.4 Let $N = 5$ and $Q = 3$, then,

$$\boldsymbol{K}_{N,Q} = \begin{bmatrix}
1 & 0 & 0 & 0 & 0 & 0 & 0 & 0 & 0 & 0 & 0 & 0 & 0 & 0 & 0 \\
0 & 0 & 0 & 0 & 0 & 1 & 0 & 0 & 0 & 0 & 0 & 0 & 0 & 0 & 0 \\
0 & 0 & 0 & 0 & 0 & 0 & 0 & 0 & 0 & 0 & 1 & 0 & 0 & 0 & 0 \\
0 & 1 & 0 & 0 & 0 & 0 & 0 & 0 & 0 & 0 & 0 & 0 & 0 & 0 & 0 \\
0 & 0 & 0 & 0 & 0 & 0 & 1 & 0 & 0 & 0 & 0 & 0 & 0 & 0 & 0 \\
0 & 0 & 0 & 0 & 0 & 0 & 0 & 0 & 0 & 0 & 0 & 1 & 0 & 0 & 0 \\
0 & 0 & 1 & 0 & 0 & 0 & 0 & 0 & 0 & 0 & 0 & 0 & 0 & 0 & 0 \\
0 & 0 & 0 & 0 & 0 & 0 & 0 & 1 & 0 & 0 & 0 & 0 & 0 & 0 & 0 \\
0 & 0 & 0 & 0 & 0 & 0 & 0 & 0 & 0 & 0 & 0 & 0 & 1 & 0 & 0 \\
0 & 0 & 0 & 1 & 0 & 0 & 0 & 0 & 0 & 0 & 0 & 0 & 0 & 0 & 0 \\
0 & 0 & 0 & 0 & 0 & 0 & 0 & 0 & 1 & 0 & 0 & 0 & 0 & 0 & 0 \\
0 & 0 & 0 & 0 & 0 & 0 & 0 & 0 & 0 & 0 & 0 & 0 & 0 & 1 & 0 \\
0 & 0 & 0 & 0 & 1 & 0 & 0 & 0 & 0 & 0 & 0 & 0 & 0 & 0 & 0 \\
0 & 0 & 0 & 0 & 0 & 0 & 0 & 0 & 0 & 1 & 0 & 0 & 0 & 0 & 0 \\
0 & 0 & 0 & 0 & 0 & 0 & 0 & 0 & 0 & 0 & 0 & 0 & 0 & 0 & 1
\end{bmatrix}. \tag{2.33}$$

In Exercise 2.7, the reader is asked to write a `MATLAB` program for finding $\boldsymbol{K}_{N,Q}$ for any given positive integers N and K.

Definition 2.10 (Diagonalization Operator) *Let* $\boldsymbol{a} \in \mathbb{C}^{N\times 1}$, *and let the* i*-th vector component of* $\boldsymbol{a}$ *be denoted by* a_i, *where* $i \in \{0, 1, \ldots, N-1\}$. *The diagonalization operator*

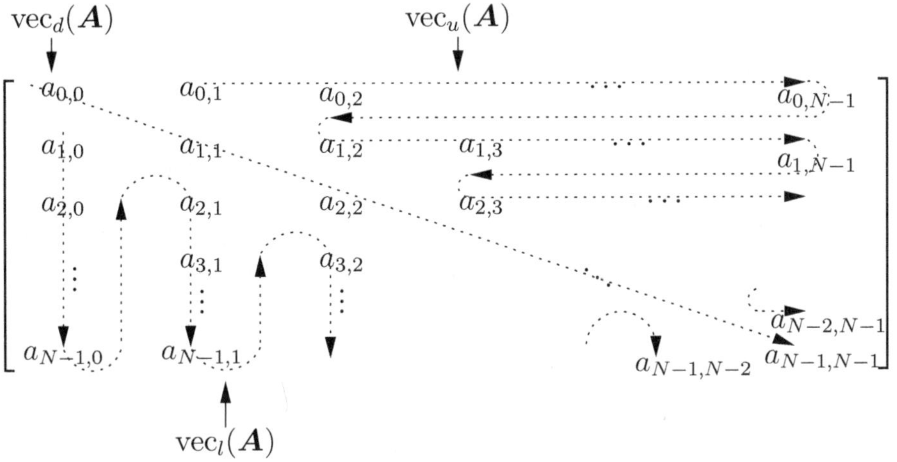

Figure 2.1 The way the three operators $\text{vec}_d(\cdot)$, $\text{vec}_l(\cdot)$, and $\text{vec}_u(\cdot)$ are returning their elements from the matrix $A \in \mathbb{C}^{N\times N}$. The operator $\text{vec}_d(\cdot)$ returns the elements on the line along the main diagonal, starting in the upper left corner and going down along the main diagonal; the operator $\text{vec}_l(\cdot)$ returns elements along the curve below the main diagonal following the order indicated in the figure; and the operator $\text{vec}_u(\cdot)$ returns elements along the curve above the main diagonal in the order indicated by the arrows along that curve.

$\text{diag} : \mathbb{C}^{N\times 1} \to \mathbb{C}^{N\times N}$ *is defined as*

$$\text{diag}(\boldsymbol{a}) = \begin{bmatrix} a_0 & 0 & \cdots & 0 \\ 0 & a_1 & \cdots & 0 \\ \vdots & & \ddots & \vdots \\ 0 & 0 & \cdots & a_{N-1} \end{bmatrix}. \tag{2.34}$$

Definition 2.11 (Special Vectorization Operators) *Let* $A \in \mathbb{C}^{N\times N}$.

Let the operator $\text{vec}_d : \mathbb{C}^{N\times N} \to \mathbb{C}^{N\times 1}$ *return all the elements on the main diagonal ordered from the upper left corner and going down to the lower right corner of the input matrix*

$$\text{vec}_d(A) = \left[a_{0,0}, a_{1,1}, a_{2,2}, \ldots, a_{N-1,N-1}\right]^T. \tag{2.35}$$

Let the operator $\text{vec}_l : \mathbb{C}^{N\times N} \to \mathbb{C}^{\frac{(N-1)N}{2}\times 1}$ *return all the elements* strictly below *the main diagonal taken in the same column-wise order as the ordinary* vec*-operator*

$$\text{vec}_l(A) = \left[a_{1,0}, a_{2,0}, \ldots, a_{N-1,0}, a_{2,1}, a_{3,1}, \ldots, a_{N-1,1}, a_{3,2}, \ldots, a_{N-1,N-2}\right]^T. \tag{2.36}$$

Let the operator $\text{vec}_u : \mathbb{C}^{N\times N} \to \mathbb{C}^{\frac{(N-1)N}{2}\times 1}$ *return all the elements* strictly above *the main diagonal taken in a* row-wise *order going from left to right, starting with the first row, then the second, and so on*

$$\mathrm{vec}_u(\boldsymbol{A}) = \left[a_{0,1}, a_{0,2}, \ldots, a_{0,N-1}, a_{1,2}, a_{1,3}, \ldots, a_{1,N-1}, a_{2,3}, \ldots, a_{N-2,N-1}\right]^T. \tag{2.37}$$

For the matrix $\boldsymbol{A} \in \mathbb{C}^{N\times N}$, Figure 2.1 shows how the three special vectorization operators $\mathrm{vec}_d(\cdot)$, $\mathrm{vec}_l(\cdot)$, and $\mathrm{vec}_u(\cdot)$ pick out the elements of $\boldsymbol{A}$ and return them into *column* vectors. The operator $\mathrm{vec}_d(\cdot)$ was also studied in Brewer (1978, Eq. (7)); the two other operators $\mathrm{vec}_l(\cdot)$ and $\mathrm{vec}_u(\cdot)$ were defined in Hjørungnes and Palomar (2008*a* and 2008*b*).

If $\boldsymbol{a} \in \mathbb{C}^{N\times 1}$, then

$$\mathrm{vec}_d(\mathrm{diag}(\boldsymbol{a})) = \boldsymbol{a}. \tag{2.38}$$

Hence, the operator $\mathrm{vec}_d(\cdot)$ is the left-inverse of the operator $\mathrm{diag}(\cdot)$. If $\boldsymbol{D} \in \mathbb{C}^{N\times N}$ is a diagonal matrix, then

$$\mathrm{diag}(\mathrm{vec}_d(\boldsymbol{D})) = \boldsymbol{D}, \tag{2.39}$$

but this formula is *not* valid for *non-diagonal* matrices. For diagonal matrices, the operator diag is the inverse of the operator vec_d; however, this is not true for non-diagonal matrices.

Example 2.5 Let $N = 3$, then the matrix $\boldsymbol{A} \in \mathbb{C}^{N\times N}$ can be written as

$$\boldsymbol{A} = \begin{bmatrix} a_{0,0} & a_{0,1} & a_{0,2} \\ a_{1,0} & a_{1,1} & a_{1,2} \\ a_{2,0} & a_{2,1} & a_{2,2} \end{bmatrix}, \tag{2.40}$$

where $(\boldsymbol{A})_{k,l} = a_{k,l} \in \mathbb{C}$ is the element in row k and column l. By using the $\mathrm{vec}(\cdot)$, $\mathrm{vec}_d(\cdot)$, $\mathrm{vec}_l(\cdot)$, and $\mathrm{vec}_u(\cdot)$ operators on $\boldsymbol{A}$, it is found that

$$\mathrm{vec}(\boldsymbol{A}) = \begin{bmatrix} a_{0,0} \\ a_{1,0} \\ a_{2,0} \\ a_{0,1} \\ a_{1,1} \\ a_{2,1} \\ a_{0,2} \\ a_{1,2} \\ a_{2,2} \end{bmatrix}, \quad \mathrm{vec}_d(\boldsymbol{A}) = \begin{bmatrix} a_{0,0} \\ a_{1,1} \\ a_{2,2} \end{bmatrix}, \quad \mathrm{vec}_l(\boldsymbol{A}) = \begin{bmatrix} a_{1,0} \\ a_{2,0} \\ a_{2,1} \end{bmatrix}, \quad \mathrm{vec}_u(\boldsymbol{A}) = \begin{bmatrix} a_{0,1} \\ a_{0,2} \\ a_{1,2} \end{bmatrix}. \tag{2.41}$$

From the example above and the definition of the operators $\text{vec}_d(\cdot)$, $\text{vec}_l(\cdot)$, and $\text{vec}_u(\cdot)$, a clear connection can be seen between the four vectorization operators $\text{vec}(\cdot)$, $\text{vec}_d(\cdot)$, $\text{vec}_l(\cdot)$, and $\text{vec}_u(\cdot)$. These connections can be found by defining three matrices, as in the following definition:

Definition 2.12 (Matrices $\boldsymbol{L}_d$, $\boldsymbol{L}_l$, and $\boldsymbol{L}_u$) *Let $\boldsymbol{A} \in \mathbb{C}^{N\times N}$. Three* unique *matrices $\boldsymbol{L}_d \in \mathbb{Z}_2^{N^2\times N}$, $\boldsymbol{L}_l \in \mathbb{Z}_2^{N^2\times \frac{N(N-1)}{2}}$, and $\boldsymbol{L}_u \in \mathbb{Z}_2^{N^2\times \frac{N(N-1)}{2}}$ contain zeros everywhere except for $+1$ at one place in each column; these matrices can be used to build up* $\text{vec}(\boldsymbol{A})$, *where $\boldsymbol{A} \in \mathbb{C}^{N\times N}$ is arbitrary, in the following way:*

$$\begin{aligned}\text{vec}\left(\boldsymbol{A}\right) &= \boldsymbol{L}_d \,\text{vec}_d\left(\boldsymbol{A}\right) + \boldsymbol{L}_l \,\text{vec}_l\left(\boldsymbol{A}\right) + \boldsymbol{L}_u \,\text{vec}_u\left(\boldsymbol{A}\right) \\ &= \left[\boldsymbol{L}_d, \boldsymbol{L}_l, \boldsymbol{L}_u\right] \begin{bmatrix} \text{vec}_d\left(\boldsymbol{A}\right) \\ \text{vec}_l\left(\boldsymbol{A}\right) \\ \text{vec}_u\left(\boldsymbol{A}\right) \end{bmatrix},\end{aligned} \tag{2.42}$$

where the terms $\boldsymbol{L}_d \,\text{vec}_d\left(\boldsymbol{A}\right)$, $\boldsymbol{L}_l \,\text{vec}_l\left(\boldsymbol{A}\right)$, and $\boldsymbol{L}_u \,\text{vec}_u\left(\boldsymbol{A}\right)$ take care of the diagonal, strictly below diagonal, and strictly above diagonal elements of $\boldsymbol{A}$, respectively.

To show how the three matrices $\boldsymbol{L}_d$, $\boldsymbol{L}_l$, and $\boldsymbol{L}_u$ can be found, the following two examples are presented.

Example 2.6 This example is related to Example 2.5, where we studied $\boldsymbol{A} \in \mathbb{C}^{3\times 3}$ given in (2.40) and the four vectorization operators applied to $\boldsymbol{A}$, as shown in (2.41).

By comparing (2.41) and (2.42), the matrices $\boldsymbol{L}_d$, $\boldsymbol{L}_l$, and $\boldsymbol{L}_u$ are found as

$$\boldsymbol{L}_d = \begin{bmatrix} 1 & 0 & 0 \\ 0 & 0 & 0 \\ 0 & 0 & 0 \\ 0 & 0 & 0 \\ 0 & 1 & 0 \\ 0 & 0 & 0 \\ 0 & 0 & 0 \\ 0 & 0 & 0 \\ 0 & 0 & 1 \end{bmatrix}, \quad \boldsymbol{L}_l = \begin{bmatrix} 0 & 0 & 0 \\ 1 & 0 & 0 \\ 0 & 1 & 0 \\ 0 & 0 & 0 \\ 0 & 0 & 0 \\ 0 & 0 & 1 \\ 0 & 0 & 0 \\ 0 & 0 & 0 \\ 0 & 0 & 0 \end{bmatrix}, \quad \boldsymbol{L}_u = \begin{bmatrix} 0 & 0 & 0 \\ 0 & 0 & 0 \\ 0 & 0 & 0 \\ 1 & 0 & 0 \\ 0 & 0 & 0 \\ 0 & 0 & 0 \\ 0 & 1 & 0 \\ 0 & 0 & 1 \\ 0 & 0 & 0 \end{bmatrix}. \tag{2.43}$$

Example 2.7 Let $N = 4$, then,

$$
\boldsymbol{L}_d = \begin{bmatrix} 1&0&0&0\\0&0&0&0\\0&0&0&0\\0&0&0&0\\0&0&0&0\\0&1&0&0\\0&0&0&0\\0&0&0&0\\0&0&0&0\\0&0&0&0\\0&0&1&0\\0&0&0&0\\0&0&0&0\\0&0&0&0\\0&0&0&0\\0&0&0&1 \end{bmatrix},\
\boldsymbol{L}_l = \begin{bmatrix} 0&0&0&0&0&0\\1&0&0&0&0&0\\0&1&0&0&0&0\\0&0&1&0&0&0\\0&0&0&0&0&0\\0&0&0&0&0&0\\0&0&0&1&0&0\\0&0&0&0&1&0\\0&0&0&0&0&0\\0&0&0&0&0&0\\0&0&0&0&0&0\\0&0&0&0&0&1\\0&0&0&0&0&0\\0&0&0&0&0&0\\0&0&0&0&0&0\\0&0&0&0&0&0 \end{bmatrix},\
\boldsymbol{L}_u = \begin{bmatrix} 0&0&0&0&0&0\\0&0&0&0&0&0\\0&0&0&0&0&0\\0&0&0&0&0&0\\1&0&0&0&0&0\\0&0&0&0&0&0\\0&0&0&0&0&0\\0&0&0&0&0&0\\0&1&0&0&0&0\\0&0&0&1&0&0\\0&0&0&0&0&0\\0&0&0&0&0&0\\0&0&1&0&0&0\\0&0&0&0&1&0\\0&0&0&0&0&1\\0&0&0&0&0&0 \end{bmatrix}. \tag{2.44}
$$

In Exercise 2.12, `MATLAB` programs should be developed for calculating the three matrices $\boldsymbol{L}_d$, $\boldsymbol{L}_l$, and $\boldsymbol{L}_u$. The matrix $\boldsymbol{L}_d$ has also been considered in Magnus and Neudecker (1988, Problem 4, p. 64) and is called the reduction matrix in Payaró and Palomar (2009, Appendix A). The two matrices $\boldsymbol{L}_l$ and $\boldsymbol{L}_u$ were introduced in Hjørungnes and Palomar (2008*a* and 2008*b*).

To find and identify Hessians of complex-valued vectors and matrices, the following definition (related to the definition in Magnus & Neudecker (1988, pp. 107–108)) is needed:

Definition 2.13 (Block Vectorization Operator) *Let $\boldsymbol{C} \in \mathbb{C}^{N\times NM}$ be the matrix given by*

$$
\boldsymbol{C} = [\boldsymbol{C}_0 \ \boldsymbol{C}_1 \ \cdots \ \boldsymbol{C}_{M-1}], \tag{2.45}
$$

where each of the block matrices $\boldsymbol{C}_i$ is complex valued and the square *of size $N \times N$, where $i \in \{0, 1, \ldots, M-1\}$. Then the* block vectorization operator *is denoted by* vecb$(\cdot)$, *and it returns the $NM \times N$ matrix given by*

$$
\operatorname{vecb}(\boldsymbol{C}) = \begin{bmatrix} \boldsymbol{C}_0 \\ \boldsymbol{C}_1 \\ \vdots \\ \boldsymbol{C}_{M-1} \end{bmatrix}. \tag{2.46}
$$

If $\operatorname{vecb}\left(\boldsymbol{C}^T\right) = \boldsymbol{C}$, *the matrix $\boldsymbol{C}$ is called* column symmetric *(Magnus and Neudecker 1988, p. 108), or equivalently $\boldsymbol{C}_i^T = \boldsymbol{C}_i$ for all $i \in \{0, 1, \ldots, M-1\}$.*

The above definition is an extension of Magnus and Neudecker (1988, p. 108) to complex matrices, such that it can be used in connection with complex-valued Hessians. A matrix that is useful for generating symmetric matrices is the *duplication matrix*. It is defined next, together with yet another vectorization operator.

Definition 2.14 (Duplication Matrix) *Let the operator* $v : \mathbb{C}^{N\times N} \to \mathbb{C}^{\frac{(N+1)N}{2}\times 1}$ *return all the elements* on and below *the main diagonal taken in the same column-wise order as the ordinary* vec*-operator:*

$$v(\boldsymbol{A}) = \left[a_{0,0}, a_{1,0}, \ldots, a_{N-1,0}, a_{1,1}, a_{2,1}, \ldots, a_{N-1,1}, a_{3,3}, \ldots, a_{N-1,N-1}\right]^T. \quad (2.47)$$

Let $\boldsymbol{A} \in \mathbb{C}^{N\times N}$ *be symmetric, then it is possible to construct* $\text{vec}(\boldsymbol{A})$ *from* $v(\boldsymbol{A})$ *with a unique matrix of size* $N^2 \times \frac{N(N+1)}{2}$ *called the* duplication matrix*; it is denoted by* $\boldsymbol{D}_N$*, and is defined by the following relation:*

$$\boldsymbol{D}_N v(\boldsymbol{A}) = \text{vec}(\boldsymbol{A}). \quad (2.48)$$

In Exercise 2.13, an explicit formula is developed for the duplication matrix, and a MATLAB program should be found for calculating the duplication matrix $\boldsymbol{D}_N$.

Let $\boldsymbol{A} \in \mathbb{C}^{N\times N}$ be symmetric such that $\boldsymbol{A}^T = \boldsymbol{A}$. If the definition of $v(\cdot)$ in Definition 2.14 is compared with the definitions of $\text{vec}_d(\cdot)$ and $\text{vec}_l(\cdot)$ in Definition 2.11, it can be seen that $v(\boldsymbol{A})$ contains the same elements as the two operators $\text{vec}_d(\boldsymbol{A})$ and $\text{vec}_l(\boldsymbol{A})$. In the next definition, the unique matrices used to transfer between these vectorization operators are defined.

Definition 2.15 (Matrices $\boldsymbol{V}_d$, $\boldsymbol{V}_l$, and $\boldsymbol{V}$) *Let* $\boldsymbol{A} \in \mathbb{C}^{N\times N}$ *be symmetric.* Unique *matrices* $\boldsymbol{V}_d \in \mathbb{Z}_2^{\frac{N(N+1)}{2}\times N}$ *and* $\boldsymbol{V}_l \in \mathbb{Z}_2^{\frac{N(N+1)}{2}\times\frac{N(N-1)}{2}}$ *contain zeros everywhere except for* $+1$ *at one place in each column, and these matrices can be used to build up* $v(\boldsymbol{A})$*, from* $\text{vec}_d(\boldsymbol{A})$ *and* $\text{vec}_l(\boldsymbol{A})$ *in the following way:*

$$v(\boldsymbol{A}) = \boldsymbol{V}_d\,\text{vec}_d(\boldsymbol{A}) + \boldsymbol{V}_l\,\text{vec}_l(\boldsymbol{A}) = [\boldsymbol{V}_d, \ \boldsymbol{V}_l]\begin{bmatrix}\text{vec}_d(\boldsymbol{A})\\ \text{vec}_l(\boldsymbol{A})\end{bmatrix}. \quad (2.49)$$

The square permutation matrix $\boldsymbol{V} \in \mathbb{Z}_2^{\frac{N(N+1)}{2}\times\frac{N(N+1)}{2}}$ *is defined by*

$$\boldsymbol{V} = [\boldsymbol{V}_d, \ \boldsymbol{V}_l]. \quad (2.50)$$

Because the matrix $\boldsymbol{V}$ *is a permutation matrix, it follows from* $\boldsymbol{V}^T\boldsymbol{V} = \boldsymbol{I}_{\frac{N(N+1)}{2}}$ *that*

$$\boldsymbol{V}_d^T\boldsymbol{V}_d = \boldsymbol{I}_N, \quad (2.51)$$

$$\boldsymbol{V}_l^T\boldsymbol{V}_l = \boldsymbol{I}_{\frac{(N-1)N}{2}}, \quad (2.52)$$

$$\boldsymbol{V}_d^T\boldsymbol{V}_l = \boldsymbol{0}_{N\times\frac{(N-1)N}{2}}. \quad (2.53)$$

Definition 2.16 (Standard Basis) *Let the* standard basis *in* $\mathbb{C}^{N\times 1}$ *be denoted by* $\boldsymbol{e}_i$*, where* $i \in \{0, 1, \ldots, N-1\}$*. The standard basis in* $\mathbb{C}^{N\times N}$ *is denoted by* $\boldsymbol{E}_{i,j} \in \mathbb{C}^{N\times N}$ *and is* *defined as*

$$\boldsymbol{E}_{i,j} = \boldsymbol{e}_i \boldsymbol{e}_j^T, \tag{2.54}$$

where $i, j \in \{0, 1, \ldots, N-1\}$.

2.5 Useful Manipulation Formulas

In this section, several useful manipulation formulas are presented. Although many of these results are well known in the literature, they are included here to make the text more complete.

A classical result from linear algebra is that if $\boldsymbol{A} \in \mathbb{C}^{N \times Q}$, then (Horn & Johnson 1985, p. 13)

$$\text{rank}\,(\boldsymbol{A}) + \dim_{\mathbb{C}}\,(\mathcal{N}(\boldsymbol{A})) = Q. \tag{2.55}$$

The following lemma states Hadamard's inequality (Magnus & Neudecker 1988), and it will be used in Chapter 6 to derive the water-filling solution of the capacity of MIMO channels.

Lemma 2.1 *Let* $\boldsymbol{A} \in \mathbb{C}^{N \times N}$ *be a positive definite matrix given by*

$$\boldsymbol{A} = \begin{bmatrix} \boldsymbol{B} & \boldsymbol{c} \\ \boldsymbol{c}^H & a_{N-1,N-1} \end{bmatrix}, \tag{2.56}$$

where $\boldsymbol{c} \in \mathbb{C}^{(N-1)\times 1}$, $\boldsymbol{B} \in \mathbb{C}^{(N-1)\times(N-1)}$, *and* $a_{N-1,N-1}$ *represent a positive scalar. Then*

$$\det(\boldsymbol{A}) \leq a_{N-1,N-1} \det(\boldsymbol{B}), \tag{2.57}$$

with equality if and only if $\boldsymbol{c} = \boldsymbol{0}_{(N-1)\times 1}$. *By repeated application of* (2.57), *it follows that if* $\boldsymbol{A} \in \mathbb{C}^{N \times N}$ *is a positive definite matrix, then*

$$\det(\boldsymbol{A}) \leq \prod_{k=0}^{N-1} a_{k,k}, \tag{2.58}$$

with equality if and only if $\boldsymbol{A}$ *is diagonal.*

Proof Because $\boldsymbol{A}$ is positive definite, $\boldsymbol{B}$ is positive definite and $a_{N-1,N-1}$ is a positive scalar. The matrix $\boldsymbol{B}^{-1}$ is also positive definite. Let $\boldsymbol{P} \in \mathbb{C}^{N \times N}$ be given as

$$\boldsymbol{P} = \begin{bmatrix} \boldsymbol{I}_{N-1}, & \boldsymbol{0}_{(N-1)\times 1} \\ -\boldsymbol{c}^H \boldsymbol{B}^{-1}, & 1 \end{bmatrix}. \tag{2.59}$$

It follows that $\det(\boldsymbol{P}) = 1$. By multiplying out, it follows that

$$\boldsymbol{P}\boldsymbol{A} = \begin{bmatrix} \boldsymbol{B}, & \boldsymbol{c} \\ \boldsymbol{0}_{1\times(N-1)}, & \alpha \end{bmatrix}, \tag{2.60}$$

where $\alpha = a_{N-1,N-1} - \boldsymbol{c}^H \boldsymbol{B}^{-1}\boldsymbol{c}$. By taking the determinant of both sides of (2.60), it follows that

$$\det(\boldsymbol{P}\boldsymbol{A}) = \det(\boldsymbol{A}) = \alpha \det(\boldsymbol{B}). \tag{2.61}$$

Because $\boldsymbol{B}^{-1}$ is positive definite, it follows that $\boldsymbol{c}^H \boldsymbol{B}^{-1}\boldsymbol{c} \geq 0$. From the definition of α, it now follows that $\alpha \leq a_{N-1,N-1}$. Putting these results together leads to the inequality in (2.57), where equality holds if and only if $\boldsymbol{c}^H \boldsymbol{B}^{-1}\boldsymbol{c} = 0$, which is equivalent to $\boldsymbol{c} = \boldsymbol{0}_{(N-1)\times 1}$. ■

The following lemma contains some of the results found in Bernstein (2005, pp. 44–45).

Lemma 2.2 *Let* $\boldsymbol{A} \in \mathbb{C}^{N\times N}$, $\boldsymbol{B} \in \mathbb{C}^{N\times M}$, $\boldsymbol{C} \in \mathbb{C}^{M\times N}$, *and* $\boldsymbol{D} \in \mathbb{C}^{M\times M}$. *If* $\boldsymbol{A}$ *is nonsingular, then*

$$\begin{bmatrix} \boldsymbol{A} & \boldsymbol{B} \\ \boldsymbol{C} & \boldsymbol{D} \end{bmatrix} = \begin{bmatrix} \boldsymbol{I}_N & \boldsymbol{0}_{N\times M} \\ \boldsymbol{C}\boldsymbol{A}^{-1} & \boldsymbol{I}_M \end{bmatrix} \begin{bmatrix} \boldsymbol{A} & \boldsymbol{0}_{N\times M} \\ \boldsymbol{0}_{M\times N} & \boldsymbol{D} - \boldsymbol{C}\boldsymbol{A}^{-1}\boldsymbol{B} \end{bmatrix} \begin{bmatrix} \boldsymbol{I}_N & \boldsymbol{A}^{-1}\boldsymbol{B} \\ \boldsymbol{0}_{M\times N} & \boldsymbol{I}_M \end{bmatrix}. \tag{2.62}$$

This result leads to

$$\det\left(\begin{bmatrix} \boldsymbol{A} & \boldsymbol{B} \\ \boldsymbol{C} & \boldsymbol{D} \end{bmatrix}\right) = \det(\boldsymbol{A}) \det\left(\boldsymbol{D} - \boldsymbol{C}\boldsymbol{A}^{-1}\boldsymbol{B}\right). \tag{2.63}$$

If $\boldsymbol{D}$ *is nonsingular, then*

$$\begin{bmatrix} \boldsymbol{A} & \boldsymbol{B} \\ \boldsymbol{C} & \boldsymbol{D} \end{bmatrix} = \begin{bmatrix} \boldsymbol{I}_N & \boldsymbol{B}\boldsymbol{D}^{-1} \\ \boldsymbol{0}_{M\times N} & \boldsymbol{I}_M \end{bmatrix} \begin{bmatrix} \boldsymbol{A} - \boldsymbol{B}\boldsymbol{D}^{-1}\boldsymbol{C} & \boldsymbol{0}_{N\times M} \\ \boldsymbol{0}_{M\times N} & \boldsymbol{D} \end{bmatrix} \begin{bmatrix} \boldsymbol{I}_N & \boldsymbol{0}_{N\times M} \\ \boldsymbol{D}^{-1}\boldsymbol{C} & \boldsymbol{I}_M \end{bmatrix}. \tag{2.64}$$

Hence,

$$\det\left(\begin{bmatrix} \boldsymbol{A} & \boldsymbol{B} \\ \boldsymbol{C} & \boldsymbol{D} \end{bmatrix}\right) = \det\left(\boldsymbol{A} - \boldsymbol{B}\boldsymbol{D}^{-1}\boldsymbol{C}\right) \det(\boldsymbol{D}). \tag{2.65}$$

If both $\boldsymbol{A}$ *and* $\boldsymbol{D}$ *are nonsingular, it follows from* (2.63) *and* (2.65) *that* $\boldsymbol{D} - \boldsymbol{C}\boldsymbol{A}^{-1}\boldsymbol{B}$ *is nonsingular, if and only if* $\boldsymbol{A} - \boldsymbol{B}\boldsymbol{D}^{-1}\boldsymbol{C}$ *is nonsingular.*

Proof The results in (2.62) and (2.64) are obtained by block matrix multiplication of the right-hand sides of these two equations. All other results in the lemma are direct consequences of (2.62) and (2.64). ■

The following lemma (Kailath, Sayed, & Hassibi 2000, p. 729) is called the *matrix inversion lemma* and is used many times in signal processing and communications (Sayed 2003; Barry, Lee, & Messerschmitt 2004).

Lemma 2.3 (Matrix Inversion Lemma) *Let* $\boldsymbol{A} \in \mathbb{C}^{N\times N}$, $\boldsymbol{B} \in \mathbb{C}^{N\times M}$, $\boldsymbol{C} \in \mathbb{C}^{M\times M}$, *and* $\boldsymbol{D} \in \mathbb{C}^{M\times N}$. *If* $\boldsymbol{A}$, $\boldsymbol{C}$, *and* $\boldsymbol{A} + \boldsymbol{B}\boldsymbol{C}\boldsymbol{D}$ *are invertible, then* $\boldsymbol{C}^{-1} + \boldsymbol{D}\boldsymbol{A}^{-1}\boldsymbol{B}$ *is invertible and*

$$[\boldsymbol{A} + \boldsymbol{B}\boldsymbol{C}\boldsymbol{D}]^{-1} = \boldsymbol{A}^{-1} - \boldsymbol{A}^{-1}\boldsymbol{B}\left[\boldsymbol{C}^{-1} + \boldsymbol{D}\boldsymbol{A}^{-1}\boldsymbol{B}\right]^{-1}\boldsymbol{D}\boldsymbol{A}^{-1}. \tag{2.66}$$

The reader is asked to prove this lemma in Exercise 2.17.

To reformulate expressions, the following lemmas are useful.

Lemma 2.4 *Let* $A \in \mathbb{C}^{N\times M}$ *and* $B \in \mathbb{C}^{M\times N}$*, then*

$$\det(I_N + AB) = \det(I_M + BA). \tag{2.67}$$

Proof This result can be shown by taking the determinant of both sides of the following identity:

$$\begin{bmatrix} I_N + AB, & A \\ 0_{M\times N}, & I_M \end{bmatrix}\begin{bmatrix} I_N, & 0_{N\times M} \\ -B, & I_M \end{bmatrix} = \begin{bmatrix} I_N, & 0_{N\times M} \\ -B, & I_M \end{bmatrix}\begin{bmatrix} I_N, & A \\ 0_{M\times N}, & I_M + BA \end{bmatrix}, \tag{2.68}$$

which are two ways of expressing the matrix $\begin{bmatrix} I_N, & A \\ -B, & I_M \end{bmatrix}$.

Alternatively, this lemma can be shown by means of (2.63) and (2.65). ■

Lemma 2.5 *Let* $A \in \mathbb{C}^{N\times M}$ *and* $B \in \mathbb{C}^{M\times N}$*. The* $N \times N$ *matrix* $I_N + AB$ *is invertible if and only if the* $M \times M$ *matrix* $I_M + BA$ *is invertible. If these two matrices are invertible, then*

$$B(I_N + AB)^{-1} = (I_M + BA)^{-1} B. \tag{2.69}$$

Proof From (2.67), it follows that $I_N + AB$ is invertible if and only if $I_M + BA$ is invertible.

By multiplying out both sides, it can be seen that the following relation holds:

$$B(I_N + AB) = (I_M + BA) B. \tag{2.70}$$

Right-multiplying the above equation with $(I_N + AB)^{-1}$ and left-multiplying with $(I_M + BA)^{-1}$ lead to (2.69). ■

The following lemma can be used to show that it is difficult to parameterize the set of all orthogonal matrices; it is found in Bernstein (2005, Corollary 11.2.4).

Lemma 2.6 *Let* $A \in \mathbb{C}^{N\times N}$*, then*

$$\det(\exp(A)) = \exp(\mathrm{Tr}\{A\}). \tag{2.71}$$

The rest of this section consists of several subsections that contain results of different categories. Subsection 2.5.1 shows several results of the Moore-Penrose inverse that will be useful when its complex differential is derived in Chapter 3. In Subsection 2.5.2, results involving the trace operator are collected. Useful material with the Kronecker and Hadamard products is presented in Subsection 2.5.3. Results that will be used to identify second-order derivatives are formulated around complex quadratic forms in

Subsection 2.5.4. Several lemmas that will be useful for finding generalized complex-valued matrix derivatives in Chapter 6 are provided in Subsection 2.5.5.

2.5.1 Moore-Penrose Inverse

Lemma 2.7 *Let $A \in \mathbb{C}^{N\times Q}$ and $B \in \mathbb{C}^{Q\times R}$, then the following properties are valid for the Moore-Penrose inverse:*

$$A^{+} = A^{-1} \text{ for nonsingular } A, \tag{2.72}$$

$$\left(A^{+}\right)^{+} = A, \tag{2.73}$$

$$\left(A^{H}\right)^{+} = \left(A^{+}\right)^{H}, \tag{2.74}$$

$$A^{H} = A^{H}AA^{+} = A^{+}AA^{H}, \tag{2.75}$$

$$A^{+} = A^{H}\left(A^{+}\right)^{H}A^{+} = A^{+}\left(A^{+}\right)^{H}A^{H}, \tag{2.76}$$

$$\left(A^{H}A\right)^{+} = A^{+}\left(A^{+}\right)^{H}, \tag{2.77}$$

$$\left(AA^{H}\right)^{+} = \left(A^{+}\right)^{H}A^{+}, \tag{2.78}$$

$$A^{+} = \left(A^{H}A\right)^{+}A^{H} = A^{H}\left(AA^{H}\right)^{+}, \tag{2.79}$$

$$A^{+} = \left(A^{H}A\right)^{-1}A^{H} \text{ if } A \text{ has full column rank}, \tag{2.80}$$

$$A^{+} = A^{H}\left(AA^{H}\right)^{-1} \text{ if } A \text{ has full row rank}, \tag{2.81}$$

$$AB = 0_{N\times R} \Leftrightarrow B^{+}A^{+} = 0_{R\times N}. \tag{2.82}$$

Proof Equations (2.72), (2.73), and (2.74) can be proved by direct insertion into the definition of the Moore-Penrose inverse.

The first part of (2.75) can be proved as follows:

$$A^{H} = A^{H}\left(A^{H}\right)^{+}A^{H} = A^{H}\left(AA^{+}\right)^{H} = A^{H}AA^{+}, \tag{2.83}$$

where the results from (2.22) and (2.74) were used. The second part of (2.75) can be proved in a similar way

$$A^{H} = A^{H}\left(A^{H}\right)^{+}A^{H} = \left(A^{+}A\right)^{H}A^{H} = A^{+}AA^{H}. \tag{2.84}$$

The first part of (2.76) can be shown by

$$A^{+} = A^{+}AA^{+} = \left(A^{H}\left(A^{H}\right)^{+}\right)^{H}A^{+} = A^{H}\left(A^{+}\right)^{H}A^{+}, \tag{2.85}$$

where (2.22) was utilized in the last equality above.
The second part of (2.76) can be proved in an analogous manner

$$A^{+} = A^{+}AA^{+} = A^{+}\left(\left(A^{+}\right)^{H}A^{H}\right)^{H} = A^{+}\left(A^{H}\right)^{+}A^{H}, \tag{2.86}$$

where (2.23) was used in the last equality.

Equations (2.77) and (2.78) can be proved by using the results from (2.75) and (2.76) in the definition of the Moore-Penrose inverse.

Equation (2.79) follows from (2.76), (2.77), and (2.78).

Equations (2.80) and (2.81) follow from (2.72) and (2.79), together with the following fact: $\text{rank}\,(A) = \text{rank}\left(A^H A\right) = \text{rank}\left(AA^H\right)$ (Horn & Johnson 1985, Section 0.4.6).

Now, (2.82) will be shown. First, it is shown that $\boldsymbol{AB} = \mathbf{0}_{N\times R}$ implies that $\boldsymbol{B}^+\boldsymbol{A}^+ = \mathbf{0}_{R\times N}$. Assume that $\boldsymbol{AB} = \mathbf{0}_{N\times R}$. From (2.79), it follows that

$$\boldsymbol{B}^+\boldsymbol{A}^+ = \left(\boldsymbol{B}^H\boldsymbol{B}\right)^+\boldsymbol{B}^H\boldsymbol{A}^H\left(\boldsymbol{A}\boldsymbol{A}^H\right)^+. \tag{2.87}$$

$\boldsymbol{AB} = \mathbf{0}_{N\times R}$ leads to $\boldsymbol{B}^H\boldsymbol{A}^H = \mathbf{0}_{R\times N}$, then (2.87) yields $\boldsymbol{B}^+\boldsymbol{A}^+ = \mathbf{0}_{R\times N}$. Second, it will be shown that $\boldsymbol{B}^+\boldsymbol{A}^+ = \mathbf{0}_{R\times N}$ implies that $\boldsymbol{AB} = \mathbf{0}_{N\times R}$. Assume that $\boldsymbol{B}^+\boldsymbol{A}^+ = \mathbf{0}_{R\times N}$. Using the implication just proved (i.e., $\boldsymbol{CD} = \mathbf{0}_{M\times P}$), then $\boldsymbol{D}^+\boldsymbol{C}^+ = \mathbf{0}_{P\times M}$, where M and P are positive integers given by the size of the matrices $\boldsymbol{C}$ and $\boldsymbol{D}$, gives $\left(\boldsymbol{A}^+\right)^+\left(\boldsymbol{B}^+\right)^+ = \mathbf{0}_{N\times R}$, the desired result follows from (2.73). ■

Lemma 2.8 *Let* $\boldsymbol{A} \in \mathbb{C}^{N\times Q}$*, then these equalities follow:*

$$\mathcal{R}(\boldsymbol{A}) = \mathcal{R}\left(\boldsymbol{A}^+\boldsymbol{A}\right), \tag{2.88}$$

$$\mathcal{C}(\boldsymbol{A}) = \mathcal{C}\left(\boldsymbol{A}\boldsymbol{A}^+\right), \tag{2.89}$$

$$\text{rank}\,(\boldsymbol{A}) = \text{rank}\left(\boldsymbol{A}^+\boldsymbol{A}\right) = \text{rank}\left(\boldsymbol{A}\boldsymbol{A}^+\right). \tag{2.90}$$

Proof From (2.24) and the definition of $\mathcal{R}(\boldsymbol{A})$, it follows that

$$\mathcal{R}(\boldsymbol{A}) = \left\{\boldsymbol{w} \in \mathbb{C}^{1\times Q} \mid \boldsymbol{w} = \boldsymbol{z}\boldsymbol{A}\left(\boldsymbol{A}^+\boldsymbol{A}\right), \text{ for some } \boldsymbol{z} \in \mathbb{C}^{1\times N}\right\} \subseteq \mathcal{R}\left(\boldsymbol{A}^+\boldsymbol{A}\right). \tag{2.91}$$

From the definition of $\mathcal{R}\left(\boldsymbol{A}^+\boldsymbol{A}\right)$, it follows that

$$\mathcal{R}\left(\boldsymbol{A}^+\boldsymbol{A}\right) = \left\{\boldsymbol{w} \in \mathbb{C}^{1\times Q} \mid \boldsymbol{w} = \boldsymbol{z}\boldsymbol{A}^+\boldsymbol{A}, \text{ for some } \boldsymbol{z} \in \mathbb{C}^{1\times Q}\right\} \subseteq \mathcal{R}(\boldsymbol{A}). \tag{2.92}$$

From (2.91) and (2.92), (2.88) follows. From (2.24) and the definition of $\mathcal{C}(\boldsymbol{A})$, it follows that

$$\mathcal{C}(\boldsymbol{A}) = \left\{\boldsymbol{w} \in \mathbb{C}^{N\times 1} \mid \boldsymbol{w} = \left(\boldsymbol{A}\boldsymbol{A}^+\right)\boldsymbol{A}\boldsymbol{z}, \text{ for some } \boldsymbol{z} \in \mathbb{C}^{Q\times 1}\right\} \subseteq \mathcal{C}\left(\boldsymbol{A}\boldsymbol{A}^+\right). \tag{2.93}$$

From the definition of $\mathcal{C}\left(\boldsymbol{A}\boldsymbol{A}^+\right)$, it follows that

$$\mathcal{C}\left(\boldsymbol{A}\boldsymbol{A}^+\right) = \left\{\boldsymbol{w} \in \mathbb{C}^{N\times 1} \mid \boldsymbol{w} = \boldsymbol{A}\boldsymbol{A}^+\boldsymbol{z}, \text{ for some } \boldsymbol{z} \in \mathbb{C}^{N\times 1}\right\} \subseteq \mathcal{C}(\boldsymbol{A}). \tag{2.94}$$

From (2.93) and (2.94), (2.89) follows. Equation (2.90) is a direct consequence of (2.88) and (2.89). ■

2.5.2 Trace Operator

From the definition of the $\text{Tr}\{\cdot\}$ operator, it follows that

$$\text{Tr}\left\{\boldsymbol{A}^T\right\} = \text{Tr}\{\boldsymbol{A}\}, \tag{2.95}$$

where $A \in \mathbb{C}^{N \times N}$. When dealing with the trace operator, the following formula is useful:

$$\mathrm{Tr}\{AB\} = \mathrm{Tr}\{BA\}, \tag{2.96}$$

where $A \in \mathbb{C}^{N \times Q}$ and $B \in \mathbb{C}^{Q \times N}$. Equation (2.96) can be proved by expressing the two sides as double sums of the components of matrices. The readers are asked to prove (2.96) in Exercise 2.9.

The $\mathrm{Tr}\{\cdot\}$ and $\mathrm{vec}(\cdot)$ operators are connected by the following formula:

$$\mathrm{Tr}\left\{A^T B\right\} = \mathrm{vec}^T(A)\,\mathrm{vec}(B), \tag{2.97}$$

where $\mathrm{vec}^T(A) = (\mathrm{vec}(A))^T$. The identity in (2.97) is shown in Exercise 2.9.

Let a_m and a_n be two complex-valued column vectors of the same size, then

$$a_m^T a_n = a_n^T a_m, \tag{2.98}$$

$$a_m^H a_n = a_n^T a_m^*. \tag{2.99}$$

For a scalar complex-valued quantity a, the following relations are obvious, but are useful for manipulating scalar expressions:

$$a = \mathrm{Tr}\{a\} = \mathrm{vec}(a). \tag{2.100}$$

The following result is well known from Harville (1997, Lemma 10.1.1 and Corollary 10.2.2):

Proposition 2.1 *If $A \in \mathbb{C}^{N \times N}$ is idempotent, then* $\mathrm{rank}(A) = \mathrm{Tr}\{A\}$. *If A, in addition, has full rank, then $A = I_N$.*

The reader is asked to prove Proposition 2.1 in Exercise 2.15.

2.5.3 Kronecker and Hadamard Products

Let $a_i \in \mathbb{C}^{N_i \times 1}$, where $i \in \{0, 1\}$, then

$$\mathrm{vec}\left(a_0 a_1^T\right) = a_1 \otimes a_0. \tag{2.101}$$

The result in (2.101) is shown in Exercise 2.14.

Lemma 2.9 *Let $A \in \mathbb{C}^{M \times N}$ and $B \in \mathbb{C}^{P \times Q}$, then*

$$(A \otimes B)^T = A^T \otimes B^T. \tag{2.102}$$

The proof of Lemma 2.9 is left for the reader in Exercise 2.16.

Lemma 2.10 (Magnus & Neudecker 1988; Harville 1997) *Let the sizes of the matrices be given such that the products AC and BD are well defined. Then*

$$(A \otimes B)(C \otimes D) = AC \otimes BD. \tag{2.103}$$

Proof Let $\boldsymbol{A} \in \mathbb{C}^{M \times N}$, $\boldsymbol{B} \in \mathbb{C}^{P \times Q}$, $\boldsymbol{C} \in \mathbb{C}^{N \times R}$, and $\boldsymbol{D} \in \mathbb{C}^{Q \times S}$. Denote element number (m, k) of the matrix $\boldsymbol{A}$ by $a_{m,k}$ and element number (k, n) of the matrix $\boldsymbol{C}$ by $c_{k,n}$. The (m, k)-th block matrix of size $P \times Q$ of the matrix $\boldsymbol{A} \otimes \boldsymbol{B}$ is $a_{m,k}\boldsymbol{B}$, and the (k, n)-th block matrix of size $Q \times S$ of the matrix $\boldsymbol{C} \otimes \boldsymbol{D}$ is $c_{k,n}\boldsymbol{D}$. Thus, the (m, n)-th block matrix of size $P \times S$ of the matrix $(\boldsymbol{A} \otimes \boldsymbol{B})(\boldsymbol{C} \otimes \boldsymbol{D})$ is given by

$$\sum_{k=0}^{N-1} a_{m,k}\boldsymbol{B}c_{k,n}\boldsymbol{D} = \left(\sum_{k=0}^{N-1} a_{m,k}c_{k,n}\right)\boldsymbol{B}\boldsymbol{D}, \tag{2.104}$$

which is equal to the (m, n)-th element of $\boldsymbol{A}\boldsymbol{C}$ times the $P \times S$ block $\boldsymbol{B}\boldsymbol{D}$, which is the (m, n)-th block of size $P \times S$ of the matrix $\boldsymbol{A}\boldsymbol{C} \otimes \boldsymbol{B}\boldsymbol{D}$. ■

To extract the vec(·) of an inner matrix from the vec(·) of a multiple-matrix product, the following result is very useful:

Lemma 2.11 *Let the sizes of the matrices* $\boldsymbol{A}$*,* $\boldsymbol{B}$*, and* $\boldsymbol{C}$ *be such that the matrix product* $\boldsymbol{A}\boldsymbol{B}\boldsymbol{C}$ *is well defined, then*

$$\operatorname{vec}(\boldsymbol{A}\boldsymbol{B}\boldsymbol{C}) = \left(\boldsymbol{C}^T \otimes \boldsymbol{A}\right)\operatorname{vec}(\boldsymbol{B}). \tag{2.105}$$

Proof Let $\boldsymbol{B} \in \mathbb{C}^{N \times Q}$, and let $\boldsymbol{B}_{:,k}$ denote column[5] number k of the matrix $\boldsymbol{B}$, and let $\boldsymbol{e}_k$ denote the standard basis vectors of size $Q \times 1$, where $k \in \{0, 1, \ldots, Q-1\}$. Then the matrix $\boldsymbol{B}$ can be expressed as $\boldsymbol{B} = \sum_{k=0}^{Q-1} \boldsymbol{B}_{:,k}\boldsymbol{e}_k^T$. By using (2.101) and (2.103), the following expression is obtained:

$$\begin{aligned}
\operatorname{vec}(\boldsymbol{A}\boldsymbol{B}\boldsymbol{C}) &= \operatorname{vec}\left(\sum_{k=0}^{Q-1} \boldsymbol{A}\boldsymbol{B}_{:,k}\boldsymbol{e}_k^T\boldsymbol{C}\right) = \sum_{k=0}^{Q-1}\operatorname{vec}\left(\left(\boldsymbol{A}\boldsymbol{B}_{:,k}\right)\left(\boldsymbol{C}^T\boldsymbol{e}_k\right)^T\right) \\
&= \sum_{k=0}^{Q-1}\left(\boldsymbol{C}^T\boldsymbol{e}_k \otimes \boldsymbol{A}\boldsymbol{B}_{:,k}\right) = \left(\boldsymbol{C}^T \otimes \boldsymbol{A}\right)\sum_{k=0}^{Q-1}\left(\boldsymbol{e}_k \otimes \boldsymbol{B}_{:,k}\right) \\
&= \left(\boldsymbol{C}^T \otimes \boldsymbol{A}\right)\sum_{k=0}^{Q-1}\operatorname{vec}\left(\boldsymbol{B}_{:,k}\boldsymbol{e}_k^T\right) = \left(\boldsymbol{C}^T \otimes \boldsymbol{A}\right)\operatorname{vec}(\boldsymbol{B}).
\end{aligned} \tag{2.106}$$

■

Let $\boldsymbol{a} \in \mathbb{C}^{N \times 1}$, then

$$\boldsymbol{a} = \operatorname{vec}(\boldsymbol{a}) = \operatorname{vec}(\boldsymbol{a}^T). \tag{2.107}$$

If $\boldsymbol{b} \in \mathbb{C}^{1 \times N}$, then

$$\boldsymbol{b} = \operatorname{vec}^T(\boldsymbol{b}) = \operatorname{vec}^T(\boldsymbol{b}^T). \tag{2.108}$$

[5] The notations $\boldsymbol{B}_{:,k}$ and $\boldsymbol{b}_k$ are used to denote the k-th column of the matrix $\boldsymbol{B}$.

The commutation matrix is denoted by $\boldsymbol{K}_{Q,N}$, and it is a permutation matrix (see Definition 2.9). It is shown in Magnus and Neudecker (1988, Section 3.7, p. 47) that

$$\boldsymbol{K}_{Q,N}^{T} = \boldsymbol{K}_{Q,N}^{-1} = \boldsymbol{K}_{N,Q}. \tag{2.109}$$

The results in (2.109) are proved in Exercise 2.6.

The following result (Magnus & Neudecker 1988, Theorem 3.9) gives the reason why the commutation matrix received its name:

Lemma 2.12 *Let $\boldsymbol{A}_i \in \mathbb{C}^{N_i \times Q_i}$ where $i \in \{0, 1\}$, then*

$$\boldsymbol{K}_{N_1,N_0} \left(\boldsymbol{A}_0 \otimes \boldsymbol{A}_1\right) = \left(\boldsymbol{A}_1 \otimes \boldsymbol{A}_0\right) \boldsymbol{K}_{Q_1,Q_0}. \tag{2.110}$$

Proof Let $\boldsymbol{X} \in \mathbb{C}^{Q_1 \times Q_0}$ be an arbitrary matrix. By utilizing (2.105) and (2.31), it can be seen that

$$\begin{aligned} \boldsymbol{K}_{N_1,N_0} \left(\boldsymbol{A}_0 \otimes \boldsymbol{A}_1\right) \operatorname{vec}\left(\boldsymbol{X}\right) &= \boldsymbol{K}_{N_1,N_0} \operatorname{vec}\left(\boldsymbol{A}_1 \boldsymbol{X} \boldsymbol{A}_0^T\right) = \operatorname{vec}\left(\boldsymbol{A}_0 \boldsymbol{X}^T \boldsymbol{A}_1^T\right) \\ &= \left(\boldsymbol{A}_1 \otimes \boldsymbol{A}_0\right) \operatorname{vec}\left(\boldsymbol{X}^T\right) = \left(\boldsymbol{A}_1 \otimes \boldsymbol{A}_0\right) \boldsymbol{K}_{Q_1,Q_0} \operatorname{vec}\left(\boldsymbol{X}\right). \end{aligned} \tag{2.111}$$

Because $\boldsymbol{X}$ was chosen arbitrarily, it is possible to set $\operatorname{vec}(\boldsymbol{X}) = \boldsymbol{e}_i$, where $\boldsymbol{e}_i$ is the standard basis vector in $\mathbb{C}^{Q_0 Q_1 \times 1}$. If this choice of $\operatorname{vec}(\boldsymbol{X})$ is inserted into (2.111), it can be seen that the i-th columns of the two $N_0 N_1 \times Q_0 Q_1$ matrices $\boldsymbol{K}_{N_1,N_0}(\boldsymbol{A}_0 \otimes \boldsymbol{A}_1)$ and $(\boldsymbol{A}_1 \otimes \boldsymbol{A}_0) \boldsymbol{K}_{Q_1,Q_0}$ are identical. This holds for all $i \in \{0, 1, \ldots, Q_0 Q_1 - 1\}$. Hence, (2.110) follows. ■

The following result is also given in Magnus and Neudecker (1988, Theorem 3.10).

Lemma 2.13 *Let $\boldsymbol{A}_i \in \mathbb{C}^{N_i \times Q_i}$, then*

$$\operatorname{vec}\left(\boldsymbol{A}_0 \otimes \boldsymbol{A}_1\right) = \left(\boldsymbol{I}_{Q_0} \otimes \boldsymbol{K}_{Q_1,N_0} \otimes \boldsymbol{I}_{N_1}\right) \left(\operatorname{vec}\left(\boldsymbol{A}_0\right) \otimes \operatorname{vec}\left(\boldsymbol{A}_1\right)\right). \tag{2.112}$$

Proof Let $\boldsymbol{e}_k^{(Q_i)}$ denote the standard basis vectors of size $Q_i \times 1$. $\boldsymbol{A}_i$ can be expressed as

$$\boldsymbol{A}_i = \sum_{k_i=0}^{Q_i-1} \left(\boldsymbol{A}_i\right)_{:,k_i} \left(\boldsymbol{e}_{k_i}^{(Q_i)}\right)^T, \tag{2.113}$$

where $i \in \{0, 1\}$. The left side of (2.112) can be expressed as

$$
\begin{aligned}
\operatorname{vec}(\boldsymbol{A}_0 \otimes \boldsymbol{A}_1) &= \sum_{k_0=0}^{Q_0-1} \sum_{k_1=0}^{Q_1-1} \operatorname{vec}\left(\left[(\boldsymbol{A}_0)_{:,k_0} \left(\boldsymbol{e}_{k_0}^{(Q_0)} \right)^T \right] \otimes \left[(\boldsymbol{A}_1)_{:,k_1} \left(\boldsymbol{e}_{k_1}^{(Q_1)} \right)^T \right] \right) \\
&= \sum_{k_0=0}^{Q_0-1} \sum_{k_1=0}^{Q_1-1} \operatorname{vec}\left(\left[(\boldsymbol{A}_0)_{:,k_0} \otimes (\boldsymbol{A}_1)_{:,k_1} \right] \left[\boldsymbol{e}_{k_0}^{(Q_0)} \otimes \boldsymbol{e}_{k_1}^{(Q_1)} \right]^T \right) \\
&= \sum_{k_0=0}^{Q_0-1} \sum_{k_1=0}^{Q_1-1} \boldsymbol{e}_{k_0}^{(Q_0)} \otimes \boldsymbol{e}_{k_1}^{(Q_1)} \otimes (\boldsymbol{A}_0)_{:,k_0} \otimes (\boldsymbol{A}_1)_{:,k_1} \\
&= \sum_{k_0=0}^{Q_0-1} \sum_{k_1=0}^{Q_1-1} \left(\boldsymbol{I}_{Q_0} \boldsymbol{e}_{k_0}^{(Q_0)} \right) \otimes \left[\boldsymbol{K}_{Q_1,N_0} \left((\boldsymbol{A}_0)_{:,k_0} \otimes \boldsymbol{e}_{k_1}^{(Q_1)} \right) \right] \otimes \left(\boldsymbol{I}_{N_1} (\boldsymbol{A}_1)_{:,k_1} \right) \\
&= \sum_{k_0=0}^{Q_0-1} \sum_{k_1=0}^{Q_1-1} \left[\boldsymbol{I}_{Q_0} \otimes \boldsymbol{K}_{Q_1,N_0} \otimes \boldsymbol{I}_{N_1} \right] \left[\boldsymbol{e}_{k_0}^{(Q_0)} \otimes (\boldsymbol{A}_0)_{:,k_0} \otimes \boldsymbol{e}_{k_1}^{(Q_1)} \otimes (\boldsymbol{A}_1)_{:,k_1} \right] \\
&= \left(\boldsymbol{I}_{Q_0} \otimes \boldsymbol{K}_{Q_1,N_0} \otimes \boldsymbol{I}_{N_1} \right) \left\{ \left(\sum_{k_0=0}^{Q_0-1} \operatorname{vec}\left((\boldsymbol{A}_0)_{:,k_0} \left(\boldsymbol{e}_{k_0}^{(Q_0)} \right)^T \right) \right) \right. \\
&\qquad \left. \otimes \left(\sum_{k_1=0}^{Q_1-1} \operatorname{vec}\left((\boldsymbol{A}_1)_{:,k_1} \left(\boldsymbol{e}_{k_1}^{(Q_1)} \right)^T \right) \right) \right\} \\
&= \left(\boldsymbol{I}_{Q_0} \otimes \boldsymbol{K}_{Q_1,N_0} \otimes \boldsymbol{I}_{N_1} \right) \left(\operatorname{vec}(\boldsymbol{A}_0) \otimes \operatorname{vec}(\boldsymbol{A}_1) \right),
\end{aligned} \tag{2.114}
$$

where (2.101), (2.110), Lemma 2.9, and $\boldsymbol{K}_{1,1} = 1$ have been used. ■

Let $\boldsymbol{A}_i \in \mathbb{C}^{N \times M}$, then

$$
\operatorname{vec}(\boldsymbol{A}_0 \odot \boldsymbol{A}_1) = \operatorname{diag}(\operatorname{vec}(\boldsymbol{A}_0)) \operatorname{vec}(\boldsymbol{A}_1). \tag{2.115}
$$

The result in (2.115) is shown in Exercise 2.10.

Lemma 2.14 *Let $\boldsymbol{A} \in \mathbb{C}^{N_0 \times N_1}$, $\boldsymbol{B} \in \mathbb{C}^{N_1 \times N_2}$, $\boldsymbol{C} \in \mathbb{C}^{N_2 \times N_3}$, and $\boldsymbol{D} \in \mathbb{C}^{N_3 \times N_0}$, then*

$$
\begin{aligned}
\operatorname{Tr}\{\boldsymbol{ABCD}\} &= \operatorname{vec}^T\left(\boldsymbol{D}^T\right) \left[\boldsymbol{C}^T \otimes \boldsymbol{A}\right] \operatorname{vec}(\boldsymbol{B}) \\
&= \operatorname{vec}^T(\boldsymbol{B}) \left[\boldsymbol{C} \otimes \boldsymbol{A}^T\right] \operatorname{vec}\left(\boldsymbol{D}^T\right).
\end{aligned} \tag{2.116}
$$

Proof The first equality in (2.116) can be shown by

$$
\begin{aligned}
\operatorname{Tr}\{\boldsymbol{ABCD}\} &= \operatorname{Tr}\{\boldsymbol{D}(\boldsymbol{ABC})\} = \operatorname{vec}^T\left(\boldsymbol{D}^T\right) \operatorname{vec}(\boldsymbol{ABC}) \\
&= \operatorname{vec}^T\left(\boldsymbol{D}^T\right) \left[\boldsymbol{C}^T \otimes \boldsymbol{A}\right] \operatorname{vec}(\boldsymbol{B}),
\end{aligned} \tag{2.117}
$$

where the results from (2.105) and (2.97) were used. The second equality in (2.116) follows by using the transpose operator on the first equality in the same equation and Lemma 2.9. ■

2.5.4 Complex Quadratic Forms

Lemma 2.15 *Let* $\boldsymbol{A}, \boldsymbol{B} \in \mathbb{C}^{N \times N}$. $\boldsymbol{z}^T \boldsymbol{A} \boldsymbol{z} = \boldsymbol{z}^T \boldsymbol{B} \boldsymbol{z}, \ \forall\, \boldsymbol{z} \in \mathbb{C}^{N \times 1}$ *is equivalent to* $\boldsymbol{A} + \boldsymbol{A}^T = \boldsymbol{B} + \boldsymbol{B}^T$.

Proof Let $(\boldsymbol{A})_{k,l} = a_{k,l}$ and $(\boldsymbol{B})_{k,l} = b_{k,l}$. Assume that $\boldsymbol{z}^T \boldsymbol{A} \boldsymbol{z} = \boldsymbol{z}^T \boldsymbol{B} \boldsymbol{z}, \ \forall\, \boldsymbol{z} \in \mathbb{C}^{N \times 1}$, and set $\boldsymbol{z} = \boldsymbol{e}_k$, where $k \in \{0, 1, \ldots, N-1\}$. Then

$$\boldsymbol{e}_k^T \boldsymbol{A} \boldsymbol{e}_k = \boldsymbol{e}_k^T \boldsymbol{B} \boldsymbol{e}_k, \tag{2.118}$$

gives that $a_{k,k} = b_{k,k}$ for all $k \in \{0, 1, \ldots, N-1\}$. Setting $\boldsymbol{z} = \boldsymbol{e}_k + \boldsymbol{e}_l$ leads to

$$\left(\boldsymbol{e}_k^T + \boldsymbol{e}_l^T\right) \boldsymbol{A}(\boldsymbol{e}_k + \boldsymbol{e}_l) = \left(\boldsymbol{e}_k^T + \boldsymbol{e}_l^T\right) \boldsymbol{B}(\boldsymbol{e}_k + \boldsymbol{e}_l), \tag{2.119}$$

which results in $a_{k,k} + a_{l,l} + a_{k,l} + a_{l,k} = b_{k,k} + b_{l,l} + b_{k,l} + b_{l,k}$. Eliminating equal terms from this equation gives $a_{k,l} + a_{l,k} = b_{k,l} + b_{l,k}$, which can be written as $\boldsymbol{A} + \boldsymbol{A}^T = \boldsymbol{B} + \boldsymbol{B}^T$.

Assuming that $\boldsymbol{A} + \boldsymbol{A}^T = \boldsymbol{B} + \boldsymbol{B}^T$, it follows that

$$\begin{aligned} \boldsymbol{z}^T \boldsymbol{A} \boldsymbol{z} &= \frac{1}{2}\left(\boldsymbol{z}^T \boldsymbol{A} \boldsymbol{z} + \boldsymbol{z}^T \boldsymbol{A}^T \boldsymbol{z}\right) = \frac{1}{2} \boldsymbol{z}^T\left(\boldsymbol{A} + \boldsymbol{A}^T\right) \boldsymbol{z} = \frac{1}{2} \boldsymbol{z}^T\left(\boldsymbol{B} + \boldsymbol{B}^T\right) \boldsymbol{z} \\ &= \frac{1}{2}\left(\boldsymbol{z}^T \boldsymbol{B} \boldsymbol{z} + \boldsymbol{z}^T \boldsymbol{B}^T \boldsymbol{z}\right) = \frac{1}{2}\left(\boldsymbol{z}^T \boldsymbol{B} \boldsymbol{z} + \boldsymbol{z}^T \boldsymbol{B} \boldsymbol{z}\right) = \boldsymbol{z}^T \boldsymbol{B} \boldsymbol{z}, \end{aligned} \tag{2.120}$$

for all $\boldsymbol{z} \in \mathbb{C}^{N \times 1}$. ■

Corollary 2.1 *Let* $\boldsymbol{A} \in \mathbb{C}^{N \times N}$. $\boldsymbol{z}^T \boldsymbol{A} \boldsymbol{z} = 0, \ \forall\, \boldsymbol{z} \in \mathbb{C}^{N \times 1}$ *is equivalent to* $\boldsymbol{A}^T = -\boldsymbol{A}$ *(i.e., A is skew-symmetric) (Bernstein 2005, p. 81).*

Proof Set $\boldsymbol{B} = \boldsymbol{0}_{N \times N}$ in Lemma 2.15, then the corollary follows. ■

Lemma 2.15 and Corollary 2.1 are also valid for *real-valued* vectors and *complex-valued* matrices as stated in the following lemma and corollary:

Lemma 2.16 *Let* $\boldsymbol{A}, \boldsymbol{B} \in \mathbb{C}^{N \times N}$. $\boldsymbol{x}^T \boldsymbol{A} \boldsymbol{x} = \boldsymbol{x}^T \boldsymbol{B} \boldsymbol{x}, \ \forall\, \boldsymbol{x} \in \mathbb{R}^{N \times 1}$ *is equivalent to* $\boldsymbol{A} + \boldsymbol{A}^T = \boldsymbol{B} + \boldsymbol{B}^T$.

Proof Let $(\boldsymbol{A})_{k,l} = a_{k,l}$ and $(\boldsymbol{B})_{k,l} = b_{k,l}$. Assume that $\boldsymbol{x}^T \boldsymbol{A} \boldsymbol{x} = \boldsymbol{x}^T \boldsymbol{B} \boldsymbol{x}, \ \forall\, \boldsymbol{x} \in \mathbb{R}^{N \times 1}$, and set $\boldsymbol{x} = \boldsymbol{e}_k$ where $k \in \{0, 1, \ldots, N-1\}$. Then

$$\boldsymbol{e}_k^T \boldsymbol{A} \boldsymbol{e}_k = \boldsymbol{e}_k^T \boldsymbol{B} \boldsymbol{e}_k, \tag{2.121}$$

gives that $a_{k,k} = b_{k,k}$ for all $k \in \{0, 1, \ldots, N-1\}$. Setting $\boldsymbol{x} = \boldsymbol{e}_k + \boldsymbol{e}_l$ leads to

$$\left(\boldsymbol{e}_k^T + \boldsymbol{e}_l^T\right) \boldsymbol{A}(\boldsymbol{e}_k + \boldsymbol{e}_l) = \left(\boldsymbol{e}_k^T + \boldsymbol{e}_l^T\right) \boldsymbol{B}(\boldsymbol{e}_k + \boldsymbol{e}_l), \tag{2.122}$$

which results in $a_{k,k} + a_{l,l} + a_{k,l} + a_{l,k} = b_{k,k} + b_{l,l} + b_{k,l} + b_{l,k}$. Eliminating equal terms from this equation gives $a_{k,l} + a_{l,k} = b_{k,l} + b_{l,k}$, which can be written $\boldsymbol{A} + \boldsymbol{A}^T = \boldsymbol{B} + \boldsymbol{B}^T$.

Assuming that $A + A^T = B + B^T$, it follows that

$$\begin{aligned} x^T A x &= \frac{1}{2}\left(x^T A x + x^T A^T x\right) = \frac{1}{2} x^T \left(A + A^T\right) x = \frac{1}{2} x^T \left(B + B^T\right) x \\ &= \frac{1}{2}\left(x^T B x + x^T B^T x\right) = \frac{1}{2}\left(x^T B x + x^T B x\right) = x^T B x, \end{aligned} \tag{2.123}$$

for all $x \in \mathbb{R}^{N\times 1}$. ∎

Corollary 2.2 *Let $A \in \mathbb{C}^{N\times N}$. $x^T A x = 0,\ \forall\, x \in \mathbb{R}^{N\times 1}$ is equivalent to $A^T = -A$ (i.e., A is skew-symmetric) (Bernstein 2005, p. 81).*

Proof Set $B = 0_{N\times N}$ in Lemma 2.16, then the corollary follows. ∎

Lemma 2.17 *Let $A, B \in \mathbb{C}^{N\times N}$. $z^H A z = z^H B z,\ \forall\, z \in \mathbb{C}^{N\times 1}$ is equivalent to $A = B$.*

Proof Let $(A)_{k,l} = a_{k,l}$ and $(B)_{k,l} = b_{k,l}$. Assume that $z^H A z = z^H B z,\ \forall\, z \in \mathbb{C}^{N\times 1}$, and set $z = e_k$ where $k \in \{0, 1, \ldots, N-1\}$. This gives in the same way as in the proof of Lemma 2.15 that $a_{k,k} = b_{k,k}$, for all $k \in \{0, 1, \ldots, N-1\}$. Also in the same way as in the proof of Lemma 2.15, setting $z = e_k + e_l$ leads to $A + A^T = B + B^T$. Next, set $z = e_k + \jmath e_l$, then manipulations of the expressions give $A - A^T = B - B^T$. The equations $A + A^T = B + B^T$ and $A - A^T = B - B^T$ imply that $A = B$.

If $A = B$, then it follows that $z^H A z = z^H B z$ for all $z \in \mathbb{C}^{N\times 1}$. ∎

The next lemma shows a result that might seem surprising.

Lemma 2.18 *Let $A, B \in \mathbb{C}^{N\times N}$. The expression $x^T A x = x^T B x,\ \forall\, x \in \mathbb{R}^{N\times 1}$ is equivalent to $z^T A z = z^T B z,\ \forall\, z \in \mathbb{C}^{N\times 1}$.*

Proof This result follows from Lemmas 2.15 and 2.16. ∎

Lemma 2.19 *Let $A, B \in \mathbb{C}^{MN\times N}$ where N and M are positive integers. If*

$$\left[I_M \otimes z^T\right] A z = \left[I_M \otimes z^T\right] B z, \tag{2.124}$$

for all $z \in \mathbb{C}^{N\times 1}$, then

$$A + \operatorname{vecb}\left(A^T\right) = B + \operatorname{vecb}\left(B^T\right). \tag{2.125}$$

Proof Let the matrix A and B be given by

$$A = \begin{bmatrix} A_0 \\ A_1 \\ \vdots \\ A_{M-1} \end{bmatrix}, \tag{2.126}$$

and

$$B = \begin{bmatrix} B_0 \\ B_1 \\ \vdots \\ B_{M-1} \end{bmatrix}, \tag{2.127}$$

where $A_i \in \mathbb{C}^{N\times N}$ and $B_i \in \mathbb{C}^{N\times N}$ for all $i \in \{0, 1, \ldots, M-1\}$.

Row number i of (2.124) can be expressed as

$$z^T A_i z = z^T B_i z, \tag{2.128}$$

for all $z \in \mathbb{C}^{N\times 1}$ and for all $i \in \{0, 1, \ldots, M-1\}$. By using Lemma 2.15 on (2.128), it follows that

$$A_i + A_i^T = B_i + B_i^T, \tag{2.129}$$

for all $i \in \{0, 1, \ldots, M-1\}$. By applying the block vectorization operator, the results inside the M results in (2.129) can be written as in (2.125). ■

2.5.5 Results for Finding Generalized Matrix Derivatives

In this subsection, several results will be presented that will be used in Chapter 6 to find generalized complex-valued matrix derivatives.

Lemma 2.20 *Let $A \in \mathbb{C}^{N\times N}$. From Definition 2.11, it follows that*

$$\mathrm{vec}_l\left(A^T\right) = \mathrm{vec}_u(A). \tag{2.130}$$

Lemma 2.21 *The following relation holds for the matrices in Definition 2.12:*

$$L_d L_d^T + L_l L_l^T + L_u L_u^T = I_{N^2}. \tag{2.131}$$

Proof From (2.42), it follows that the $N^2 \times N^2$ matrix $[L_d, L_l, L_u]$ is a permutation matrix. Hence, its inverse is given by its transposed

$$[L_d, L_l, L_u][L_d, L_l, L_u]^T = I_{N^2}. \tag{2.132}$$

By multiplying out the left-hand side as a block matrix, the lemma follows. ■

Lemma 2.22 *For the matrices defined in Definition 2.12, the following relations hold:*

$$L_d^T L_d = I_N, \tag{2.133}$$

$$L_l^T L_l = I_{\frac{N(N-1)}{2}}, \tag{2.134}$$

$$L_u^T L_u = I_{\frac{N(N-1)}{2}}, \tag{2.135}$$

$$L_d^T L_l = 0_{N\times\frac{N(N-1)}{2}}, \tag{2.136}$$

$$L_d^T L_u = 0_{N\times\frac{N(N-1)}{2}}, \tag{2.137}$$

$$L_l^T L_u = 0_{\frac{N(N-1)}{2}\times\frac{N(N-1)}{2}}. \tag{2.138}$$

Proof Because the three matrices L_d, L_l, and L_u are given by nonoverlapping parts of a permutation matrix, the above relations follow. ■

Lemma 2.23 *Let $A \in \mathbb{C}^{N\times N}$, then*

$$\mathrm{vec}\left(I_N \odot A\right) = L_d \,\mathrm{vec}_d(A), \tag{2.139}$$

where $\odot$ denotes the Hadamard product (see Definition 2.7).

Proof This follows by using the diagonal matrix $\boldsymbol{I}_N \odot \boldsymbol{A}$ in (2.42). Because $\boldsymbol{I}_N \odot \boldsymbol{A}$ is diagonal, it follows that $\text{vec}_l(\boldsymbol{I}_N \odot \boldsymbol{A}) = \text{vec}_u(\boldsymbol{I}_N \odot \boldsymbol{A}) = \boldsymbol{0}_{\frac{N(N-1)}{2}\times 1}$ and $\text{vec}_d(\boldsymbol{I}_N \odot \boldsymbol{A}) = \text{vec}_d(\boldsymbol{A})$. Inserting these results into (2.42) leads to (2.139). ■

Lemma 2.24 *Let $\boldsymbol{A} \in \mathbb{C}^{N\times N}$, then*

$$\boldsymbol{L}_d^T \,\text{vec}\,(\boldsymbol{A}) = \text{vec}_d\,(\boldsymbol{A})\,, \tag{2.140}$$

$$\boldsymbol{L}_l^T \,\text{vec}\,(\boldsymbol{A}) = \text{vec}_l\,(\boldsymbol{A})\,, \tag{2.141}$$

$$\boldsymbol{L}_u^T \,\text{vec}\,(\boldsymbol{A}) = \text{vec}_u\,(\boldsymbol{A})\,. \tag{2.142}$$

Proof Multiplying (2.42) from the left by $\boldsymbol{L}_d^T$ and using Lemma 2.22 result in (2.140). In a similar manner, (2.141) and (2.142) follow. ■

Lemma 2.25 *The following relation holds between the matrices defined in Definition 2.12:*

$$\boldsymbol{K}_{N,N} = \boldsymbol{L}_d\boldsymbol{L}_d^T + \boldsymbol{L}_l\boldsymbol{L}_u^T + \boldsymbol{L}_u\boldsymbol{L}_l^T\,. \tag{2.143}$$

Proof Using the operators defined earlier and the commutation matrix, we get for the matrix $\boldsymbol{A} \in \mathbb{C}^{N\times N}$

$$\begin{aligned}
\boldsymbol{K}_{N,N}\,\text{vec}\,(\boldsymbol{A}) &= \text{vec}\left(\boldsymbol{A}^T\right) \\
&= \boldsymbol{L}_d\,\text{vec}_d\,(\boldsymbol{A}) + \boldsymbol{L}_l\,\text{vec}_u\,(\boldsymbol{A}) + \boldsymbol{L}_u\,\text{vec}_l\,(\boldsymbol{A}) \\
&= \boldsymbol{L}_d\boldsymbol{L}_d^T\,\text{vec}\,(\boldsymbol{A}) + \boldsymbol{L}_l\boldsymbol{L}_u^T\,\text{vec}\,(\boldsymbol{A}) + \boldsymbol{L}_u\boldsymbol{L}_l^T\,\text{vec}\,(\boldsymbol{A}) \\
&= \left[\boldsymbol{L}_d\boldsymbol{L}_d^T + \boldsymbol{L}_l\boldsymbol{L}_u^T + \boldsymbol{L}_u\boldsymbol{L}_l^T\right]\text{vec}\,(\boldsymbol{A})\,.
\end{aligned} \tag{2.144}$$

Because this holds for any $\boldsymbol{A} \in \mathbb{C}^{N\times N}$, the lemma follows by setting $\text{vec}(\boldsymbol{A})$ equal to the i-th standard vector in $\mathbb{C}^{N^2\times 1}$ for all $i \in \{0, 1, \ldots, N^2-1\}$. ■

Lemma 2.26 *Let $\boldsymbol{A} \in \mathbb{C}^{N\times N}$, then*

$$(\boldsymbol{K}_{N,N} \odot \boldsymbol{I}_{N^2})\,\text{vec}\,(\boldsymbol{A}) = \boldsymbol{L}_d\,\text{vec}_d\,(\boldsymbol{A})\,. \tag{2.145}$$

This can also be expressed as

$$\boldsymbol{L}_d\boldsymbol{L}_d^T = \boldsymbol{I}_{N^2} \odot \boldsymbol{K}_{N,N}\,. \tag{2.146}$$

Proof From (2.140), it follows that $\boldsymbol{L}_d\,\text{vec}_d\,(\boldsymbol{A}) = \boldsymbol{L}_d\boldsymbol{L}_d^T\,\text{vec}\,(\boldsymbol{A})$. By studying (2.143) and using the knowledge that $\boldsymbol{L}_l$ and $\boldsymbol{L}_u$ have distinct columns taken from an $N^2 \times N^2$ permutation matrix, it is seen that the term $\boldsymbol{L}_d\boldsymbol{L}_d^T$ contains all the diagonal elements of $\boldsymbol{K}_{N,N}$. By taking the Hadamard product on each side of (2.143) with $\boldsymbol{I}_{N^2}$, the result in (2.145) follows. The result in (2.146) is a consequence of (2.140), together with $\boldsymbol{L}_d\,\text{vec}_d\,(\boldsymbol{A}) = \boldsymbol{L}_d\boldsymbol{L}_d^T\,\text{vec}\,(\boldsymbol{A})$. ■

Lemma 2.27 *The following relations hold:*

$$\boldsymbol{K}_{N,N}\boldsymbol{L}_d = \boldsymbol{L}_d, \tag{2.147}$$

$$\boldsymbol{K}_{N,N}\boldsymbol{L}_u = \boldsymbol{L}_l, \tag{2.148}$$

$$\boldsymbol{K}_{N,N}\boldsymbol{L}_l = \boldsymbol{L}_u, \tag{2.149}$$

$$\boldsymbol{K}_{N,N}\boldsymbol{D}_N = \boldsymbol{D}_N. \tag{2.150}$$

Proof Note that $\boldsymbol{K}_{N,N}^T = \boldsymbol{K}_{N,N}$ (Magnus & Neudecker 1988, pp. 46–48). Since $\text{vec}_d(\boldsymbol{A}) = \boldsymbol{L}_d^T \text{vec}(\boldsymbol{A})$ and $\text{vec}_d(\boldsymbol{A}^T) = \boldsymbol{L}_d^T \text{vec}(\boldsymbol{A}^T) = \boldsymbol{L}_d^T \boldsymbol{K}_{N,N} \text{vec}(\boldsymbol{A})$ are equal for all $\boldsymbol{A} \in \mathbb{C}^{N\times N}$, it follows that (2.147) holds. Because $\text{vec}_l(\boldsymbol{A}) = \boldsymbol{L}_l^T \text{vec}(\boldsymbol{A})$ and $\text{vec}_u(\boldsymbol{A}^T) = \boldsymbol{L}_u^T \text{vec}(\boldsymbol{A}^T) = \boldsymbol{L}_u^T \boldsymbol{K}_{N,N} \text{vec}(\boldsymbol{A})$ are equal for all $\boldsymbol{A} \in \mathbb{C}^{N\times N}$, (2.148) and (2.149) are true. Let $\boldsymbol{B} \in \mathbb{C}^{N\times N}$ be symmetric. Because

$$\boldsymbol{K}_{N,N}\boldsymbol{D}_N v(\boldsymbol{B}) = \boldsymbol{K}_{N,N} \text{vec}(\boldsymbol{B}) = \text{vec}(\boldsymbol{B}) = \boldsymbol{D}_N v(\boldsymbol{B}), \tag{2.151}$$

it follows that (2.150) is valid. ■

Lemma 2.28 *Let* $\boldsymbol{A} \in \mathbb{C}^{N\times N}$ *and* $(\boldsymbol{A})_{i,j} = a_{i,j}$, *then*

$$\boldsymbol{A} = \sum_{i=0}^{N-1}\sum_{j=0}^{N-1} a_{i,j}\boldsymbol{E}_{i,j} = \sum_{i=0}^{N-1} a_{i,i}\boldsymbol{E}_{i,i} + \sum_{j=0}^{N-2}\sum_{i=j+1}^{N-1} a_{i,j}\boldsymbol{E}_{i,j} + \sum_{i=0}^{N-2}\sum_{j=i+1}^{N-1} a_{i,j}\boldsymbol{E}_{i,j}, \tag{2.152}$$

where $\boldsymbol{E}_{i,j}$ *is given in Definition 2.16, the sum* $\sum_{i=0}^{N-1} a_{i,i}\boldsymbol{E}_{i,i}$ *takes care of all the elements on the main diagonal, the sum* $\sum_{j=0}^{N-2}\sum_{i=j+1}^{N-1} a_{i,j}\boldsymbol{E}_{i,j}$ *considers all elements strictly below the main diagonal, and the sum* $\sum_{i=0}^{N-2}\sum_{j=i+1}^{N-1} a_{i,j}\boldsymbol{E}_{i,j}$ *contains all terms strictly above the main diagonal.*

Proof This result follows directly from the way matrices are built up. ■

Lemma 2.29 *The* $N^2 \times N$ *matrix* $\boldsymbol{L}_d$ *has the following properties:*

$$\boldsymbol{L}_d = \left[\text{vec}\left(\boldsymbol{e}_0\boldsymbol{e}_0^T\right), \text{vec}\left(\boldsymbol{e}_1\boldsymbol{e}_1^T\right), \ldots, \text{vec}\left(\boldsymbol{e}_{N-1}\boldsymbol{e}_{N-1}^T\right)\right], \tag{2.153}$$

$$\text{rank}(\boldsymbol{L}_d) = N, \tag{2.154}$$

$$\boldsymbol{L}_d^+ = \boldsymbol{L}_d^T, \tag{2.155}$$

$$(\boldsymbol{L}_d)_{i+jN,k} = \delta_{i,j,k}, \quad \forall i, j, k \in \{0, 1, \ldots, N-1\}, \tag{2.156}$$

where $\delta_{i,j,k}$ *denotes the Kronecker delta function with three integer-valued input arguments, which is* $+1$ *when* all *input arguments are equal and* 0 *otherwise.*

Proof First, (2.153) is shown by taking the vec(·) operator on the diagonal elements of $\boldsymbol{A}$:

$$\begin{aligned}
\boldsymbol{L}_d \operatorname{vec}_d(\boldsymbol{A}) &= \operatorname{vec}\left(\sum_{i=0}^{N-1} a_{i,i}\boldsymbol{E}_{i,i}\right) = \sum_{i=0}^{N-1} a_{i,i}\operatorname{vec}(\boldsymbol{E}_{i,i}) = \sum_{i=0}^{N-1} a_{i,i}\operatorname{vec}\left(\boldsymbol{e}_i\boldsymbol{e}_i^T\right) \\
&= \sum_{i=0}^{N-1} \boldsymbol{e}_i \otimes \boldsymbol{e}_i\, a_{i,i} = [\boldsymbol{e}_0 \otimes \boldsymbol{e}_0, \boldsymbol{e}_1 \otimes \boldsymbol{e}_1, \ldots, \boldsymbol{e}_{N-1} \otimes \boldsymbol{e}_{N-1}] \operatorname{vec}_d(\boldsymbol{A}) \\
&= \left[\operatorname{vec}\left(\boldsymbol{e}_0\boldsymbol{e}_0^T\right), \operatorname{vec}\left(\boldsymbol{e}_1\boldsymbol{e}_1^T\right), \ldots, \operatorname{vec}\left(\boldsymbol{e}_{N-1}\boldsymbol{e}_{N-1}^T\right)\right] \operatorname{vec}_d(\boldsymbol{A}),
\end{aligned} \tag{2.157}$$

which shows that (2.153) holds, where (2.215) from Exercise 2.14 has been used.

From (2.153), (2.154) follows directly.

Because $\boldsymbol{L}_d$ has full column rank, (2.155) follows from (2.80) and (2.133).

Let $i, j, k \in \{0, 1, \ldots, N-1\}$, then (2.156) can be shown as follows:

$$(\boldsymbol{L}_d)_{i+jN,k} = (\boldsymbol{e}_k \otimes \boldsymbol{e}_k)_{i+jN} = \delta_{j,k}(\boldsymbol{e}_k)_i = \delta_{j,k}\delta_{i,k} = \delta_{i,j,k}, \tag{2.158}$$

where $\delta_{k,l}$ denotes the Kronecker delta function with two integer-valued input arguments, i.e., $\delta_{k,l} = 1$, when $k = l$ and $\delta_{k,l} = 0$ when $k \neq l$. ■

Lemma 2.30 *The $N^2 \times \frac{N(N-1)}{2}$ matrix $\boldsymbol{L}_l$ from Definition 2.12 satisfies the following properties:*

$$\boldsymbol{L}_l = \left[\operatorname{vec}\left(\boldsymbol{e}_1\boldsymbol{e}_0^T\right), \operatorname{vec}\left(\boldsymbol{e}_2\boldsymbol{e}_0^T\right), \ldots, \operatorname{vec}\left(\boldsymbol{e}_{N-1}\boldsymbol{e}_0^T\right), \operatorname{vec}\left(\boldsymbol{e}_2\boldsymbol{e}_1^T\right), \ldots, \operatorname{vec}\left(\boldsymbol{e}_{N-1}\boldsymbol{e}_{N-2}^T\right)\right], \tag{2.159}$$

$$\operatorname{rank}(\boldsymbol{L}_l) = \frac{N(N-1)}{2}, \tag{2.160}$$

$$\boldsymbol{L}_l^+ = \boldsymbol{L}_l^T, \tag{2.161}$$

$$(\boldsymbol{L}_l)_{i+jN,k+lN-\frac{l^2+3l+2}{2}} = \delta_{j,l}\cdot\delta_{k,i}, \tag{2.162}$$

where $i, j, k, l \in \{0, 1, \ldots, N-1\}$, and $k > l$.

Proof Equation (2.159) can be derived by using the vec(·) operator on the terms of $\boldsymbol{A} \in \mathbb{C}^{N\times N}$, which are located strictly below the main diagonal

$$\begin{aligned}
\boldsymbol{L}_l \operatorname{vec}_l(\boldsymbol{A}) &= \operatorname{vec}\left(\sum_{j=0}^{N-2}\sum_{i=j+1}^{N-1} a_{i,j}\boldsymbol{E}_{i,j}\right) = \sum_{j=0}^{N-2}\sum_{i=j+1}^{N-1} a_{i,j}\operatorname{vec}\left(\boldsymbol{E}_{i,j}\right) \\
&= \sum_{j=0}^{N-2}\sum_{i=j+1}^{N-1} a_{i,j}\operatorname{vec}\left(\boldsymbol{e}_i\boldsymbol{e}_j^T\right) = \sum_{j=0}^{N-2}\sum_{i=j+1}^{N-1} \boldsymbol{e}_j \otimes \boldsymbol{e}_i\, a_{i,j}
\end{aligned}$$

$$= [e_0 \otimes e_1, e_0 \otimes e_2, \ldots, e_0 \otimes e_{N-1}, e_1 \otimes e_2, \ldots, e_{N-2} \otimes e_{N-1}] \begin{bmatrix} a_{1,0} \\ a_{2,0} \\ \vdots \\ a_{N-1,0} \\ a_{2,1} \\ \vdots \\ a_{N-1,N-2} \end{bmatrix}$$

$$= [e_0 \otimes e_1, e_0 \otimes e_2, \ldots, e_0 \otimes e_{N-1}, e_1 \otimes e_2, \ldots, e_{N-2} \otimes e_{N-1}] \operatorname{vec}_l(A). \quad (2.163)$$

Because (2.163) is valid for all $A \in \mathbb{C}^{N \times N}$, the result in (2.159) follows by setting $\operatorname{vec}_l(A)$ equal to the i-th standard basis vector in $\mathbb{C}^{\frac{N(N-1)}{2} \times 1}$ for all $i \in \{0, 1, \ldots, N-1\}$.

The result in (2.160) follows directly by the fact that the columns of L_l are given by different columns of a permutation matrix.

From (2.160), it follows that the $N^2 \times \frac{N(N-1)}{2}$ matrix L_l has full column rank, then (2.161) follows from (2.80) and (2.134).

It remains to show (2.162). The number of columns of L_l is $\frac{N(N-1)}{2}$; hence, the element that should be decided is $(L_l)_{i+jN,q}$, where $i, j \in \{0, 1, \ldots, N-1\}$ and $q \in \left\{0, 1, \ldots, \frac{N(N-1)}{2} - 1\right\}$. The one-dimensional index $q \in \{0, 1, \ldots, \frac{N(N-1)}{2} - 1\}$ runs through all elements strictly below the main diagonal of an $N \times N$ matrix when moving from column to column from the upper elements and down each column in the same order as used when the operator $\operatorname{vec}(\cdot)$ is applied on an $N \times N$ matrix. By studying the one-dimensional index q carefully, it is seen that the first column of L_l corresponds to $q = 0$ for elements in row number 1 and column number 0, where the numbering of the rows and columns starts with 0. The first element in the first column of an $N \times N$ matrix is *not* numbered by q because this element is not located strictly below the main diagonal. Let the row number for generating the index q be denoted by k, and let the column be number l of an $N \times N$ matrix, where $k, l \in \{0, 1, \ldots, N-1\}$. For elements strictly below the main diagonal, it is required that $k > l$. By studying the number of columns in L_l that should be generated by going along the columns of an $N \times N$ matrix to the element in row number k and column number l, it is seen that the index q can be expressed as in terms of k and l as

$$q = k + lN - \sum_{p=0}^{l} (p+1) = k + lN - \frac{l^2 + 3l + 2}{2}, \quad (2.164)$$

where the sum $\sum_{p=0}^{l} (p+1)$ represents the elements among the first l columns that should not be indexed by q because they are located above or on the main diagonal. The expression in (2.162) is found as follows:

$$\begin{aligned} (L_l)_{i+jN,k+lN-\frac{l^2+3l+2}{2}} &= \left(\operatorname{vec}\left(e_k e_l^T\right)\right)_{i+jN} = (e_l \otimes e_k)_{i+jN} \\ &= \delta_{j,l} (e_k)_i = \delta_{j,l} \delta_{k,i}, \end{aligned} \quad (2.165)$$

which was going to be shown. ■

Lemma 2.31 *The* $N^2 \times \frac{N(N-1)}{2}$ *matrix* $\boldsymbol{L}_u$ *defined in Definition 2.12 satisfies the following properties:*

$$\boldsymbol{L}_u = \left[\operatorname{vec}\left(\boldsymbol{e}_0\boldsymbol{e}_1^T\right), \operatorname{vec}\left(\boldsymbol{e}_0\boldsymbol{e}_2^T\right), \ldots, \operatorname{vec}\left(\boldsymbol{e}_0\boldsymbol{e}_{N-1}^T\right), \operatorname{vec}\left(\boldsymbol{e}_1\boldsymbol{e}_2^T\right), \ldots, \operatorname{vec}\left(\boldsymbol{e}_{N-2}\boldsymbol{e}_{N-1}^T\right)\right], \tag{2.166}$$

$$\operatorname{rank}(\boldsymbol{L}_u) = \frac{N(N-1)}{2} \tag{2.167}$$

$$\boldsymbol{L}_u^+ = \boldsymbol{L}_u^T, \tag{2.168}$$

$$(\boldsymbol{L}_u)_{i+jN,\, l+kN-\frac{k^2+3k+2}{2}} = \delta_{l,j} \cdot \delta_{k,i}, \tag{2.169}$$

where $i, j, k, l \in \{0, 1, \ldots, N-1\}$ *and* $l > k$.

Proof Equation (2.166) can be derived by using the vec(·) operator on the terms of $\boldsymbol{A} \in \mathbb{C}^{N\times N}$, which are located strictly above the main diagonal:

$$\begin{aligned}
&\boldsymbol{L}_u \operatorname{vec}_u(\boldsymbol{A}) \\
&= \operatorname{vec}\left(\sum_{i=0}^{N-2}\sum_{j=i+1}^{N-1} a_{i,j}\boldsymbol{E}_{i,j}\right) = \sum_{i=0}^{N-2}\sum_{j=i+1}^{N-1} a_{i,j}\operatorname{vec}\left(\boldsymbol{E}_{i,j}\right) \\
&= \sum_{i=0}^{N-2}\sum_{j=i+1}^{N-1} a_{i,j}\operatorname{vec}\left(\boldsymbol{e}_i\boldsymbol{e}_j^T\right) = \sum_{i=0}^{N-2}\sum_{j=i+1}^{N-1} \boldsymbol{e}_j \otimes \boldsymbol{e}_i\, a_{i,j} \\
&= \left[\boldsymbol{e}_1 \otimes \boldsymbol{e}_0, \boldsymbol{e}_2 \otimes \boldsymbol{e}_0, \ldots, \boldsymbol{e}_{N-1} \otimes \boldsymbol{e}_0, \boldsymbol{e}_2 \otimes \boldsymbol{e}_1, \ldots, \boldsymbol{e}_{N-1} \otimes \boldsymbol{e}_{N-2}\right] \begin{bmatrix} a_{0,1} \\ a_{0,2} \\ \vdots \\ a_{0,N-1} \\ a_{1,2} \\ \vdots \\ a_{N-2,N-1} \end{bmatrix} \\
&= \left[\boldsymbol{e}_1 \otimes \boldsymbol{e}_0, \boldsymbol{e}_2 \otimes \boldsymbol{e}_0, \ldots, \boldsymbol{e}_{N-1} \otimes \boldsymbol{e}_0, \boldsymbol{e}_2 \otimes \boldsymbol{e}_1, \ldots, \boldsymbol{e}_{N-1} \otimes \boldsymbol{e}_{N-2}\right] \operatorname{vec}_u(\boldsymbol{A}).
\end{aligned} \tag{2.170}$$

The equation in (2.166) now follows from (2.170) because (2.170) is valid for all $\boldsymbol{A} \in \mathbb{C}^{N\times N}$. The i-th columns of each side of (2.166) are shown to be equal by setting $\operatorname{vec}_u(\boldsymbol{A})$ equal to the i-th standard unit vector in $\mathbb{C}^{\frac{N(N-1)}{2}\times 1}$.

The result in (2.167) follows from (2.166).

From (2.167), it follows that the matrix $\boldsymbol{L}_u$ has full column rank; hence, it follows from (2.80) and (2.135) that (2.168) holds.

The matrix $\boldsymbol{L}_u$ has size $N^2 \times \frac{N(N-1)}{2}$, such the task is to specify the elements $(\boldsymbol{L}_u)_{i+jN,q}$, where $i, j \in \{0, 1, \ldots, N-1\}$ and $q \in \{0, 1, \ldots, \frac{N(N-1)}{2} - 1\}$ specify the column of $\boldsymbol{L}_u$. Here, q is the number of elements that is strictly above the main diagonal when the elements of an $N \times N$ matrix are visited in a row-wise manner, starting from the first row, and going from left to right until the element in row number k and column number l. For elements strictly above the main diagonal, it is required that $l > k$. Using the same logic as in the proof of (2.162), the column numbering of $\boldsymbol{L}_u$ can be found as

$$q = l + kN - \sum_{p=0}^{k}(p+1) = l + kN - \frac{k^2+3k+2}{2}, \tag{2.171}$$

where the term $\sum_{p=0}^{k}(p+1)$ gives the number of elements that have been visited when traversing the rows from left to right, which should not be counted until the element in row number k and column number l is reached, meaning that they are located on or below the main diagonal. The expression in (2.169) can be shown as follows:

$$\begin{aligned}(\boldsymbol{L}_u)_{i+jN,l+kN-\frac{k^2+3k+2}{2}} &= \left(\operatorname{vec}\left(\boldsymbol{e}_k\boldsymbol{e}_l^T\right)\right)_{i+jN} = (\boldsymbol{e}_l \otimes \boldsymbol{e}_k)_{i+jN} \\ &= \delta_{l,j}\,(\boldsymbol{e}_k)_i = \delta_{l,j}\delta_{k,i},\end{aligned} \tag{2.172}$$

which is the same as in (2.169). ■

Proposition 2.2 *Let* $\boldsymbol{A} \in \mathbb{C}^{N\times N}$*, then,*

$$\operatorname{vec}_d(\boldsymbol{A}) = (\boldsymbol{A} \odot \boldsymbol{I}_N)\,\boldsymbol{1}_{N\times 1}. \tag{2.173}$$

Proof This result follows directly from the definition of $\operatorname{vec}_d(\cdot)$ and by multiplying out the right side of (2.173)

$$(\boldsymbol{A} \odot \boldsymbol{I}_N)\,\boldsymbol{1}_{N\times 1} = \begin{bmatrix} a_{0,0} \\ a_{1,1} \\ \vdots \\ a_{N-1,N-1} \end{bmatrix}, \tag{2.174}$$

which is equal to $\operatorname{vec}_d(\boldsymbol{A})$. ■

The duplication matrix is well known from the literature (Magnus & Neudecker 1988, pp. 48–53), and in the next lemma, the connections between the duplication matrix and the matrices $\boldsymbol{L}_d$, $\boldsymbol{L}_l$, and $\boldsymbol{L}_u$, defined in Definition 2.12, are shown.

Lemma 2.32 *The following relations hold between the three special matrices* $\boldsymbol{L}_d$*,* $\boldsymbol{L}_l$*, and* $\boldsymbol{L}_u$ *and the duplication matrix* $\boldsymbol{D}_N$*:*

$$\boldsymbol{D}_N = \boldsymbol{L}_d\boldsymbol{V}_d^T + (\boldsymbol{L}_l + \boldsymbol{L}_u)\,\boldsymbol{V}_l^T, \tag{2.175}$$

$$\boldsymbol{L}_d = \boldsymbol{D}_N\boldsymbol{V}_d, \tag{2.176}$$

$$\boldsymbol{L}_l + \boldsymbol{L}_u = \boldsymbol{D}_N\boldsymbol{V}_l, \tag{2.177}$$

$$\boldsymbol{V}_d = \boldsymbol{D}_N^+\boldsymbol{L}_d, \tag{2.178}$$

$$\boldsymbol{V}_l = \boldsymbol{D}_N^+\left(\boldsymbol{L}_l + \boldsymbol{L}_u\right), \tag{2.179}$$

where the two matrices $\boldsymbol{V}_d$ *and* $\boldsymbol{V}_l$ *are defined in Definition 2.15.*

Proof Let $A \in \mathbb{C}^{N\times N}$ be symmetric. For a symmetric A, it follows that $\text{vec}_l(A) = \text{vec}_u(A)$. Using this result in (2.42) yields

$$\begin{aligned}\text{vec}(A) &= L_d \text{vec}_d(A) + L_l \text{vec}_l(A) + L_u \text{vec}_u(A)\\ &= L_d \text{vec}_d(A) + (L_l + L_u)\text{vec}_l(A)\\ &= [L_d,\ L_l + L_u]\begin{bmatrix}\text{vec}_d(A)\\ \text{vec}_l(A)\end{bmatrix}.\end{aligned} \tag{2.180}$$

Alternatively, vec (A) can be expressed by (2.48) as follows:

$$\text{vec}(A) = D_N v(A) = D_N [V_d,\ V_l]\begin{bmatrix}\text{vec}_d(A)\\ \text{vec}_l(A)\end{bmatrix}, \tag{2.181}$$

where (2.49) was used. Because the right-hand sides of (2.180) and (2.181) are identical for all symmetric matrices A, it follows that

$$[L_d,\ L_l + L_u] = D_N [V_d,\ V_l]. \tag{2.182}$$

Right-multiplying the above equation by $[V_d,\ V_l]^T$ leads to (2.175). Multiplying out the right-hand side of (2.182) gives $D_N [V_d,\ V_l] = [D_N V_d,\ D_N V_l]$. By comparing this block matrix with the block matrix on the left-hand side of (2.182), the results in (2.176) and (2.177) follow.

The duplication matrix D_N has size $N^2 \times \frac{N(N+1)}{2}$ and is left invertible by its Moore-Penrose inverse, which is given by Magnus and Neudecker (1988, p. 49):

$$D_N^+ = \left(D_N^T D_N\right)^{-1} D_N^T. \tag{2.183}$$

By left-multiplying (2.48) by D_N^+, the following relation holds:

$$v(A) = D_N^+ \text{vec}(A). \tag{2.184}$$

Because $D_N^+ D_N = I_N$, (2.178) and (2.179) follow by left-multiplying (2.176) and (2.177) by D_N^+, respectively. ■

2.6 Exercises

2.1 Let $f : \mathbb{C} \times \mathbb{C} \to \mathbb{C}$ be given by

$$f(z, z^*) = u(x, y) + \jmath v(x, y), \tag{2.185}$$

where $z = x + \jmath y$, $\text{Re}\{f(z, z^*)\} = u(x, y)$, and $\text{Im}\{f(z, z^*)\} = v(x, y)$. Show that (2.6) is equivalent to the traditional formulation of the Cauchy-Riemann equations

$$\frac{\partial u}{\partial x} = \frac{\partial v}{\partial y}, \tag{2.186}$$

$$\frac{\partial u}{\partial y} = -\frac{\partial v}{\partial x}. \tag{2.187}$$

2.2 Functions that are going to be maximized or minimized must be real-valued. The results of this exercise show that in engineering problems of practical interests, the objective functions that are interesting do *not* satisfy the Cauchy-Riemann equations.

Let the function $f : \mathbb{C} \times \mathbb{C} \rightarrow \mathbb{R}$ be given by

$$f(z, z^*) = u(x, y) + \jmath v(x, y), \tag{2.188}$$

where $z = x + \jmath y$, $\text{Re}\{f(z, z^*)\} = u(x, y)$, and $\text{Im}\{f(z, z^*)\} = v(x, y)$ are real-valued; hence, $v(x, y) = 0$, and assume that it satisfies the Cauchy-Riemann equations. Show that f is then a constant function.

2.3 Decide whether the following functions are analytic or non-analytic:

$$f(z) = z^*, \tag{2.189}$$

$$f(z) = \sin(z), \tag{2.190}$$

$$f(z) = \exp(\jmath z), \tag{2.191}$$

$$f(z) = |z|, \tag{2.192}$$

$$f(z) = \frac{1}{z}, \tag{2.193}$$

$$f(z) = \text{Re}\{z\}, \tag{2.194}$$

$$f(z) = \text{Im}\{z\}, \tag{2.195}$$

$$f(z) = \text{Re}\{z\} + \jmath \, \text{Im}\{z\}, \tag{2.196}$$

$$f(z) = \text{Re}\{z\} - \jmath \, \text{Im}\{z\}, \tag{2.197}$$

$$f(z) = \ln(z), \tag{2.198}$$

where the principal value (Kreyszig 1988, p. 754) of $\ln(z)$ is used in this book.

2.4 Let $\boldsymbol{z} \in \mathbb{C}^{N\times 1}$ be an arbitrary complex-valued vector. Show that the Moore-Penrose inverse of $\boldsymbol{z}$ is given by

$$\boldsymbol{z}^{+} = \begin{cases} \frac{\boldsymbol{z}^H}{\|\boldsymbol{z}\|^2}, & \text{if } \boldsymbol{z} \neq \boldsymbol{0}_{N\times 1}, \\ \boldsymbol{0}_{1\times N}, & \text{if } \boldsymbol{z} = \boldsymbol{0}_{N\times 1}. \end{cases} \tag{2.199}$$

2.5 Assume that $\boldsymbol{A} \in \mathbb{C}^{N\times N}$ and $\boldsymbol{B} \in \mathbb{C}^{N\times N}$ commute (i.e., $\boldsymbol{AB} = \boldsymbol{BA}$). Show that

$$\exp(\boldsymbol{A}) \exp(\boldsymbol{B}) = \exp(\boldsymbol{B}) \exp(\boldsymbol{A}) = \exp(\boldsymbol{A} + \boldsymbol{B}). \tag{2.200}$$

2.6 Show that the following properties are valid for the commutation matrix:

$$\boldsymbol{K}_{Q,N}^T = \boldsymbol{K}_{Q,N}^{-1} = \boldsymbol{K}_{N,Q}, \tag{2.201}$$

$$\boldsymbol{K}_{1,N} = \boldsymbol{K}_{N,1} = \boldsymbol{I}_N, \tag{2.202}$$

$$\boldsymbol{K}_{Q,N} = \sum_{j=0}^{N-1} \sum_{i=0}^{Q-1} \boldsymbol{E}_{i,j} \otimes \boldsymbol{E}_{i,j}^T, \tag{2.203}$$

$$\left[\boldsymbol{K}_{Q,N}\right]_{i+jN,k+lQ} = \delta_{i,l}\delta_{j,k}, \tag{2.204}$$

where $\boldsymbol{E}_{i,j}$ of size $Q \times N$ contains only 0s except for $+1$ in the (i, j)-th position.

2.7 Write a MATLAB program that finds the commutation matrix $\boldsymbol{K}_{N,Q}$ without using `for`- or `while`- loops. (Hint: One useful MATLAB function to avoid loops is `find`.)

2.8 Show that

$$\text{Tr}\left\{\boldsymbol{A}^T \boldsymbol{B}\right\} = \text{vec}^T(\boldsymbol{A})\,\text{vec}(\boldsymbol{B}), \tag{2.205}$$

where the matrices $\boldsymbol{A} \in \mathbb{C}^{N\times M}$ and $\boldsymbol{B} \in \mathbb{C}^{N\times M}$.

2.9 Show that

$$\text{Tr}\{\boldsymbol{AB}\} = \text{Tr}\{\boldsymbol{BA}\}, \tag{2.206}$$

where the matrices $\boldsymbol{A} \in \mathbb{C}^{M\times N}$ and $\boldsymbol{B} \in \mathbb{C}^{N\times M}$.

2.10 Show that

$$\text{vec}(\boldsymbol{A} \odot \boldsymbol{B}) = \text{diag}(\text{vec}(\boldsymbol{A}))\,\text{vec}(\boldsymbol{B}), \tag{2.207}$$

where $\boldsymbol{A}, \boldsymbol{B} \in \mathbb{C}^{N\times M}$.

2.11 Let $\boldsymbol{A} \in \mathbb{C}^{M\times N}$, $\boldsymbol{B} \in \mathbb{C}^{P\times Q}$. Use Lemma 2.13 to show that

$$\text{vec}(\boldsymbol{A} \otimes \boldsymbol{B}) = [\boldsymbol{I}_N \otimes \boldsymbol{G}]\,\text{vec}(\boldsymbol{A}) = [\boldsymbol{H} \otimes \boldsymbol{I}_P]\,\text{vec}(\boldsymbol{B}), \tag{2.208}$$

where $\boldsymbol{G} \in \mathbb{C}^{QMP\times M}$ and $\boldsymbol{H} \in \mathbb{C}^{QMN\times Q}$ are given by

$$\boldsymbol{G} = \left[\boldsymbol{K}_{Q,M} \otimes \boldsymbol{I}_P\right]\left[\boldsymbol{I}_M \otimes \text{vec}(\boldsymbol{B})\right], \tag{2.209}$$

$$\boldsymbol{H} = \left[\boldsymbol{I}_N \otimes \boldsymbol{K}_{Q,M}\right]\left[\text{vec}(\boldsymbol{A}) \otimes \boldsymbol{I}_Q\right]. \tag{2.210}$$

2.12 Write MATLAB programs that find the matrices $\boldsymbol{L}_d$, $\boldsymbol{L}_l$, and $\boldsymbol{L}_u$ without using any `for`- or `while`- loops. (Hint: One useful MATLAB function to avoid loops is `find`.)

2.13 Let the identity matrix $\boldsymbol{I}_{\frac{N(N+1)}{2}}$ have columns that are indexed as follows:

$$\boldsymbol{I}_{\frac{N(N+1)}{2}} = \left[\boldsymbol{u}_{0,0}, \boldsymbol{u}_{1,0}, \cdots, \boldsymbol{u}_{N-1,0}, \boldsymbol{u}_{1,1}, \cdots, \boldsymbol{u}_{N-1,1}, \boldsymbol{u}_{2,2}, \cdots, \boldsymbol{u}_{N-1,N-1}\right], \tag{2.211}$$

where the vector $\boldsymbol{u}_{i,j} \in \mathbb{R}^{\frac{N(N+1)}{2} \times 1}$ contains 0s everywhere except in component number $jN + i + 1 - \frac{1}{2}(j+1)j$.
Show[6] that the duplication matrix $\boldsymbol{D}_N$ of size $N^2 \times \frac{N(N+1)}{2}$ can be expressed as

$$\begin{aligned} \boldsymbol{D}_N &= \sum_{i \geq j} \operatorname{vec}\left(\boldsymbol{T}_{i,j}\right) \boldsymbol{u}_{i,j}^T \\ &= \left[\operatorname{vec}(\boldsymbol{T}_{0,0}), \operatorname{vec}(\boldsymbol{T}_{1,0}), \cdots, \operatorname{vec}(\boldsymbol{T}_{N-1,0}), \operatorname{vec}(\boldsymbol{T}_{1,1}), \cdots, \operatorname{vec}(\boldsymbol{T}_{N-1,N-1})\right], \end{aligned} \tag{2.212}$$

where $\boldsymbol{u}_{i,j}$ is defined above, and where $\boldsymbol{T}_{i,j}$ is an $N \times N$ matrix defined as

$$\boldsymbol{T}_{i,j} = \begin{cases} \boldsymbol{E}_{i,j} + \boldsymbol{E}_{j,i}, & \text{if } i \neq j, \\ \boldsymbol{E}_{i,i}, & \text{if } i = j, \end{cases} \tag{2.213}$$

where $\boldsymbol{E}_{i,j}$ is found in Definition 2.16.
By using Definitions 2.15 and (2.175), show that

$$\boldsymbol{D}_N \boldsymbol{D}_N^T = \boldsymbol{I}_{N^2} + \boldsymbol{K}_{N,N} - \boldsymbol{K}_{N,N} \odot \boldsymbol{I}_{N^2}. \tag{2.214}$$

By means of (2.212), write a MATLAB program for finding the duplication matrix $\boldsymbol{D}_N$ without any `for-` or `while-` loops.

2.14 Let $\boldsymbol{a}_i \in \mathbb{C}^{N_i \times 1}$, where $i \in \{0, 1\}$, then show that

$$\operatorname{vec}\left(\boldsymbol{a}_0 \boldsymbol{a}_1^T\right) = \boldsymbol{a}_1 \otimes \boldsymbol{a}_0, \tag{2.215}$$

by using the definitions of the vec operator and the Kronecker product. Show also that the following is valid:

$$\boldsymbol{a}_0 \otimes \boldsymbol{a}_1^T = \boldsymbol{a}_0 \boldsymbol{a}_1^T = \boldsymbol{a}_1^T \otimes \boldsymbol{a}_0. \tag{2.216}$$

2.15 Show that Proposition 2.1 holds.

2.16 Let $\boldsymbol{A} \in \mathbb{C}^{M \times N}$ and $\boldsymbol{B} \in \mathbb{C}^{P \times Q}$. Show that

$$(\boldsymbol{A} \otimes \boldsymbol{B})^T = \boldsymbol{A}^T \otimes \boldsymbol{B}^T. \tag{2.217}$$

2.17 Let $\boldsymbol{A} \in \mathbb{C}^{N \times N}$, $\boldsymbol{B} \in \mathbb{C}^{N \times M}$, $\boldsymbol{C} \in \mathbb{C}^{M \times M}$, and $\boldsymbol{D} \in \mathbb{C}^{M \times N}$. Use Lemma 2.2 to show that if $\boldsymbol{A} \in \mathbb{C}^{N \times N}$, $\boldsymbol{C} \in \mathbb{C}^{M \times M}$, and $\boldsymbol{A} + \boldsymbol{B}\boldsymbol{C}\boldsymbol{D}$ are invertible, then $\boldsymbol{C}^{-1} + \boldsymbol{D}\boldsymbol{A}^{-1}\boldsymbol{B}$ is invertible. Show that the matrix inversion lemma stated in Lemma 2.3 is valid by showing (2.66).

2.18 Write a MATLAB program that implements the operator $v : \mathbb{C}^{N \times N} \to \mathbb{C}^{\frac{N(N+1)}{2} \times 1}$ without any loops. By using the program that implements the operator $v(\cdot)$, write a

[6] The following result is formulated in Magnus (1988, Theorem 4.3).

`MATLAB` program that finds the three matrices $\boldsymbol{V}_d$, $\boldsymbol{V}_l$, and $\boldsymbol{V}$ without using any `for-` or `while-` loops.

2.19 Given three positive integers M, N, and P, let $\boldsymbol{A} \in \mathbb{C}^{M \times N}$ and $\boldsymbol{B} \in \mathbb{C}^{NP \times P}$ be column symmetric. Show that the matrix $\boldsymbol{C} \triangleq [\boldsymbol{A} \otimes \boldsymbol{I}_P]\,\boldsymbol{B}$ is column symmetric, that is,

$$\operatorname{vecb}\left(\boldsymbol{C}^T\right) = \boldsymbol{C}. \tag{2.218}$$

3 Theory of Complex-Valued Matrix Derivatives

3.1 Introduction

A theory developed for finding derivatives with respect to *real-valued* matrices with independent elements was presented in Magnus and Neudecker (1988) for scalar, vector, and matrix functions. There, the matrix derivatives with respect to a real-valued matrix variable are found by means of the differential of the function. This theory is extended in this chapter to the case where the function depends on a complex-valued matrix variable and its complex conjugate, when all the elements of the matrix are independent. It will be shown how the complex differential of the function can be used to identify the derivative of the function with respect to both the complex-valued input matrix variable and its complex conjugate. This is a natural extension of the real-valued vector derivatives in Kreutz-Delgado (2008)[1] and the real-valued matrix derivatives in Magnus and Neudecker (1988) to the case of complex-valued matrix derivatives. The complex-valued input variable and its complex conjugate should be treated as independent when finding complex matrix derivatives. For scalar complex-valued functions that depend on a complex-valued *vector* and its complex conjugate, a theory for finding derivatives with respect to complex-valued vectors, when all the vector components are independent, was given in Brandwood (1983). This was extended to a systematic and simple way of finding derivatives of scalar, vector, and matrix functions with respect to complex-valued *matrices* when the matrix elements are independent (Hjørungnes & Gesbert 2007*a*). In this chapter, the definition of the complex-valued matrix derivative will be given, and a procedure will be presented for how to obtain the complex-valued matrix derivative. Central to this procedure is the complex differential of a function, because in the complex-valued matrix definition, the first issue is to find the complex differential of the function at hand.

The organization of the rest of this chapter is as follows: Section 3.2 contains an introduction to the area of complex differentials, where several ways for finding the complex differential are presented, together with the derivation of many useful complex differentials. The most important complex differentials are collected into Table 3.1

[1] Derivatives with respect to real-valued and complex-valued vectors were studied in Kreutz-Delgado (2008; 2009, June 25), respectively. Derivatives of a scalar function with respect to real-valued or complex-valued column vectors were organized as *row vectors* in Kreutz-Delgado (2008; 2009, June 25). The definition given in this chapter is a natural generalization of the definitions used in Kreutz-Delgado (2008; 2009, June 25).

and are easy for the reader to locate. In Section 3.3, the definition of complex-valued matrix derivatives is given together with a procedure that can be used to find complex-valued matrix derivatives. Fundamental results – including topics such as the chain rule of complex-valued matrix derivatives, conditions for finding stationary points for scalar real-valued functions, the direction in which a scalar real-valued function has its maximum and minimum rates of change, and the steepest descent method – are stated in Section 3.4. Section 3.5 presents exercises related to the material presented in this chapter. Some of these exercises can be directly applied in signal processing and communications.

3.2 Complex Differentials

Just as in the real-valued case (Magnus & Neudecker 1988), the symbol d will be used to denote the complex differential. The complex differential has the same size as the matrix it is applied to and can be found component-wise (i.e., $(d\boldsymbol{Z})_{k,l} = d\,(\boldsymbol{Z})_{k,l}$). Let $z = x + \jmath y \in \mathbb{C}$ represent a complex scalar variable, where $\mathrm{Re}\{z\} = x$ and $\mathrm{Im}\{z\} = y$. The following four relations hold between the real and imaginary parts of z and its complex conjugate z^*:

$$z = x + \jmath y, \tag{3.1}$$

$$z^* = x - \jmath y, \tag{3.2}$$

$$x = \frac{z + z^*}{2}, \tag{3.3}$$

$$y = \frac{z - z^*}{2\jmath}. \tag{3.4}$$

For complex differentials (Fong 2006), these four relations can be formulated as follows:

$$dz = dx + \jmath dy, \tag{3.5}$$

$$dz^* = dx - \jmath dy, \tag{3.6}$$

$$dx = \frac{dz + dz^*}{2}, \tag{3.7}$$

$$dy = \frac{dz - dz^*}{2\jmath}. \tag{3.8}$$

In studying (3.5) and (3.6), the following relation holds:

$$dz^* = (dz)^*. \tag{3.9}$$

Let us consider the scalar function $f : \mathbb{C} \times \mathbb{C} \to \mathbb{C}$ denoted by $f(z, z^*)$. Because the function f can be considered as a function of the two complex-valued variables z and z^*, both of which depend on the x and y through (3.1) and (3.2), the function f can also be seen as a function that depends on the two real-valued variables x and y. If f is considered as a function of the two independent real-valued variables x and y, the differential of f can be expressed as follows (Edwards & Penney 1986):

$$df = \frac{\partial f}{\partial x} dx + \frac{\partial f}{\partial y} dy, \tag{3.10}$$

where $\frac{\partial f}{\partial x}$ and $\frac{\partial f}{\partial y}$ are the partial derivatives of f with respect to x and y, respectively. By inserting the differential expression of dx and dy from (3.7) and (3.8) into (3.10), the following expression is found:

$$df = \frac{\partial f}{\partial x} \frac{dz + dz^*}{2} + \frac{\partial f}{\partial y} \frac{dz - dz^*}{2\jmath} = \frac{1}{2}\left(\frac{\partial f}{\partial x} - \jmath \frac{\partial f}{\partial y}\right) dz + \frac{1}{2}\left(\frac{\partial f}{\partial x} + \jmath \frac{\partial f}{\partial y}\right) dz^*. \tag{3.11}$$

A complex-valued expression (Fong 2006, Eq. (1.4)) similar to the one in (3.10) is also valid when z and z^* are treated as two independent variables:

$$df = \frac{\partial f}{\partial z} dz + \frac{\partial f}{\partial z^*} dz^*. \tag{3.12}$$

If (3.11) and (3.12) are compared, it is seen that

$$\frac{\partial f}{\partial z} = \frac{1}{2}\left(\frac{\partial f}{\partial x} - \jmath \frac{\partial f}{\partial y}\right), \tag{3.13}$$

and

$$\frac{\partial f}{\partial z^*} = \frac{1}{2}\left(\frac{\partial f}{\partial x} + \jmath \frac{\partial f}{\partial y}\right), \tag{3.14}$$

which are in agreement with the formal derivatives defined in Definition 2.2 (see (2.11) and (2.12)).

The above analysis can be extended to a scalar complex-valued function that depends on a complex-valued matrix variable $\boldsymbol{Z}$ and its complex conjugate $\boldsymbol{Z}^*$. Let us study the scalar function $f : \mathbb{C}^{N\times Q} \times \mathbb{C}^{N\times Q} \to \mathbb{C}$ denoted by $f(\boldsymbol{Z}, \boldsymbol{Z}^*)$. The complex-valued matrix variables $\boldsymbol{Z}$ and $\boldsymbol{Z}^*$ can also be expressed:

$$\boldsymbol{Z} = \boldsymbol{X} + \jmath \boldsymbol{Y}, \tag{3.15}$$

$$\boldsymbol{Z}^* = \boldsymbol{X} - \jmath \boldsymbol{Y}, \tag{3.16}$$

where $\mathrm{Re}\{\boldsymbol{Z}\} = \boldsymbol{X}$ and $\mathrm{Im}\{\boldsymbol{Z}\} = \boldsymbol{Y}$. The relations in (3.15) and (3.16) are equivalent to (2.2) and (2.3), respectively. The complex differential of a matrix is found by using the differential operator on each element of the matrix; hence, (2.2), (2.3), (2.4), and (2.5) can be carried over to differential forms, in the following way:

$$d\boldsymbol{Z} = d\,\mathrm{Re}\{\boldsymbol{Z}\} + \jmath d\,\mathrm{Im}\{\boldsymbol{Z}\} = d\boldsymbol{X} + \jmath d\boldsymbol{Y}, \tag{3.17}$$

$$d\boldsymbol{Z}^* = d\,\mathrm{Re}\{\boldsymbol{Z}\} - \jmath d\,\mathrm{Im}\{\boldsymbol{Z}\} = d\boldsymbol{X} - \jmath d\boldsymbol{Y}, \tag{3.18}$$

$$d\,\mathrm{Re}\{\boldsymbol{Z}\} = d\boldsymbol{X} = \tfrac{1}{2}\left(d\boldsymbol{Z} + d\boldsymbol{Z}^*\right), \tag{3.19}$$

$$d\,\mathrm{Im}\{\boldsymbol{Z}\} = d\boldsymbol{Y} = \tfrac{1}{2\jmath}\left(d\boldsymbol{Z} - d\boldsymbol{Z}^*\right). \tag{3.20}$$

Given all components within the two real-valued $N \times Q$ matrices $\boldsymbol{X}$ and $\boldsymbol{Y}$, the differential of f might be expressed in terms of the independent real-valued variables $x_{k,l}$

and $y_{k,l}$ or the independent (when considering complex derivatives) complex-valued variables $z_{k,l}$ and $z_{k,l}^*$ in the following way:

$$df = \sum_{k=0}^{Q-1}\sum_{l=0}^{N-1} \frac{\partial f}{\partial x_{k,l}} dx_{k,l} + \sum_{k=0}^{Q-1}\sum_{l=0}^{N-1} \frac{\partial f}{\partial y_{k,l}} dy_{k,l} \tag{3.21}$$

$$= \sum_{k=0}^{Q-1}\sum_{l=0}^{N-1} \frac{\partial f}{\partial z_{k,l}} dz_{k,l} + \sum_{k=0}^{Q-1}\sum_{l=0}^{N-1} \frac{\partial f}{\partial z_{k,l}^*} dz_{k,l}^*, \tag{3.22}$$

where $\frac{\partial f}{\partial x_{k,l}}$, $\frac{\partial f}{\partial y_{k,l}}$, $\frac{\partial f}{\partial z_{k,l}}$, and $\frac{\partial f}{\partial z_{k,l}^*}$ are the derivatives of f with respect to $x_{k,l}$, $y_{k,l}$, $z_{k,l}$, and $z_{k,l}^*$, respectively. The NQ formal derivatives of $\frac{\partial f}{\partial z_{k,l}}$ and the NQ formal derivatives of $\frac{\partial f}{\partial z_{k,l}^*}$ can be organized into matrices in several ways; later in this and in the next chapter, we will see several alternative definitions for the derivatives of a scalar function f with respect to complex-valued matrices $\boldsymbol{Z}$ and $\boldsymbol{Z}^*$.

This section contains three subsections. In Subsection 3.2.1, a procedure that can often be used to find the complex differentials is presented. Several basic complex differentials that are essential for finding complex derivatives are presented in Subsection 3.2.2, together with their derivations. Two lemmas are presented in Subsection 3.2.3; these will be used to identify both first- and second-order derivatives in this and later chapters.

3.2.1 Procedure for Finding Complex Differentials

Let the two input complex-valued matrix variables be denoted $\boldsymbol{Z}_0 \in \mathbb{C}^{N\times Q}$ and $\boldsymbol{Z}_1 \in \mathbb{C}^{N\times Q}$, where all elements of these two matrices are independent. It is assumed that these two complex-valued matrix variables can be treated independently when finding complex-valued matrix derivatives.

A procedure that can often be used to find the differentials of a complex matrix function $\boldsymbol{F} : \mathbb{C}^{N\times Q} \times \mathbb{C}^{N\times Q} \to \mathbb{C}^{M\times P}$, denoted by $\boldsymbol{F}(\boldsymbol{Z}_0, \boldsymbol{Z}_1)$, is to calculate the difference

$$\boldsymbol{F}(\boldsymbol{Z}_0 + d\boldsymbol{Z}_0, \boldsymbol{Z}_1 + d\boldsymbol{Z}_1) - \boldsymbol{F}(\boldsymbol{Z}_0, \boldsymbol{Z}_1)$$
$$= \text{First-order}(d\boldsymbol{Z}_0, d\boldsymbol{Z}_1) + \text{Higher-order}(d\boldsymbol{Z}_0, d\boldsymbol{Z}_1), \tag{3.23}$$

where First-order$(\cdot,\cdot)$ returns the terms that depend on $d\boldsymbol{Z}_0$ or $d\boldsymbol{Z}_1$ of the *first order*, and Higher-order$(\cdot,\cdot)$ returns the terms that depend on the *higher-order* terms of $d\boldsymbol{Z}_0$ and $d\boldsymbol{Z}_1$. The differential is then given by First-order$(\cdot,\cdot)$ as

$$d\boldsymbol{F} = \text{First-order}(\boldsymbol{F}(\boldsymbol{Z}_0 + d\boldsymbol{Z}_0, \boldsymbol{Z}_1 + d\boldsymbol{Z}_1) - \boldsymbol{F}(\boldsymbol{Z}_0, \boldsymbol{Z}_1)). \tag{3.24}$$

This procedure will be used several times in this chapter.

3.2.2 Basic Complex Differential Properties

Some of the basic properties of complex differentials are presented in this subsection.

Proposition 3.1 *Let $\boldsymbol{A} \in \mathbb{C}^{M\times P}$ be a constant matrix that is not dependent on the complex matrix variable $\boldsymbol{Z}$ or $\boldsymbol{Z}^*$. The complex differential of a constant matrix $\boldsymbol{A}$ is* *given by*

$$dA = \mathbf{0}_{M\times P}. \tag{3.25}$$

Proof Let the function used in (3.23) be given as $F(Z_0, Z_1) = A$. By obtaining the difference in (3.23), it is found that

$$F(Z_0 + dZ_0, Z_1 + dZ_1) - F(Z_0, Z_1) = A - A = \mathbf{0}_{M\times P}. \tag{3.26}$$

Here, both first-order and higher-order terms are equal to the zero matrix $\mathbf{0}_{M\times P}$. Hence, (3.25) follows. ■

Proposition 3.2 *Let $A \in \mathbb{C}^{M\times N}$, $Z \in \mathbb{C}^{N\times Q}$, and $B \in \mathbb{C}^{Q\times P}$, where A and B are independent of Z and Z^*. Then*

$$d(AZB) = A(dZ)B. \tag{3.27}$$

Proof The procedure presented in Subsection 3.2.1 is now followed. Let the function used in Subsection 3.2.1 be given by $F(Z_0, Z_1) = AZ_0B$. The difference on the left-hand side of (3.23) can be written as

$$\begin{aligned} F(Z_0 + dZ_0, Z_1 + dZ_1) - F(Z_0, Z_1) &= A(Z_0 + dZ_0)B - AZ_0B \\ &= A(dZ_0)B. \end{aligned} \tag{3.28}$$

It is seen that the left-hand side of (3.28) contains only one first-order term. By choosing the two complex-valued matrix variables Z_0 and Z_1 in (3.28) as $Z_0 = Z$ and $Z_1 = Z^*$, it is seen that (3.27) follows. ■

Corollary 3.1 *Let $a \in \mathbb{C}$ be a constant that is independent of $Z \in \mathbb{C}^{N\times Q}$ and $Z^* \in \mathbb{C}^{N\times Q}$. Then*

$$d(aZ) = a\,dZ, \tag{3.29}$$

Proof If we set $A = aI_N$ and $B = I_Q$ in (3.27), the result follows. ■

Proposition 3.3 *Let $Z_i \in \mathbb{C}^{N\times Q}$ for $i \in \{0, 1, \ldots, L-1\}$. The complex differential of a sum is given by*

$$d(Z_0 + Z_1) = dZ_0 + dZ_1. \tag{3.30}$$

The complex differential of a sum of L for such matrices can be expressed as

$$d\left(\sum_{k=0}^{L-1} Z_k\right) = \sum_{k=0}^{L-1} dZ_k. \tag{3.31}$$

Proof Let the function in the procedure outlined in Subsection 3.2.1 be given by $F(Z_0, Z_1) = Z_0 + Z_1$. By forming the difference in (3.23), it is found that

$$\begin{aligned} F(Z_0 + dZ_0, Z_1 + dZ_1) - F(Z_0, Z_1) &= Z_0 + dZ_0 + Z_1 + dZ_1 - Z_0 - Z_1 \\ &= dZ_0 + dZ_1. \end{aligned} \tag{3.32}$$

Both terms on the right-hand side of (3.32) are of the first order in $d\boldsymbol{Z}_0$ or $d\boldsymbol{Z}_1$; hence, (3.30) follows.

By repeated application of (3.30), (3.31) follows. ■

Proposition 3.4 *If* $\boldsymbol{Z} \in \mathbb{C}^{N\times N}$, *then*

$$d(\text{Tr}\{\boldsymbol{Z}\}) = \text{Tr}\{d\boldsymbol{Z}\}. \tag{3.33}$$

Proof If the procedure in Subsection 3.2.1 should be followed, it is first adapted to scalar functions. Let the function f be defined as $f : \mathbb{C}^{N\times N} \times \mathbb{C}^{N\times N} \to \mathbb{C}$, where it is given by $f(\boldsymbol{Z}_0, \boldsymbol{Z}_1) = \text{Tr}\{\boldsymbol{Z}_0\}$. The left-hand side of (3.23) can be written as

$$\begin{aligned} \boldsymbol{F}(\boldsymbol{Z}_0 + d\boldsymbol{Z}_0, \boldsymbol{Z}_1 + d\boldsymbol{Z}_1) - \boldsymbol{F}(\boldsymbol{Z}_0, \boldsymbol{Z}_1) &= \text{Tr}\{\boldsymbol{Z}_0 + d\boldsymbol{Z}_0\} - \text{Tr}\{\boldsymbol{Z}_0\} \\ &= \text{Tr}\{d\boldsymbol{Z}_0\}. \end{aligned} \tag{3.34}$$

The right-hand side of (3.34) contains only first-order terms in $d\boldsymbol{Z}_0$. By choosing $\boldsymbol{Z}_0 = \boldsymbol{Z}$ and $\boldsymbol{Z}_1 = \boldsymbol{Z}^*$ in (3.34), the result in (3.33) follows. ■

Proposition 3.5 *Let* $\boldsymbol{Z}_0 \in \mathbb{C}^{M\times N}$ *and* $\boldsymbol{Z}_1 \in \mathbb{C}^{N\times P}$, *such that the matrix product* $\boldsymbol{Z}_0\boldsymbol{Z}_1$ *is well defined. Then*[2]

$$d\boldsymbol{Z}_0\boldsymbol{Z}_1 = (d\boldsymbol{Z}_0)\boldsymbol{Z}_1 + \boldsymbol{Z}_0 d\boldsymbol{Z}_1. \tag{3.35}$$

Proof To find the complex differential of $\boldsymbol{Z}_0\boldsymbol{Z}_1$, the procedure outlined in (3.23) is followed. Let $\boldsymbol{F}(\boldsymbol{Z}_0, \boldsymbol{Z}_1) = \boldsymbol{Z}_0\boldsymbol{Z}_1$. First, the left-hand side of (3.23) is written as

$$\boldsymbol{F}(\boldsymbol{Z}_0 + d\boldsymbol{Z}_0, \boldsymbol{Z}_1 + d\boldsymbol{Z}_1) - \boldsymbol{F}(\boldsymbol{Z}_0, \boldsymbol{Z}_1) = \boldsymbol{Z}_0 d\boldsymbol{Z}_1 + (d\boldsymbol{Z}_0)\boldsymbol{Z}_1 + (d\boldsymbol{Z}_0)d\boldsymbol{Z}_1.$$

The complex differential of $\boldsymbol{F}(\boldsymbol{Z}_0, \boldsymbol{Z}_1)$ can be identified as all the first-order terms on $d\boldsymbol{Z}_0$ or $d\boldsymbol{Z}_1$; therefore, $d\boldsymbol{Z}_0\boldsymbol{Z}_1 = \boldsymbol{Z}_0 d\boldsymbol{Z}_1 + (d\boldsymbol{Z}_0)\boldsymbol{Z}_1$. ■

Proposition 3.6 *Let* $\boldsymbol{Z}_0 \in \mathbb{C}^{N\times Q}$ *and* $\boldsymbol{Z}_1 \in \mathbb{C}^{M\times P}$. *The complex differential of the Kronecker product is given by*

$$d(\boldsymbol{Z}_0 \otimes \boldsymbol{Z}_1) = (d\boldsymbol{Z}_0) \otimes \boldsymbol{Z}_1 + \boldsymbol{Z}_0 \otimes d\boldsymbol{Z}_1. \tag{3.36}$$

Proof The procedure in Subsection 3.2.1 is followed, so let $\boldsymbol{F} : \mathbb{C}^{N\times Q} \times \mathbb{C}^{M\times P} \to \mathbb{C}^{NM\times QP}$ be given by $\boldsymbol{F}(\boldsymbol{Z}_0, \boldsymbol{Z}_1) = \boldsymbol{Z}_0 \otimes \boldsymbol{Z}_1$, expanding the difference of the left-hand side of (3.23):

$$\begin{aligned} \boldsymbol{F}(\boldsymbol{Z}_0 + d\boldsymbol{Z}_0, \boldsymbol{Z}_1 + d\boldsymbol{Z}_1) - \boldsymbol{F}(\boldsymbol{Z}_0, \boldsymbol{Z}_1) &= (\boldsymbol{Z}_0 + d\boldsymbol{Z}_0) \otimes (\boldsymbol{Z}_1 + d\boldsymbol{Z}_1) - \boldsymbol{Z}_0 \otimes \boldsymbol{Z}_1 \\ &= \boldsymbol{Z}_0 \otimes d\boldsymbol{Z}_1 + (d\boldsymbol{Z}_0) \otimes \boldsymbol{Z}_1 + (d\boldsymbol{Z}_0) \otimes d\boldsymbol{Z}_1, \end{aligned} \tag{3.37}$$

where it was used so that the Kronecker product follows the distributive law.[3] Three addends are present on the right-hand side of (3.37); the first two are of the first order in $d\boldsymbol{Z}_0$ and $d\boldsymbol{Z}_1$, and the third addend is of the second order. Because the differential

[2] In this book, the following notation is used when taking differentials of *matrix products*: $d\boldsymbol{Z}_0\boldsymbol{Z}_1 = d(\boldsymbol{Z}_0\boldsymbol{Z}_1)$.

[3] Let $\boldsymbol{A}, \boldsymbol{B} \in \mathbb{C}^{N\times Q}$ and $\boldsymbol{C}, \boldsymbol{D} \in \mathbb{C}^{M\times P}$, then $(\boldsymbol{A} + \boldsymbol{B}) \otimes (\boldsymbol{C} + \boldsymbol{D}) = \boldsymbol{A} \otimes \boldsymbol{C} + \boldsymbol{A} \otimes \boldsymbol{D} + \boldsymbol{B} \otimes \boldsymbol{C} + \boldsymbol{B} \otimes \boldsymbol{D}$. This is shown in Horn and Johnson (1991, Section 4.2).

of $\boldsymbol{F}$ is equal to the first-order terms in $d\boldsymbol{Z}_0$ and $d\boldsymbol{Z}_1$ in (3.37), the result in (3.36) follows. ■

Proposition 3.7 *Let* $\boldsymbol{Z}_0 \in \mathbb{C}^{N\times Q}$ *for* $i \in \{0, 1\}$. *The complex differential of the Hadamard product is given by*

$$d(\boldsymbol{Z}_0 \odot \boldsymbol{Z}_1) = (d\boldsymbol{Z}_0) \odot \boldsymbol{Z}_1 + \boldsymbol{Z}_0 \odot d\boldsymbol{Z}_1. \tag{3.38}$$

Proof Let $\boldsymbol{F} : \mathbb{C}^{N\times Q} \times \mathbb{C}^{N\times Q} \to \mathbb{C}^{N\times Q}$ be given as $\boldsymbol{F}(\boldsymbol{Z}_0, \boldsymbol{Z}_1) = \boldsymbol{Z}_0 \odot \boldsymbol{Z}_1$. The difference on the left-hand side of (3.23) can be written as

$$\begin{aligned} \boldsymbol{F}(\boldsymbol{Z}_0 + d\boldsymbol{Z}_0, \boldsymbol{Z}_1 + d\boldsymbol{Z}_1) - \boldsymbol{F}(\boldsymbol{Z}_0, \boldsymbol{Z}_1) &= (\boldsymbol{Z}_0 + d\boldsymbol{Z}_0) \odot (\boldsymbol{Z}_1 + d\boldsymbol{Z}_1) - \boldsymbol{Z}_0 \odot \boldsymbol{Z}_1 \\ &= \boldsymbol{Z}_0 \odot d\boldsymbol{Z}_1 + (d\boldsymbol{Z}_0) \odot \boldsymbol{Z}_1 + (d\boldsymbol{Z}_0) \odot d\boldsymbol{Z}_1. \end{aligned} \tag{3.39}$$

Among the three addends on the right-hand side in (3.39), the first two are first order in $d\boldsymbol{Z}_0$ and $d\boldsymbol{Z}_1$, and the third addend is second order. Hence, (3.38) follows. ■

Proposition 3.8 *Let* $\boldsymbol{Z} \in \mathbb{C}^{N\times N}$ *be invertible. Then the complex differential of the inverse matrix* $\boldsymbol{Z}^{-1}$ *is given by*

$$d\boldsymbol{Z}^{-1} = -\boldsymbol{Z}^{-1}(d\boldsymbol{Z})\boldsymbol{Z}^{-1}. \tag{3.40}$$

Proof Because $\boldsymbol{Z} \in \mathbb{C}^{N\times N}$ is invertible, the following relation is satisfied:

$$\boldsymbol{Z}\boldsymbol{Z}^{-1} = \boldsymbol{I}_N. \tag{3.41}$$

By applying the differential operator d on both sides of (3.41) and using the results from (3.25) and (3.35), it is found that

$$(d\boldsymbol{Z})\boldsymbol{Z}^{-1} + \boldsymbol{Z}d\boldsymbol{Z}^{-1} = d\boldsymbol{I}_N = \boldsymbol{0}_{N\times N}. \tag{3.42}$$

Solving $d\boldsymbol{Z}^{-1}$ from this equation yields (3.40). ■

Proposition 3.9 *Let* reshape$(\cdot)$ *be any* linear reshaping operator[4] *of the input matrix. The complex differential of the operator* reshape$(\cdot)$ *is given by*

$$d\,\text{reshape}(\boldsymbol{Z}) = \text{reshape}(d\boldsymbol{Z}). \tag{3.43}$$

Proof Because reshape$(\cdot)$ is a *linear* operator, it follows that

$$\begin{aligned} \text{reshape}(\boldsymbol{Z} + d\boldsymbol{Z}) - \text{reshape}(\boldsymbol{Z}) &= \text{reshape}(\boldsymbol{Z}) + \text{reshape}(d\boldsymbol{Z}) - \text{reshape}(\boldsymbol{Z}) \\ &= \text{reshape}(d\boldsymbol{Z}). \end{aligned} \tag{3.44}$$

By using the procedure from Subsection 3.2.1, and because the right-hand side of (3.44) contains only one first-order term, the result in (3.43) follows. ■

[4] The size of the output vector/matrix might be different from the input, so reshape$(\cdot)$ performs linear reshaping of its input argument. Hence, the reshape$(\cdot)$ operator might delete certain input components, keep all input components, and/or make multiple copies of certain input components.

The differentiation rule of the reshaping operator reshape($\cdot$) in Table 3.1 is valid for any *linear reshaping* operator reshape($\cdot$) of a matrix; examples of such operators include transpose $(\cdot)^T$ and vec($\cdot$).

Proposition 3.10 *Let $\boldsymbol{Z} \in \mathbb{C}^{N\times Q}$. Then the complex differential of the matrix $\boldsymbol{Z}^*$ is given by*

$$d\boldsymbol{Z}^* = (d\boldsymbol{Z})^*. \tag{3.45}$$

Proof Because the differential operator of a matrix operates on each component of the matrix, and because of (3.9), the expression in (3.45) is valid. ■

Proposition 3.11 *Let $\boldsymbol{Z} \in \mathbb{C}^{N\times Q}$. Then the differential of the complex Hermitian of $\boldsymbol{Z}$ is given by*

$$d\boldsymbol{Z}^H = (d\boldsymbol{Z})^H. \tag{3.46}$$

Proof Because the Hermitian operator is given by the complex conjugate transpose, this result follows from (3.43) when using reshape($\cdot$) as the $(\cdot)^T$ plus (3.45)

$$d\boldsymbol{Z}^H = d\left(\boldsymbol{Z}^*\right)^T = \left(d\boldsymbol{Z}^*\right)^T = \left((d\boldsymbol{Z})^*\right)^T = (d\boldsymbol{Z})^H. \tag{3.47}$$

■

Proposition 3.12 *Let $\boldsymbol{Z} \in \mathbb{C}^{N\times N}$. Then the complex differential of the determinant is given by*

$$d\det(\boldsymbol{Z}) = \mathrm{Tr}\left\{\boldsymbol{C}^T(\boldsymbol{Z})d\boldsymbol{Z}\right\}. \tag{3.48}$$

where the matrix $\boldsymbol{C}(\boldsymbol{Z}) \in \mathbb{C}^{N\times N}$ contains the cofactors[5] *denoted by $c_{k,l}(\boldsymbol{Z})$ of $\boldsymbol{Z}$. If $\boldsymbol{Z} \in \mathbb{C}^{N\times N}$ is invertible, then the complex differential of the determinant is given by*

$$d\det(\boldsymbol{Z}) = \det(\boldsymbol{Z})\,\mathrm{Tr}\left\{\boldsymbol{Z}^{-1}d\boldsymbol{Z}\right\}. \tag{3.49}$$

Proof Let $c_{k,l}(\boldsymbol{Z})$ be the cofactor of $z_{k,l} \triangleq (\boldsymbol{Z})_{k,l}$, where $\boldsymbol{Z} \in \mathbb{C}^{N\times N}$. The determinant can be formulated along any row or column, and if the column number l is considered, the determinant can be written as

$$\det(\boldsymbol{Z}) = \sum_{k=0}^{N-1} c_{k,l}(\boldsymbol{Z})z_{k,l}, \tag{3.50}$$

where the cofactor $c_{k,l}(\boldsymbol{Z})$ is independent of $z_{k,l}$ and $z_{k,l}^*$.

If $f : \mathbb{C}^{N\times Q} \times \mathbb{C}^{N\times Q} \to \mathbb{C}$ is a scalar complex-valued function denoted by $f(\boldsymbol{Z}, \boldsymbol{Z}^*)$, then the connection between the differential of f, the derivatives of f with respect to all the components of $\boldsymbol{Z}$ and $\boldsymbol{Z}^*$, and the differentials of the components of $\boldsymbol{Z}$ and $\boldsymbol{Z}^*$ can be written as in (3.22). If (3.22) is used on the function $f(\boldsymbol{Z}, \boldsymbol{Z}^*) = \det(\boldsymbol{Z})$, where

[5] A cofactor $c_{k,l}(\boldsymbol{Z})$ of $\boldsymbol{Z} \in \mathbb{C}^{N\times N}$ is equal to $(-1)^{k+l}$ times the (k, l)-th minor of $\boldsymbol{Z}$, denoted by $m_{k,l}(\boldsymbol{Z})$. The minor $m_{k,l}(\boldsymbol{Z})$ is equal to the determinant of the $(N-1)\times(N-1)$ submatrix of $\boldsymbol{Z}$ found by deleting its k-th row and l-th column.

$N = Q$, then it is found that the derivatives of f with respect to $z_{k,l}$ and $z_{k,l}^*$ are given by

$$\frac{\partial}{\partial z_{k,l}} f = c_{k,l}(\boldsymbol{Z}), \tag{3.51}$$

$$\frac{\partial}{\partial z_{k,l}^*} f = 0, \tag{3.52}$$

where (3.50) has been utilized. Inserting (3.51) and (3.52) into (3.22) leads to

$$d \det(\boldsymbol{Z}) = \sum_{k=0}^{N-1} \sum_{l=0}^{N-1} c_{k,l}(\boldsymbol{Z}) dz_{k,l}. \tag{3.53}$$

The following identity is valid for square matrices $\boldsymbol{A} \in \mathbb{C}^{N \times N}$ and $\boldsymbol{B} \in \mathbb{C}^{N \times N}$:

$$\operatorname{Tr}\{\boldsymbol{A}\boldsymbol{B}\} = \sum_{p=0}^{N-1} \sum_{q=0}^{N-1} a_{p,q} b_{q,p}. \tag{3.54}$$

If (3.54) is used on the expression in (3.53), the following expression for the differential $\det(\boldsymbol{Z})$ is found:

$$d \det(\boldsymbol{Z}) = \operatorname{Tr}\left\{\boldsymbol{C}^T(\boldsymbol{Z}) d\boldsymbol{Z}\right\}, \tag{3.55}$$

where $\boldsymbol{C}(\boldsymbol{Z}) \in \mathbb{C}^{N \times N}$ is the matrix of cofactors of $\boldsymbol{Z}$ such that $c_{k,l}(\boldsymbol{Z}) = (\boldsymbol{C}(\boldsymbol{Z}))_{k,l}$. Therefore, (3.48) holds.

Assume now that $\boldsymbol{Z}$ is invertible. The following formula is valid for invertible matrices (Kreyszig 1988, p. 411):

$$\boldsymbol{C}^T(\boldsymbol{Z}) = \boldsymbol{Z}^{\#} = \det(\boldsymbol{Z})\boldsymbol{Z}^{-1}. \tag{3.56}$$

When (3.56) is used in (3.48), it follows that the differential of $\det(\boldsymbol{Z})$ can be written as

$$d \det(\boldsymbol{Z}) = \operatorname{Tr}\{\boldsymbol{Z}^{\#} d\boldsymbol{Z}\} = \det(\boldsymbol{Z}) \operatorname{Tr}\left\{\boldsymbol{Z}^{-1} d\boldsymbol{Z}\right\}, \tag{3.57}$$

which completes the last part of the proposition. ■

Proposition 3.13 *Let $\boldsymbol{Z} \in \mathbb{C}^{N \times N}$ be nonsingular. Then the complex differential of the adjoint of $\boldsymbol{Z}$ can be expressed as*

$$d\boldsymbol{Z}^{\#} = \det(\boldsymbol{Z}) \left[\operatorname{Tr}\left\{\boldsymbol{Z}^{-1}(d\boldsymbol{Z})\right\} \boldsymbol{Z}^{-1} - \boldsymbol{Z}^{-1}(d\boldsymbol{Z})\boldsymbol{Z}^{-1}\right]. \tag{3.58}$$

Proof For invertible matrices $\boldsymbol{Z}^{\#} = \det(\boldsymbol{Z})\boldsymbol{Z}^{-1}$, using the complex differential operator on this matrix relation and the results in (3.35), (3.40), and (3.49) yields

$$\begin{aligned} d\boldsymbol{Z}^{\#} &= (d \det(\boldsymbol{Z}))\boldsymbol{Z}^{-1} + \det(\boldsymbol{Z}) d\boldsymbol{Z}^{-1} \\ &= \det(\boldsymbol{Z}) \operatorname{Tr}\left\{\boldsymbol{Z}^{-1} d\boldsymbol{Z}\right\} \boldsymbol{Z}^{-1} - \det(\boldsymbol{Z})\boldsymbol{Z}^{-1} (d\boldsymbol{Z}) \boldsymbol{Z}^{-1} \\ &= \det(\boldsymbol{Z}) \left[\operatorname{Tr}\left\{\boldsymbol{Z}^{-1} d\boldsymbol{Z}\right\} \boldsymbol{Z}^{-1} - \boldsymbol{Z}^{-1} (d\boldsymbol{Z}) \boldsymbol{Z}^{-1}\right], \end{aligned} \tag{3.59}$$

which is the desired result. ■

The following differential is important when finding derivatives of the mutual information and the capacity of an MIMO channel, because the capacity is given by the logarithm of a determinant expression (Telatar 1995).

Proposition 3.14 *Let $\boldsymbol{Z} \in \mathbb{C}^{N\times N}$ be invertible with a determinant that is not both real and negative. Then the differential of the natural logarithm of the determinant is given by*

$$d\ln(\det(\boldsymbol{Z})) = \operatorname{Tr}\left\{\boldsymbol{Z}^{-1} d\boldsymbol{Z}\right\}. \tag{3.60}$$

Proof In this book, the principal value is used for $\ln(z)$, and its derivative is given by Kreyszig (1988, p. 755):

$$\frac{\partial \ln(z)}{\partial z} = \frac{1}{z}. \tag{3.61}$$

Hence, the complex differential of $\ln(z)$ is given by

$$d\ln(z) = \frac{dz}{z}, \tag{3.62}$$

when the variable z is not located on the negative real axis or in the origin.

Assume that $\det(\boldsymbol{Z})$ is not both real and non-positive. Then

$$d\ln(\det(\boldsymbol{Z})) = \frac{d\det(\boldsymbol{Z})}{\det(\boldsymbol{Z})} = \frac{\det(\boldsymbol{Z})\operatorname{Tr}\left\{\boldsymbol{Z}^{-1}d\boldsymbol{Z}\right\}}{\det(\boldsymbol{Z})} = \operatorname{Tr}\left\{\boldsymbol{Z}^{-1}d\boldsymbol{Z}\right\}, \tag{3.63}$$

where (3.49) was used to find $d\det(\boldsymbol{Z})$. ■

The differential of the *real-valued* Moore-Penrose inverse is given in Magnus and Neudecker (1988) and Harville (1997), but the complex-valued version is derived next.

Proposition 3.15 (Differential of the Moore-Penrose Inverse) *Let $\boldsymbol{Z} \in \mathbb{C}^{N\times Q}$. Then the complex differential of $\boldsymbol{Z}^{+}$ is given by*

$$\begin{aligned} d\boldsymbol{Z}^{+} &= -\boldsymbol{Z}^{+}(d\boldsymbol{Z})\boldsymbol{Z}^{+} + \boldsymbol{Z}^{+}(\boldsymbol{Z}^{+})^{H}(d\boldsymbol{Z}^{H})\left(\boldsymbol{I}_N - \boldsymbol{Z}\boldsymbol{Z}^{+}\right) \\ &\quad + \left(\boldsymbol{I}_Q - \boldsymbol{Z}^{+}\boldsymbol{Z}\right)(d\boldsymbol{Z}^{H})(\boldsymbol{Z}^{+})^{H}\boldsymbol{Z}^{+}. \end{aligned} \tag{3.64}$$

Proof Equation (2.25) leads to $d\boldsymbol{Z}^{+} = d\boldsymbol{Z}^{+}\boldsymbol{Z}\boldsymbol{Z}^{+} = (d\boldsymbol{Z}^{+}\boldsymbol{Z})\boldsymbol{Z}^{+} + \boldsymbol{Z}^{+}\boldsymbol{Z}d\boldsymbol{Z}^{+}$. If $\boldsymbol{Z}d\boldsymbol{Z}^{+}$ is found from $d\boldsymbol{Z}\boldsymbol{Z}^{+} = (d\boldsymbol{Z})\boldsymbol{Z}^{+} + \boldsymbol{Z}d\boldsymbol{Z}^{+}$, and is inserted in the expression for $d\boldsymbol{Z}^{+}$, then it is found that

$$\begin{aligned} d\boldsymbol{Z}^{+} &= (d\boldsymbol{Z}^{+}\boldsymbol{Z})\boldsymbol{Z}^{+} + \boldsymbol{Z}^{+}(d\boldsymbol{Z}\boldsymbol{Z}^{+} - (d\boldsymbol{Z})\boldsymbol{Z}^{+}) \\ &= (d\boldsymbol{Z}^{+}\boldsymbol{Z})\boldsymbol{Z}^{+} + \boldsymbol{Z}^{+}d\boldsymbol{Z}\boldsymbol{Z}^{+} - \boldsymbol{Z}^{+}(d\boldsymbol{Z})\boldsymbol{Z}^{+}. \end{aligned} \tag{3.65}$$

It is seen from (3.65) that it remains to express $d\boldsymbol{Z}^{+}\boldsymbol{Z}$ and $d\boldsymbol{Z}\boldsymbol{Z}^{+}$ in terms of $d\boldsymbol{Z}$ and $d\boldsymbol{Z}^{*}$. First, $d\boldsymbol{Z}^{+}\boldsymbol{Z}$ is handled as

$$\begin{aligned} d\boldsymbol{Z}^{+}\boldsymbol{Z} &= d\boldsymbol{Z}^{+}\boldsymbol{Z}\boldsymbol{Z}^{+}\boldsymbol{Z} = (d\boldsymbol{Z}^{+}\boldsymbol{Z})\boldsymbol{Z}^{+}\boldsymbol{Z} + \boldsymbol{Z}^{+}\boldsymbol{Z}(d\boldsymbol{Z}^{+}\boldsymbol{Z}) \\ &= (\boldsymbol{Z}^{+}\boldsymbol{Z}(d\boldsymbol{Z}^{+}\boldsymbol{Z}))^{H} + \boldsymbol{Z}^{+}\boldsymbol{Z}(d\boldsymbol{Z}^{+}\boldsymbol{Z}). \end{aligned} \tag{3.66}$$

The expression $\boldsymbol{Z}(d\boldsymbol{Z}^{+}\boldsymbol{Z})$ can be found from $d\boldsymbol{Z} = d\boldsymbol{Z}\boldsymbol{Z}^{+}\boldsymbol{Z} = (d\boldsymbol{Z})\boldsymbol{Z}^{+}\boldsymbol{Z} + \boldsymbol{Z}(d\boldsymbol{Z}^{+}\boldsymbol{Z})$, and it is given by $\boldsymbol{Z}(d\boldsymbol{Z}^{+}\boldsymbol{Z}) = d\boldsymbol{Z} - (d\boldsymbol{Z})\boldsymbol{Z}^{+}\boldsymbol{Z} = (d\boldsymbol{Z})(\boldsymbol{I}_Q - \boldsymbol{Z}^{+}\boldsymbol{Z})$. If this expression

Table 3.1 Important complex differentials.

Function	Differential of function
$\boldsymbol{A}$	$\boldsymbol{0}$
$a\boldsymbol{Z}$	$ad\boldsymbol{Z}$
$\boldsymbol{AZB}$	$\boldsymbol{A}(d\boldsymbol{Z})\boldsymbol{B}$
$\boldsymbol{Z}_0 + \boldsymbol{Z}_1$	$d\boldsymbol{Z}_0 + d\boldsymbol{Z}_1$
$\mathrm{Tr}\{\boldsymbol{Z}\}$	$\mathrm{Tr}\{d\boldsymbol{Z}\}$
$\boldsymbol{Z}_0\boldsymbol{Z}_1$	$(d\boldsymbol{Z}_0)\boldsymbol{Z}_1 + \boldsymbol{Z}_0 d\boldsymbol{Z}_1$
$\boldsymbol{Z}_0 \otimes \boldsymbol{Z}_1$	$(d\boldsymbol{Z}_0) \otimes \boldsymbol{Z}_1 + \boldsymbol{Z}_0 \otimes d\boldsymbol{Z}_1$
$\boldsymbol{Z}_0 \odot \boldsymbol{Z}_1$	$(d\boldsymbol{Z}_0) \odot \boldsymbol{Z}_1 + \boldsymbol{Z}_0 \odot d\boldsymbol{Z}_1$
$\boldsymbol{Z}^{-1}$	$-\boldsymbol{Z}^{-1}(d\boldsymbol{Z})\boldsymbol{Z}^{-1}$
$\det(\boldsymbol{Z})$	$\det(\boldsymbol{Z})\,\mathrm{Tr}\left\{\boldsymbol{Z}^{-1}d\boldsymbol{Z}\right\}$
$\ln(\det(\boldsymbol{Z}))$	$\mathrm{Tr}\left\{\boldsymbol{Z}^{-1}d\boldsymbol{Z}\right\}$
$\mathrm{reshape}(\boldsymbol{Z})$	$\mathrm{reshape}(d\boldsymbol{Z})$
$\boldsymbol{Z}^*$	$(d\boldsymbol{Z})^*$
$\boldsymbol{Z}^H$	$(d\boldsymbol{Z})^H$
$\boldsymbol{Z}^\#$	$\left[\mathrm{Tr}\left\{\boldsymbol{Z}^{-1}(d\boldsymbol{Z})\right\}\boldsymbol{Z}^{-1} - \boldsymbol{Z}^{-1}(d\boldsymbol{Z})\boldsymbol{Z}^{-1}\right]$
$\boldsymbol{Z}^+$	$-\boldsymbol{Z}^+(d\boldsymbol{Z})\boldsymbol{Z}^+ + \boldsymbol{Z}^+(\boldsymbol{Z}^+)^H(d\boldsymbol{Z}^H)\left(\boldsymbol{I}_N - \boldsymbol{Z}\boldsymbol{Z}^+\right) + \left(\boldsymbol{I}_Q - \boldsymbol{Z}^+\boldsymbol{Z}\right)(d\boldsymbol{Z}^H)(\boldsymbol{Z}^+)^H\boldsymbol{Z}^+$
$e^z = \exp(z)$	$e^z dz$
$\ln(z)$	$\frac{dz}{z}$

is inserted into (3.66), it is found that

$$\begin{aligned} d\boldsymbol{Z}^+\boldsymbol{Z} &= (\boldsymbol{Z}^+(d\boldsymbol{Z})(\boldsymbol{I}_Q - \boldsymbol{Z}^+\boldsymbol{Z}))^H + \boldsymbol{Z}^+(d\boldsymbol{Z})(\boldsymbol{I}_Q - \boldsymbol{Z}^+\boldsymbol{Z}) \\ &= (\boldsymbol{I}_Q - \boldsymbol{Z}^+\boldsymbol{Z})\left(d\boldsymbol{Z}^H\right)(\boldsymbol{Z}^+)^H + \boldsymbol{Z}^+(d\boldsymbol{Z})(\boldsymbol{I}_Q - \boldsymbol{Z}^+\boldsymbol{Z}). \end{aligned} \tag{3.67}$$

Second, it can be shown in a similar manner that

$$d\boldsymbol{Z}\boldsymbol{Z}^+ = (\boldsymbol{I}_N - \boldsymbol{Z}\boldsymbol{Z}^+)(d\boldsymbol{Z})\boldsymbol{Z}^+ + (\boldsymbol{Z}^+)^H\left(d\boldsymbol{Z}^H\right)(\boldsymbol{I}_N - \boldsymbol{Z}\boldsymbol{Z}^+). \tag{3.68}$$

If the expressions for $d\boldsymbol{Z}^+\boldsymbol{Z}$ and $d\boldsymbol{Z}\boldsymbol{Z}^+$ are inserted into (3.65), then (3.64) is obtained. ■

If $\boldsymbol{Z} \in \mathbb{C}^{N\times N}$ is invertible, then the Moore-Penrose inverse reduces into the normal matrix inverse. It is seen from (3.64) that the differential of the Moore-Penrose inverse reduces to the differential of the inverse matrices in (3.40) if the matrix is invertible.

Several of the most important properties of complex differentials are summarized in Table 3.1, assuming $\boldsymbol{A}$, $\boldsymbol{B}$, and a to be constants, and $\boldsymbol{Z}$, $\boldsymbol{Z}_0$, and $\boldsymbol{Z}_1$ to be complex-valued matrix variables. The complex differential of the complex exponential function of a scalar argument e^z and the complex differential of the principal value of $\ln(z)$ are also included in Table 3.1.

3.2.3 Results Used to Identify First- and Second-Order Derivatives

The two real-valued matrix variables $\mathrm{Re}\{\boldsymbol{Z}\} = \boldsymbol{X}$ and $\mathrm{Im}\{\boldsymbol{Z}\} = \boldsymbol{Y}$ are independent of each other, and, hence, are their differentials. Although the complex variables $\boldsymbol{Z}$ and $\boldsymbol{Z}^*$

are related, their differentials are linearly independent in the way of the next lemma. This lemma is very important for identifying first-order complex-valued matrix derivatives from the complex differential of the function under consideration. The idea of identifying the first-order complex-valued matrix derivatives from the complex differential is the key procedure for finding matrix derivatives.

Lemma 3.1 *Let* $\boldsymbol{Z} \in \mathbb{C}^{N\times Q}$ *and* $\boldsymbol{A}_i \in \mathbb{C}^{M\times NQ}$. *If*

$$\boldsymbol{A}_0 d\,\mathrm{vec}(\boldsymbol{Z}) + \boldsymbol{A}_1 d\,\mathrm{vec}(\boldsymbol{Z}^*) = \boldsymbol{0}_{M\times 1}, \tag{3.69}$$

for all $d\boldsymbol{Z} \in \mathbb{C}^{N\times Q}$, *then* $\boldsymbol{A}_i = \boldsymbol{0}_{M\times NQ}$ *for* $i \in \{0, 1\}$.

Proof Let $\boldsymbol{A}_i \in \mathbb{C}^{M\times NQ}$ be an arbitrary complex-valued function of $\boldsymbol{Z} \in \mathbb{C}^{N\times Q}$ and $\boldsymbol{Z}^* \in \mathbb{C}^{N\times Q}$. By using the vec$(\cdot)$ operator on (3.17) and (3.18), it follows that $d\,\mathrm{vec}(\boldsymbol{Z}) = d\,\mathrm{vec}(\mathrm{Re}\{\boldsymbol{Z}\}) + \jmath d\,\mathrm{vec}(\mathrm{Im}\{\boldsymbol{Z}\})$ and $d\,\mathrm{vec}(\boldsymbol{Z}^*) = d\,\mathrm{vec}(\mathrm{Re}\{\boldsymbol{Z}\}) - \jmath d\,\mathrm{vec}(\mathrm{Im}\{\boldsymbol{Z}\})$. If these two expressions are substituted into the expression of the lemma statement given by $\boldsymbol{A}_0 d\,\mathrm{vec}(\boldsymbol{Z}) + \boldsymbol{A}_1 d\,\mathrm{vec}(\boldsymbol{Z}^*) = \boldsymbol{0}_{M\times 1}$, then it follows that

$$\boldsymbol{A}_0(d\,\mathrm{vec}(\mathrm{Re}\{\boldsymbol{Z}\}) + \jmath d\,\mathrm{vec}(\mathrm{Im}\{\boldsymbol{Z}\})) + \boldsymbol{A}_1(d\,\mathrm{vec}(\mathrm{Re}\{\boldsymbol{Z}\}) - \jmath d\,\mathrm{vec}(\mathrm{Im}\{\boldsymbol{Z}\})) = \boldsymbol{0}_{M\times 1}. \tag{3.70}$$

This is equivalent to

$$(\boldsymbol{A}_0 + \boldsymbol{A}_1) d\,\mathrm{vec}(\mathrm{Re}\{\boldsymbol{Z}\}) + \jmath(\boldsymbol{A}_0 - \boldsymbol{A}_1) d\,\mathrm{vec}(\mathrm{Im}\{\boldsymbol{Z}\}) = \boldsymbol{0}_{M\times 1}. \tag{3.71}$$

Because the differentials $d\,\mathrm{Re}\{\boldsymbol{Z}\}$ and $d\,\mathrm{Im}\{\boldsymbol{Z}\}$ are independent, so are the differentials $d\,\mathrm{vec}(\mathrm{Re}\{\boldsymbol{Z}\})$ and $d\,\mathrm{vec}(\mathrm{Im}\{\boldsymbol{Z}\})$. Therefore, $\boldsymbol{A}_0 + \boldsymbol{A}_1 = \boldsymbol{0}_{M\times NQ}$ and $\boldsymbol{A}_0 - \boldsymbol{A}_1 = \boldsymbol{0}_{M\times NQ}$. Hence, it follows that $\boldsymbol{A}_0 = \boldsymbol{A}_1 = \boldsymbol{0}_{M\times NQ}$. ∎

The next lemma is important for identifying second-order complex-valued matrix derivatives. These derivatives are treated in detail in Chapter 5, and they are called Hessians.

Lemma 3.2 *Let* $\boldsymbol{Z} \in \mathbb{C}^{N\times Q}$ *and* $\boldsymbol{B}_i \in \mathbb{C}^{NQ\times NQ}$. *If*

$$\left(d\,\mathrm{vec}^T(\boldsymbol{Z})\right)\boldsymbol{B}_0 d\,\mathrm{vec}(\boldsymbol{Z}) + \left(d\,\mathrm{vec}^T(\boldsymbol{Z}^*)\right)\boldsymbol{B}_1 d\,\mathrm{vec}(\boldsymbol{Z}) + \left(d\,\mathrm{vec}^T(\boldsymbol{Z}^*)\right)\boldsymbol{B}_2 d\,\mathrm{vec}(\boldsymbol{Z}^*) = 0, \tag{3.72}$$

for all $d\boldsymbol{Z} \in \mathbb{C}^{N\times Q}$, *then* $\boldsymbol{B}_0 = -\boldsymbol{B}_0^T$, $\boldsymbol{B}_1 = \boldsymbol{0}_{NQ\times NQ}$, *and* $\boldsymbol{B}_2 = -\boldsymbol{B}_2^T$ *(i.e.,* $\boldsymbol{B}_0$ *and* $\boldsymbol{B}_2$ *are skew-symmetric).*

Proof Inserting the expressions $d\,\mathrm{vec}(\boldsymbol{Z}) = d\,\mathrm{vec}(\mathrm{Re}\{\boldsymbol{Z}\}) + \jmath d\,\mathrm{vec}(\mathrm{Im}\{\boldsymbol{Z}\})$ and $d\,\mathrm{vec}(\boldsymbol{Z}^*) = d\,\mathrm{vec}(\mathrm{Re}\{\boldsymbol{Z}\}) - \jmath d\,\mathrm{vec}(\mathrm{Im}\{\boldsymbol{Z}\})$ into the second-order differential expression given in the lemma leads to

$$\begin{aligned}
&\left[d\,\mathrm{vec}^T(\mathrm{Re}\{\boldsymbol{Z}\})\right]\left[\boldsymbol{B}_0 + \boldsymbol{B}_1 + \boldsymbol{B}_2\right] d\,\mathrm{vec}(\mathrm{Re}\{\boldsymbol{Z}\}) \\
&\quad + \left[d\,\mathrm{vec}^T(\mathrm{Im}\{\boldsymbol{Z}\})\right]\left[-\boldsymbol{B}_0 + \boldsymbol{B}_1 - \boldsymbol{B}_2\right] d\,\mathrm{vec}(\mathrm{Im}\{\boldsymbol{Z}\}) \\
&\quad + \left[d\,\mathrm{vec}^T(\mathrm{Re}\{\boldsymbol{Z}\})\right]\left[\jmath\left(\boldsymbol{B}_0 + \boldsymbol{B}_0^T\right) + \jmath\left(\boldsymbol{B}_1 - \boldsymbol{B}_1^T\right)\right. \\
&\quad \left. - \jmath\left(\boldsymbol{B}_2 + \boldsymbol{B}_2^T\right)\right] d\,\mathrm{vec}(\mathrm{Im}\{\boldsymbol{Z}\}) = 0.
\end{aligned} \tag{3.73}$$

Equation (3.73) is valid for all $d\boldsymbol{Z}$; furthermore, the differentials of $d\,\mathrm{vec}(\mathrm{Re}\{\boldsymbol{Z}\})$ and $d\,\mathrm{vec}(\mathrm{Im}\{\boldsymbol{Z}\})$ are independent. If $d\,\mathrm{vec}(\mathrm{Im}\{\boldsymbol{Z}\})$ is set to the zero vector, then it follows from (3.73) and Corollary 2.2 (which is valid for real-valued vectors) that

$$\boldsymbol{B}_0 + \boldsymbol{B}_1 + \boldsymbol{B}_2 = -\boldsymbol{B}_0^T - \boldsymbol{B}_1^T - \boldsymbol{B}_2^T. \tag{3.74}$$

In the same way, by setting $d\,\mathrm{vec}(\mathrm{Re}\{\boldsymbol{Z}\})$ to the zero vector, it follows from (3.73) and Corollary 2.2 that

$$-\boldsymbol{B}_0 + \boldsymbol{B}_1 - \boldsymbol{B}_2 = \boldsymbol{B}_0^T - \boldsymbol{B}_1^T + \boldsymbol{B}_2^T. \tag{3.75}$$

Because of the skew-symmetry in (3.74) and (3.75) and the linear independence of $d\,\mathrm{vec}(\mathrm{Re}\{\boldsymbol{Z}\})$ and $d\,\mathrm{vec}(\mathrm{Im}\{\boldsymbol{Z}\})$, it follows from (3.73) and Corollary 2.2 that

$$\left(\boldsymbol{B}_0 + \boldsymbol{B}_0^T\right) + \left(\boldsymbol{B}_1 - \boldsymbol{B}_1^T\right) - \left(\boldsymbol{B}_2 + \boldsymbol{B}_2^T\right) = \boldsymbol{0}_{NQ\times NQ}. \tag{3.76}$$

Equations (3.74), (3.75), and (3.76) lead to $\boldsymbol{B}_0 = -\boldsymbol{B}_0^T$, $\boldsymbol{B}_1 = -\boldsymbol{B}_1^T$, and $\boldsymbol{B}_2 = -\boldsymbol{B}_2^T$. Because the matrices $\boldsymbol{B}_0$ and $\boldsymbol{B}_2$ are skew-symmetric, Corollary 2.1 (which is valid for complex-valued matrices) reduces the equation stated inside the lemma formulation

$$\left(d\,\mathrm{vec}^T(\boldsymbol{Z})\right)\boldsymbol{B}_0 d\,\mathrm{vec}(\boldsymbol{Z}) + \left(d\,\mathrm{vec}^T(\boldsymbol{Z}^*)\right)\boldsymbol{B}_1 d\,\mathrm{vec}(\boldsymbol{Z}) + \left(d\,\mathrm{vec}^T(\boldsymbol{Z}^*)\right)\boldsymbol{B}_2 d\,\mathrm{vec}(\boldsymbol{Z}^*) = 0, \tag{3.77}$$

into $\left(d\,\mathrm{vec}^T(\boldsymbol{Z}^*)\right)\boldsymbol{B}_1 d\,\mathrm{vec}(\boldsymbol{Z}) = 0$. Then Lemma 2.17 results in $\boldsymbol{B}_1 = \boldsymbol{0}_{NQ\times NQ}$. ■

3.3 Derivative with Respect to Complex Matrices

The most general definition of the derivative is given here, from which the definitions for less general cases follow. They will be given later in an identification table.

Definition 3.1 (Derivative wrt. Complex Matrices) *Let* $\boldsymbol{F} : \mathbb{C}^{N\times Q} \times \mathbb{C}^{N\times Q} \rightarrow \mathbb{C}^{M\times P}$*. Then the derivative of the matrix function* $\boldsymbol{F}(\boldsymbol{Z}, \boldsymbol{Z}^*) \in \mathbb{C}^{M\times P}$ *with respect to* $\boldsymbol{Z} \in \mathbb{C}^{N\times Q}$ *is denoted by* $\mathcal{D}_{\boldsymbol{Z}}\boldsymbol{F}$*, and the derivative of the matrix function* $\boldsymbol{F}(\boldsymbol{Z}, \boldsymbol{Z}^*) \in \mathbb{C}^{M\times P}$ *with respect to* $\boldsymbol{Z}^* \in \mathbb{C}^{N\times Q}$ *is denoted by* $\mathcal{D}_{\boldsymbol{Z}^*}\boldsymbol{F}$*. The size of both these derivatives is* $MP \times NQ$*. The derivatives* $\mathcal{D}_{\boldsymbol{Z}}\boldsymbol{F}$ *and* $\mathcal{D}_{\boldsymbol{Z}^*}\boldsymbol{F}$ *are defined by the following differential expression:*

$$d\,\mathrm{vec}(\boldsymbol{F}) = (\mathcal{D}_{\boldsymbol{Z}}\boldsymbol{F})\,d\,\mathrm{vec}(\boldsymbol{Z}) + (\mathcal{D}_{\boldsymbol{Z}^*}\boldsymbol{F})\,d\,\mathrm{vec}(\boldsymbol{Z}^*). \tag{3.78}$$

$\mathcal{D}_{\boldsymbol{Z}}\boldsymbol{F}(\boldsymbol{Z}, \boldsymbol{Z}^*)$ *and* $\mathcal{D}_{\boldsymbol{Z}^*}\boldsymbol{F}(\boldsymbol{Z}, \boldsymbol{Z}^*)$ *are called the* Jacobian matrices *of* $\boldsymbol{F}$ *with respect to the matrix* $\boldsymbol{Z}$ *and* $\boldsymbol{Z}^*$*, respectively.*

Notice that Definition 3.1 is a generalization of formal derivatives, given in Definition 2.2, for the case of matrix functions that depend on complex-valued matrix variables. For scalar functions of scalar variables, Definitions 2.2 and 3.1 return the same result, and the reason for this can be found in Section 3.2.

Table 3.2 shows how the derivatives of the different types of functions in Table 2.2 can be identified from the differentials of these functions. By subtracting the differential

Table 3.2 Identification table.

Function type	Differential	Derivative wrt. z, $\boldsymbol{z}$, or $\boldsymbol{Z}$	Derivative wrt. z^*, $\boldsymbol{z}^*$, or $\boldsymbol{Z}^*$	Size of derivatives
$f(z, z^*)$	$df = a_0 dz + a_1 dz^*$	$\mathcal{D}_z f(z, z^*) = a_0$	$\mathcal{D}_{z^*} f(z, z^*) = a_1$	1×1
$f(\boldsymbol{z}, \boldsymbol{z}^*)$	$df = \boldsymbol{a}_0 d\boldsymbol{z} + \boldsymbol{a}_1 d\boldsymbol{z}^*$	$\mathcal{D}_{\boldsymbol{z}} f(\boldsymbol{z}, \boldsymbol{z}^*) = \boldsymbol{a}_0$	$\mathcal{D}_{\boldsymbol{z}^*} f(\boldsymbol{z}, \boldsymbol{z}^*) = \boldsymbol{a}_1$	$1 \times N$
$f(\boldsymbol{Z}, \boldsymbol{Z}^*)$	$df = \text{vec}^T(A_0) d\,\text{vec}(\boldsymbol{Z}) + \text{vec}^T(A_1) d\,\text{vec}(\boldsymbol{Z}^*)$	$\mathcal{D}_{\boldsymbol{Z}} f(\boldsymbol{Z}, \boldsymbol{Z}^*) = \text{vec}^T(A_0)$	$\mathcal{D}_{\boldsymbol{Z}^*} f(\boldsymbol{Z}, \boldsymbol{Z}^*) = \text{vec}^T(A_1)$	$1 \times NQ$
$f(\boldsymbol{Z}, \boldsymbol{Z}^*)$	$df = \text{Tr}\left\{A_0^T d\boldsymbol{Z} + A_1^T d\boldsymbol{Z}^*\right\}$	$\frac{\partial}{\partial \boldsymbol{Z}} f(\boldsymbol{Z}, \boldsymbol{Z}^*) = A_0$	$\frac{\partial}{\partial \boldsymbol{Z}^*} f(\boldsymbol{Z}, \boldsymbol{Z}^*) = A_1$	$N \times Q$
$\boldsymbol{f}(z, z^*)$	$d\boldsymbol{f} = \boldsymbol{b}_0 dz + \boldsymbol{b}_1 dz^*$	$\mathcal{D}_z \boldsymbol{f}(z, z^*) = \boldsymbol{b}_0$	$\mathcal{D}_{z^*} \boldsymbol{f}(z, z^*) = \boldsymbol{b}_1$	$M \times 1$
$\boldsymbol{f}(\boldsymbol{z}, \boldsymbol{z}^*)$	$d\boldsymbol{f} = \boldsymbol{B}_0 d\boldsymbol{z} + \boldsymbol{B}_1 d\boldsymbol{z}^*$	$\mathcal{D}_{\boldsymbol{z}} \boldsymbol{f}(\boldsymbol{z}, \boldsymbol{z}^*) = \boldsymbol{B}_0$	$\mathcal{D}_{\boldsymbol{z}^*} \boldsymbol{f}(\boldsymbol{z}, \boldsymbol{z}^*) = \boldsymbol{B}_1$	$M \times N$
$\boldsymbol{f}(\boldsymbol{Z}, \boldsymbol{Z}^*)$	$d\boldsymbol{f} = \boldsymbol{\beta}_0 d\,\text{vec}(\boldsymbol{Z}) + \boldsymbol{\beta}_1 d\,\text{vec}(\boldsymbol{Z}^*)$	$\mathcal{D}_{\boldsymbol{Z}} \boldsymbol{f}(\boldsymbol{Z}, \boldsymbol{Z}^*) = \boldsymbol{\beta}_0$	$\mathcal{D}_{\boldsymbol{Z}^*} \boldsymbol{f}(\boldsymbol{Z}, \boldsymbol{Z}^*) = \boldsymbol{\beta}_1$	$M \times NQ$
$\boldsymbol{F}(z, z^*)$	$d\,\text{vec}(\boldsymbol{F}) = \boldsymbol{c}_0 dz + \boldsymbol{c}_1 dz^*$	$\mathcal{D}_z \boldsymbol{F}(z, z^*) = \boldsymbol{c}_0$	$\mathcal{D}_{z^*} \boldsymbol{F}(z, z^*) = \boldsymbol{c}_1$	$MP \times 1$
$\boldsymbol{F}(\boldsymbol{z}, \boldsymbol{z}^*)$	$d\,\text{vec}(\boldsymbol{F}) = \boldsymbol{C}_0 d\boldsymbol{z} + \boldsymbol{C}_1 d\boldsymbol{z}^*$	$\mathcal{D}_{\boldsymbol{z}} \boldsymbol{F}(\boldsymbol{z}, \boldsymbol{z}^*) = \boldsymbol{C}_0$	$\mathcal{D}_{\boldsymbol{z}^*} \boldsymbol{F}(\boldsymbol{z}, \boldsymbol{z}^*) = \boldsymbol{C}_1$	$MP \times N$
$\boldsymbol{F}(\boldsymbol{Z}, \boldsymbol{Z}^*)$	$d\,\text{vec}(\boldsymbol{F}) = \boldsymbol{\zeta}_0 d\,\text{vec}(\boldsymbol{Z}) + \boldsymbol{\zeta}_1 d\,\text{vec}(\boldsymbol{Z}^*)$	$\mathcal{D}_{\boldsymbol{Z}} \boldsymbol{F}(\boldsymbol{Z}, \boldsymbol{Z}^*) = \boldsymbol{\zeta}_0$	$\mathcal{D}_{\boldsymbol{Z}^*} \boldsymbol{F}(\boldsymbol{Z}, \boldsymbol{Z}^*) = \boldsymbol{\zeta}_1$	$MP \times NQ$

Adapted from Hjørungnes and Gesbert (2007*a*), © 2007 IEEE.

in (3.78) from the corresponding differential in the last line of Table 3.2, it follows that

$$(\boldsymbol{\zeta}_0 - \mathcal{D}_{\boldsymbol{Z}}\boldsymbol{F}(\boldsymbol{Z},\boldsymbol{Z}^*))\,d\,\mathrm{vec}(\boldsymbol{Z}) + (\boldsymbol{\zeta}_1 - \mathcal{D}_{\boldsymbol{Z}^*}\boldsymbol{F}(\boldsymbol{Z},\boldsymbol{Z}^*))\,d\,\mathrm{vec}(\boldsymbol{Z}^*) = \boldsymbol{0}_{MP\times 1}. \tag{3.79}$$

The derivatives in Table 3.2 then follow by applying Lemma 3.1 on this equation. Table 3.2 is an extension of the corresponding table given in Magnus and Neudecker (1988), which is valid in the real variable case. Table 3.2 shows that $z \in \mathbb{C}$, $\boldsymbol{z} \in \mathbb{C}^{N\times 1}$, $\boldsymbol{Z} \in \mathbb{C}^{N\times Q}$, $f \in \mathbb{C}$, $\boldsymbol{f} \in \mathbb{C}^{M\times 1}$, and $\boldsymbol{F} \in \mathbb{C}^{M\times P}$. Furthermore, $a_i \in \mathbb{C}$, $\boldsymbol{a}_i \in \mathbb{C}^{1\times N}$, $\boldsymbol{A}_i \in \mathbb{C}^{N\times Q}$, $\boldsymbol{b}_i \in \mathbb{C}^{M\times 1}$, $\boldsymbol{B}_i \in \mathbb{C}^{M\times N}$, $\boldsymbol{\beta}_i \in \mathbb{C}^{M\times NQ}$, $\boldsymbol{c}_i \in \mathbb{C}^{MP\times 1}$, $\boldsymbol{C}_i \in \mathbb{C}^{MP\times N}$, and $\boldsymbol{\zeta}_i \in \mathbb{C}^{MP\times NQ}$, and each of these might be a function of z, $\boldsymbol{z}$, $\boldsymbol{Z}$, z^*, $\boldsymbol{z}^*$, or $\boldsymbol{Z}^*$, but *not* on the differential operator d. For example, in the most general matrix case, then in the expression $d\,\mathrm{vec}(\boldsymbol{F}) = \boldsymbol{\zeta}_0 d\,\mathrm{vec}(\boldsymbol{Z}) + \boldsymbol{\zeta}_1 d\,\mathrm{vec}(\boldsymbol{Z}^*)$, the two matrices $\boldsymbol{\zeta}_0$ and $\boldsymbol{\zeta}_1$ are not dependent on the differential operator d. For scalar functions of the type $f(\boldsymbol{Z},\boldsymbol{Z}^*)$, two alternative definitions for the derivatives are given. The notation $\frac{\partial}{\partial \boldsymbol{Z}} f(\boldsymbol{Z},\boldsymbol{Z}^*)$ and $\frac{\partial}{\partial \boldsymbol{Z}^*} f(\boldsymbol{Z},\boldsymbol{Z}^*)$ will be defined in Subsection 4.2.3 in the next chapter.

Definition 3.2 (Formal Derivatives of Vector Functions wrt. Vectors) *If* $\boldsymbol{f} : \mathbb{C}^{N\times 1} \times \mathbb{C}^{N\times 1} \to \mathbb{C}^{M\times 1}$, *then the two* formal derivatives of a vector function *with respect to the two row vector variables* $\boldsymbol{z}^T$ *and* $\boldsymbol{z}^H$ *are denoted by* $\frac{\partial}{\partial \boldsymbol{z}^T}\boldsymbol{f}(\boldsymbol{z},\boldsymbol{z}^*)$ *and* $\frac{\partial}{\partial \boldsymbol{z}^H}\boldsymbol{f}(\boldsymbol{z},\boldsymbol{z}^*)$. *These two formal derivatives are sized as* $M \times N$, *and they are defined as*

$$\frac{\partial}{\partial \boldsymbol{z}^T}\boldsymbol{f}(\boldsymbol{z},\boldsymbol{z}^*) = \begin{bmatrix} \frac{\partial}{\partial z_0} f_0 & \cdots & \frac{\partial}{\partial z_{N-1}} f_0 \\ \vdots & & \vdots \\ \frac{\partial}{\partial z_0} f_{M-1} & \cdots & \frac{\partial}{\partial z_{N-1}} f_{M-1} \end{bmatrix}, \tag{3.80}$$

and

$$\frac{\partial}{\partial \boldsymbol{z}^H}\boldsymbol{f}(\boldsymbol{z},\boldsymbol{z}^*) = \begin{bmatrix} \frac{\partial}{\partial z_0^*} f_0 & \cdots & \frac{\partial}{\partial z_{N-1}^*} f_0 \\ \vdots & & \vdots \\ \frac{\partial}{\partial z_0^*} f_{M-1} & \cdots & \frac{\partial}{\partial z_{N-1}^*} f_{M-1} \end{bmatrix}, \tag{3.81}$$

where z_i *and* f_i *is the component number* i *of the vectors* $\boldsymbol{z}$ *and* $\boldsymbol{f}$, *respectively.*

Notice that $\frac{\partial}{\partial \boldsymbol{z}^T}\boldsymbol{f} = \mathcal{D}_{\boldsymbol{z}}\boldsymbol{f}$ and $\frac{\partial}{\partial \boldsymbol{z}^H}\boldsymbol{f} = \mathcal{D}_{\boldsymbol{z}^*}\boldsymbol{f}$. Using the formal derivative notation in Definition 3.2, the derivatives of the function $\boldsymbol{F}(\boldsymbol{Z},\boldsymbol{Z}^*)$, in Definition 3.1, are

$$\mathcal{D}_{\boldsymbol{Z}}\boldsymbol{F}(\boldsymbol{Z},\boldsymbol{Z}^*) = \frac{\partial\,\mathrm{vec}(\boldsymbol{F}(\boldsymbol{Z},\boldsymbol{Z}^*))}{\partial\,\mathrm{vec}^T(\boldsymbol{Z})}, \tag{3.82}$$

$$\mathcal{D}_{\boldsymbol{Z}^*}\boldsymbol{F}(\boldsymbol{Z},\boldsymbol{Z}^*) = \frac{\partial\,\mathrm{vec}(\boldsymbol{F}(\boldsymbol{Z},\boldsymbol{Z}^*))}{\partial\,\mathrm{vec}^T(\boldsymbol{Z}^*)}. \tag{3.83}$$

This is a generalization of the real-matrix variable case studied thoroughly in Magnus and Neudecker (1988) to the complex-matrix variable case.

Definition 3.3 (Formal Derivative of Matrix Functions wrt. Scalars) *If* $\boldsymbol{F} : \mathbb{C}^{N\times Q} \times \mathbb{C}^{N\times Q} \to \mathbb{C}^{M\times P}$, *then the* formal derivative of the matrix function $\boldsymbol{F} \in \mathbb{C}^{M\times P}$ with respect to the scalar $z \in \mathbb{C}$ *is defined as*

$$\frac{\partial \boldsymbol{F}}{\partial z} = \begin{bmatrix} \frac{\partial f_{0,0}}{\partial z} & \cdots & \frac{\partial f_{0,P-1}}{\partial z} \\ \vdots & \ddots & \vdots \\ \frac{\partial f_{M-1,0}}{\partial z} & \cdots & \frac{\partial f_{M-1,P-1}}{\partial z} \end{bmatrix}, \tag{3.84}$$

where $\frac{\partial \boldsymbol{F}}{\partial z}$ *has size* $M \times P$ *and* $f_{i,j}$ *is the* (i, j)*-th component function of* $\boldsymbol{F}$*, where* $i \in \{0, 1, \ldots, M-1\}$ *and* $j \in \{0, 1, \ldots, P-1\}$.

By using Definitions 3.2 and 3.3, it is possible to find the following alternative expression for the derivative of the matrix function $\boldsymbol{F} \in \mathbb{C}^{M\times P}$ with respect to the matrix $\boldsymbol{Z}$:

$$\mathcal{D}_{\boldsymbol{Z}} \boldsymbol{F}(\boldsymbol{Z}, \boldsymbol{Z}^*) = \frac{\partial \operatorname{vec}(\boldsymbol{F}(\boldsymbol{Z}, \boldsymbol{Z}^*))}{\partial \operatorname{vec}^T(\boldsymbol{Z})} = \begin{bmatrix} \frac{\partial f_{0,0}}{\partial z_{0,0}} & \frac{\partial f_{0,0}}{\partial z_{1,0}} & \cdots & \frac{\partial f_{0,0}}{\partial z_{N-1,Q-1}} \\ \frac{\partial f_{1,0}}{\partial z_{0,0}} & \frac{\partial f_{1,0}}{\partial z_{1,0}} & \cdots & \frac{\partial f_{1,0}}{\partial z_{N-1,Q-1}} \\ \vdots & \vdots & \ddots & \vdots \\ \frac{\partial f_{M-1,P-1}}{\partial z_{0,0}} & \frac{\partial f_{M-1,P-1}}{\partial z_{1,0}} & \cdots & \frac{\partial f_{M-1,P-1}}{\partial z_{N-1,Q-1}} \end{bmatrix}$$

$$= \left[\frac{\partial \operatorname{vec}(\boldsymbol{F})}{\partial z_{0,0}} \; \frac{\partial \operatorname{vec}(\boldsymbol{F})}{\partial z_{1,0}} \cdots \frac{\partial \operatorname{vec}(\boldsymbol{F})}{\partial z_{N-1,Q-1}} \right] = \sum_{n=0}^{N-1} \sum_{q=0}^{Q-1} \frac{\partial \operatorname{vec}(\boldsymbol{F})}{\partial z_{n,q}} \operatorname{vec}^T\left(\boldsymbol{E}_{n,q}\right) \tag{3.85}$$

$$= \sum_{n=0}^{N-1} \sum_{q=0}^{Q-1} \operatorname{vec}\left(\frac{\partial \boldsymbol{F}}{\partial z_{n,q}}\right) \operatorname{vec}^T\left(\boldsymbol{E}_{n,q}\right), \tag{3.86}$$

where $z_{i,j}$ is the (i, j)-th component of $\boldsymbol{Z}$, and where $\boldsymbol{E}_{n,q}$ is an $N \times Q$ matrix containing only 0s except for $+1$ at the (n, q)-th position. The notation $\boldsymbol{E}_{n,q}$ is here a natural generalization of the square matrices given in Definition 2.16 to nonsquare matrices. Using (3.85), it follows that

$$\mathcal{D}_{\boldsymbol{Z}^*} \boldsymbol{F}(\boldsymbol{Z}, \boldsymbol{Z}^*) = \sum_{n=0}^{N-1} \sum_{q=0}^{Q-1} \operatorname{vec}\left(\frac{\partial \boldsymbol{F}}{\partial z_{n,q}^*}\right) \operatorname{vec}^T\left(\boldsymbol{E}_{n,q}\right). \tag{3.87}$$

The following lemma shows how to find the derivatives of the complex conjugate of a matrix function when the derivatives of the matrix function are already known.

Lemma 3.3 *Let the derivatives of* $\boldsymbol{F} : \mathbb{C}^{N\times Q} \times \mathbb{C}^{N\times Q} \to \mathbb{C}^{M\times P}$ *with respect to the two complex-valued variables* $\boldsymbol{Z}$ *and* $\boldsymbol{Z}^*$ *be known and given by* $\mathcal{D}_{\boldsymbol{Z}} \boldsymbol{F}$ *and* $\mathcal{D}_{\boldsymbol{Z}^*} \boldsymbol{F}$*, respectively. The derivatives of the matrix function* $\boldsymbol{F}^*$ *with respect to* $\boldsymbol{Z}$ *and* $\boldsymbol{Z}^*$ *are given by*

$$\mathcal{D}_{\boldsymbol{Z}} \boldsymbol{F}^* = (\mathcal{D}_{\boldsymbol{Z}^*} \boldsymbol{F})^*, \tag{3.88}$$

$$\mathcal{D}_{\boldsymbol{Z}^*} \boldsymbol{F}^* = (\mathcal{D}_{\boldsymbol{Z}} \boldsymbol{F})^*. \tag{3.89}$$

Proof By taking the complex conjugation of both sides of (3.78), it is found that

$$\begin{aligned} d \operatorname{vec}(\boldsymbol{F}^*) &= (\mathcal{D}_{\boldsymbol{Z}} \boldsymbol{F})^* \, d \operatorname{vec}(\boldsymbol{Z}^*) + (\mathcal{D}_{\boldsymbol{Z}^*} \boldsymbol{F})^* \, d \operatorname{vec}(\boldsymbol{Z}) \\ &= (\mathcal{D}_{\boldsymbol{Z}^*} \boldsymbol{F})^* \, d \operatorname{vec}(\boldsymbol{Z}) + (\mathcal{D}_{\boldsymbol{Z}} \boldsymbol{F})^* \, d \operatorname{vec}(\boldsymbol{Z}^*). \end{aligned} \tag{3.90}$$

By using Definition 3.1, it is seen that (3.88) and (3.89) follow. ■

Table 3.3 Procedure for finding the derivatives with respect to complex-valued matrix variables.

Step 1:	Compute the differential $d\,\text{vec}(\boldsymbol{F})$.
Step 2:	Manipulate the expression into the form given (3.78).
Step 3:	The matrices $\mathcal{D}_{\boldsymbol{Z}}\boldsymbol{F}(\boldsymbol{Z},\boldsymbol{Z}^*)$ and $\mathcal{D}_{\boldsymbol{Z}^*}\boldsymbol{F}(\boldsymbol{Z},\boldsymbol{Z}^*)$ can now be read out by using Definition 3.1.

To find the derivative of a product of two functions, the following lemma can be used:

Lemma 3.4 *Let* $\boldsymbol{F} : \mathbb{C}^{N\times Q} \times \mathbb{C}^{N\times Q} \to \mathbb{C}^{M\times P}$ *be given by*

$$\boldsymbol{F}(\boldsymbol{Z},\boldsymbol{Z}^*) = \boldsymbol{G}(\boldsymbol{Z},\boldsymbol{Z}^*)\boldsymbol{H}(\boldsymbol{Z},\boldsymbol{Z}^*), \tag{3.91}$$

where $\boldsymbol{G} : \mathbb{C}^{N\times Q} \times \mathbb{C}^{N\times Q} \to \mathbb{C}^{M\times R}$ *and* $\boldsymbol{H} : \mathbb{C}^{N\times Q} \times \mathbb{C}^{N\times Q} \to \mathbb{C}^{R\times P}$. *Then the following relations hold:*

$$\mathcal{D}_{\boldsymbol{Z}}\boldsymbol{F} = \left(\boldsymbol{H}^T \otimes \boldsymbol{I}_M\right)\mathcal{D}_{\boldsymbol{Z}}\boldsymbol{G} + (\boldsymbol{I}_P \otimes \boldsymbol{G})\,\mathcal{D}_{\boldsymbol{Z}}\boldsymbol{H}, \tag{3.92}$$

$$\mathcal{D}_{\boldsymbol{Z}^*}\boldsymbol{F} = \left(\boldsymbol{H}^T \otimes \boldsymbol{I}_M\right)\mathcal{D}_{\boldsymbol{Z}^*}\boldsymbol{G} + (\boldsymbol{I}_P \otimes \boldsymbol{G})\,\mathcal{D}_{\boldsymbol{Z}^*}\boldsymbol{H}. \tag{3.93}$$

Proof The complex differential of $\boldsymbol{F}$ can be expressed as

$$d\boldsymbol{F} = \boldsymbol{I}_M\,(d\boldsymbol{G})\,\boldsymbol{H} + \boldsymbol{G}\,(d\boldsymbol{H})\,\boldsymbol{I}_P. \tag{3.94}$$

By using the definition of the derivative of $\boldsymbol{G}$ and $\boldsymbol{H}$ after applying the vec$(\cdot)$, it is found that

$$\begin{aligned} d\,\text{vec}(\boldsymbol{F}) &= \left(\boldsymbol{H}^T \otimes \boldsymbol{I}_M\right) d\,\text{vec}(\boldsymbol{G}) + (\boldsymbol{I}_P \otimes \boldsymbol{G})\, d\,\text{vec}(\boldsymbol{H}) \\ &= \left(\boldsymbol{H}^T \otimes \boldsymbol{I}_M\right) \left[(\mathcal{D}_{\boldsymbol{Z}}\boldsymbol{G})\, d\,\text{vec}(\boldsymbol{Z}) + (\mathcal{D}_{\boldsymbol{Z}^*}\boldsymbol{G})\, d\,\text{vec}(\boldsymbol{Z}^*)\right] \\ &\quad + (\boldsymbol{I}_P \otimes \boldsymbol{G}) \left[(\mathcal{D}_{\boldsymbol{Z}}\boldsymbol{H})\, d\,\text{vec}(\boldsymbol{Z}) + (\mathcal{D}_{\boldsymbol{Z}^*}\boldsymbol{H})\, d\,\text{vec}(\boldsymbol{Z}^*)\right] \\ &= \left[\left(\boldsymbol{H}^T \otimes \boldsymbol{I}_M\right)\mathcal{D}_{\boldsymbol{Z}}\boldsymbol{G} + (\boldsymbol{I}_P \otimes \boldsymbol{G})\,\mathcal{D}_{\boldsymbol{Z}}\boldsymbol{H}\right] d\,\text{vec}(\boldsymbol{Z}) \\ &\quad + \left[\left(\boldsymbol{H}^T \otimes \boldsymbol{I}_M\right)\mathcal{D}_{\boldsymbol{Z}^*}\boldsymbol{G} + (\boldsymbol{I}_P \otimes \boldsymbol{G})\,\mathcal{D}_{\boldsymbol{Z}^*}\boldsymbol{H}\right] d\,\text{vec}(\boldsymbol{Z}^*). \end{aligned} \tag{3.95}$$

The derivatives of $\boldsymbol{F}$ with respect to $\boldsymbol{Z}$ and $\boldsymbol{Z}^*$ can now be identified as in (3.92) and (3.93), respectively. ∎

3.3.1 Procedure for Finding Complex-Valued Matrix Derivatives

Finding the derivative of the complex matrix function $\boldsymbol{F}$ with respect to the complex matrices $\boldsymbol{Z}$ and $\boldsymbol{Z}^*$ can be achieved using the three-step procedure shown in Table 3.3.

For less general function types, as given in Table 2.2, a similar procedure can be used. In Chapter 4, many examples will be given for how this procedure can be used to find complex-valued matrix derivatives for all the cases shown in Table 3.2.

3.4 Fundamental Results on Complex-Valued Matrix Derivatives

In this section, some fundamental results are presented. All of these results are important when solving practical optimization problems involving differentiation with respect to a complex-valued matrix. These results include the chain rule, conditions for finding stationary points for a scalar real-valued function that depends on complex matrices, and in which direction a scalar real-valued function has the minimum or maximum rate of change. It will be shown how this result should be used in the *steepest ascent or descent method* (Luenberger 1973). For certain types of functions, the same procedure used for the real-valued matrix case (Magnus & Neudecker 1988) can be used; this result is also stated in a theorem.

The rest of this section is organized as follows: In Subsection 3.4.1, the chain rule will be formulated and it can be used to find complicated derivatives. Subsection 3.4.2 presents several topics for scalar real-valued functions, including three equivalent ways to find stationary points, the relationship between the complex-valued derivative of such functions with respect to the input matrix variable and its complex conjugate, and the directions in which such functions have maximum and minimum rates of change. When the function has only *one* independent input matrix variable, the relation between the theory presented in Magnus and Neudecker (1988) and the complex-valued matrix derivatives is presented in Subsection 3.4.3.

3.4.1 Chain Rule

One big advantage of the way the derivative is defined in Definition 3.1 compared with other definitions (see discussion about notation on Magnus and Neudecker (1988, pp. 171–173) is that the chain rule is valid in a very simple form. The chain rule is formulated in the following theorem.

Theorem 3.1 (Chain Rule) *Let* $(\mathcal{S}_0, \mathcal{S}_1) \subseteq \mathbb{C}^{N\times Q} \times \mathbb{C}^{N\times Q}$*, and let* $\boldsymbol{F} : \mathcal{S}_0 \times \mathcal{S}_1 \to \mathbb{C}^{M\times P}$ *be differentiable with respect to its first and second arguments at an interior point* $(\boldsymbol{Z}, \boldsymbol{Z}^*)$ *in the set* $\mathcal{S}_0 \times \mathcal{S}_1$*. Let* $\mathcal{T}_0 \times \mathcal{T}_1 \subseteq \mathbb{C}^{M\times P} \times \mathbb{C}^{M\times P}$ *be such that* $(\boldsymbol{F}(\boldsymbol{Z}, \boldsymbol{Z}^*), \boldsymbol{F}^*(\boldsymbol{Z}, \boldsymbol{Z}^*)) \in \mathcal{T}_0 \times \mathcal{T}_1$ *for all* $(\boldsymbol{Z}, \boldsymbol{Z}^*) \in \mathcal{S}_0 \times \mathcal{S}_1$*. Assume that* $\boldsymbol{G} : \mathcal{T}_0 \times \mathcal{T}_1 \to \mathbb{C}^{R\times S}$ *is differentiable at an inner point* $(\boldsymbol{F}(\boldsymbol{Z}, \boldsymbol{Z}^*), \boldsymbol{F}^*(\boldsymbol{Z}, \boldsymbol{Z}^*)) \in \mathcal{T}_0 \times \mathcal{T}_1$*. Define the composite function* $\boldsymbol{H} : \mathcal{S}_0 \times \mathcal{S}_1 \to \mathbb{C}^{R\times S}$ *by*

$$\boldsymbol{H}(\boldsymbol{Z}, \boldsymbol{Z}^*) = \boldsymbol{G}(\boldsymbol{F}(\boldsymbol{Z}, \boldsymbol{Z}^*), \boldsymbol{F}^*(\boldsymbol{Z}, \boldsymbol{Z}^*)). \tag{3.96}$$

The derivatives $\mathcal{D}_{\boldsymbol{Z}}\boldsymbol{H}$ *and* $\mathcal{D}_{\boldsymbol{Z}^*}\boldsymbol{H}$ *are as follows:*

$$\mathcal{D}_{\boldsymbol{Z}}\boldsymbol{H} = (\mathcal{D}_{\boldsymbol{F}}\boldsymbol{G})(\mathcal{D}_{\boldsymbol{Z}}\boldsymbol{F}) + (\mathcal{D}_{\boldsymbol{F}^*}\boldsymbol{G})(\mathcal{D}_{\boldsymbol{Z}}\boldsymbol{F}^*), \tag{3.97}$$

$$\mathcal{D}_{\boldsymbol{Z}^*}\boldsymbol{H} = (\mathcal{D}_{\boldsymbol{F}}\boldsymbol{G})(\mathcal{D}_{\boldsymbol{Z}^*}\boldsymbol{F}) + (\mathcal{D}_{\boldsymbol{F}^*}\boldsymbol{G})(\mathcal{D}_{\boldsymbol{Z}^*}\boldsymbol{F}^*). \tag{3.98}$$

Proof From Definition 3.1, it follows that

$$d\,\mathrm{vec}(\boldsymbol{H}) = d\,\mathrm{vec}(\boldsymbol{G}) = (\mathcal{D}_{\boldsymbol{F}}\boldsymbol{G})\,d\,\mathrm{vec}(\boldsymbol{F}) + (\mathcal{D}_{\boldsymbol{F}^*}\boldsymbol{G})\,d\,\mathrm{vec}(\boldsymbol{F}^*). \tag{3.99}$$

The complex differentials of vec($\boldsymbol{F}$) and vec($\boldsymbol{F}^*$) are given by

$$d\,\text{vec}(\boldsymbol{F}) = (\mathcal{D}_{\boldsymbol{Z}}\boldsymbol{F})\,d\,\text{vec}(\boldsymbol{Z}) + (\mathcal{D}_{\boldsymbol{Z}^*}\boldsymbol{F})\,d\,\text{vec}(\boldsymbol{Z}^*), \tag{3.100}$$

$$d\,\text{vec}(\boldsymbol{F}^*) = (\mathcal{D}_{\boldsymbol{Z}}\boldsymbol{F}^*)\,d\,\text{vec}(\boldsymbol{Z}) + (\mathcal{D}_{\boldsymbol{Z}^*}\boldsymbol{F}^*)\,d\,\text{vec}(\boldsymbol{Z}^*). \tag{3.101}$$

By substituting the results from (3.100) and (3.101) into (3.99), and then using the definition of the derivatives with respect to $\boldsymbol{Z}$ and $\boldsymbol{Z}^*$, the theorem follows. ■

3.4.2 Scalar Real-Valued Functions

In this subsection, several results will be presented for scalar real-valued functions. Topics such as necessary conditions for optimality (stationary points), the relation between the derivative of a real-valued function with respect to the input matrix variable and its complex conjugate, and the direction of maximum and minimum rate of change will be treated.

The next theorem shows that when working on scalar real-valued functions that are dependent on complex matrices, three equivalent ways can be used to identify stationary points.[6]

Theorem 3.2 *Let* $f : \mathbb{C}^{N\times Q} \times \mathbb{C}^{N\times Q} \to \mathbb{R}$. *A stationary point of the function* $f(\boldsymbol{Z}, \boldsymbol{Z}^*) = g(\boldsymbol{X}, \boldsymbol{Y})$, *where* $g : \mathbb{R}^{N\times Q} \times \mathbb{R}^{N\times Q} \to \mathbb{R}$ *and* $\boldsymbol{Z} = \boldsymbol{X} + \jmath\boldsymbol{Y}$ *is then found by one of the following three equivalent conditions:*

$$\mathcal{D}_{\boldsymbol{X}}g(\boldsymbol{X}, \boldsymbol{Y}) = \boldsymbol{0}_{1\times NQ} \;\wedge\; \mathcal{D}_{\boldsymbol{Y}}g(\boldsymbol{X}, \boldsymbol{Y}) = \boldsymbol{0}_{1\times NQ}, \tag{3.102}$$

$$\mathcal{D}_{\boldsymbol{Z}}f(\boldsymbol{Z}, \boldsymbol{Z}^*) = \boldsymbol{0}_{1\times NQ}, \tag{3.103}$$

or

$$\mathcal{D}_{\boldsymbol{Z}^*}f(\boldsymbol{Z}, \boldsymbol{Z}^*) = \boldsymbol{0}_{1\times NQ}. \tag{3.104}$$

In (3.102), *the symbol* $\wedge$ *means that both of the equations stated in* (3.102) *must be satisfied at the same time.*

Proof In optimization theory (Magnus & Neudecker 1988), a stationary point is defined as a point where the derivatives with respect to all independent variables vanish. Because $\text{Re}\{\boldsymbol{Z}\} = \boldsymbol{X}$ and $\text{Im}\{\boldsymbol{Z}\} = \boldsymbol{Y}$ contain only independent variables, (3.102) gives a stationary point by definition. By using the chain rule, in Theorem 3.1, on both sides of the equation $f(\boldsymbol{Z}, \boldsymbol{Z}^*) = g(\boldsymbol{X}, \boldsymbol{Y})$ and taking the derivative with respect to $\boldsymbol{X}$ and $\boldsymbol{Y}$, the following two equations are obtained:

$$(\mathcal{D}_{\boldsymbol{Z}}f)(\mathcal{D}_{\boldsymbol{X}}\boldsymbol{Z}) + (\mathcal{D}_{\boldsymbol{Z}^*}f)(\mathcal{D}_{\boldsymbol{X}}\boldsymbol{Z}^*) = \mathcal{D}_{\boldsymbol{X}}g, \tag{3.105}$$

$$(\mathcal{D}_{\boldsymbol{Z}}f)(\mathcal{D}_{\boldsymbol{Y}}\boldsymbol{Z}) + (\mathcal{D}_{\boldsymbol{Z}^*}f)(\mathcal{D}_{\boldsymbol{Y}}\boldsymbol{Z}^*) = \mathcal{D}_{\boldsymbol{Y}}g. \tag{3.106}$$

From (3.17) and (3.18), it follows directly that $\mathcal{D}_{\boldsymbol{X}}\boldsymbol{Z} = \mathcal{D}_{\boldsymbol{X}}\boldsymbol{Z}^* = \boldsymbol{I}_{NQ}$ and $\mathcal{D}_{\boldsymbol{Y}}\boldsymbol{Z} = -\mathcal{D}_{\boldsymbol{Y}}\boldsymbol{Z}^* = \jmath\boldsymbol{I}_{NQ}$. If these results are inserted into (3.105) and (3.106), these two

[6] Notice that a stationary point can be a local minimum, a local maximum, or a saddle point.

equations can be formulated into a block matrix form in the following way:

$$\begin{bmatrix} \mathcal{D}_X g \\ \mathcal{D}_Y g \end{bmatrix} = \begin{bmatrix} 1 & 1 \\ \jmath & -\jmath \end{bmatrix} \begin{bmatrix} \mathcal{D}_Z f \\ \mathcal{D}_{Z^*} f \end{bmatrix}. \tag{3.107}$$

This equation is equivalent to the following matrix equation:

$$\begin{bmatrix} \mathcal{D}_Z f \\ \mathcal{D}_{Z^*} f \end{bmatrix} = \begin{bmatrix} \frac{1}{2} & -\frac{\jmath}{2} \\ \frac{1}{2} & \frac{\jmath}{2} \end{bmatrix} \begin{bmatrix} \mathcal{D}_X g \\ \mathcal{D}_Y g \end{bmatrix}. \tag{3.108}$$

Because $\mathcal{D}_X g \in \mathbb{R}^{1\times NQ}$ and $\mathcal{D}_Y g \in \mathbb{R}^{1\times NQ}$, it is seen from (3.108) that the three relations (3.102), (3.103), and (3.104) are equivalent. ■

Notice that (3.107) and (3.108) are multivariable generalizations of the corresponding scalar Wirtinger and partial derivatives given in (2.11), (2.12), (2.13), and (2.14).

The next theorem gives a simplified way of finding the derivative of a scalar real-valued function with respect to $\boldsymbol{Z}$ when the derivative with respect to $\boldsymbol{Z}^*$ is already known.

Theorem 3.3 *Let* $f : \mathbb{C}^{N\times Q} \times \mathbb{C}^{N\times Q} \to \mathbb{R}$. *Then the following holds:*

$$\mathcal{D}_{Z^*} f = (\mathcal{D}_Z f)^* . \tag{3.109}$$

Proof Because $f \in \mathbb{R}$, it is possible to write the df in the following two ways:

$$df = (\mathcal{D}_Z f)\, d\,\text{vec}(\boldsymbol{Z}) + (\mathcal{D}_{Z^*} f)\, d\,\text{vec}(\boldsymbol{Z}^*), \tag{3.110}$$

$$df = df^* = (\mathcal{D}_Z f)^*\, d\,\text{vec}(\boldsymbol{Z}^*) + (\mathcal{D}_{Z^*} f)^*\, d\,\text{vec}(\boldsymbol{Z}), \tag{3.111}$$

where $df = df^*$ because $f \in \mathbb{R}$. By subtracting (3.110) from (3.111) and then applying Lemma 3.1, it follows that $\mathcal{D}_Z f = (\mathcal{D}_{Z^*} f)^*$, which is equivalent to (3.109). ■

Let $f : \mathbb{C}^{M\times Q} \times \mathbb{C}^{M\times Q} \to \mathbb{R}$ be denoted $f(\boldsymbol{Z}, \boldsymbol{Z}^*)$, where $\boldsymbol{Z}$ contains independent matrix elements. By using the result from Theorem 3.3, (3.110) can be rewritten as

$$\begin{aligned} df &= (\mathcal{D}_Z f)\, d\,\text{vec}(\boldsymbol{Z}) + (\mathcal{D}_{Z^*} f)\, d\,\text{vec}(\boldsymbol{Z}^*) \\ &= (\mathcal{D}_Z f)\, d\,\text{vec}(\boldsymbol{Z}) + (\mathcal{D}_Z f)^*\, d\,\text{vec}(\boldsymbol{Z}^*) \\ &= (\mathcal{D}_Z f)\, d\,\text{vec}(\boldsymbol{Z}) + ((\mathcal{D}_Z f)\, d\,\text{vec}(\boldsymbol{Z}))^* \\ &= 2\,\text{Re}\,\{(\mathcal{D}_Z f)\, d\,\text{vec}(\boldsymbol{Z})\}. \end{aligned} \tag{3.112}$$

This expression will be used in the proof of the next theorem.

In engineering, we are often interested in maximizing or minimizing a real-valued scalar variable, so it is important to find the direction where the function is increasing and decreasing fastest. The following theorem gives an answer to this question and can be applied in the widely used steepest ascent and descent methods.

Theorem 3.4 *Let* $f : \mathbb{C}^{N\times Q} \times \mathbb{C}^{N\times Q} \to \mathbb{R}$. *The directions where the function* f *has the maximum and minimum rate of change with respect to* $\text{vec}(\boldsymbol{Z})$ *are given by* $[\mathcal{D}_{Z^*} f(\boldsymbol{Z}, \boldsymbol{Z}^*)]^T$ *and* $-[\mathcal{D}_{Z^*} f(\boldsymbol{Z}, \boldsymbol{Z}^*)]^T$, *respectively.*

Proof From Theorem 3.3 and (3.112), it follows that

$$df = 2\,\mathrm{Re}\{(\mathcal{D}_{\boldsymbol{Z}} f)\, d\,\mathrm{vec}(\boldsymbol{Z})\} = 2\,\mathrm{Re}\left\{(\mathcal{D}_{\boldsymbol{Z}^*} f)^*\, d\,\mathrm{vec}(\boldsymbol{Z})\right\}. \tag{3.113}$$

Let $\boldsymbol{a}_i \in \mathbb{C}^{K\times 1}$, where $i \in \{0, 1\}$. Then

$$\mathrm{Re}\left\{\boldsymbol{a}_0^H \boldsymbol{a}_1\right\} = \left\langle \begin{bmatrix} \mathrm{Re}\{\boldsymbol{a}_0\} \\ \mathrm{Im}\{\boldsymbol{a}_0\} \end{bmatrix}, \begin{bmatrix} \mathrm{Re}\{\boldsymbol{a}_1\} \\ \mathrm{Im}\{\boldsymbol{a}_1\} \end{bmatrix} \right\rangle, \tag{3.114}$$

where $\langle \cdot, \cdot \rangle$ is the ordinary Euclidean inner product (Young 1990) between real vectors in $\mathbb{R}^{2K\times 1}$. By using this inner product, the differential of f can be written as

$$df = 2\left\langle \begin{bmatrix} \mathrm{Re}\left\{(\mathcal{D}_{\boldsymbol{Z}^*} f)^T\right\} \\ \mathrm{Im}\left\{(\mathcal{D}_{\boldsymbol{Z}^*} f)^T\right\} \end{bmatrix}, \begin{bmatrix} \mathrm{Re}\{d\,\mathrm{vec}(\boldsymbol{Z})\} \\ \mathrm{Im}\{d\,\mathrm{vec}(\boldsymbol{Z})\} \end{bmatrix} \right\rangle. \tag{3.115}$$

By applying the Cauchy-Schwartz inequality (Young 1990) for inner products, it can be shown that the maximum value of df occurs when $d\,\mathrm{vec}(\boldsymbol{Z}) = \alpha\,(\mathcal{D}_{\boldsymbol{Z}^*} f)^T$ for $\alpha > 0$, and from this, it follows that the minimum rate of change occurs when $d\,\mathrm{vec}(\boldsymbol{Z}) = -\beta\,(\mathcal{D}_{\boldsymbol{Z}^*} f)^T$, for $\beta > 0$. ■

Remark *Let* $g : \mathbb{C}^{K\times 1} \times \mathbb{C}^{K\times 1} \to \mathbb{R}$ *be given by*

$$g(\boldsymbol{a}_0, \boldsymbol{a}_1) = 2\,\mathrm{Re}\left\{\boldsymbol{a}_0^T \boldsymbol{a}_1\right\}. \tag{3.116}$$

If $K = 2$ *and* $\boldsymbol{a}_0 = [1, \jmath]^T$, *then* $g(\boldsymbol{a}_0, \boldsymbol{a}_0) = 0$ *despite the fact that* $[1, \jmath]^T \neq \boldsymbol{0}_{2\times 1}$. *Therefore, the function g defined in* (3.116) *is* not *an inner product, and a Cauchy-Schwartz inequality is not valid for this function. By examining the proof of Theorem 3.4, it can be seen that the reason why* $[\mathcal{D}_{\boldsymbol{Z}} f(\boldsymbol{Z}, \boldsymbol{Z}^*)]^T$ *is* not *the direction of maximum change of rate is that the function g in* (3.116) *is* not *an inner product.*

If a real-valued function f is being optimized with respect to the variable $\boldsymbol{Z}$ *by means of the steepest ascent or descent method, it follows from Theorem 3.4 that the updating term must be proportional to* $\mathcal{D}_{\boldsymbol{Z}^*} f(\boldsymbol{Z}, \boldsymbol{Z}^*)$, *and* not $\mathcal{D}_{\boldsymbol{Z}} f(\boldsymbol{Z}, \boldsymbol{Z}^*)$. *The update equation for optimizing the real-valued function in Theorem 3.4 by means of the steepest ascent or descent method can be expressed as*

$$\mathrm{vec}^T(\boldsymbol{Z}_{k+1}) = \mathrm{vec}^T(\boldsymbol{Z}_k) + \mu \mathcal{D}_{\boldsymbol{Z}^*} f(\boldsymbol{Z}_k, \boldsymbol{Z}_k^*), \tag{3.117}$$

where μ *is a* real positive constant if it is a maximization problem *or a* real negative constant if it is a minimization problem, *and* $\boldsymbol{Z}_k \in \mathbb{C}^{N\times Q}$ *is the value of the unknown matrix after k iterations. In* (3.117), *the size of* $\mathrm{vec}^T(\boldsymbol{Z}_k)$ *and* $\mathcal{D}_{\boldsymbol{Z}^*} f(\boldsymbol{Z}_k, \boldsymbol{Z}_k^*)$ *is* $1 \times NQ$.

Example 3.1 Let $f : \mathbb{C} \times \mathbb{C} \to \mathbb{R}$ be given by

$$f(z, z^*) = |z|^2 = z z^*, \tag{3.118}$$

such that f can be used to express the squared Euclidean distance. It is possible to visualize this function over the complex plane z by a contour plot like the one shown in

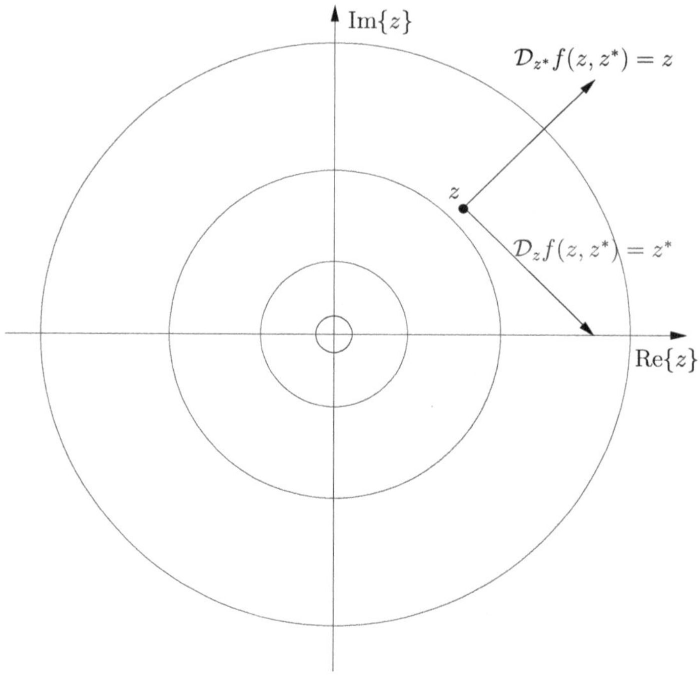

Figure 3.1 Contour plot of the function $f(z, z^*) = |z|^2$ taken from Example 3.1. The location of an arbitrary point z is shown by $\bullet$, and the two vectors $\mathcal{D}_{z^*} f(z, z^*) = z$ and $\mathcal{D}_z f(z, z^*) = z^*$ are drawn from the point z.

Figure 3.1, and this is a widely used function in engineering. The formal derivatives of this function are

$$\mathcal{D}_{z^*} f(z, z^*) = z, \tag{3.119}$$

$$\mathcal{D}_z f(z, z^*) = z^*. \tag{3.120}$$

These two derivatives are shown with two vectors (arrows) in Figure 3.1 out of the point z, which is marked with $\bullet$. It is seen from Figure 3.1 that the function f is increasing faster along the upper vector $\mathcal{D}_{z^*} f(z, z^*) = z$ than along the lower vector $\mathcal{D}_z f(z, z^*) = z^*$. The function f is maximally increasing in the direction of $\mathcal{D}_{z^*} f(z, z^*) = z$ when the starting position is z. This simple example can be used for remembering the important general components of Theorem 3.4.

3.4.3 One Independent Input Matrix Variable

In this subsection, the case in which the input variable of the functions is just *one* matrix variable with independent matrix components will be studied. It will be shown that the same procedure applied in the real-valued case (Magnus & Neudecker 1988) can be used for this case.

Theorem 3.5 *Let $F : \mathbb{C}^{N\times Q} \times \mathbb{C}^{N\times Q} \to \mathbb{C}^{M\times P}$ and $G : \mathbb{C}^{N\times Q} \to \mathbb{C}^{M\times P}$, where the differentials of Z_0 and Z_1 are assumed to be independent. If $F(Z_0, Z_1) = G(Z_0)$, then $\mathcal{D}_Z F(Z, Z^*) = \mathcal{D}_Z G(Z)$ can be obtained by the procedure given in Magnus and Neudecker (1988) for finding the derivative of the function G and $\mathcal{D}_{Z^*} F(Z, Z^*) = 0_{MP\times NQ}$.*

If $F(Z_0, Z_1) = G(Z_1)$, where the differentials of Z_0 and Z_1 are independent, then $\mathcal{D}_Z F(Z, Z^*) = 0_{MP\times NQ}$, and $\mathcal{D}_{Z^*} F(Z, Z^*) = \mathcal{D}_Z G(Z)|_{Z=Z^*}$ can be obtained by the procedure given in Magnus and Neudecker (1988).

Proof Assume that $F(Z_0, Z_1) = G(Z_0)$ where Z_0 and Z_1 have independent differentials. Applying the vec$(\cdot)$ and the differential operator to this equation leads to

$$d\,\text{vec}\,(F) = \left(\mathcal{D}_{Z_0} F\right) d\,\text{vec}\,(Z_0) + \left(\mathcal{D}_{Z_1} F\right) d\,\text{vec}\,(Z_1) = \left(\mathcal{D}_{Z_0} G\right) d\,\text{vec}\,(Z_0). \tag{3.121}$$

By setting $Z_0 = Z$ and $Z_1 = Z^*$ and using Lemma 3.1, it is seen that $\mathcal{D}_{Z^*} F = 0_{MP\times NQ}$ and $\mathcal{D}_Z F(Z, Z^*) = \mathcal{D}_Z G(Z)$. Because the last equation depends on only one matrix variable, Z, the same techniques as given in Magnus and Neudecker (1988) can be used. The first part of the theorem is then proved, and the second part can be shown in a similar way. ■

3.5 Exercises

3.1 Let $Z \in \mathbb{C}^{N\times N}$, and let perm : $\mathbb{C}^{N\times N} \to \mathbb{C}$ denote the *permanent* function of a complex-valued input matrix, that is,

$$\text{perm}(Z) \triangleq \sum_{k=0}^{N-1} m_{k,l}(Z) z_{k,l}, \tag{3.122}$$

where $m_{k,l}(Z)$ represents the (k, l)-th minor of Z, which is equal to the determinant of the matrix found from Z by deleting its k-th row and l-th column. Show that the differential of perm(Z) is given by

$$d\,\text{perm}(Z) = \text{Tr}\left\{M^T(Z) dZ\right\}, \tag{3.123}$$

where the $N \times N$ matrix $M(Z)$ contains the minors of Z, that is, $(M(Z))_{k,l} = m_{k,l}(Z)$.

3.2 When $z \in \mathbb{C}$ is a scalar, it follows from the product rule that $dz^k = kz^{k-1}dz$, where $k \in \mathbb{N}$. In this exercise, matrix versions of this result are derived. Let $Z \in \mathbb{C}^{N\times N}$ be a square matrix. Show that

$$dZ^k = \sum_{l=1}^{k} Z^{l-1}(dZ)Z^{k-l}, \tag{3.124}$$

where $k \in \mathbb{N}$. Use (3.125) to show that

$$d\,\text{Tr}\left\{Z^k\right\} = k\,\text{Tr}\left\{Z^{k-1} dZ\right\}. \tag{3.125}$$

3.3 Show that the complex differential of $\exp(\boldsymbol{Z})$, where $\boldsymbol{Z} \in \mathbb{C}^{N\times N}$ and $\exp(\cdot)$ is the exponential matrix function given in Definition 2.5, can be expressed as

$$d\exp(\boldsymbol{Z}) = \sum_{k=0}^{\infty} \frac{1}{(k+1)} \sum_{i=0}^{k} \boldsymbol{Z}^i (d\boldsymbol{Z}) \boldsymbol{Z}^{k-i}. \tag{3.126}$$

3.4 Show that the complex differential of $\operatorname{Tr}\{\exp(\boldsymbol{Z})\}$, where $\boldsymbol{Z} \in \mathbb{C}^{N\times N}$ and $\exp(\cdot)$ is given in Definition 2.5, is given by

$$d\operatorname{Tr}\{\exp(\boldsymbol{Z})\} = \operatorname{Tr}\{\exp(\boldsymbol{Z})\, d\boldsymbol{Z}\}. \tag{3.127}$$

From this result, show that the derivative of $\operatorname{Tr}\{\exp(\boldsymbol{Z})\}$ with respect to $\boldsymbol{Z}$ is given by

$$\mathcal{D}_{\boldsymbol{Z}} \operatorname{Tr}\{\exp(\boldsymbol{Z})\} = \operatorname{vec}^T\left(\exp\left(\boldsymbol{Z}^T\right)\right). \tag{3.128}$$

3.5 Let $t \in \mathbb{R}$ and $\boldsymbol{A} \in \mathbb{C}^{N\times N}$. Show that

$$\frac{d}{dt}\exp(t\boldsymbol{A}) = \boldsymbol{A}\exp(t\boldsymbol{A}) = \exp(t\boldsymbol{A})\,\boldsymbol{A}. \tag{3.129}$$

3.6 Let $\boldsymbol{Z} \in \mathbb{C}^{N\times Q}$, and let $\boldsymbol{A} \in \mathbb{C}^{M\times N}$ and $\boldsymbol{B} \in \mathbb{C}^{Q\times M}$ be two matrices that are independent of $\boldsymbol{Z}$ and $\boldsymbol{Z}^*$. Show that the complex differential of $\operatorname{Tr}\left\{\boldsymbol{A}\boldsymbol{Z}^+\boldsymbol{B}\right\}$ can be expressed as

$$\begin{aligned} d\operatorname{Tr}\left\{\boldsymbol{A}\boldsymbol{Z}^+\boldsymbol{B}\right\} = &\operatorname{Tr}\left\{\boldsymbol{A}\left(-\boldsymbol{Z}^+(d\boldsymbol{Z})\boldsymbol{Z}^+ + \boldsymbol{Z}^+(\boldsymbol{Z}^+)^H(d\boldsymbol{Z}^H)\left(\boldsymbol{I}_N - \boldsymbol{Z}\boldsymbol{Z}^+\right)\right.\right. \\ &\left.\left. + \left(\boldsymbol{I}_Q - \boldsymbol{Z}^+\boldsymbol{Z}\right)(d\boldsymbol{Z}^H)(\boldsymbol{Z}^+)^H\boldsymbol{Z}^+\right)\boldsymbol{B}\right\}. \end{aligned} \tag{3.130}$$

Assume that $N = Q$ and $\boldsymbol{Z} \in \mathbb{C}^{N\times N}$ is nonsingular. Use (3.130) to find an expression for $d\operatorname{Tr}\left\{\boldsymbol{A}\boldsymbol{Z}^{-1}\boldsymbol{B}\right\}$.

3.7 Let $a \in \mathbb{C} \setminus \{0\}$ and $\boldsymbol{Z} \in \mathbb{C}^{N\times Q}$, and let the function $\boldsymbol{F} : \mathbb{C}^{N\times Q} \times \mathbb{C}^{N\times Q} \to \mathbb{C}^{N\times Q}$ be given by

$$\boldsymbol{F}(\boldsymbol{Z}, \boldsymbol{Z}^*) = a\boldsymbol{Z}. \tag{3.131}$$

Let $\boldsymbol{G} : \mathbb{C}^{N\times Q} \times \mathbb{C}^{N\times Q} \to \mathbb{C}^{R\times R}$ be denoted by $\boldsymbol{G}(\boldsymbol{F}, \boldsymbol{F}^*)$, and let $\boldsymbol{H} : \mathbb{C}^{N\times Q} \times \mathbb{C}^{N\times Q} \to \mathbb{C}^{R\times R}$ be a composed function given as

$$\boldsymbol{H}(\boldsymbol{Z}, \boldsymbol{Z}^*) = \boldsymbol{G}(\boldsymbol{F}(\boldsymbol{Z}, \boldsymbol{Z}^*), \boldsymbol{F}^*(\boldsymbol{Z}, \boldsymbol{Z}^*)). \tag{3.132}$$

By means of the chain rule, show that

$$\mathcal{D}_{\boldsymbol{F}}\boldsymbol{G}|_{\boldsymbol{F}=\boldsymbol{F}(\boldsymbol{Z},\boldsymbol{Z}^*)} = \frac{1}{a}\mathcal{D}_{\boldsymbol{Z}}\boldsymbol{H}. \tag{3.133}$$

3.8 In a MIMO system, the signals are transmitted over a channel where both the transmitter and the receiver are equipped with multiple antennas. Let the numbers of transmit and receive antennas be M_t and M_r, respectively, and let the memoryless fixed MIMO transfer channel be denoted $\boldsymbol{H}$ (see Figure 3.2). Assume that the channel is contaminated with white zero-mean complex circularly symmetric Gaussian-distributed additive noise $\boldsymbol{n} \in \mathbb{C}^{M_r\times 1}$ with covariance matrix given by the identity matrix: $\mathbb{E}\left[\boldsymbol{n}\boldsymbol{n}^H\right] = \boldsymbol{I}_{M_r}$, where $\mathbb{E}[\cdot]$ denotes the expected value operator. The mutual information, denoted I,

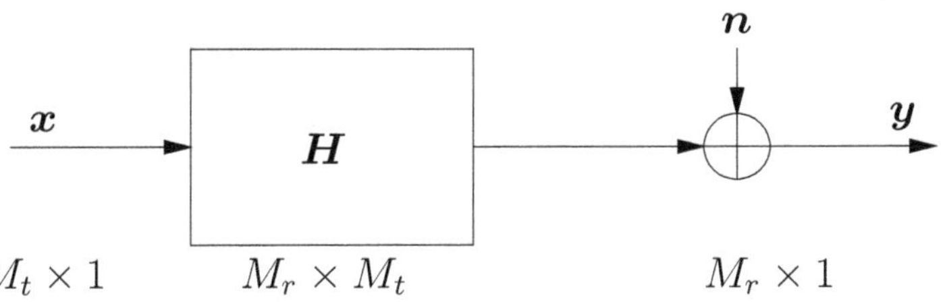

Figure 3.2 MIMO channel with input vector $\boldsymbol{x} \in \mathbb{C}^{M_t \times 1}$, additive Gaussian noise $\boldsymbol{n} \in \mathbb{C}^{M_r \times 1}$, output vector $\boldsymbol{y} \in \mathbb{C}^{M_r \times 1}$, and memoryless fixed known transfer function $\boldsymbol{H} \in \mathbb{C}^{M_r \times M_t}$.

between the channel input, which is assumed to be zero-mean complex circularly symmetric Gaussian-distributed vector $\boldsymbol{x} \in \mathbb{C}^{M_t \times 1}$, and the channel output vector $\boldsymbol{y} \in \mathbb{C}^{M_r \times 1}$ of the MIMO channel was derived in Telatar (1995) as

$$I = \ln\left(\det\left(\boldsymbol{I}_{M_r} + \boldsymbol{H}\boldsymbol{Q}\boldsymbol{H}^H\right)\right), \tag{3.134}$$

where $\boldsymbol{Q} \triangleq \mathbb{E}\left[\boldsymbol{x}\boldsymbol{x}^H\right] \in \mathbb{C}^{M_t \times M_t}$ is the covariance matrix of $\boldsymbol{x} \in \mathbb{C}^{M_t \times 1}$, which is assumed to be independent of the channel noise $\boldsymbol{n}$. Consider $I : \mathbb{C}^{M_r \times M_t} \times \mathbb{C}^{M_r \times M_t} \to \mathbb{R}$ as a function of $\boldsymbol{H}$ and $\boldsymbol{H}^*$, such that this function is denoted $I(\boldsymbol{H}, \boldsymbol{H}^*)$. Show that the complex differential of $I(\boldsymbol{H}, \boldsymbol{H}^*)$ can be expressed as

$$\begin{aligned} dI(\boldsymbol{H}, \boldsymbol{H}^*) = {} & \operatorname{Tr}\left\{\boldsymbol{Q}\boldsymbol{H}^H\left(\boldsymbol{I}_{M_r} + \boldsymbol{H}\boldsymbol{Q}\boldsymbol{H}^H\right)^{-1} d\boldsymbol{H}\right\} \\ & + \operatorname{Tr}\left\{\boldsymbol{Q}^T\boldsymbol{H}^T\left(\boldsymbol{I}_{M_r} + \boldsymbol{H}\boldsymbol{Q}\boldsymbol{H}^H\right)^{-T} d\boldsymbol{H}^*\right\}. \end{aligned} \tag{3.135}$$

Based on (3.135), show that the derivatives of $I(\boldsymbol{H}, \boldsymbol{H}^*)$ with respect to both $\boldsymbol{H}$ and $\boldsymbol{H}^*$ are given as

$$\mathcal{D}_{\boldsymbol{H}} I(\boldsymbol{H}, \boldsymbol{H}^*) = \operatorname{vec}^T\left(\left(\boldsymbol{I}_{M_r} + \boldsymbol{H}\boldsymbol{Q}\boldsymbol{H}^H\right)^{-T} \boldsymbol{H}^*\boldsymbol{Q}^T\right), \tag{3.136}$$

$$\mathcal{D}_{\boldsymbol{H}^*} I(\boldsymbol{H}, \boldsymbol{H}^*) = \operatorname{vec}^T\left(\left(\boldsymbol{I}_{M_r} + \boldsymbol{H}\boldsymbol{Q}\boldsymbol{H}^H\right)^{-1} \boldsymbol{H}\boldsymbol{Q}\right), \tag{3.137}$$

respectively. Explain why these results are in agreement with Palomar and Verdú (2006, Theorem 1).

3.9 Let $\boldsymbol{A}^H = \boldsymbol{A} \in \mathbb{C}^{N \times N}$ be given. Consider the function $f : \mathbb{C}^{N \times 1} \times \mathbb{C}^{N \times 1} \to \mathbb{R}$ defined as

$$f(\boldsymbol{z}, \boldsymbol{z}^*) = \frac{\boldsymbol{z}^H\boldsymbol{A}\boldsymbol{z}}{\boldsymbol{z}^H\boldsymbol{z}}, \tag{3.138}$$

and f is defined for $\boldsymbol{z} \neq \boldsymbol{0}_{N \times 1}$. The expression in (3.138) is called the Rayleigh quotient (Strang 1988). By using the theory presented in this chapter, show that df can be expressed as

$$df = \left[\frac{\boldsymbol{z}^H\boldsymbol{A}}{\boldsymbol{z}^H\boldsymbol{z}} - \frac{\boldsymbol{z}^H\boldsymbol{A}\boldsymbol{z}}{\left(\boldsymbol{z}^H\boldsymbol{z}\right)^2}\boldsymbol{z}^H\right] d\boldsymbol{z} + \left[\frac{\boldsymbol{z}^T\boldsymbol{A}^T}{\boldsymbol{z}^H\boldsymbol{z}} - \frac{\boldsymbol{z}^H\boldsymbol{A}\boldsymbol{z}}{\left(\boldsymbol{z}^H\boldsymbol{z}\right)^2}\boldsymbol{z}^T\right] d\boldsymbol{z}^*. \tag{3.139}$$

By using Table 3.2, show that the derivatives of f with respect to z and z^* can be identified as

$$\mathcal{D}_z f = \frac{z^H A}{z^H z} - \frac{z^H A z}{\left(z^H z\right)^2} z^H, \tag{3.140}$$

$$\mathcal{D}_{z^*} f = \frac{z^T A^T}{z^H z} - \frac{z^H A z}{\left(z^H z\right)^2} z^T. \tag{3.141}$$

By studying the necessary conditions for optimality (i.e., $\mathcal{D}_{z^*} f = \mathbf{0}_{1\times N}$), show that the maximum and minimum values of f are given by the maximum and minimum eigenvalue of A.

3.10 Consider the following function $f : \mathbb{C}^{N\times 1} \times \mathbb{C}^{N\times 1} \rightarrow \mathbb{R}$ given by

$$f(z, z^*) = \sigma_d^2 - z^H p - p^H z + z^H R z, \tag{3.142}$$

where $\sigma_d^2 > 0$, $p \in \mathbb{C}^{N\times 1}$, and $R \in \mathbb{C}^{N\times N}$ are independent of both z and z^*. The function given in (3.142) represents the mean square error (MSE) between the output of a finite impulse response (FIR) filter of length N with complex-valued coefficients collected into the vector $z \in \mathbb{C}^{N\times 1}$ and the desired output signal (Haykin 2002, Chapter 2). In (3.142), σ_d^2 represents the variance of the desired output signal, $R^H = R$ is the autocorrelation matrix of the input signal of size $N \times N$, and p is the cross-correlation vector between the input vector and the desired scalar output signal. Show that the values of the FIR filter coefficient z that is minimizing the function in (3.142) must satisfy

$$R z = p. \tag{3.143}$$

These are called the *Wiener-Hopf equations*.

Using the steepest descent method, show that the update equation for minimizing f defined in (3.142) is given by

$$z_{k+1} = z_k + \mu \left(p - R z_k\right), \tag{3.144}$$

where μ is a positive step size and k is the iteration index.

3.11 Consider the linear model shown in Figure 3.2, where the output of the channel $y \in \mathbb{C}^{M_r\times 1}$ is given by

$$y = Hx + n, \tag{3.145}$$

where $H \in \mathbb{C}^{M_r\times M_t}$ is a fixed MIMO transfer function and the input signal $x \in \mathbb{C}^{M_t\times 1}$ is uncorrelated with the additive noise vector $n \in \mathbb{C}^{M_r\times 1}$. All signals are assumed to have zero-mean. The three vectors x, n, and y have autocorrelation matrices given by

$$R_x = \mathbb{E}\left[x x^H\right], \tag{3.146}$$

$$R_n = \mathbb{E}\left[n n^H\right], \tag{3.147}$$

$$R_y = \mathbb{E}\left[y y^H\right] = R_n + H R_x H^H, \tag{3.148}$$

respectively. Assume that a linear complex-valued receiver filter $\boldsymbol{Z} \in \mathbb{C}^{M_t \times M_r}$ is applied to the received signal $\boldsymbol{y}$ such that the output of the receiver filter is $\boldsymbol{Z}\boldsymbol{y} \in \mathbb{C}^{M_t \times 1}$. Show that the MSE, denoted $f : \mathbb{C}^{M_t \times M_r} \times \mathbb{C}^{M_t \times M_r} \to \mathbb{R}$, between the output of the receiver filter $\boldsymbol{Z}\boldsymbol{y}$ and the original signal $\boldsymbol{x}$ defined as $f(\boldsymbol{Z}, \boldsymbol{Z}^*) = \mathbb{E}\left[\|\boldsymbol{Z}\boldsymbol{y} - \boldsymbol{x}\|^2\right]$ can be expressed as

$$f(\boldsymbol{Z}, \boldsymbol{Z}^*) = \operatorname{Tr}\left\{\boldsymbol{Z}\left[\boldsymbol{H}\boldsymbol{R}_x\boldsymbol{H}^H + \boldsymbol{R}_n\right]\boldsymbol{Z}^H - \boldsymbol{Z}\boldsymbol{H}\boldsymbol{R}_x - \boldsymbol{R}_x\boldsymbol{H}^H\boldsymbol{Z}^H + \boldsymbol{R}_x\right\}. \quad (3.149)$$

Show that the value of the filter coefficient $\boldsymbol{Z}$ that is minimizing the MSE function $f(\boldsymbol{Z}, \boldsymbol{Z}^*)$ is satisfying

$$\boldsymbol{Z} = \boldsymbol{R}_x\boldsymbol{H}^H\left[\boldsymbol{H}\boldsymbol{R}_x\boldsymbol{H}^H + \boldsymbol{R}_n\right]^{-1}. \quad (3.150)$$

The minimum MSE receiver filter in (3.150) is called the *Wiener filter* (Sayed 2008).

3.12 Some of the results from this and the previous exercise are presented in Kailath et al. (2000, Section 3.4) and Sayed (2003, Section 2.6).

Use the matrix inversion lemma in Lemma 2.3 to show that the minimum MSE receiver filter in (3.150) can be reformulated as

$$\boldsymbol{Z} = \left[\boldsymbol{R}_x^{-1} + \boldsymbol{H}^H\boldsymbol{R}_n^{-1}\boldsymbol{H}\right]^{-1}\boldsymbol{H}^H\boldsymbol{R}_n^{-1}. \quad (3.151)$$

Show that the minimum value using the minimum MSE filter in (3.150) can be expressed as

$$\operatorname{Tr}\left\{\left[\boldsymbol{R}_x^{-1} + \boldsymbol{H}^H\boldsymbol{R}_n^{-1}\boldsymbol{H}\right]^{-1}\right\} \triangleq g(\boldsymbol{H}, \boldsymbol{H}^*), \quad (3.152)$$

where the function $g : \mathbb{C}^{M_r \times M_t} \times \mathbb{C}^{M_r \times M_t} \to \mathbb{R}$ has been defined to be equal to this minimum MSE value. Show that the complex differential of g can be expressed as

$$dg = -\operatorname{Tr}\left\{\left[\boldsymbol{R}_x^{-1} + \boldsymbol{H}^H\boldsymbol{R}_n^{-1}\boldsymbol{H}\right]^{-2}\boldsymbol{H}^H\boldsymbol{R}_n^{-1}d\boldsymbol{H}\right.$$
$$\left. + \left[\boldsymbol{R}_x^{-T} + \boldsymbol{H}^T\boldsymbol{R}_n^{-T}\boldsymbol{H}^*\right]^{-2}\boldsymbol{H}^T\boldsymbol{R}_n^{-T}d\boldsymbol{H}^*\right\}. \quad (3.153)$$

Show that the derivatives of g with respect to $\boldsymbol{H}$ and $\boldsymbol{H}^*$ can be expressed as

$$\mathcal{D}_{\boldsymbol{H}}g = -\operatorname{vec}^T\left(\boldsymbol{R}_n^{-T}\boldsymbol{H}^*\left[\boldsymbol{R}_x^{-T} + \boldsymbol{H}^T\boldsymbol{R}_n^{-T}\boldsymbol{H}^*\right]^{-2}\right), \quad (3.154)$$

$$\mathcal{D}_{\boldsymbol{H}^*}g = -\operatorname{vec}^T\left(\boldsymbol{R}_n^{-1}\boldsymbol{H}\left[\boldsymbol{R}_x^{-1} + \boldsymbol{H}^H\boldsymbol{R}_n^{-1}\boldsymbol{H}\right]^{-2}\right), \quad (3.155)$$

respectively. It is observed from the above equations that $(\mathcal{D}_{\boldsymbol{H}^*}g)^* = \mathcal{D}_{\boldsymbol{H}}g$, which is in agreement with Theorem 3.3.

4 Development of Complex-Valued Derivative Formulas

4.1 Introduction

The definition of a complex-valued matrix derivative was given in Chapter 3 (see Definition 3.1). In this chapter, it will be shown how the complex-valued matrix derivatives can be found for all nine different types of functions given in Table 2.2. Three different choices are given for the complex-valued *input variables* of the functions, namely, scalar, vector, or matrix; in addition, three possibilities for the type of *output* that functions return, again, could be scalar, vector, or matrix. The derivative can be identified through the complex differential by using Table 3.2. In this chapter, it will be shown how the theory introduced in Chapters 2 and 3 can be used to find complex-valued matrix derivatives through examples. Many results are collected inside tables to make them more accessible.

The rest of this chapter is organized as follows: The simplest case, when the output of a function is a complex-valued scalar, is treated in Section 4.2, which contains three subsections (4.2.1, 4.2.2, and 4.2.3) when the input variables are scalars, vectors, and matrices, respectively. Section 4.3 looks at the case of vector functions; it contains Subsections 4.3.1, 4.3.2, and 4.3.3, which treat the three cases of complex-valued scalar, vector, and matrix input variables, respectively. Matrix functions are considered in Section 4.4, which contains three subsections. The three cases of complex-valued matrix functions with scalar, vector, and matrix inputs are treated in Subsections 4.4.1, 4.4.2, and 4.4.3, respectively. The chapter ends with Section 4.5, which consists of 10 exercises.

4.2 Complex-Valued Derivatives of Scalar Functions

4.2.1 Complex-Valued Derivatives of $f(z, z^*)$

If the variables z and z^* are treated as independent variables, then the derivatives $\mathcal{D}_z f(z, z^*)$ and $\mathcal{D}_{z^*} f(z, z^*)$ can be found as for scalar functions having two independent variables. The case of scalar function of scalar independent variables is treated extensively in the literature for scalar input variables (see, for example, Kreyszig 1988, and Edwards & Penney 1986). See also Example 2.2 for how this can be done. To make the reader more familiar with how to treat the variables z and z^* independently, some examples are given below.

Example 4.1 By examining Definition 2.2 of the *formal derivatives*, the operators of finding the derivative with respect to z and z^* can be expressed, respectively, as

$$\frac{\partial}{\partial z} = \frac{1}{2}\left(\frac{\partial}{\partial x} - \jmath\frac{\partial}{\partial y}\right), \tag{4.1}$$

and

$$\frac{\partial}{\partial z^*} = \frac{1}{2}\left(\frac{\partial}{\partial x} + \jmath\frac{\partial}{\partial y}\right), \tag{4.2}$$

where $z = x + \jmath y$, $\mathrm{Re}\{z\} = x$, and $\mathrm{Im}\{z\} = y$. To show that the two operators in (4.1) and (4.2) are in agreement with the fact that z and z^* can be treated as independent variables when finding derivatives, we can try to use the operators in (4.1) and (4.2) to find the derivative of z and z^*, that is

$$\frac{\partial z^*}{\partial z} = \frac{1}{2}\left(\frac{\partial}{\partial x} - \jmath\frac{\partial}{\partial y}\right)(x - \jmath y) = \frac{1}{2}(1 - 1) = 0, \tag{4.3}$$

and

$$\frac{\partial z}{\partial z^*} = \frac{1}{2}\left(\frac{\partial}{\partial x} + \jmath\frac{\partial}{\partial y}\right)(x + \jmath y) = \frac{1}{2}(1 - 1) = 0, \tag{4.4}$$

which are expected because z and z^* should be treated as independent variables, as shown by Lemma 3.1. The derivative of z and z^* with respect to itself can be found in a similar way, that is,

$$\frac{\partial z}{\partial z} = \frac{1}{2}\left(\frac{\partial}{\partial x} - \jmath\frac{\partial}{\partial y}\right)(x + \jmath y) = \frac{1}{2}(1 + 1) = 1, \tag{4.5}$$

and

$$\frac{\partial z^*}{\partial z^*} = \frac{1}{2}\left(\frac{\partial}{\partial x} + \jmath\frac{\partial}{\partial y}\right)(x - \jmath y) = \frac{1}{2}(1 + 1) = 1. \tag{4.6}$$

The derivative of the real (x) and imaginary parts (y) of z with respect to z and z^* can be found as

$$\frac{\partial x}{\partial z} = \frac{\partial}{\partial z}\left(\frac{1}{2}(z + z^*)\right) = \frac{1}{2}, \tag{4.7}$$

$$\frac{\partial x}{\partial z^*} = \frac{\partial}{\partial z^*}\left(\frac{1}{2}(z + z^*)\right) = \frac{1}{2}, \tag{4.8}$$

$$\frac{\partial y}{\partial z} = \frac{\partial}{\partial z}\left(\frac{1}{2\jmath}(z - z^*)\right) = \frac{1}{2\jmath}, \tag{4.9}$$

$$\frac{\partial y}{\partial z^*} = \frac{\partial}{\partial z^*}\left(\frac{1}{2\jmath}(z - z^*)\right) = \frac{\jmath}{2} = -\frac{1}{2\jmath}. \tag{4.10}$$

Table 4.1 Complex-valued derivatives of functions of the type $f(z, z^*)$.

$f(z, z^*)$	$\frac{\partial f}{\partial x}$	$\frac{\partial f}{\partial y}$	$\frac{\partial f}{\partial z}$	$\frac{\partial f}{\partial z^*}$
$\text{Re}\{z\} = x$	1	0	$\frac{1}{2}$	$\frac{1}{2}$
$\text{Im}\{z\} = y$	0	1	$\frac{1}{2\jmath}$	$-\frac{1}{2\jmath}$
z	1	$\jmath$	1	0
z^*	1	$-\jmath$	0	1

These results and others are collected in Table 4.1. The derivative of z^* with respect to x can be found as follows:

$$\frac{\partial z^*}{\partial x} = \frac{\partial (x - \jmath y)}{\partial x} = 1. \tag{4.11}$$

The remaining results in Table 4.1 can be derived in a similar fashion.

Example 4.2 Let $f : \mathbb{C} \times \mathbb{C} \to \mathbb{R}$ be defined as

$$f(z, z^*) = \sqrt{zz^*} = |z| = \sqrt{x^2 + y^2}, \tag{4.12}$$

such that f represents the squared Euclidean distance from the origin to z. Assume that $z \neq 0$ in this example. By treating z and z^* as independent variables, the derivative of f with respect to z and z^* can be calculated as

$$\frac{\partial f}{\partial z} = \frac{\partial \sqrt{zz^*}}{\partial z} = \frac{z^*}{2\sqrt{z^* z}} = \frac{z^*}{2|z|} = \frac{1}{2} e^{-\jmath \angle z}, \tag{4.13}$$

$$\frac{\partial f}{\partial z^*} = \frac{\partial \sqrt{zz^*}}{\partial z^*} = \frac{z}{2\sqrt{z^* z}} = \frac{z}{2|z|} = \frac{1}{2} e^{\jmath \angle z}, \tag{4.14}$$

where the function $\angle(\cdot) : \mathbb{C}\backslash\{0\} \to (-\pi, \pi]$ is the principal value of the argument (Kreyszig 1988, Section 12.2) of the input. It is seen that (4.13) and (4.14) are in agreement with Theorem 3.3.

These derivatives can be calculated alternatively by using Definition 2.2. This is done by first finding the derivatives of f with respect to the real (x) and imaginary parts (y) of $z = x + \jmath y$, and then inserting the result into (2.11) and (2.12). First, the derivatives of f with respect to x and y are found:

$$\frac{\partial f}{\partial x} = \frac{x}{\sqrt{x^2 + y^2}} = \frac{\text{Re}\{z\}}{|z|}, \tag{4.15}$$

and

$$\frac{\partial f}{\partial y} = \frac{y}{\sqrt{x^2 + y^2}} = \frac{\text{Im}\{z\}}{|z|}. \tag{4.16}$$

Inserting the results from (4.15) and (4.16) into both (2.11) and (2.12) gives

$$\frac{\partial f}{\partial z} = \frac{1}{2}\left(\frac{\partial f}{\partial x} - \jmath \frac{\partial f}{\partial y}\right) = \frac{1}{2}\left(\frac{\mathrm{Re}\{z\}}{|z|} - \jmath \frac{\mathrm{Im}\{z\}}{|z|}\right) = \frac{z^*}{2|z|} = \frac{1}{2}e^{-\jmath \angle z}, \quad (4.17)$$

and

$$\frac{\partial f}{\partial z^*} = \frac{1}{2}\left(\frac{\partial f}{\partial x} + \jmath \frac{\partial f}{\partial y}\right) = \frac{1}{2}\left(\frac{\mathrm{Re}\{z\}}{|z|} + \jmath \frac{\mathrm{Im}\{z\}}{|z|}\right) = \frac{z}{2|z|} = \frac{1}{2}e^{\jmath \angle z}. \quad (4.18)$$

Hence, (4.17) and (4.18) are in agreement with the results found in (4.13) and (4.14), respectively. However, it is seen that it is more involved to find the derivatives $\frac{\partial f}{\partial z}$ and $\frac{\partial f}{\partial z^*}$ by using Definition 2.2 than by treating z and z^* independently.

Example 4.3 Let $f : \mathbb{C} \times \mathbb{C} \to \mathbb{R}$ be defined as

$$f(z, z^*) = \angle z = \arctan \frac{\mathrm{Im}\{z\}}{\mathrm{Re}\{z\}} = \arctan \frac{y}{x}, \quad (4.19)$$

where arctan(·) is the inverse tangent function (Edwards & Penney 1986). Expressed in polar coordinates, z is given by

$$z = |z|e^{\jmath \angle z}. \quad (4.20)$$

The input argument of 0 is not defined, so it is assumed that $z \neq 0$ in this example. Two alternative methods are presented to find the derivative of f with respect to z and z^*. By treating z and z^* as independent variables, it is found that

$$\frac{\partial f}{\partial z^*} = \frac{1}{1 + \frac{\mathrm{Im}^2\{z\}}{\mathrm{Re}^2\{z\}}} \frac{-\frac{1}{2\jmath}\mathrm{Re}\{z\} - \frac{1}{2}\mathrm{Im}\{z\}}{\mathrm{Re}^2\{z\}} = \frac{\jmath}{2} \frac{\mathrm{Re}\{z\} + \jmath\,\mathrm{Im}\{z\}}{\mathrm{Re}^2\{z\} + \mathrm{Im}^2\{z\}} = \frac{\jmath z}{2|z|^2} = \frac{\jmath}{2z^*}. \quad (4.21)$$

By using (3.109), it is found that

$$\frac{\partial f}{\partial z} = \left(\frac{\partial f}{\partial z^*}\right)^* = -\frac{\jmath}{2z}. \quad (4.22)$$

If $\frac{\partial f}{\partial z^*}$ is found by the use of the operator given in (4.2), the derivatives of f with respect to x and y are found first:

$$\frac{\partial f}{\partial x} = \frac{1}{1 + \left(\frac{y}{x}\right)^2} \frac{(-y)}{x^2} = \frac{-y}{x^2 + y^2} = -\frac{\mathrm{Im}\{z\}}{|z|^2}, \quad (4.23)$$

and

$$\frac{\partial f}{\partial y} = \frac{1}{1 + \left(\frac{y}{x}\right)^2} \frac{1}{x} = \frac{x}{x^2 + y^2} = \frac{\mathrm{Re}\{z\}}{|z|^2}. \quad (4.24)$$

Inserting (4.23) and (4.24) into (2.12) yields

$$\frac{\partial f}{\partial z^*} = \frac{1}{2}\left(\frac{\partial f}{\partial x} + \jmath \frac{\partial f}{\partial y}\right) = \frac{1}{2}\left(-\frac{\text{Im}\{z\}}{|z|^2} + \jmath \frac{\text{Re}\{z\}}{|z|^2}\right)$$
$$= \frac{1}{2}\frac{\jmath}{|z|^2}(\text{Re}\{z\} + \jmath\, \text{Im}\{z\}) = \frac{\jmath z}{2|z|^2} = \frac{\jmath}{2z^*}. \tag{4.25}$$

It is seen that (4.21) and (4.25) give the same result; however, it is observed that direct calculation by treating z and z^* as independent variables is easier because it requires fewer calculations.

Example 4.4 When optimizing a communication system where multilevel phase shift keying (PSK) symbols are in used as the signal alphabet (Hjørungnes 2005), the derivative of $|\angle z|$ with respect to z^* might be needed. In this example, we will study this case by using the chain rule.

Let $h : \mathbb{C} \times \mathbb{C} \to \mathbb{R}$ be given by

$$h(z, z^*) = g(f(z, z^*)) = |\angle z|, \tag{4.26}$$

where the function $g : \mathbb{R} \to \mathbb{R}$ is defined by

$$g(x) = |x|, \tag{4.27}$$

and $f : \mathbb{C} \times \mathbb{C} \to \mathbb{R}$ by

$$f(z, z^*) = \angle z = \arctan\left(\frac{y}{x}\right). \tag{4.28}$$

By using the chain rule (Theorem 3.1), we find that

$$\frac{\partial h(z, z^*)}{\partial z^*} = \left.\frac{\partial g(x)}{\partial x}\right|_{x=f(z,z^*)} \frac{\partial f(z, z^*)}{\partial z^*}. \tag{4.29}$$

From real-valued calculus, we know that $\frac{\partial |x|}{\partial x} = \frac{|x|}{x}$, and $\frac{\partial f(z,z^*)}{\partial z^*}$ was derived in (4.21). Putting these results together gives

$$\frac{\partial h(z, z^*)}{\partial z^*} = \frac{|\angle z|}{\angle z}\frac{\jmath}{2z^*}. \tag{4.30}$$

4.2.2 Complex-Valued Derivatives of $f(\boldsymbol{z}, \boldsymbol{z}^*)$

Let $\boldsymbol{a} \in \mathbb{C}^{N\times 1}$, $\boldsymbol{A} \in \mathbb{C}^{N\times N}$, and $\boldsymbol{z} \in \mathbb{C}^{N\times 1}$. Some examples of functions of the type $f(\boldsymbol{z}, \boldsymbol{z}^*)$ include $\boldsymbol{a}^T\boldsymbol{z}$, $\boldsymbol{a}^T\boldsymbol{z}^*$, $\boldsymbol{z}^T\boldsymbol{a}$, $\boldsymbol{z}^H\boldsymbol{a}$, $\boldsymbol{z}^T\boldsymbol{A}\boldsymbol{z}$, $\boldsymbol{z}^H\boldsymbol{A}\boldsymbol{z}$, and $\boldsymbol{z}^H\boldsymbol{A}\boldsymbol{z}^*$. The complex differentials and derivatives of these functions are shown in Table 4.2.

Two examples for how the results in Table 4.2 can be derived are given in the sequel.

Table 4.2 Complex-valued derivatives of functions of the type $f(z, z^*)$.

$f(z, z^*)$	Differential df	$\mathcal{D}_z f(z, z^*)$	$\mathcal{D}_{z^*} f(z, z^*)$
$a^T z = z^T a$	$a^T dz$	a^T	$0_{1\times N}$
$a^T z^* = z^H a$	$a^T dz^*$	$0_{1\times N}$	a^T
$z^T A z$	$z^T \left(A + A^T\right) dz$	$z^T \left(A + A^T\right)$	$0_{1\times N}$
$z^H A z$	$z^H A dz + z^T A^T dz^*$	$z^H A$	$z^T A^T$
$z^H A z^*$	$z^H \left(A + A^T\right) dz^*$	$0_{1\times N}$	$z^H \left(A + A^T\right)$

Example 4.5 Let $f : \mathbb{C}^{N\times 1} \times \mathbb{C}^{N\times 1} \to \mathbb{C}$ be given by

$$f(z, z^*) = z^H a = a^T z^*, \tag{4.31}$$

where $a \in \mathbb{C}^{N\times 1}$ is a vector independent of z and z^*. To find the derivative of this function, the procedure outlined in Section 3.3 is followed, where the first step is to find the complex differential of f. This complex differential should be manipulated into the form corresponding to the function type $f(z, z^*)$ given in Table 3.2. The complex differential of this function can be written as

$$df = \left(dz^H\right) a = a^T dz^*, \tag{4.32}$$

where the complex differential rules in (3.27) and (3.43) were applied. It is seen from the second line of Table 3.2 that now the complex differential of f is in the appropriate form. Therefore, we can identify the derivatives of f with respect to z and z^* as

$$\mathcal{D}_z f = 0_{1\times N}, \tag{4.33}$$

$$\mathcal{D}_{z^*} f = a^T, \tag{4.34}$$

respectively. These results are included in Table 4.2.

The procedure for finding the derivatives is always to reformulate the complex differential of the current functional type into the corresponding form in Table 3.2, and then to read out the derivatives directly.

Example 4.6 Let $f : \mathbb{C}^{N\times 1} \times \mathbb{C}^{N\times 1} \to \mathbb{C}$ be given by

$$f(z, z^*) = z^H A z. \tag{4.35}$$

This function frequently appears in array signal processing (Jonhson & Dudgeon 1993). The differential of this function can be expressed as

$$df = (dz^H) A z + z^H A dz = z^H A dz + z^T A^T dz^*, \tag{4.36}$$

where (3.27), (3.35), and (3.43) were utilized. From (4.36), the derivatives of $z^H A z$ with respect to z and z^* follow from Table 3.2, and the results are included in Table 4.2.

The other lines of Table 4.2 can be derived in a similar fashion. Some of the results included in Table 4.2 can also be found in Brandwood (1983), and they are used, for example, in Jaffer and Jones (1995) for designing complex FIR filters with respect to a weighted least-squares design criterion, and in Huang and Benesty (2003), for doing adaptive blind multichannel identification in the frequency domain.

4.2.3 Complex-Valued Derivatives of $f(\boldsymbol{Z}, \boldsymbol{Z}^*)$

Examples of functions of the type $f(\boldsymbol{Z}, \boldsymbol{Z}^*)$ are $\text{Tr}\{\boldsymbol{Z}\}$, $\text{Tr}\{\boldsymbol{Z}^*\}$, $\text{Tr}\{\boldsymbol{A}\boldsymbol{Z}\}$, $\det(\boldsymbol{Z})$, $\text{Tr}\{\boldsymbol{Z}\boldsymbol{A}_0\boldsymbol{Z}^T\boldsymbol{A}_1\}$, $\text{Tr}\{\boldsymbol{Z}\boldsymbol{A}_0\boldsymbol{Z}\boldsymbol{A}_1\}$, $\text{Tr}\{\boldsymbol{Z}\boldsymbol{A}_0\boldsymbol{Z}^H\boldsymbol{A}_1\}$, $\text{Tr}\{\boldsymbol{Z}\boldsymbol{A}_0\boldsymbol{Z}^*\boldsymbol{A}_1\}$, $\text{Tr}\{\boldsymbol{Z}^p\}$, $\text{Tr}\{\boldsymbol{A}\boldsymbol{Z}^{-1}\}$, $\det(\boldsymbol{A}_0\boldsymbol{Z}\boldsymbol{A}_1)$, $\det(\boldsymbol{Z}^2)$, $\det(\boldsymbol{Z}\boldsymbol{Z}^T)$, $\det(\boldsymbol{Z}\boldsymbol{Z}^*)$, $\det(\boldsymbol{Z}\boldsymbol{Z}^H)$, $\det(\boldsymbol{Z}^p)$, $\lambda(\boldsymbol{Z})$, and $\lambda^*(\boldsymbol{Z})$, where $\boldsymbol{Z} \in \mathbb{C}^{N\times Q}$ or possibly $\boldsymbol{Z} \in \mathbb{C}^{N\times N}$, if this is required for the functions to be defined. The sizes of $\boldsymbol{A}$, $\boldsymbol{A}_0$, and $\boldsymbol{A}_1$ are chosen such that these functions are well defined. The operators $\text{Tr}\{\cdot\}$ and $\det(\cdot)$ are defined in Section 2.4, and $\lambda(\boldsymbol{Z})$ returns an eigenvalue of $\boldsymbol{Z}$.

For functions of the type $f(\boldsymbol{Z}, \boldsymbol{Z}^*)$, it is also common to arrange the formal derivatives $\frac{\partial}{\partial z_{k,l}} f$ and $\frac{\partial}{\partial z_{k,l}^*} f$ in an alternative way (Magnus & Neudecker 1988, Section 9.2) than in the expressions $\mathcal{D}_{\boldsymbol{Z}} f(\boldsymbol{Z}, \boldsymbol{Z}^*)$ and $\mathcal{D}_{\boldsymbol{Z}^*} f(\boldsymbol{Z}, \boldsymbol{Z}^*)$. The notation for one alternative way of organizing all the formal derivatives is $\frac{\partial}{\partial \boldsymbol{Z}} f$ and $\frac{\partial}{\partial \boldsymbol{Z}^*} f$. In this alternative way, the formal derivatives of the elements of the matrix $\boldsymbol{Z} \in \mathbb{C}^{N\times Q}$ are arranged as

$$\frac{\partial}{\partial \boldsymbol{Z}} f = \begin{bmatrix} \frac{\partial}{\partial z_{0,0}} f & \cdots & \frac{\partial}{\partial z_{0,Q-1}} f \\ \vdots & & \vdots \\ \frac{\partial}{\partial z_{N-1,0}} f & \cdots & \frac{\partial}{\partial z_{N-1,Q-1}} f \end{bmatrix}, \tag{4.37}$$

$$\frac{\partial}{\partial \boldsymbol{Z}^*} f = \begin{bmatrix} \frac{\partial}{\partial z_{0,0}^*} f & \cdots & \frac{\partial}{\partial z_{0,Q-1}^*} f \\ \vdots & & \vdots \\ \frac{\partial}{\partial z_{N-1,0}^*} f & \cdots & \frac{\partial}{\partial z_{N-1,Q-1}^*} f \end{bmatrix}. \tag{4.38}$$

The quantities $\frac{\partial}{\partial \boldsymbol{Z}} f$ and $\frac{\partial}{\partial \boldsymbol{Z}^*} f$ are called the *gradient* of f with respect to $\boldsymbol{Z}$ and $\boldsymbol{Z}^*$. Equations (4.37) and (4.38) are generalizations to the complex case of one of the ways to define the derivative of real scalar functions with respect to real matrices, as described in Magnus and Neudecker (1988). Notice that the way of arranging the formal derivatives in (4.37) and (4.38) is different from the way given in (3.82) and (3.83). The connection between these two alternative ways to arrange the derivatives of a scalar function with respect to matrices is now elaborated.

From Table 3.2, it is observed that the derivatives of f can be identified from two alternative expressions of df. These two alternative ways for expressing df are equal and can be put together as

$$df = \text{vec}^T(\boldsymbol{A}_0) d\,\text{vec}(\boldsymbol{Z}) + \text{vec}^T(\boldsymbol{A}_1) d\,\text{vec}(\boldsymbol{Z}^*) \tag{4.39}$$

$$= \text{Tr}\left\{\boldsymbol{A}_0^T d\boldsymbol{Z} + \boldsymbol{A}_1^T d\boldsymbol{Z}^*\right\}, \tag{4.40}$$

where $\boldsymbol{A}_0 \in \mathbb{C}^{N\times Q}$ and $\boldsymbol{A}_1 \in \mathbb{C}^{N\times Q}$ depend on $\boldsymbol{Z} \in \mathbb{C}^{N\times Q}$ and $\boldsymbol{Z}^* \in \mathbb{C}^{N\times Q}$ in general. The traditional way of identifying the derivatives of f with respect to $\boldsymbol{Z}$ and $\boldsymbol{Z}^*$ can be read out from (4.39) in the following way:

$$\mathcal{D}_{\boldsymbol{Z}} f = \operatorname{vec}^T(\boldsymbol{A}_0), \tag{4.41}$$

$$\mathcal{D}_{\boldsymbol{Z}^*} f = \operatorname{vec}^T(\boldsymbol{A}_1). \tag{4.42}$$

In an alternative way, from (4.40), two gradients of f with respect to $\boldsymbol{Z}$ and $\boldsymbol{Z}^*$ are identified as

$$\frac{\partial}{\partial \boldsymbol{Z}} f = \boldsymbol{A}_0, \tag{4.43}$$

$$\frac{\partial}{\partial \boldsymbol{Z}^*} f = \boldsymbol{A}_1. \tag{4.44}$$

The size of the gradient of f with respect to $\boldsymbol{Z}$ and $\boldsymbol{Z}^*$ (i.e., $\frac{\partial}{\partial \boldsymbol{Z}} f$ and $\frac{\partial}{\partial \boldsymbol{Z}^*} f$) is $N \times Q$, and the size of $\mathcal{D}_{\boldsymbol{Z}} f(\boldsymbol{Z}, \boldsymbol{Z}^*)$ and $\mathcal{D}_{\boldsymbol{Z}^*} f(\boldsymbol{Z}, \boldsymbol{Z}^*)$ is $1 \times NQ$, so these two ways of organizing the formal derivatives are different, although their components are the same. By comparing (4.41) and (4.42) to (4.43) and (4.44), respectively, it is seen that the connection between the two ways of defining the derivatives is given by

$$\mathcal{D}_{\boldsymbol{Z}} f(\boldsymbol{Z}, \boldsymbol{Z}^*) = \operatorname{vec}^T\left(\frac{\partial}{\partial \boldsymbol{Z}} f(\boldsymbol{Z}, \boldsymbol{Z}^*)\right), \tag{4.45}$$

$$\mathcal{D}_{\boldsymbol{Z}^*} f(\boldsymbol{Z}, \boldsymbol{Z}^*) = \operatorname{vec}^T\left(\frac{\partial}{\partial \boldsymbol{Z}^*} f(\boldsymbol{Z}, \boldsymbol{Z}^*)\right). \tag{4.46}$$

At some places in the literature (Haykin 2002; Palomar and Verdú 2006), an alternative notation is used for the gradient of scalar functions $f : \mathbb{C}^{N\times Q} \times \mathbb{C}^{N\times Q} \to \mathbb{C}$. This alternative notation used for the gradient of f with respect to $\boldsymbol{Z}$ and $\boldsymbol{Z}^*$ is

$$\nabla_{\boldsymbol{Z}^*} f \triangleq \frac{\partial}{\partial \boldsymbol{Z}} f, \tag{4.47}$$

$$\nabla_{\boldsymbol{Z}} f \triangleq \frac{\partial}{\partial \boldsymbol{Z}^*} f, \tag{4.48}$$

respectively. Because it is easy to forget that the derivation should be done with respect to $\boldsymbol{Z}^*$ when the notation $\nabla_{\boldsymbol{Z}}$ is used (and vice versa for $\boldsymbol{Z}$ and $\nabla_{\boldsymbol{Z}^*}$), the notations $\nabla_{\boldsymbol{Z}}$ and $\nabla_{\boldsymbol{Z}^*}$ will not be used in this book.

From Theorem 3.4, it is seen that for a scalar real-valued function $f : \mathbb{C}^{N\times Q} \times \mathbb{C}^{N\times Q} \to \mathbb{R}$, the direction with respect to $\operatorname{vec}(\boldsymbol{Z})$ where the function decreases fastest is $-[\mathcal{D}_{\boldsymbol{Z}^*} f(\boldsymbol{Z}, \boldsymbol{Z}^*)]^T$. When using the vec operator, the steepest descent method is expressed in (3.117). If the notation introduced in (4.38) is utilized, it can be seen that the steepest descent method (3.117) can be reformulated as

$$\boldsymbol{Z}_{k+1} = \boldsymbol{Z}_k + \mu \left.\frac{\partial}{\partial \boldsymbol{Z}^*} f(\boldsymbol{Z}, \boldsymbol{Z}^*)\right|_{\boldsymbol{Z}=\boldsymbol{Z}_k}, \tag{4.49}$$

where μ and $\boldsymbol{Z}_k$ play the same roles as in (3.117), k represents the number of iterations, and (4.46) was used.

Example 4.7 Let $\boldsymbol{Z}_i \in \mathbb{C}^{N_i \times Q_i}$, where $i \in \{0, 1\}$, and let the function $f : \mathbb{C}^{N_0 \times Q_0} \times \mathbb{C}^{N_1 \times Q_1} \to \mathbb{C}$ be given by

$$f(\boldsymbol{Z}_0, \boldsymbol{Z}_1) = \operatorname{Tr}\{\boldsymbol{Z}_0 \boldsymbol{A}_0 \boldsymbol{Z}_1 \boldsymbol{A}_1\}, \tag{4.50}$$

where $\boldsymbol{A}_0$ and $\boldsymbol{A}_1$ are independent of $\boldsymbol{Z}_0$ and $\boldsymbol{Z}_1$. For the matrix product within the trace to be well defined, $\boldsymbol{A}_0 \in \mathbb{C}^{Q_0 \times N_1}$ and $\boldsymbol{A}_1 \in \mathbb{C}^{Q_1 \times N_0}$. The differential of this function can be expressed as

$$\begin{aligned} df &= \operatorname{Tr}\{(d\boldsymbol{Z}_0)\boldsymbol{A}_0 \boldsymbol{Z}_1 \boldsymbol{A}_1 + \boldsymbol{Z}_0 \boldsymbol{A}_0 (d\boldsymbol{Z}_1)\boldsymbol{A}_1\} \\ &= \operatorname{Tr}\{\boldsymbol{A}_0 \boldsymbol{Z}_1 \boldsymbol{A}_1 (d\boldsymbol{Z}_0) + \boldsymbol{A}_1 \boldsymbol{Z}_0 \boldsymbol{A}_0 (d\boldsymbol{Z}_1)\}, \end{aligned} \tag{4.51}$$

where (2.96) and (2.97) have been used. From (4.51), it is possible to find the differentials of $\operatorname{Tr}\{\boldsymbol{Z}\boldsymbol{A}_0\boldsymbol{Z}^T\boldsymbol{A}_1\}$, $\operatorname{Tr}\{\boldsymbol{Z}\boldsymbol{A}_0\boldsymbol{Z}\boldsymbol{A}_1\}$, $\operatorname{Tr}\{\boldsymbol{Z}\boldsymbol{A}_0\boldsymbol{Z}^H\boldsymbol{A}_1\}$, and $\operatorname{Tr}\{\boldsymbol{Z}\boldsymbol{A}_0\boldsymbol{Z}^*\boldsymbol{A}_1\}$. The differentials of these four functions are

$$d\operatorname{Tr}\{\boldsymbol{Z}\boldsymbol{A}_0\boldsymbol{Z}^T\boldsymbol{A}_1\} = \operatorname{Tr}\left\{\left(\boldsymbol{A}_0\boldsymbol{Z}^T\boldsymbol{A}_1 + \boldsymbol{A}_0^T\boldsymbol{Z}^T\boldsymbol{A}_1^T\right)d\boldsymbol{Z}\right\}, \tag{4.52}$$

$$d\operatorname{Tr}\{\boldsymbol{Z}\boldsymbol{A}_0\boldsymbol{Z}\boldsymbol{A}_1\} = \operatorname{Tr}\{(\boldsymbol{A}_0\boldsymbol{Z}\boldsymbol{A}_1 + \boldsymbol{A}_1\boldsymbol{Z}\boldsymbol{A}_0)\,d\boldsymbol{Z}\}, \tag{4.53}$$

$$d\operatorname{Tr}\{\boldsymbol{Z}\boldsymbol{A}_0\boldsymbol{Z}^H\boldsymbol{A}_1\} = \operatorname{Tr}\left\{\boldsymbol{A}_0\boldsymbol{Z}^H\boldsymbol{A}_1 d\boldsymbol{Z} + \boldsymbol{A}_0^T\boldsymbol{Z}^T\boldsymbol{A}_1^T d\boldsymbol{Z}^*\right\}, \tag{4.54}$$

$$d\operatorname{Tr}\{\boldsymbol{Z}\boldsymbol{A}_0\boldsymbol{Z}^*\boldsymbol{A}_1\} = \operatorname{Tr}\left\{\boldsymbol{A}_0\boldsymbol{Z}^*\boldsymbol{A}_1 d\boldsymbol{Z} + \boldsymbol{A}_1\boldsymbol{Z}\boldsymbol{A}_0 d\boldsymbol{Z}^*\right\}, \tag{4.55}$$

where (2.95) and (2.96) have been used several times. These four differential expressions are now in the same form as (4.40), such that the derivatives of these four functions with respect to $\boldsymbol{Z}$ and $\boldsymbol{Z}^*$ can be found; they are included in Table 4.3.

Example 4.8 Let $f : \mathbb{C}^{N \times N} \to \mathbb{C}$ be given by $f(\boldsymbol{Z}) = \operatorname{Tr}\{\boldsymbol{A}\boldsymbol{Z}^{-1}\}$ where $\boldsymbol{Z} \in \mathbb{C}^{N \times N}$ is nonsingular. The differential of this function can be expressed as

$$df = \operatorname{Tr}\left\{\boldsymbol{A}d\boldsymbol{Z}^{-1}\right\} = -\operatorname{Tr}\left\{\boldsymbol{A}\boldsymbol{Z}^{-1}(d\boldsymbol{Z})\boldsymbol{Z}^{-1}\right\} = -\operatorname{Tr}\left\{\boldsymbol{Z}^{-1}\boldsymbol{A}\boldsymbol{Z}^{-1}d\boldsymbol{Z}\right\}, \tag{4.56}$$

where (3.40) was utilized. The derivative of f with respect to $\boldsymbol{Z}$ and $\boldsymbol{Z}^*$ can now be identified; these results are included in Table 4.3.

Example 4.9 Let $f : \mathbb{C}^{N \times N} \to \mathbb{C}$ be given by $f(\boldsymbol{Z}) = \operatorname{Tr}\{\boldsymbol{Z}^p\}$ where $p \in \mathbb{N}$ is a positive integer. By means of (3.33) and repeated application of (3.35), the differential of this function is then given by

$$df = \operatorname{Tr}\left\{\sum_{i=1}^{p} \boldsymbol{Z}^{i-1}(d\boldsymbol{Z})\boldsymbol{Z}^{p-i}\right\} = \sum_{i=1}^{p} \operatorname{Tr}\left\{\boldsymbol{Z}^{p-1}d\boldsymbol{Z}\right\} = p\operatorname{Tr}\left\{\boldsymbol{Z}^{p-1}d\boldsymbol{Z}\right\}. \tag{4.57}$$

From this equation, it is possible to find the derivatives of the function $\operatorname{Tr}\{\boldsymbol{Z}^p\}$ with respect to $\boldsymbol{Z}$ and $\boldsymbol{Z}^*$; the results are included in Table 4.3.

Table 4.3 Complex-valued derivatives of functions of the type $f(\boldsymbol{Z}, \boldsymbol{Z}^*)$.

$f(\boldsymbol{Z}, \boldsymbol{Z}^*)$	$\frac{\partial}{\partial \boldsymbol{Z}} f$	$\frac{\partial}{\partial \boldsymbol{Z}^*} f$
$\mathrm{Tr}\{\boldsymbol{Z}\}$	$\boldsymbol{I}_N$	$\boldsymbol{0}_{N\times N}$
$\mathrm{Tr}\{\boldsymbol{Z}^*\}$	$\boldsymbol{0}_{N\times N}$	$\boldsymbol{I}_N$
$\mathrm{Tr}\{\boldsymbol{A}\boldsymbol{Z}\}$	$\boldsymbol{A}^T$	$\boldsymbol{0}_{N\times Q}$
$\mathrm{Tr}\{\boldsymbol{Z}\boldsymbol{A}_0\boldsymbol{Z}^T\boldsymbol{A}_1\}$	$\boldsymbol{A}_1^T\boldsymbol{Z}\boldsymbol{A}_0^T + \boldsymbol{A}_1\boldsymbol{Z}\boldsymbol{A}_0$	$\boldsymbol{0}_{N\times Q}$
$\mathrm{Tr}\{\boldsymbol{Z}\boldsymbol{A}_0\boldsymbol{Z}\boldsymbol{A}_1\}$	$\boldsymbol{A}_1^T\boldsymbol{Z}^T\boldsymbol{A}_0^T + \boldsymbol{A}_0^T\boldsymbol{Z}^T\boldsymbol{A}_1^T$	$\boldsymbol{0}_{N\times Q}$
$\mathrm{Tr}\{\boldsymbol{Z}\boldsymbol{A}_0\boldsymbol{Z}^H\boldsymbol{A}_1\}$	$\boldsymbol{A}_1^T\boldsymbol{Z}^*\boldsymbol{A}_0^T$	$\boldsymbol{A}_1\boldsymbol{Z}\boldsymbol{A}_0$
$\mathrm{Tr}\{\boldsymbol{Z}\boldsymbol{A}_0\boldsymbol{Z}^*\boldsymbol{A}_1\}$	$\boldsymbol{A}_1^T\boldsymbol{Z}^H\boldsymbol{A}_0^T$	$\boldsymbol{A}_0^T\boldsymbol{Z}^T\boldsymbol{A}_1^T$
$\mathrm{Tr}\{\boldsymbol{A}\boldsymbol{Z}^{-1}\}$	$-\left(\boldsymbol{Z}^T\right)^{-1}\boldsymbol{A}^T\left(\boldsymbol{Z}^T\right)^{-1}$	$\boldsymbol{0}_{N\times N}$
$\mathrm{Tr}\{\boldsymbol{Z}^p\}$	$p\left(\boldsymbol{Z}^T\right)^{p-1}$	$\boldsymbol{0}_{N\times N}$
$\det(\boldsymbol{Z})$	$\det(\boldsymbol{Z})\left(\boldsymbol{Z}^T\right)^{-1}$	$\boldsymbol{0}_{N\times N}$
$\det(\boldsymbol{A}_0\boldsymbol{Z}\boldsymbol{A}_1)$	$\det(\boldsymbol{A}_0\boldsymbol{Z}\boldsymbol{A}_1)\boldsymbol{A}_0^T\left(\boldsymbol{A}_1^T\boldsymbol{Z}^T\boldsymbol{A}_0^T\right)^{-1}\boldsymbol{A}_1^T$	$\boldsymbol{0}_{N\times Q}$
$\det(\boldsymbol{Z}^2)$	$2\det^2(\boldsymbol{Z})\left(\boldsymbol{Z}^T\right)^{-1}$	$\boldsymbol{0}_{N\times N}$
$\det(\boldsymbol{Z}\boldsymbol{Z}^T)$	$2\det(\boldsymbol{Z}\boldsymbol{Z}^T)\left(\boldsymbol{Z}\boldsymbol{Z}^T\right)^{-1}\boldsymbol{Z}$	$\boldsymbol{0}_{N\times Q}$
$\det(\boldsymbol{Z}\boldsymbol{Z}^*)$	$\det(\boldsymbol{Z}\boldsymbol{Z}^*)(\boldsymbol{Z}^H\boldsymbol{Z}^T)^{-1}\boldsymbol{Z}^H$	$\det(\boldsymbol{Z}\boldsymbol{Z}^*)\boldsymbol{Z}^T\left(\boldsymbol{Z}^H\boldsymbol{Z}^T\right)^{-1}$
$\det(\boldsymbol{Z}\boldsymbol{Z}^H)$	$\det(\boldsymbol{Z}\boldsymbol{Z}^H)(\boldsymbol{Z}^*\boldsymbol{Z}^T)^{-1}\boldsymbol{Z}^*$	$\det(\boldsymbol{Z}\boldsymbol{Z}^H)\left(\boldsymbol{Z}\boldsymbol{Z}^H\right)^{-1}\boldsymbol{Z}$
$\det(\boldsymbol{Z}^p)$	$p\det^p(\boldsymbol{Z})\left(\boldsymbol{Z}^T\right)^{-1}$	$\boldsymbol{0}_{N\times N}$
$\lambda(\boldsymbol{Z})$	$\frac{\boldsymbol{v}_0^*\boldsymbol{u}_0^T}{\boldsymbol{v}_0^H\boldsymbol{u}_0}$	$\boldsymbol{0}_{N\times N}$
$\lambda^*(\boldsymbol{Z})$	$\boldsymbol{0}_{N\times N}$	$\frac{\boldsymbol{v}_0\boldsymbol{u}_0^H}{\boldsymbol{v}_0^T\boldsymbol{u}_0^*}$

Example 4.10 Let $f : \mathbb{C}^{N\times Q} \to \mathbb{C}$ be given by

$$f(\boldsymbol{Z}) = \det(\boldsymbol{A}_0\boldsymbol{Z}\boldsymbol{A}_1), \tag{4.58}$$

where $\boldsymbol{Z} \in \mathbb{C}^{N\times Q}$, $\boldsymbol{A}_0 \in \mathbb{C}^{M\times N}$, and $\boldsymbol{A}_1 \in \mathbb{C}^{Q\times M}$, where M is a positive integer. The matrices $\boldsymbol{A}_0$ and $\boldsymbol{A}_1$ are assumed to be independent of $\boldsymbol{Z}$. The complex differential of f can be expressed as

$$df = \det(\boldsymbol{A}_0\boldsymbol{Z}\boldsymbol{A}_1)\,\mathrm{Tr}\left\{\boldsymbol{A}_1\left(\boldsymbol{A}_0\boldsymbol{Z}\boldsymbol{A}_1\right)^{-1}\boldsymbol{A}_0 d\boldsymbol{Z}\right\}. \tag{4.59}$$

From (4.59), the derivatives of $\det(\boldsymbol{A}_0\boldsymbol{Z}\boldsymbol{A}_1)$ with respect to $\boldsymbol{Z}$ and $\boldsymbol{Z}^*$ follow; the results are included in Table 4.3.

Example 4.11 Let $f : \mathbb{C}^{N\times Q} \times \mathbb{C}^{Q\times N} \to \mathbb{C}$ be defined as

$$f(\boldsymbol{Z}_0, \boldsymbol{Z}_1) = \det(\boldsymbol{Z}_0\boldsymbol{Z}_1), \tag{4.60}$$

where it is assumed that $\boldsymbol{Z}_0\boldsymbol{Z}_1$ is nonsingular. Notice that $\boldsymbol{Z}_0 \in \mathbb{C}^{N\times Q}$ and $\boldsymbol{Z}_1 \in \mathbb{C}^{Q\times N}$, such that, in general, the matrices $\boldsymbol{Z}_0$ and $\boldsymbol{Z}_1$ have different sizes. The complex differential of this function can be calculated by means of (3.35) and (3.49) as

$$\begin{aligned} df &= \det(\boldsymbol{Z}_0\boldsymbol{Z}_1)\operatorname{Tr}\left\{(\boldsymbol{Z}_0\boldsymbol{Z}_1)^{-1}d(\boldsymbol{Z}_0\boldsymbol{Z}_1)\right\} \\ &= \det(\boldsymbol{Z}_0\boldsymbol{Z}_1)\operatorname{Tr}\left\{\boldsymbol{Z}_1(\boldsymbol{Z}_0\boldsymbol{Z}_1)^{-1}d\boldsymbol{Z}_0 + (\boldsymbol{Z}_0\boldsymbol{Z}_1)^{-1}\boldsymbol{Z}_0 d\boldsymbol{Z}_1\right\}. \end{aligned} \tag{4.61}$$

From (4.61), it is possible to find the complex differentials and derivatives of the functions $\det\left(\boldsymbol{Z}^2\right)$, $\det\left(\boldsymbol{Z}\boldsymbol{Z}^T\right)$, $\det(\boldsymbol{Z}\boldsymbol{Z}^*)$, and $\det\left(\boldsymbol{Z}\boldsymbol{Z}^H\right)$

$$d\det\left(\boldsymbol{Z}^2\right) = 2\left[\det(\boldsymbol{Z})\right]^2\operatorname{Tr}\left\{\boldsymbol{Z}^{-1}d\boldsymbol{Z}\right\}, \tag{4.62}$$

$$d\det\left(\boldsymbol{Z}\boldsymbol{Z}^T\right) = 2\det(\boldsymbol{Z}\boldsymbol{Z}^T)\operatorname{Tr}\left\{\boldsymbol{Z}^T\left(\boldsymbol{Z}\boldsymbol{Z}^T\right)^{-1}d\boldsymbol{Z}\right\}, \tag{4.63}$$

$$d\det(\boldsymbol{Z}\boldsymbol{Z}^*) = \det(\boldsymbol{Z}\boldsymbol{Z}^*)\operatorname{Tr}\left\{\boldsymbol{Z}^*(\boldsymbol{Z}\boldsymbol{Z}^*)^{-1}d\boldsymbol{Z} + (\boldsymbol{Z}\boldsymbol{Z}^*)^{-1}\boldsymbol{Z}d\boldsymbol{Z}^*\right\}, \tag{4.64}$$

$$d\det\left(\boldsymbol{Z}\boldsymbol{Z}^H\right) = \det(\boldsymbol{Z}\boldsymbol{Z}^H)\operatorname{Tr}\left\{\boldsymbol{Z}^H(\boldsymbol{Z}\boldsymbol{Z}^H)^{-1}d\boldsymbol{Z} + \boldsymbol{Z}^T\left(\boldsymbol{Z}^*\boldsymbol{Z}^T\right)^{-1}d\boldsymbol{Z}^*\right\}. \tag{4.65}$$

From these four complex differentials, the derivatives of these four determinant functions can be identified; they are included in Table 4.3, assuming that the inverse matrices involved exist.

Example 4.12 Let $f : \mathbb{C}^{N\times N} \to \mathbb{C}$ be defined as

$$f(\boldsymbol{Z}) = \det(\boldsymbol{Z}^p), \tag{4.66}$$

where $p \in \mathbb{N}$ is a positive integer and $\boldsymbol{Z} \in \mathbb{C}^{N\times N}$ is assumed to be nonsingular. From (3.49), the complex differential of f can be expressed as

$$df = d\left(\det(\boldsymbol{Z})\right)^p = p\left(\det(\boldsymbol{Z})\right)^{p-1} d\det(\boldsymbol{Z}) = p\left(\det(\boldsymbol{Z})\right)^p \operatorname{Tr}\{\boldsymbol{Z}^{-1}d\boldsymbol{Z}\}. \tag{4.67}$$

The derivatives of f with respect to $\boldsymbol{Z}$ and $\boldsymbol{Z}^*$ can be identified from (4.67) and they are included in Table 4.3. The result for $p = 1$ is also explicitly included in Table 4.3, and it can alternatively be derived from (4.59).

Example 4.13 Let $f(\boldsymbol{Z}, \boldsymbol{Z}^*) = \operatorname{Tr}\left\{\boldsymbol{F}(\boldsymbol{Z}, \boldsymbol{Z}^*)\right\}$, where $\boldsymbol{F} : \mathbb{C}^{N\times Q} \times \mathbb{C}^{N\times Q} \to \mathbb{C}^{M\times M}$. It is assumed that the complex differential of $\operatorname{vec}(\boldsymbol{F})$ can be expressed as in (3.78). Then, it follows from (2.97), (3.43), and (3.78) that the complex differential of f can be written as

$$df = \operatorname{vec}^T(\boldsymbol{I}_M)\left[(\mathcal{D}_{\boldsymbol{Z}}\boldsymbol{F})d\operatorname{vec}(\boldsymbol{Z}) + (\mathcal{D}_{\boldsymbol{Z}^*}\boldsymbol{F})d\operatorname{vec}(\boldsymbol{Z}^*)\right]. \tag{4.68}$$

From this equation, $\mathcal{D}_{\boldsymbol{Z}}f$ and $\mathcal{D}_{\boldsymbol{Z}^*}f$ follow:

$$\mathcal{D}_{\boldsymbol{Z}}f = \operatorname{vec}^T(\boldsymbol{I}_M)\mathcal{D}_{\boldsymbol{Z}}\boldsymbol{F}, \tag{4.69}$$

$$\mathcal{D}_{\boldsymbol{Z}^*}f = \operatorname{vec}^T(\boldsymbol{I}_M)\mathcal{D}_{\boldsymbol{Z}^*}\boldsymbol{F}. \tag{4.70}$$

When the derivatives of $\boldsymbol{F}$ are already known, the above expressions are useful for finding the derivatives of $f(\boldsymbol{Z}, \boldsymbol{Z}^*) = \operatorname{Tr}\left\{\boldsymbol{F}(\boldsymbol{Z}, \boldsymbol{Z}^*)\right\}$.

Example 4.14 Let λ_0 be a simple eigenvalue[1] of $\boldsymbol{Z}_0 \in \mathbb{C}^{N\times N}$, and let $\boldsymbol{u}_0 \in \mathbb{C}^{N\times 1}$ be the normalized corresponding eigenvector, such that $\boldsymbol{Z}_0\boldsymbol{u}_0 = \lambda_0\boldsymbol{u}_0$. Let $\lambda : \mathbb{C}^{N\times N} \to \mathbb{C}$ and $\boldsymbol{u} : \mathbb{C}^{N\times N} \to \mathbb{C}^{N\times 1}$ be defined such that

$$\boldsymbol{Z}\boldsymbol{u}(\boldsymbol{Z}) = \lambda(\boldsymbol{Z})\boldsymbol{u}(\boldsymbol{Z}), \tag{4.71}$$

$$\boldsymbol{u}_0^H\boldsymbol{u}(\boldsymbol{Z}) = 1, \tag{4.72}$$

$$\lambda(\boldsymbol{Z}_0) = \lambda_0, \tag{4.73}$$

$$\boldsymbol{u}(\boldsymbol{Z}_0) = \boldsymbol{u}_0. \tag{4.74}$$

Let the normalized left eigenvector of $\boldsymbol{Z}_0$ corresponding to λ_0 be denoted $\boldsymbol{v}_0 \in \mathbb{C}^{N\times 1}$ (i.e., $\boldsymbol{v}_0^H\boldsymbol{Z}_0 = \lambda_0\boldsymbol{v}_0^H$), or, equivalently $\boldsymbol{Z}_0^H\boldsymbol{v}_0 = \lambda_0^*\boldsymbol{v}_0$. To find the complex differential of $\lambda(\boldsymbol{Z})$ at $\boldsymbol{Z} = \boldsymbol{Z}_0$, take the complex differential of both sides of (4.71) evaluated at $\boldsymbol{Z} = \boldsymbol{Z}_0$

$$(d\boldsymbol{Z})\,\boldsymbol{u}_0 + \boldsymbol{Z}_0 d\boldsymbol{u} = (d\lambda)\,\boldsymbol{u}_0 + \lambda_0 d\boldsymbol{u}. \tag{4.75}$$

Premultiplying (4.75) by $\boldsymbol{v}_0^H$ gives

$$\boldsymbol{v}_0^H\,(d\boldsymbol{Z})\,\boldsymbol{u}_0 = (d\lambda)\,\boldsymbol{v}_0^H\boldsymbol{u}_0. \tag{4.76}$$

From Horn and Johnson (1985, Lemma 6.3.10), it follows that $\boldsymbol{v}_0^H\boldsymbol{u}_0 \neq 0$, and, hence,

$$d\lambda = \frac{\boldsymbol{v}_0^H\,(d\boldsymbol{Z})\,\boldsymbol{u}_0}{\boldsymbol{v}_0^H\boldsymbol{u}_0} = \operatorname{Tr}\left\{\frac{\boldsymbol{u}_0\boldsymbol{v}_0^H}{\boldsymbol{v}_0^H\boldsymbol{u}_0}d\boldsymbol{Z}\right\}. \tag{4.77}$$

This result is included in Table 4.3, and it will be used later when the derivatives of the eigenvector $\boldsymbol{u}$ and the Hessian of λ are found. The complex differential of λ^* at $\boldsymbol{Z}_0$ can now also be found by complex conjugating (4.77)

$$d\lambda^* = \frac{\boldsymbol{v}_0^T\,(d\boldsymbol{Z}^*)\,\boldsymbol{u}_0^*}{\boldsymbol{v}_0^T\boldsymbol{u}_0^*} = \operatorname{Tr}\left\{\frac{\boldsymbol{u}_0^*\boldsymbol{v}_0^T}{\boldsymbol{v}_0^T\boldsymbol{u}_0^*}d\boldsymbol{Z}^*\right\}. \tag{4.78}$$

These results are derived in Magnus and Neudecker (1988, Section 8.9). The derivatives of $\lambda(\boldsymbol{Z})$ and $\lambda^*(\boldsymbol{Z})$ at $\boldsymbol{Z}_0$ with respect to $\boldsymbol{Z}$ and $\boldsymbol{Z}^*$ can be found from the complex differentials in (4.77) and (4.78); these are included in Table 4.3.

---

[1] The matrix $\boldsymbol{Z}_0 \in \mathbb{C}^{N\times N}$ has in general N different complex eigenvalues. The roots of the characteristic equation (i.e., the eigenvalues), need not be distinct. The number of times an eigenvalue appears is equal to its algebraic multiplicity. If one eigenvalue appears only once, it is called a *simple eigenvalue* (Horn & Johnson 1985).

4.3 Complex-Valued Derivatives of Vector Functions

4.3.1 Complex-Valued Derivatives of $\boldsymbol{f}(z, z^*)$

Examples of functions of the type $\boldsymbol{f}(z, z^*)$ are $\boldsymbol{a}z$, $\boldsymbol{a}zz^*$, and $\boldsymbol{a}f(z, z^*)$, where $\boldsymbol{a} \in \mathbb{C}^{M\times 1}$ and $z \in \mathbb{C}$. These functions can be differentiated by finding the complex differentials of the scalar functions z, zz^*, and $f(z, z^*)$, respectively.

Example 4.15 Let $\boldsymbol{f}(z, z^*) = \boldsymbol{a}f(z, z^*)$, then the complex differential of this function is given by

$$d\boldsymbol{f} = \boldsymbol{a}df = \boldsymbol{a}\left(\mathcal{D}_z f(z, z^*)\right) dz + \boldsymbol{a}\left(\mathcal{D}_{z^*} f(z, z^*)\right) dz^*, \tag{4.79}$$

where df was found from Table 3.2. From (4.79), it follows that $\mathcal{D}_z \boldsymbol{f} = \boldsymbol{a}\mathcal{D}_z f(z, z^*)$ and $\mathcal{D}_{z^*} \boldsymbol{f} = \boldsymbol{a}\mathcal{D}_{z^*} f(z, z^*)$. The derivatives of the vector functions $\boldsymbol{a}z$ and $\boldsymbol{a}zz^*$ follow from these results.

4.3.2 Complex-Valued Derivatives of $\boldsymbol{f}(\boldsymbol{z}, \boldsymbol{z}^*)$

Examples of functions of the type $\boldsymbol{f}(\boldsymbol{z}, \boldsymbol{z}^*)$ are $\boldsymbol{A}\boldsymbol{z}$, $\boldsymbol{A}\boldsymbol{z}^*$, and $\boldsymbol{F}(\boldsymbol{z}, \boldsymbol{z}^*)\boldsymbol{a}$, where $\boldsymbol{z} \in \mathbb{C}^{N\times 1}$, $\boldsymbol{A} \in \mathbb{C}^{M\times N}$, $\boldsymbol{F} \in \mathbb{C}^{M\times P}$, and $\boldsymbol{a} \in \mathbb{C}^{P\times 1}$.

Example 4.16 Let $\boldsymbol{f} : \mathbb{C}^{N\times 1} \times \mathbb{C}^{N\times 1} \to \mathbb{C}^{M\times 1}$ be given by $\boldsymbol{f}(\boldsymbol{z}, \boldsymbol{z}^*) = \boldsymbol{F}(\boldsymbol{z}, \boldsymbol{z}^*)\boldsymbol{a}$, where $\boldsymbol{F} : \mathbb{C}^{N\times 1} \times \mathbb{C}^{N\times 1} \to \mathbb{C}^{M\times P}$. The complex differential of $\boldsymbol{f}$ is computed as

$$\begin{aligned} d\boldsymbol{f} &= d\,\mathrm{vec}(\boldsymbol{f}) = d\,\mathrm{vec}(\boldsymbol{F}(\boldsymbol{z}, \boldsymbol{z}^*)\boldsymbol{a}) = \left(\boldsymbol{a}^T \otimes \boldsymbol{I}_M\right) d\,\mathrm{vec}(\boldsymbol{F}) \\ &= \left(\boldsymbol{a}^T \otimes \boldsymbol{I}_M\right) \left[\left(\mathcal{D}_{\boldsymbol{z}} \boldsymbol{F}(\boldsymbol{z}, \boldsymbol{z}^*)\right) d\boldsymbol{z} + \left(\mathcal{D}_{\boldsymbol{z}^*} \boldsymbol{F}(\boldsymbol{z}, \boldsymbol{z}^*)\right) d\boldsymbol{z}^*\right], \end{aligned} \tag{4.80}$$

where (2.105) and Table 3.2 were used. From (4.80), the derivatives of $\boldsymbol{f}$ with respect to $\boldsymbol{z}$ and $\boldsymbol{z}^*$ follow:

$$\mathcal{D}_{\boldsymbol{z}} \boldsymbol{f} = \left(\boldsymbol{a}^T \otimes \boldsymbol{I}_M\right) \mathcal{D}_{\boldsymbol{z}} \boldsymbol{F}(\boldsymbol{z}, \boldsymbol{z}^*), \tag{4.81}$$

$$\mathcal{D}_{\boldsymbol{z}^*} \boldsymbol{f} = \left(\boldsymbol{a}^T \otimes \boldsymbol{I}_M\right) \mathcal{D}_{\boldsymbol{z}^*} \boldsymbol{F}(\boldsymbol{z}, \boldsymbol{z}^*). \tag{4.82}$$

4.3.3 Complex-Valued Derivatives of $\boldsymbol{f}(\boldsymbol{Z}, \boldsymbol{Z}^*)$

Examples of functions of the type $\boldsymbol{f}(\boldsymbol{Z}, \boldsymbol{Z}^*)$ are $\boldsymbol{Z}\boldsymbol{a}$, $\boldsymbol{Z}^T\boldsymbol{a}$, $\boldsymbol{Z}^*\boldsymbol{a}$, $\boldsymbol{Z}^H\boldsymbol{a}$, $\boldsymbol{F}(\boldsymbol{Z}, \boldsymbol{Z}^*)\boldsymbol{a}$, $\boldsymbol{u}(\boldsymbol{Z})$ (eigenvector), $\boldsymbol{u}^*(\boldsymbol{Z})$ (eigenvector), and $\boldsymbol{v}^H(\boldsymbol{Z})$ (left eigenvector), where the sizes of $\boldsymbol{a}$, $\boldsymbol{Z}$, and $\boldsymbol{F}$ are chosen such that the functions are well defined. The complex differentials of $\boldsymbol{Z}\boldsymbol{a}$, $\boldsymbol{Z}^T\boldsymbol{a}$, $\boldsymbol{Z}^*\boldsymbol{a}$, and $\boldsymbol{Z}^H\boldsymbol{a}$ follow from the complex differential of $\boldsymbol{F}(\boldsymbol{Z}, \boldsymbol{Z}^*)\boldsymbol{a}$, and the complex differential of $\boldsymbol{F}(\boldsymbol{Z}, \boldsymbol{Z}^*)\boldsymbol{a}$ can be found in an analogous manner as in (4.80).

Example 4.17 The complex differential of the eigenvector $\boldsymbol{u}(\boldsymbol{Z})$ is now found at $\boldsymbol{Z} = \boldsymbol{Z}_0$. The derivation here is similar to the one in Magnus and Neudecker (1988, Section 8.9), where the same result for $d\boldsymbol{u}$ at $\boldsymbol{Z} = \boldsymbol{Z}_0$ was derived; however, additional details are included here. See the discussion around (4.71) to (4.74) for an introduction to the eigenvalue and eigenvector notation. Let $\boldsymbol{Y}_0 = \lambda_0 \boldsymbol{I}_N - \boldsymbol{Z}_0$, then it follows from (4.75) that

$$\begin{aligned} \boldsymbol{Y}_0 d\boldsymbol{u} &= (d\boldsymbol{Z})\,\boldsymbol{u}_0 - (d\lambda)\,\boldsymbol{u}_0 = (d\boldsymbol{Z})\,\boldsymbol{u}_0 - \frac{\boldsymbol{v}_0^H (d\boldsymbol{Z})\boldsymbol{u}_0}{\boldsymbol{v}_0^H \boldsymbol{u}_0}\boldsymbol{u}_0 \\ &= \left(\boldsymbol{I}_N - \frac{\boldsymbol{u}_0 \boldsymbol{v}_0^H}{\boldsymbol{v}_0^H \boldsymbol{u}_0}\right)(d\boldsymbol{Z})\,\boldsymbol{u}_0, \end{aligned} \tag{4.83}$$

where (4.77) was utilized. Premultiplying (4.83) with $\boldsymbol{Y}_0^+$ (where $(\cdot)^+$ is the Moore-Penrose inverse from Definition 2.4) results in

$$\boldsymbol{Y}_0^+ \boldsymbol{Y}_0 d\boldsymbol{u} = \boldsymbol{Y}_0^+ \left(\boldsymbol{I}_N - \frac{\boldsymbol{u}_0 \boldsymbol{v}_0^H}{\boldsymbol{v}_0^H \boldsymbol{u}_0}\right)(d\boldsymbol{Z})\,\boldsymbol{u}_0. \tag{4.84}$$

Because λ_0 is a simple eigenvalue, $\dim_{\mathbb{C}}(\mathcal{N}(\boldsymbol{Y}_0)) = 1$ (Horn & Johnson 1985), where $\mathcal{N}(\cdot)$ denotes the null space (see Section 2.4). Hence, it follows from (2.55) that $\text{rank}(\boldsymbol{Y}_0) = N - \dim_{\mathbb{C}}(\mathcal{N}(\boldsymbol{Y}_0)) = N - 1$. From $\boldsymbol{Y}_0 \boldsymbol{u}_0 = \boldsymbol{0}_{N\times 1}$, it follows from (2.82) that $\boldsymbol{u}_0^+ \boldsymbol{Y}_0^+ = \boldsymbol{0}_{1\times N}$. It can be shown by direct insertion in Definition 2.4 of the Moore-Penrose inverse that the inverse of the normalized eigenvector $\boldsymbol{u}_0$ is given by $\boldsymbol{u}_0^+ = \boldsymbol{u}_0^H$ (see Exercise 2.4 for the Moore-Penrose inverse of an arbitrary complex-valued vector). From these results, it follows that $\boldsymbol{u}_0^H \boldsymbol{Y}_0^+ = \boldsymbol{0}_{1\times N}$. Set $\boldsymbol{C}_0 = \boldsymbol{Y}_0^+ \boldsymbol{Y}_0 + \boldsymbol{u}_0 \boldsymbol{u}_0^H$, then it can be shown from the two facts $\boldsymbol{u}_0^H \boldsymbol{Y}_0^+ = \boldsymbol{0}_{1\times N}$ and $\boldsymbol{Y}_0 \boldsymbol{u}_0 = \boldsymbol{0}_{N\times 1}$ that $\boldsymbol{C}_0^2 = \boldsymbol{C}_0$ (i.e., $\boldsymbol{C}_0$ is idempotent). It can be shown by the direct use of Definition 2.4 that the matrix $\boldsymbol{Y}_0^+ \boldsymbol{Y}_0$ is also idempotent. With the use of Proposition 2.1, it is found that

$$\begin{aligned} \text{rank}(\boldsymbol{C}_0) &= \text{Tr}\{\boldsymbol{C}_0\} = \text{Tr}\left\{\boldsymbol{Y}_0^+ \boldsymbol{Y}_0 + \boldsymbol{u}_0 \boldsymbol{u}_0^H\right\} = \text{Tr}\left\{\boldsymbol{Y}_0^+ \boldsymbol{Y}_0\right\} + \text{Tr}\left\{\boldsymbol{u}_0 \boldsymbol{u}_0^H\right\} \\ &= \text{rank}\left(\boldsymbol{Y}_0^+ \boldsymbol{Y}_0\right) + 1 = \text{rank}(\boldsymbol{Y}_0) + 1 = N - 1 + 1 = N, \end{aligned} \tag{4.85}$$

where (2.90) was used. From Proposition 2.1 and (4.85), it follows that $\boldsymbol{C}_0 = \boldsymbol{I}_N$. Using the complex differential operator on both sides of the normalization in (4.72) yields $\boldsymbol{u}_0^H d\boldsymbol{u} = 0$. Using these results, it follows that

$$\boldsymbol{Y}_0^+ \boldsymbol{Y}_0 d\boldsymbol{u} = \left(\boldsymbol{I}_N - \boldsymbol{u}_0 \boldsymbol{u}_0^H\right) d\boldsymbol{u} = d\boldsymbol{u} - \boldsymbol{u}_0 \boldsymbol{u}_0^H d\boldsymbol{u} = d\boldsymbol{u}. \tag{4.86}$$

Equations (4.84) and (4.86) lead to

$$d\boldsymbol{u} = (\lambda_0 \boldsymbol{I}_N - \boldsymbol{Z}_0)^+ \left(\boldsymbol{I}_N - \frac{\boldsymbol{u}_0 \boldsymbol{v}_0^H}{\boldsymbol{v}_0^H \boldsymbol{u}_0}\right)(d\boldsymbol{Z})\,\boldsymbol{u}_0. \tag{4.87}$$

From (4.87), it is possible to find the derivative of the eigenvector function $\boldsymbol{u}(\boldsymbol{Z})$ evaluated at $\boldsymbol{Z}_0$ with respect to the matrix $\boldsymbol{Z}$ in the following way:

$$d\boldsymbol{u} = \operatorname{vec}(d\boldsymbol{u}) = \operatorname{vec}\left((\lambda_0 \boldsymbol{I}_N - \boldsymbol{Z}_0)^+ \left(\boldsymbol{I}_N - \frac{\boldsymbol{u}_0 \boldsymbol{v}_0^H}{\boldsymbol{v}_0^H \boldsymbol{u}_0}\right)(d\boldsymbol{Z})\, \boldsymbol{u}_0\right)$$

$$= \left(\boldsymbol{u}_0^T \otimes \left[(\lambda_0 \boldsymbol{I}_N - \boldsymbol{Z}_0)^+ \left(\boldsymbol{I}_N - \frac{\boldsymbol{u}_0 \boldsymbol{v}_0^H}{\boldsymbol{v}_0^H \boldsymbol{u}_0}\right)\right]\right) d \operatorname{vec}(\boldsymbol{Z}), \tag{4.88}$$

where (2.105) was used. From (4.88), it follows that

$$\mathcal{D}_{\boldsymbol{Z}} \boldsymbol{u} = \boldsymbol{u}_0^T \otimes \left[(\lambda_0 \boldsymbol{I}_N - \boldsymbol{Z}_0)^+ \left(\boldsymbol{I}_N - \frac{\boldsymbol{u}_0 \boldsymbol{v}_0^H}{\boldsymbol{v}_0^H \boldsymbol{u}_0}\right)\right]. \tag{4.89}$$

The complex differential and the derivative of $\boldsymbol{u}^*$ follow with the use of (3.45), (4.88), and (4.89).

Example 4.18 The left eigenvector function $\boldsymbol{v} : \mathbb{C}^{N \times N} \to \mathbb{C}^{N \times 1}$ with the argument $\boldsymbol{Z} \in \mathbb{C}^{N \times N}$, denoted $\boldsymbol{v}(\boldsymbol{Z})$, is defined through the following four relations:

$$\boldsymbol{v}^H(\boldsymbol{Z})\boldsymbol{Z} = \lambda(\boldsymbol{Z})\boldsymbol{v}^H, \tag{4.90}$$

$$\boldsymbol{v}_0^H \boldsymbol{v}(\boldsymbol{Z}) = 1, \tag{4.91}$$

$$\lambda(\boldsymbol{Z}_0) = \lambda_0, \tag{4.92}$$

$$\boldsymbol{v}(\boldsymbol{Z}_0) = \boldsymbol{v}_0. \tag{4.93}$$

The complex differential of $\boldsymbol{v}(\boldsymbol{Z})$ at $\boldsymbol{Z} = \boldsymbol{Z}_0$ can be found, using a procedure similar to the one used in Example 4.17 for finding $d\boldsymbol{u}$ at $\boldsymbol{Z} = \boldsymbol{Z}_0$, leading to

$$d\boldsymbol{v}^H = \boldsymbol{v}_0^H (d\boldsymbol{Z}) \left(\boldsymbol{I}_N - \frac{\boldsymbol{u}_0 \boldsymbol{v}_0^H}{\boldsymbol{v}_0^H \boldsymbol{u}_0}\right)(\lambda_0 \boldsymbol{I}_N - \boldsymbol{Z}_0)^+. \tag{4.94}$$

In general, it is hard to work with derivatives of eigenvalues and eigenvectors because the derivatives depend on the algebraic multiplicity of the corresponding eigenvalue. For this reason, it is better to try to rewrite the objective function such that the eigenvalues and eigenvectors do not appear explicitly. Two such cases are given in communication problems in Hjørungnes and Gesbert (2007*c* and 2007*d*), and the latter is explained in detail in Section 7.4.

4.4 Complex-Valued Derivatives of Matrix Functions

4.4.1 Complex-Valued Derivatives of $\boldsymbol{F}(z, z^*)$

Examples of functions of the type $\boldsymbol{F}(z, z^*)$ are $\boldsymbol{A}z$, $\boldsymbol{A}zz^*$, and $\boldsymbol{A}f(z, z^*)$, where $\boldsymbol{A} \in \mathbb{C}^{M \times P}$ is independent of z and z^*. These functions can be differentiated by finding the complex differentials of the scalar functions z, zz^*, and $f(z, z^*)$.

Example 4.19 Let $\boldsymbol{F} : \mathbb{C} \times \mathbb{C} \to \mathbb{C}^{M \times P}$ be given by

$$\boldsymbol{F}(z, z^*) = \boldsymbol{A} f(z, z^*), \tag{4.95}$$

where $f : \mathbb{C} \times \mathbb{C} \to \mathbb{C}$ has derivatives that can be identified from

$$df = (\mathcal{D}_z f)\, dz + (\mathcal{D}_{z^*} f)\, dz^*, \tag{4.96}$$

and where $\boldsymbol{A} \in \mathbb{C}^{M \times P}$ is independent of z and z^*. The complex differential of vec($\boldsymbol{F}$) can be expressed as

$$d\,\text{vec}(\boldsymbol{F}) = \text{vec}(\boldsymbol{A})\, df = \text{vec}(\boldsymbol{A})\,(\mathcal{D}_z f)\, dz + \text{vec}(\boldsymbol{A})\,(\mathcal{D}_{z^*} f)\, dz^*. \tag{4.97}$$

Now, the derivatives of $\boldsymbol{F}$ with respect to z and z^* can be identified as

$$\mathcal{D}_z \boldsymbol{F} = \text{vec}(\boldsymbol{A})\, \mathcal{D}_z f, \tag{4.98}$$

$$\mathcal{D}_{z^*} \boldsymbol{F} = \text{vec}(\boldsymbol{A})\, \mathcal{D}_{z^*} f. \tag{4.99}$$

In more complicated examples than those shown above, the complex differential of vec($\boldsymbol{F}$) should be reformulated directly, possibly component-wise, to be put into a form such that the derivatives can be identified (i.e., following the general procedure outlined in Section 3.3).

4.4.2 Complex-Valued Derivatives of $\boldsymbol{F}(\boldsymbol{z}, \boldsymbol{z}^*)$

Examples of functions of the type $\boldsymbol{F}(\boldsymbol{z}, \boldsymbol{z}^*)$ are $\boldsymbol{z}\boldsymbol{z}^T$ and $\boldsymbol{z}\boldsymbol{z}^H$, where $\boldsymbol{z} \in \mathbb{C}^{N \times 1}$.

Example 4.20 Let $\boldsymbol{F} : \mathbb{C}^{N \times 1} \times \mathbb{C}^{N \times 1} \to \mathbb{C}^{N \times N}$ be given by $\boldsymbol{F}(\boldsymbol{z}, \boldsymbol{z}^*) = \boldsymbol{z}\boldsymbol{z}^H$. The complex differential of the $\boldsymbol{F}$ can be expressed as

$$d\boldsymbol{F} = (d\boldsymbol{z})\boldsymbol{z}^H + \boldsymbol{z}\, d\boldsymbol{z}^H. \tag{4.100}$$

And from this equation, it follows that

$$\begin{aligned} d\,\text{vec}(\boldsymbol{F}) &= \left[\boldsymbol{z}^* \otimes \boldsymbol{I}_N\right] d\,\text{vec}(\boldsymbol{z}) + \left[\boldsymbol{I}_N \otimes \boldsymbol{z}\right] d\,\text{vec}(\boldsymbol{z}^H) \\ &= \left[\boldsymbol{z}^* \otimes \boldsymbol{I}_N\right] d\boldsymbol{z} + \left[\boldsymbol{I}_N \otimes \boldsymbol{z}\right] d\boldsymbol{z}^*. \end{aligned} \tag{4.101}$$

Hence, the derivatives of $\boldsymbol{F}(\boldsymbol{z}, \boldsymbol{z}^*) = \boldsymbol{z}\boldsymbol{z}^H$ with respect to $\boldsymbol{z}$ and $\boldsymbol{z}^*$ are given by

$$\mathcal{D}_{\boldsymbol{z}} \boldsymbol{F} = \boldsymbol{z}^* \otimes \boldsymbol{I}_N, \tag{4.102}$$

$$\mathcal{D}_{\boldsymbol{z}^*} \boldsymbol{F} = \boldsymbol{I}_N \otimes \boldsymbol{z}. \tag{4.103}$$

Table 4.4 Complex-valued derivatives of functions of the type $F(Z, Z^*)$.

$F(Z, Z^*)$	$\mathcal{D}_Z F(Z, Z^*)$	$\mathcal{D}_{Z^*} F(Z, Z^*)$
Z	I_{NQ}	$0_{NQ\times NQ}$
Z^T	$K_{N,Q}$	$0_{NQ\times NQ}$
Z^*	$0_{NQ\times NQ}$	I_{NQ}
Z^H	$0_{NQ\times NQ}$	$K_{N,Q}$
ZZ^T	$(I_{N^2} + K_{N,N})(Z \otimes I_N)$	$0_{N^2\times NQ}$
$Z^T Z$	$\left(I_{Q^2} + K_{Q,Q}\right)\left(I_Q \otimes Z^T\right)$	$0_{Q^2\times NQ}$
ZZ^H	$Z^* \otimes I_N$	$K_{N,N}(Z \otimes I_N)$
Z^{-1}	$-(Z^T)^{-1} \otimes Z^{-1}$	$0_{N^2\times N^2}$
Z^p	$\sum_{i=1}^{p} \left((Z^T)^{p-i} \otimes Z^{i-1}\right)$	$0_{N^2\times N^2}$
$Z \otimes Z$	$A(Z) + B(Z)$	$0_{N^2Q^2\times NQ}$
$Z \otimes Z^*$	$A(Z^*)$	$B(Z)$
$Z^* \otimes Z^*$	$0_{N^2Q^2\times NQ}$	$A(Z^*) + B(Z^*)$
$Z \odot Z$	$2\,\mathrm{diag}(\mathrm{vec}(Z))$	$0_{NQ\times NQ}$
$Z \odot Z^*$	$\mathrm{diag}(\mathrm{vec}(Z^*))$	$\mathrm{diag}(\mathrm{vec}(Z))$
$Z^* \odot Z^*$	$0_{NQ\times NQ}$	$2\,\mathrm{diag}(\mathrm{vec}(Z^*))$
$\exp(Z)$	$\sum_{k=0}^{\infty} \frac{1}{(k+1)!} \sum_{i=0}^{k} \left(Z^T\right)^{k-i} \otimes Z^i$	$0_{N^2\times N^2}$
$\exp(Z^*)$	$0_{N^2\times N^2}$	$\sum_{k=0}^{\infty} \frac{1}{(k+1)!} \sum_{i=0}^{k} \left(\left(Z^H\right)^{k-i} \otimes (Z^*)^i\right)$
$\exp\left(Z^H\right)$	$0_{N^2\times N^2}$	$\sum_{k=0}^{\infty} \frac{1}{(k+1)!} \sum_{i=0}^{k} \left((Z^*)^{k-i} \otimes (Z^H)^i\right) K_{N,N}$

4.4.3 Complex-Valued Derivatives of $F(Z, Z^*)$

Examples of functions of the form $F(Z, Z^*)$ are $Z, Z^T, Z^*, Z^H, ZZ^T, Z^T Z, ZZ^H, Z^{-1}, Z^+, Z^\#, Z^p, Z \otimes Z, Z \otimes Z^*, Z^* \otimes Z^*, Z \odot Z, Z \odot Z^*, Z^* \odot Z^*, \exp(Z), \exp(Z^*)$, and $\exp(Z^H)$, where $Z \in \mathbb{C}^{N\times Q}$ or possibly $Z \in \mathbb{C}^{N\times N}$, if this is required for the function to be defined.

Example 4.21 If $F(Z) = Z \in \mathbb{C}^{N\times Q}$, then

$$d\,\mathrm{vec}(F) = d\,\mathrm{vec}(Z) = I_{NQ} d\,\mathrm{vec}(Z). \tag{4.104}$$

From this expression of the complex differentials of vec(F), the derivatives of $F(Z) = Z$ with respect to Z and Z^* can be identified from the last line of Table 3.2. These derivatives are included in Table 4.4.

Example 4.22 Let $\boldsymbol{F} : \mathbb{C}^{N\times Q} \times \mathbb{C}^{Q\times M} \to \mathbb{C}^{N\times M}$, where

$$\boldsymbol{F}(\boldsymbol{Z}_0, \boldsymbol{Z}_1) = \boldsymbol{Z}_0\boldsymbol{Z}_1, \tag{4.105}$$

for $\boldsymbol{Z}_0 \in \mathbb{C}^{N\times Q}$ and $\boldsymbol{Z}_1 \in \mathbb{C}^{Q\times M}$, such that the sizes of $\boldsymbol{Z}_0$ and $\boldsymbol{Z}_1$ are different in general. The operator vec(·) applied to the complex differential of $\boldsymbol{F}$, see (3.35), yields

$$\begin{aligned} d\,\mathrm{vec}(\boldsymbol{F}) &= \mathrm{vec}\left((d\boldsymbol{Z}_0)\boldsymbol{Z}_1\right) + \mathrm{vec}\left(\boldsymbol{Z}_0(d\boldsymbol{Z}_1)\right) \\ &= \left(\boldsymbol{Z}_1^T \otimes \boldsymbol{I}_N\right) d\,\mathrm{vec}(\boldsymbol{Z}_0) + \left(\boldsymbol{I}_M \otimes \boldsymbol{Z}_0\right) d\,\mathrm{vec}(\boldsymbol{Z}_1). \end{aligned} \tag{4.106}$$

From this result, the complex differentials of $\boldsymbol{Z}\boldsymbol{Z}^T$, $\boldsymbol{Z}^T\boldsymbol{Z}$, and $\boldsymbol{Z}\boldsymbol{Z}^H$ can be derived and are given by

$$d\boldsymbol{Z}\boldsymbol{Z}^T = \left(\boldsymbol{I}_{N^2} + \boldsymbol{K}_{N,N}\right)\left(\boldsymbol{Z} \otimes \boldsymbol{I}_N\right) d\,\mathrm{vec}(\boldsymbol{Z}), \tag{4.107}$$

$$d\boldsymbol{Z}^T\boldsymbol{Z} = \left(\boldsymbol{I}_{Q^2} + \boldsymbol{K}_{Q,Q}\right)\left(\boldsymbol{I}_Q \otimes \boldsymbol{Z}^T\right) d\,\mathrm{vec}(\boldsymbol{Z}), \tag{4.108}$$

$$d\boldsymbol{Z}\boldsymbol{Z}^H = \left(\boldsymbol{Z}^* \otimes \boldsymbol{I}_N\right) d\,\mathrm{vec}(\boldsymbol{Z}) + \boldsymbol{K}_{N,N}\left(\boldsymbol{Z} \otimes \boldsymbol{I}_N\right) d\,\mathrm{vec}(\boldsymbol{Z}^*), \tag{4.109}$$

where $\boldsymbol{K}_{Q,N}$ is given in Definition 2.9. The derivatives of these three functions can now be identified; they are included in Table 4.4.

Example 4.23 Let $\boldsymbol{Z} \in \mathbb{C}^{N\times N}$ be a nonsingular matrix, and $\boldsymbol{F} : \mathbb{C}^{N\times N} \to \mathbb{C}^{N\times N}$ be given by

$$\boldsymbol{F}(\boldsymbol{Z}) = \boldsymbol{Z}^{-1}. \tag{4.110}$$

By using (2.105) and (3.40), it follows that

$$d\,\mathrm{vec}(\boldsymbol{F}) = -\left(\left(\boldsymbol{Z}^T\right)^{-1} \otimes \boldsymbol{Z}^{-1}\right) d\,\mathrm{vec}(\boldsymbol{Z}). \tag{4.111}$$

From this result, the derivatives of $\boldsymbol{F}$ with respect to $\boldsymbol{Z}$ and $\boldsymbol{Z}^*$ are identified and included in Table 4.4.

Example 4.24 Let $\boldsymbol{F} : \mathbb{C}^{N\times N} \to \mathbb{C}^{N\times N}$, where

$$\boldsymbol{F}(\boldsymbol{Z}) = \boldsymbol{Z}^p, \tag{4.112}$$

for $\boldsymbol{Z} \in \mathbb{C}^{N\times N}$ and where $p \in \mathbb{N}$ is a positive integer. Hence, the function $\boldsymbol{F}(\boldsymbol{Z})$ in (4.112) represents *matrix power*. By repeated application of (3.35), it can be shown that

$$d\boldsymbol{F} = \sum_{i=1}^{p} \boldsymbol{Z}^{i-1}(d\boldsymbol{Z})\boldsymbol{Z}^{p-i}, \tag{4.113}$$

from which it follows that

$$d\,\mathrm{vec}(\boldsymbol{F}) = \sum_{i=1}^{p} \left(\left(\boldsymbol{Z}^T\right)^{p-i} \otimes \boldsymbol{Z}^{i-1} \right) d\,\mathrm{vec}(\boldsymbol{Z}). \tag{4.114}$$

Now the derivatives of $\boldsymbol{F}$ with respect to $\boldsymbol{Z}$ and $\boldsymbol{Z}^*$ can be found; they are included in Table 4.4.

Example 4.25 Let $\boldsymbol{F} : \mathbb{C}^{N_0 \times Q_0} \times \mathbb{C}^{N_1 \times Q_1} \to \mathbb{C}^{N_0 N_1 \times Q_0 Q_1}$ be given by

$$\boldsymbol{F}(\boldsymbol{Z}_0, \boldsymbol{Z}_1) = \boldsymbol{Z}_0 \otimes \boldsymbol{Z}_1, \tag{4.115}$$

where $\boldsymbol{Z}_i \in \mathbb{C}^{N_i \times Q_i}$ where $i \in \{0, 1\}$. The complex differential of this function follows from (3.36):

$$d\boldsymbol{F} = (d\boldsymbol{Z}_0) \otimes \boldsymbol{Z}_1 + \boldsymbol{Z}_0 \otimes d\boldsymbol{Z}_1. \tag{4.116}$$

Applying the vec(·) operator to (4.116) yields

$$d\,\mathrm{vec}(\boldsymbol{F}) = \mathrm{vec}\left((d\boldsymbol{Z}_0) \otimes \boldsymbol{Z}_1\right) + \mathrm{vec}\left(\boldsymbol{Z}_0 \otimes d\boldsymbol{Z}_1\right). \tag{4.117}$$

From (2.103) and (2.112), it follows that

$$\begin{aligned}
\mathrm{vec}\left((d\boldsymbol{Z}_0) \otimes \boldsymbol{Z}_1\right) &= \left(\boldsymbol{I}_{Q_0} \otimes \boldsymbol{K}_{Q_1,N_0} \otimes \boldsymbol{I}_{N_1}\right) \left[(d\,\mathrm{vec}(\boldsymbol{Z}_0)) \otimes \mathrm{vec}(\boldsymbol{Z}_1)\right] \\
&= \left(\boldsymbol{I}_{Q_0} \otimes \boldsymbol{K}_{Q_1,N_0} \otimes \boldsymbol{I}_{N_1}\right) \left[\left(\boldsymbol{I}_{N_0 Q_0} d\,\mathrm{vec}(\boldsymbol{Z}_0)\right) \otimes (\mathrm{vec}(\boldsymbol{Z}_1) 1)\right] \\
&= \left(\boldsymbol{I}_{Q_0} \otimes \boldsymbol{K}_{Q_1,N_0} \otimes \boldsymbol{I}_{N_1}\right) \left[\left(\boldsymbol{I}_{N_0 Q_0} \otimes \mathrm{vec}(\boldsymbol{Z}_1)\right) (d\,\mathrm{vec}(\boldsymbol{Z}_0) \otimes 1)\right] \\
&= \left(\boldsymbol{I}_{Q_0} \otimes \boldsymbol{K}_{Q_1,N_0} \otimes \boldsymbol{I}_{N_1}\right) \left[\boldsymbol{I}_{N_0 Q_0} \otimes \mathrm{vec}(\boldsymbol{Z}_1)\right] d\,\mathrm{vec}(\boldsymbol{Z}_0),
\end{aligned} \tag{4.118}$$

and, in a similar way, it follows that

$$\begin{aligned}
\mathrm{vec}\left(\boldsymbol{Z}_0 \otimes d\boldsymbol{Z}_1\right) &= \left(\boldsymbol{I}_{Q_0} \otimes \boldsymbol{K}_{Q_1,N_0} \otimes \boldsymbol{I}_{N_1}\right) \left[\mathrm{vec}(\boldsymbol{Z}_0) \otimes d\,\mathrm{vec}(\boldsymbol{Z}_1)\right] \\
&= \left(\boldsymbol{I}_{Q_0} \otimes \boldsymbol{K}_{Q_1,N_0} \otimes \boldsymbol{I}_{N_1}\right) \left[\mathrm{vec}(\boldsymbol{Z}_0) \otimes \boldsymbol{I}_{N_1 Q_1}\right] d\,\mathrm{vec}(\boldsymbol{Z}_1).
\end{aligned} \tag{4.119}$$

Inserting the results from (4.118) and (4.119) into (4.117) gives

$$\begin{aligned}
d\,\mathrm{vec}(\boldsymbol{F}) &= \left(\boldsymbol{I}_{Q_0} \otimes \boldsymbol{K}_{Q_1,N_0} \otimes \boldsymbol{I}_{N_1}\right) \left[\boldsymbol{I}_{N_0 Q_0} \otimes \mathrm{vec}(\boldsymbol{Z}_1)\right] d\,\mathrm{vec}(\boldsymbol{Z}_0) \\
&\quad + \left(\boldsymbol{I}_{Q_0} \otimes \boldsymbol{K}_{Q_1,N_0} \otimes \boldsymbol{I}_{N_1}\right) \left[\mathrm{vec}(\boldsymbol{Z}_0) \otimes \boldsymbol{I}_{N_1 Q_1}\right] d\,\mathrm{vec}(\boldsymbol{Z}_1).
\end{aligned} \tag{4.120}$$

Define the matrices $\boldsymbol{A}(\boldsymbol{Z}_1)$ and $\boldsymbol{B}(\boldsymbol{Z}_0)$ by

$$\boldsymbol{A}(\boldsymbol{Z}_1) \triangleq \left(\boldsymbol{I}_{Q_0} \otimes \boldsymbol{K}_{Q_1,N_0} \otimes \boldsymbol{I}_{N_1}\right) \left[\boldsymbol{I}_{N_0 Q_0} \otimes \mathrm{vec}(\boldsymbol{Z}_1)\right], \tag{4.121}$$

$$\boldsymbol{B}(\boldsymbol{Z}_0) \triangleq \left(\boldsymbol{I}_{Q_0} \otimes \boldsymbol{K}_{Q_1,N_0} \otimes \boldsymbol{I}_{N_1}\right) \left[\mathrm{vec}(\boldsymbol{Z}_0) \otimes \boldsymbol{I}_{N_1 Q_1}\right]. \tag{4.122}$$

By means of the matrices $\boldsymbol{A}(\boldsymbol{Z}_1)$ and $\boldsymbol{B}(\boldsymbol{Z}_0)$, it is then possible to rewrite the complex differential of the Kronecker product $\boldsymbol{F}(\boldsymbol{Z}_0, \boldsymbol{Z}_1) = \boldsymbol{Z}_0 \otimes \boldsymbol{Z}_1$ as

$$d\,\mathrm{vec}(\boldsymbol{F}) = \boldsymbol{A}(\boldsymbol{Z}_1) d\,\mathrm{vec}(\boldsymbol{Z}_0) + \boldsymbol{B}(\boldsymbol{Z}_0) d\,\mathrm{vec}(\boldsymbol{Z}_1). \tag{4.123}$$

From (4.123), the complex differentials of $Z \otimes Z$, $Z \otimes Z^*$, and $Z^* \otimes Z^*$ can be expressed as

$$d Z \otimes Z = (A(Z) + B(Z)) d \operatorname{vec}(Z), \tag{4.124}$$

$$d Z \otimes Z^* = A(Z^*) d \operatorname{vec}(Z) + B(Z) d \operatorname{vec}(Z^*), \tag{4.125}$$

$$d Z^* \otimes Z^* = (A(Z^*) + B(Z^*)) d \operatorname{vec}(Z^*). \tag{4.126}$$

Now, the derivatives of these three functions with respect to Z and Z^* can be identified from the last three equations above, and these derivatives are included in Table 4.4.

Example 4.26 Let $F : \mathbb{C}^{N \times Q} \times \mathbb{C}^{N \times Q} \to \mathbb{C}^{N \times Q}$ be given by

$$F(Z_0, Z_1) = Z_0 \odot Z_1. \tag{4.127}$$

The complex differential of this function follows from (3.38) and is given by

$$dF = (dZ_0) \odot Z_1 + Z_0 \odot dZ_1 = Z_1 \odot dZ_0 + Z_0 \odot dZ_1. \tag{4.128}$$

Applying the $\operatorname{vec}(\cdot)$ operator to (4.128) and using (2.115) results in

$$d \operatorname{vec}(F) = \operatorname{diag}(\operatorname{vec}(Z_1)) \, d \operatorname{vec}(Z_0) + \operatorname{diag}(\operatorname{vec}(Z_0)) \, d \operatorname{vec}(Z_1). \tag{4.129}$$

The complex differentials of $Z \odot Z$, $Z \odot Z^*$, and $Z^* \odot Z^*$ can be derived from (4.129):

$$d Z \odot Z = 2 \operatorname{diag}(\operatorname{vec}(Z)) d \operatorname{vec}(Z), \tag{4.130}$$

$$d Z \odot Z^* = \operatorname{diag}(\operatorname{vec}(Z^*)) d \operatorname{vec}(Z) + \operatorname{diag}(\operatorname{vec}(Z)) d \operatorname{vec}(Z^*), \tag{4.131}$$

$$d Z^* \odot Z^* = 2 \operatorname{diag}(\operatorname{vec}(Z^*)) d \operatorname{vec}(Z^*). \tag{4.132}$$

The derivatives of these three functions with respect to Z and Z^* can now be obtained and are included in Table 4.4.

Example 4.27 The complex differential of the exponential matrix function (see Definition 2.5) can be expressed as

$$d \exp(Z) = \sum_{k=1}^{\infty} \frac{1}{k!} d Z^k = \sum_{k=0}^{\infty} \frac{1}{(k+1)!} d Z^{k+1} = \sum_{k=0}^{\infty} \frac{1}{(k+1)!} \sum_{i=0}^{k} Z^i (dZ) Z^{k-i}, \tag{4.133}$$

where the complex differential rules in (3.25) and (3.35) have been used. Applying $\operatorname{vec}(\cdot)$ on (4.133) yields

$$d \operatorname{vec}(\exp(Z)) = \sum_{k=0}^{\infty} \frac{1}{(k+1)!} \sum_{i=0}^{k} \left(\left(Z^T\right)^{k-i} \otimes Z^i \right) d \operatorname{vec}(Z). \tag{4.134}$$

In a similar way, the complex differentials and derivatives of the functions $\exp(\boldsymbol{Z}^*)$ and $\exp(\boldsymbol{Z}^H)$ can be found to be

$$d\,\text{vec}(\exp(\boldsymbol{Z}^*)) = \sum_{k=0}^{\infty} \frac{1}{(k+1)!} \sum_{i=0}^{k} \left(\left(\boldsymbol{Z}^H\right)^{k-i} \otimes (\boldsymbol{Z}^*)^i \right) d\,\text{vec}(\boldsymbol{Z}^*), \tag{4.135}$$

$$d\,\text{vec}(\exp(\boldsymbol{Z}^H)) = \sum_{k=0}^{\infty} \frac{1}{(k+1)!} \sum_{i=0}^{k} \left((\boldsymbol{Z}^*)^{k-i} \otimes (\boldsymbol{Z}^H)^i \right) \boldsymbol{K}_{N,N} d\,\text{vec}(\boldsymbol{Z}^*). \tag{4.136}$$

The derivatives of $\exp(\boldsymbol{Z})$, $\exp(\boldsymbol{Z}^*)$, and $\exp(\boldsymbol{Z}^H)$ with respect to $\boldsymbol{Z}$ and $\boldsymbol{Z}^*$ can now be derived; they are included in Table 4.4.

Example 4.28 Let $\boldsymbol{F} : \mathbb{C}^{N\times Q} \times \mathbb{C}^{N\times Q} \to \mathbb{C}^{Q\times N}$ be given by

$$\boldsymbol{F}(\boldsymbol{Z}, \boldsymbol{Z}^*) = \boldsymbol{Z}^+, \tag{4.137}$$

where $\boldsymbol{Z} \in \mathbb{C}^{N\times Q}$. The reason for including both variables $\boldsymbol{Z}$ and $\boldsymbol{Z}^*$ in this function definition is that the complex differential of $\boldsymbol{Z}^+$ (see (3.64)) depends on both $d\boldsymbol{Z}$ and $d\boldsymbol{Z}^*$. Using the vec(·) and the differential operator d on (3.64) and utilizing (2.105) and (2.31) results in

$$\begin{aligned} d\,\text{vec}(\boldsymbol{F}) = &- \left[(\boldsymbol{Z}^+)^T \otimes \boldsymbol{Z}^+ \right] d\,\text{vec}(\boldsymbol{Z}) \\ &+ \left[\left(\boldsymbol{I}_N - (\boldsymbol{Z}^+)^T \boldsymbol{Z}^T \right) \otimes \boldsymbol{Z}^+ (\boldsymbol{Z}^+)^H \right] \boldsymbol{K}_{N,Q} d\,\text{vec}(\boldsymbol{Z}^*) \\ &+ \left[(\boldsymbol{Z}^+)^T (\boldsymbol{Z}^+)^* \otimes (\boldsymbol{I}_Q - \boldsymbol{Z}^+\boldsymbol{Z}) \right] \boldsymbol{K}_{N,Q} d\,\text{vec}(\boldsymbol{Z}^*). \end{aligned} \tag{4.138}$$

From (4.138), the derivatives $\mathcal{D}_{\boldsymbol{Z}}\boldsymbol{F}$ and $\mathcal{D}_{\boldsymbol{Z}^*}\boldsymbol{F}$ can be expressed as

$$\mathcal{D}_{\boldsymbol{Z}}\boldsymbol{F} = -(\boldsymbol{Z}^+)^T \otimes \boldsymbol{Z}^+, \tag{4.139}$$

$$\begin{aligned} \mathcal{D}_{\boldsymbol{Z}^*}\boldsymbol{F} = &\left\{ \left[\left(\boldsymbol{I}_N - (\boldsymbol{Z}^+)^T \boldsymbol{Z}^T \right) \otimes \boldsymbol{Z}^+ (\boldsymbol{Z}^+)^H \right] \right. \\ &\left. + \left[(\boldsymbol{Z}^+)^T (\boldsymbol{Z}^+)^* \otimes (\boldsymbol{I}_Q - \boldsymbol{Z}^+\boldsymbol{Z}) \right] \right\} \boldsymbol{K}_{N,Q}. \end{aligned} \tag{4.140}$$

If the matrix $\boldsymbol{Z} \in \mathbb{C}^{N\times N}$ is invertible, then $\boldsymbol{Z}^+ = \boldsymbol{Z}^{-1}$, and (4.139) and (4.140) reduce to $\mathcal{D}_{\boldsymbol{Z}}\boldsymbol{F} = -\boldsymbol{Z}^{-T} \otimes \boldsymbol{Z}^{-1}$ and $\mathcal{D}_{\boldsymbol{Z}^*}\boldsymbol{F} = \boldsymbol{0}_{N^2\times N^2}$, which is in agreement with the results found in Example 4.23 and Table 4.4.

Example 4.29 Let $\boldsymbol{F} : \mathbb{C}^{N\times N} \to \mathbb{C}^{N\times N}$ be given by $\boldsymbol{F}(\boldsymbol{Z}) = \boldsymbol{Z}^\#$ (i.e., the function $\boldsymbol{F}$ represents the adjoint matrix of the input variable $\boldsymbol{Z}$). The complex differential of this function is given in (3.58). Using the vec(·) operator on (3.58) leads to

$$d\,\text{vec}(\boldsymbol{F}) = \det(\boldsymbol{Z}) \left[\text{vec}(\boldsymbol{Z}^{-1})\, \text{vec}^T\left((\boldsymbol{Z}^{-1})^T \right) - \left[(\boldsymbol{Z}^{-1})^T \otimes \boldsymbol{Z}^{-1} \right] \right] d\,\text{vec}(\boldsymbol{Z}). \tag{4.141}$$

From this, it follows that

$$\mathcal{D}_Z F = \det(Z)\left[\operatorname{vec}(Z^{-1})\operatorname{vec}^T\left((Z^{-1})^T\right) - \left[(Z^{-1})^T \otimes Z^{-1}\right]\right]. \tag{4.142}$$

Because the expressions associated with the complex differential of the Moore-Penrose inverse and the adjoint matrices are so long, they are not included in Table 4.4.

4.5 Exercises

4.1 Use the following identity $|z|^2 = zz^*$ and the chain rule to find $\frac{\partial |z|}{\partial z}$ and $\frac{\partial |z|}{\partial z^*}$. Make sure that this alternative derivation leads to the same result as given in (4.13) and (4.14).

4.2 Show that

$$\frac{\partial |z^*|}{\partial z} = \frac{z^*}{2|z|}, \tag{4.143}$$

and

$$\frac{\partial |z^*|}{\partial z^*} = \frac{z}{2|z|}. \tag{4.144}$$

4.3 For *real-valued* scalar variables, we know that $\frac{d|x|^2}{dx} = \frac{dx^2}{dx}$, where $x \in \mathbb{R}$. Show that for the complex-valued case (i.e., $z \in \mathbb{C}$), then

$$\frac{\partial z^2}{\partial z} \neq \frac{\partial |z|^2}{\partial z}, \tag{4.145}$$

in general.

4.4 Find $\frac{\partial \angle z}{\partial z}$ by differentiating (4.20) with respect to z^*.

4.5 Show that

$$\frac{\partial \angle z^*}{\partial z^*} = -\frac{\jmath}{2z^*}, \tag{4.146}$$

and

$$\frac{\partial \angle z^*}{\partial z} = \frac{\jmath}{2z}, \tag{4.147}$$

by means of the results already derived in this chapter.

4.6 Let $\boldsymbol{A}^H = \boldsymbol{A} \in \mathbb{C}^{N \times N}$ and $\boldsymbol{B}^H = \boldsymbol{B} \in \mathbb{C}^{N \times N}$ be given constant matrices where $\boldsymbol{B}$ is positive or negative definite such that $\boldsymbol{z}^H \boldsymbol{B} \boldsymbol{z} \neq 0, \forall \boldsymbol{z} \neq \boldsymbol{0}_{N \times 1}$. Let $f : \mathbb{C}^{N \times 1} \times \mathbb{C}^{N \times 1} \rightarrow \mathbb{R}$ be given by

$$f(\boldsymbol{z}, \boldsymbol{z}^*) = \frac{\boldsymbol{z}^H \boldsymbol{A} \boldsymbol{z}}{\boldsymbol{z}^H \boldsymbol{B} \boldsymbol{z}}, \tag{4.148}$$

where f is not defined for $z = \mathbf{0}_{N\times 1}$. The expression in (4.148) is called the generalized Rayleigh quotient. Show that the df is given by

$$df = \left[\frac{z^H A}{z^H B z} - \frac{z^H A z}{\left(z^H B z\right)^2} z^H B\right] dz + \left[\frac{z^T A^T}{z^H B z} - \frac{z^H A z}{\left(z^H B z\right)^2} z^T B^T\right] dz^*. \quad (4.149)$$

From this complex differential, the derivatives of f with respect to z and z^* are identified as

$$\mathcal{D}_z f = \frac{z^H A}{z^H B z} - \frac{z^H A z}{\left(z^H B z\right)^2} z^H B, \quad (4.150)$$

$$\mathcal{D}_{z^*} f = \frac{z^T A^T}{z^H B z} - \frac{z^H A z}{\left(z^H B z\right)^2} z^T B^T. \quad (4.151)$$

By studying the equation $\mathcal{D}_{z^*} f = \mathbf{0}_{1\times N}$, show that the maximum and minimum values of f are given by the maximum and minimum eigenvalues of the generalized eigenvalue problem $Az = \lambda Bz$. See Therrien (1992, Section 2.6) for an introduction to the generalized eigenvalue problem $Az = \lambda Bz$, where λ are roots of the equation $\det(A - \lambda B) = 0$.

Assume that B is positive definite. Then B has a unique positive definite square root (Horn & Johnson 1991, p. 448). Let this square root be denoted $B^{1/2}$. Explain why

$$\lambda_{\min}(B^{-1/2} A B^{-1/2}) \leq f(z, z^*) \leq \lambda_{\max}(B^{-1/2} A B^{-1/2}), \quad (4.152)$$

where $\lambda_{\min}(\cdot)$ and $\lambda_{\max}(\cdot)$ denote the minimum and maximum eigenvalues of the matrix input argument.[2]

4.7 Show that the derivatives with respect to Z and Z^* of the function $f(Z, Z^*) = \ln(\det(Z))$, when $Z \in \mathbb{C}^{N\times N}$ is nonsingular, are given by

$$\mathcal{D}_Z f = \text{vec}^T\left(Z^{-T}\right), \quad (4.153)$$

and

$$\mathcal{D}_{Z^*} f = \mathbf{0}_{1\times N^2}. \quad (4.154)$$

4.8 Assume that $Z_0^H = Z_0$. Let λ_0 be a *simple* real eigenvalue of $Z_0 \in \mathbb{C}^{N\times N}$, and let $u_0 \in \mathbb{C}^{N\times 1}$ be the normalized corresponding eigenvector, such that $Z_0 u_0 = \lambda_0 u_0$. Let $\lambda : \mathbb{C}^{N\times N} \to \mathbb{C}$ and $u : \mathbb{C}^{N\times N} \to \mathbb{C}^{N\times 1}$ be defined such that

$$Z u(Z) = \lambda(Z) u(Z), \quad (4.155)$$

$$u_0^H u(Z) = 1, \quad (4.156)$$

$$\lambda(Z_0) = \lambda_0, \quad (4.157)$$

$$u(Z_0) = u_0. \quad (4.158)$$

[2] The eigenvalues of $B^{-1}A$ are equal to the eigenvalues of $B^{-1/2}AB^{-1/2}$ because the matrix products CD and DC have equal eigenvalues, when $C, D \in \mathbb{C}^{N\times N}$ and C is invertible. The reason for this can be seen from $\det(\lambda I_N - CD) = \det(\lambda CC^{-1} - CD) = \det(C(\lambda C^{-1} - D)) = \det((\lambda C^{-1} - D)C) = \det(\lambda I_N - DC)$.

Show that the complex differentials $d\lambda$ and $d\boldsymbol{u}$, at $\boldsymbol{Z}_0$ are given by

$$d\lambda = \boldsymbol{u}_0^H (d\boldsymbol{Z})\, \boldsymbol{u}_0, \tag{4.159}$$

$$d\boldsymbol{u} = (\lambda_0 \boldsymbol{I}_N - \boldsymbol{Z}_0)^+ (d\boldsymbol{Z})\, \boldsymbol{u}_0. \tag{4.160}$$

4.9 Let $\boldsymbol{Z} \in \mathbb{C}^{N\times N}$ have all eigenvalues with absolute value less than one. Show that

$$(\boldsymbol{I}_N - \boldsymbol{Z})^{-1} = \sum_{k=0}^{\infty} \boldsymbol{Z}^k, \tag{4.161}$$

(see Magnus & Neudecker 1988, p. 169). Furthermore, show that the derivative of $(\boldsymbol{I}_N - \boldsymbol{Z})^{-1}$ with respect to $\boldsymbol{Z}$ and $\boldsymbol{Z}^*$ can be expressed as

$$\mathcal{D}_{\boldsymbol{Z}} (\boldsymbol{I}_N - \boldsymbol{Z})^{-1} = \sum_{k=1}^{\infty} \sum_{l=1}^{k} \left(\boldsymbol{Z}^{k-l}\right)^T \otimes \boldsymbol{Z}^{l-1}, \tag{4.162}$$

$$\mathcal{D}_{\boldsymbol{Z}^*} (\boldsymbol{I}_N - \boldsymbol{Z})^{-1} = \boldsymbol{0}_{N^2\times N^2}. \tag{4.163}$$

4.10 The natural logarithm of a square complex-valued matrix $\boldsymbol{Z} \in \mathbb{C}^{N\times N}$ can be expressed as follows (Horn & Johnson 1991, p. 492):

$$\ln (\boldsymbol{I}_N - \boldsymbol{Z}) \triangleq -\sum_{k=1}^{\infty} \frac{1}{k} \boldsymbol{Z}^k, \tag{4.164}$$

and it is defined for all matrices $\boldsymbol{Z} \in \mathbb{C}^{N\times N}$ such that the absolute value of all eigenvalues is smaller than one. Show that the complex differential of $\ln (\boldsymbol{I}_N - \boldsymbol{Z})$ can be expressed as

$$d \ln (\boldsymbol{I}_N - \boldsymbol{Z}) = -\sum_{k=1}^{\infty} \frac{1}{k} \sum_{l=1}^{k} \boldsymbol{Z}^{l-1} (d\boldsymbol{Z})\, \boldsymbol{Z}^{k-l}. \tag{4.165}$$

Use the expression for $d \ln (\boldsymbol{I}_N - \boldsymbol{Z})$ to show that the derivatives of $\ln (\boldsymbol{I}_N - \boldsymbol{Z})$ with respect to $\boldsymbol{Z}$ and $\boldsymbol{Z}^*$ are given by

$$\mathcal{D}_{\boldsymbol{Z}} \ln (\boldsymbol{I}_N - \boldsymbol{Z}) = -\sum_{k=1}^{\infty} \frac{1}{k} \sum_{l=1}^{k} \left(\boldsymbol{Z}^{k-l}\right)^T \otimes \boldsymbol{Z}^{l-1}, \tag{4.166}$$

$$\mathcal{D}_{\boldsymbol{Z}^*} \ln (\boldsymbol{I}_N - \boldsymbol{Z}) = \boldsymbol{0}_{N^2\times N^2}, \tag{4.167}$$

respectively. Use (4.165) to show that

$$d \operatorname{Tr} \{\ln (\boldsymbol{I}_N - \boldsymbol{Z})\} = -\operatorname{Tr} \left\{(\boldsymbol{I}_N - \boldsymbol{Z})^{-1} d\boldsymbol{Z}\right\}. \tag{4.168}$$

4.11 Let $f : \mathbb{C}^{N\times Q} \times \mathbb{C}^{N\times Q} \to \mathbb{C}$ be given by

$$f(\boldsymbol{Z}, \boldsymbol{Z}^*) = \ln \left(\det \left(\boldsymbol{R}_n + \boldsymbol{Z}\boldsymbol{A}\boldsymbol{R}_x\boldsymbol{A}^H\boldsymbol{Z}^H\right)\right), \tag{4.169}$$

where the three matrices $\boldsymbol{R}_n \in \mathbb{C}^{N\times N}$ (positive semidefinite), $\boldsymbol{R}_x \in \mathbb{C}^{P\times P}$ (positive semidefinite), and $\boldsymbol{A} \in \mathbb{C}^{Q\times P}$ are independent of $\boldsymbol{Z}$ and $\boldsymbol{Z}^*$. Show that the derivatives of

f with respect to $\boldsymbol{Z}$ and $\boldsymbol{Z}^*$ are given by

$$\mathcal{D}_{\boldsymbol{Z}} f = \text{vec}^T \left(\boldsymbol{R}_n^{-T} \boldsymbol{Z}^* \boldsymbol{A}^* \left[(\boldsymbol{R}_x^*)^{-1} + \boldsymbol{A}^T \boldsymbol{Z}^T \boldsymbol{R}_n^{-T} \boldsymbol{Z}^* \boldsymbol{A}^* \right]^{-1} \boldsymbol{A}^T \right), \tag{4.170}$$

$$\mathcal{D}_{\boldsymbol{Z}^*} f = \text{vec}^T \left(\boldsymbol{R}_n^{-1} \boldsymbol{Z} \boldsymbol{A} \left[\boldsymbol{R}_x^{-1} + \boldsymbol{A}^H \boldsymbol{Z}^H \boldsymbol{R}_n^{-1} \boldsymbol{Z} \boldsymbol{A} \right]^{-1} \boldsymbol{A}^H \right). \tag{4.171}$$

Explain why (4.171) is in agreement with Palomar and Verdú (2006, Eq. (21)).

4.12 Let $f : \mathbb{C}^{N \times Q} \times \mathbb{C}^{N \times Q} \to \mathbb{C}$ be given by

$$f(\boldsymbol{Z}, \boldsymbol{Z}^*) = \ln \left(\det \left(\boldsymbol{R}_n + \boldsymbol{A} \boldsymbol{Z} \boldsymbol{R}_x \boldsymbol{Z}^H \boldsymbol{A}^H \right) \right), \tag{4.172}$$

where the three matrices $\boldsymbol{R}_n \in \mathbb{C}^{P \times P}$ (positive semidefinite), $\boldsymbol{R}_x \in \mathbb{C}^{Q \times Q}$ (positive semidefinite), and $\boldsymbol{A} \in \mathbb{C}^{P \times N}$ are independent of $\boldsymbol{Z}$ and $\boldsymbol{Z}^*$. Show that the derivatives of f with respect to $\boldsymbol{Z}$ and $\boldsymbol{Z}^*$ can be expressed as

$$\mathcal{D}_{\boldsymbol{Z}} f = \text{vec}^T \left(\boldsymbol{A}^T \boldsymbol{R}_n^{-T} \boldsymbol{A}^* \boldsymbol{Z}^* \left((\boldsymbol{R}_x^*)^{-1} + \boldsymbol{Z}^T \boldsymbol{A}^T \boldsymbol{R}_n^{-T} \boldsymbol{A}^* \boldsymbol{Z}^* \right)^{-1} \right), \tag{4.173}$$

$$\mathcal{D}_{\boldsymbol{Z}^*} f = \text{vec}^T \left(\boldsymbol{A}^H \boldsymbol{R}_n^{-1} \boldsymbol{A} \boldsymbol{Z} (\boldsymbol{R}_x^{-1} + \boldsymbol{Z}^H \boldsymbol{A}^H \boldsymbol{R}_n^{-1} \boldsymbol{A} \boldsymbol{Z})^{-1} \right). \tag{4.174}$$

Explain why (4.174) is in agreement with Palomar and Verdú (2006, Eq. (22)).

5 Complex Hessian Matrices for Scalar, Vector, and Matrix Functions

5.1 Introduction

This chapter provides the tools for finding Hessians (i.e., second-order derivatives) in a systematic way when the input variables are complex-valued matrices. The proposed theory is useful when solving numerous problems that involve optimization when the unknown parameter is a complex-valued matrix. In an effort to build adaptive optimization algorithms, it is important to find out if a certain value of the complex-valued parameter matrix at a stationary point[1] is a maximum, minimum, or saddle point; the Hessian can then be utilized very efficiently. The complex Hessian might also be used to accelerate the convergence of iterative optimization algorithms, to study the stability of iterative algorithms, and to study convexity and concavity of an objective function. The methods presented in this chapter are general, such that many results can be derived using the introduced framework. Complex Hessians are derived for some useful examples taken from signal processing and communications.

The problem of finding Hessians has been treated for *real-valued* matrix variables in Magnus and Neudecker (1988, Chapter 10). For complex-valued *vector* variables, the Hessian matrix is treated for scalar functions in Brookes (July 2009) and Kreutz-Delgado (2009, June 25th). Both gradients and Hessians for scalar functions that depend on complex-valued *vectors* are studied in van den Bos (1994*a*). The Hessian of real-valued functions depending on real-valued matrix variables is used in Payaró and Palomar (2009) to enhance the connection between information theory and estimation theory. A complex version of Newton's recursion formula is derived in Abatzoglou, Mendel, and Harada (1991) and Yan and Fan (2000), and there the topic of Hessian matrices is briefly treated for real scalar functions, which depend on complex-valued *vectors*. A theory for finding *complex-valued Hessian matrices* is presented in this chapter for the three cases of complex-valued scalar, vector, and matrix functions when the input variables are complex-valued matrices.

The Hessian matrix of a function is a matrix that contains the *second-order derivatives* of the function. In this chapter, the Hessian matrix will be defined; it will be also shown how it can be obtained for the three cases of complex-valued scalar, vector, and matrix

[1] Recall that a *stationary point* is a point where the derivative of the function is equal to the null vector, such that a stationary point is among the points that satisfy the necessary conditions for optimality (see Theorem 3.2).

functions. Only the case where the function f is a complex scalar function was treated in Hjørungnes and Gesbert (2007*b*). However, these results are extended to complex-valued vector and matrix functions as well in this chapter, and these results are novel. The way the Hessian is defined in this chapter is a generalization of the *real-valued case* given in Magnus and Neudecker (1988). The main contribution of this chapter lies in the proposed approach on how to obtain Hessians in a way that is both simple and systematic, based on the so-called second-order complex differential of the scalar, vector, or matrix function.

In this chapter, it is assumed that the functions are *twice differentiable* with respect to the complex-valued parameter matrix and its complex conjugate. Section 3.2 presented theory showing that these two parameter matrices have linearly independent differentials, which will also be used in this chapter when finding the Hessians through second-order complex differentials.

The rest of this chapter is organized as follows: Section 5.2 presents two alternative ways for representing the complex-valued matrix variable $\boldsymbol{Z}$ and its complex conjugate $\boldsymbol{Z}^*$. In Subsection 5.2.1, the first way of representing the complex-valued matrix variables is similar to that in previous chapters, where the two matrices $\boldsymbol{Z} \in \mathbb{C}^{N\times Q}$ and $\boldsymbol{Z}^* \in \mathbb{C}^{N\times Q}$ are used explicitly. These two matrix variables should be treated as independent when finding complex matrix derivatives. In addition, an augmented alternative representation $\mathcal{Z} \triangleq [\boldsymbol{Z}\ \boldsymbol{Z}^*] \in \mathbb{C}^{N\times 2Q}$ is presented in Subsection 5.2.2. The augmented matrix variable $\mathcal{Z}$ contains only independent differentials (see Subsection 3.2.3). The augmented representation simplifies the presentation on how to obtain complex Hessians of scalar, vector, and matrix functions. In Section 5.3, it is shown how the Hessian (second-order derivative) of a *scalar function* f can be found. Two alternative ways of finding the complex Hessian of scalar function are presented. The first way is shown in Subsection 5.3.1, where the Hessian is identified from the second-order differential when $\boldsymbol{Z}$ and $\boldsymbol{Z}^*$ are used as matrix variables. An alternative way of finding the Hessians of complex-valued scalar functions is presented in Subsection 5.3.2, based on the augmented matrix variable $\mathcal{Z}$. The way to find the Hessian for complex-valued *vector functions* is given in Section 5.4, and the case of complex-valued *matrix* functions is presented in Section 5.5. Several examples of how the complex Hessian might be calculated are presented in Section 5.6 for the three cases of scalar, vector, and matrix functions. Exercises are given in Section 5.7.

5.2 Alternative Representations of Complex-Valued Matrix Variables

5.2.1 Complex-Valued Matrix Variables $\boldsymbol{Z}$ and $\boldsymbol{Z}^*$

As in previous chapters, one way of representing complex-valued input matrix variables is by the use of two matrices $\boldsymbol{Z} \in \mathbb{C}^{N\times Q}$ and $\boldsymbol{Z}^* \in \mathbb{C}^{N\times Q}$. In this chapter, it is assumed that all the elements within $\boldsymbol{Z}$ are independent. It follows from Lemma 3.1 that the elements within $d\boldsymbol{Z}$ and $d\boldsymbol{Z}^*$ are linearly independent. Lemmas 3.1 and 3.2 are presented in Subsection 3.2.3, and they will be used in this chapter to identify the complex Hessians.

Neither of the matrices $d\boldsymbol{Z}$ nor $d\boldsymbol{Z}^*$ is a function of $\boldsymbol{Z}$ or $\boldsymbol{Z}^*$ and, hence, their differentials are the zero matrix. Mathematically, this can be formulated as

$$d^2\boldsymbol{Z} = d\left(d\boldsymbol{Z}\right) = \boldsymbol{0}_{N\times Q} = d\left(d\boldsymbol{Z}^*\right) = d^2\boldsymbol{Z}^*. \tag{5.1}$$

The representation of the input matrix variables as $\boldsymbol{Z}$ and $\boldsymbol{Z}^*$ will be used to develop a theory for finding Hessians of complex-valued *scalar* functions in Subsection 5.3.1. In the next subsection, an alternative representation of the complex-valued matrix variables will be presented. It will be used to simplify the process of finding complex Hessians of scalar, vector, and matrix functions in Subsection 5.3.1, Sections 5.4, and 5.5, respectively.

5.2.2 Augmented Complex-Valued Matrix Variables $\mathcal{Z}$

To simplify the presentation for the Hessians, an alternative representation of the complex-valued matrix variables will be defined in this subsection.

From Lemma 3.1, it is seen that *all* the components of the two matrices $d\boldsymbol{Z}$ and $d\boldsymbol{Z}^*$ are linearly independent. This motivates the definition of the *augmented complex-valued matrix variable* $\mathcal{Z}$ of size $N \times 2Q$, defined as follows:

$$\mathcal{Z} \triangleq \left[\boldsymbol{Z}, \boldsymbol{Z}^*\right] \in \mathbb{C}^{N\times 2Q}. \tag{5.2}$$

The differentials of all the components of $\mathcal{Z}$ are linearly independent (see Lemma 3.1). Hence, the matrix $\mathcal{Z}$ can be treated as a matrix that contains only independent elements when finding complex-valued matrix derivatives. This augmented matrix will be used in this chapter to develop a theory for complex-valued functions of scalars, vectors, and matrices in similar lines, as was done for the *real-valued* case in Magnus and Neudecker (1988, Chapter 10). The main reason for introducing the augmented matrix variable is to make the presentation of the complex Hessian matrices more compact and easier to follow. When dealing with the complex matrix variables $\boldsymbol{Z}$ and $\boldsymbol{Z}^*$ explicitly, four Hessian matrices have to be found instead of *one*, which is the case when the augmented matrix variable $\mathcal{Z}$ is used.

The differential of the vectorization operator of the augmented matrix variable $\mathcal{Z}$ will be used throughout this chapter and it is given by

$$d\,\text{vec}\left(\mathcal{Z}\right) = \begin{bmatrix} d\,\text{vec}\left(\boldsymbol{Z}\right) \\ d\,\text{vec}\left(\boldsymbol{Z}^*\right) \end{bmatrix}. \tag{5.3}$$

The complex-valued matrix variables $\mathcal{Z}$ and $\mathcal{Z}^*$ contain the *same matrix components*; however, the matrix elements are rearranged inside the two matrix variables. Both of the matrix variables $\mathcal{Z}$ and $\mathcal{Z}^*$ are used in the development of complex Hessians. The differential of the vectorization operator of the symbol $\mathcal{Z}^*$ is given by

$$d\,\text{vec}\left(\mathcal{Z}^*\right) = \begin{bmatrix} d\,\text{vec}\left(\boldsymbol{Z}^*\right) \\ d\,\text{vec}\left(\boldsymbol{Z}\right) \end{bmatrix}. \tag{5.4}$$

Table 5.1 Classification of scalar, vector, and matrix functions, which depend on the augmented matrix variable $\mathcal{Z} \in \mathbb{C}^{N\times 2Q}$.

Function type	$\mathcal{Z} \in \mathbb{C}^{N\times 2Q}$
Scalar function $f \in \mathbb{C}$	$f(\mathcal{Z})$ $f : \mathbb{C}^{N\times 2Q} \rightarrow \mathbb{C}$
Vector function $\boldsymbol{f} \in \mathbb{C}^{M\times 1}$	$\boldsymbol{f}(\mathcal{Z})$ $\boldsymbol{f} : \mathbb{C}^{N\times 2Q} \rightarrow \mathbb{C}^{M\times 1}$
Matrix function $\boldsymbol{F} \in \mathbb{C}^{M\times P}$	$\boldsymbol{F}(\mathcal{Z})$ $\boldsymbol{F} : \mathbb{C}^{N\times 2Q} \rightarrow \mathbb{C}^{M\times P}$

From (5.3) and (5.4), it is seen that the vectors $d\,\text{vec}(\mathcal{Z})$ and $d\,\text{vec}(\mathcal{Z}^*)$ are connected through the following relation:

$$d\,\text{vec}(\mathcal{Z}^*) = \begin{bmatrix} \boldsymbol{0}_{NQ\times NQ} & \boldsymbol{I}_{NQ} \\ \boldsymbol{I}_{NQ} & \boldsymbol{0}_{NQ\times NQ} \end{bmatrix} d\,\text{vec}(\mathcal{Z}) = \left\{ \begin{bmatrix} 0 & 1 \\ 1 & 0 \end{bmatrix} \otimes \boldsymbol{I}_{NQ} \right\} d\,\text{vec}(\mathcal{Z}). \tag{5.5}$$

This is equivalent to the following expression:

$$d\,\text{vec}^T(\mathcal{Z}) = \left(d\,\text{vec}^H(\mathcal{Z})\right) \begin{bmatrix} \boldsymbol{0}_{NQ\times NQ} & \boldsymbol{I}_{NQ} \\ \boldsymbol{I}_{NQ} & \boldsymbol{0}_{NQ\times NQ} \end{bmatrix} = \left(d\,\text{vec}^H(\mathcal{Z})\right) \left\{ \begin{bmatrix} 0 & 1 \\ 1 & 0 \end{bmatrix} \otimes \boldsymbol{I}_{NQ} \right\}, \tag{5.6}$$

which will be used later in this chapter.

The second-order differential is given by the differential of the differential of the augmented matrix variable; it is given by

$$d^2\mathcal{Z} = d(d\mathcal{Z}) = \left[d(d\boldsymbol{Z}) \;\; d(d\boldsymbol{Z}^*)\right] = \left[\boldsymbol{0}_{N\times Q} \;\; \boldsymbol{0}_{N\times Q}\right] = \boldsymbol{0}_{N\times 2Q}. \tag{5.7}$$

In a similar manner, the second-order differential of the variable $\mathcal{Z}^*$ is also the zero matrix

$$d^2\mathcal{Z}^* = d(d\mathcal{Z}^*) = \boldsymbol{0}_{N\times 2Q}. \tag{5.8}$$

Three types of functions will be studied; in this chapter, these depend on the augmented matrix variables. The three functions are scalar $f : \mathbb{C}^{N\times 2Q} \rightarrow \mathbb{C}$, vector $\boldsymbol{f} : \mathbb{C}^{N\times 2Q} \rightarrow \mathbb{C}^{M\times 1}$, and matrix $\boldsymbol{F} : \mathbb{C}^{N\times 2Q} \rightarrow \mathbb{C}^{M\times P}$. Because both matrix variables $\boldsymbol{Z}$ and $\boldsymbol{Z}^*$ are contained within the augmented matrix variable $\mathcal{Z}$, only the augmented matrix variable $\mathcal{Z}$ is used in the function definitions in Table 5.1. The complex conjugate of the augmented matrix variable $\mathcal{Z}^*$ is *not* needed in this case because it is redundant. Each of these function types is presented in Table 5.1. The theory for finding Hessians of scalar functions of the type $f(\mathcal{Z})$ is presented in Subsection 5.3.2. The way to find Hessians of vector functions $\boldsymbol{f}(\mathcal{Z})$ is presented in Section 5.4. For matrix functions $\boldsymbol{F}(\mathcal{Z})$, the theory for identifying the Hessians is presented in Section 5.5.

In the next section, scalar functions of the type $f : \mathbb{C}^{N\times Q} \times \mathbb{C}^{N\times Q} \rightarrow \mathbb{C}$ denoted by $f(\boldsymbol{Z}, \boldsymbol{Z}^*)$ are studied: hence, the traditional input matrix variables $\boldsymbol{Z} \in \mathbb{C}^{N\times Q}$ and $\boldsymbol{Z}^* \in \mathbb{C}^{N\times Q}$ are used.

5.3 Complex Hessian Matrices of Scalar Functions

This section contains the following three subsections. In Subsection 5.3.1, the complex Hessian matrix of a scalar function $f(\boldsymbol{Z}, \boldsymbol{Z}^*)$ is found when $\boldsymbol{Z}$ and $\boldsymbol{Z}^*$ are the matrix variables. Complex Hessian matrices of scalar functions $f(\boldsymbol{\mathcal{Z}})$ are studied for the case where the augmented matrix variable $\boldsymbol{\mathcal{Z}}$ is used in Subsection 5.3.2. The connection between these two approaches is explained in Subsection 5.3.3.

5.3.1 Complex Hessian Matrices of Scalar Functions Using $\boldsymbol{Z}$ and $\boldsymbol{Z}^*$

In this subsection, a systematic theory is introduced for finding the four Hessians of a complex-valued *scalar* function $f : \mathbb{C}^{N\times Q} \times \mathbb{C}^{N\times Q} \rightarrow \mathbb{C}$ with respect to a complex-valued matrix variable $\boldsymbol{Z}$ and the complex conjugate $\boldsymbol{Z}^*$ of this variable. The presentation given here follows the method proposed in Hjørungnes and Gesbert (2007*b*). In this subsection, the studied function is denoted by $f(\boldsymbol{Z}, \boldsymbol{Z}^*)$, and it is assumed to be twice differentiable in the matrix variables $\boldsymbol{Z}$ and $\boldsymbol{Z}^*$. The Hessian matrix depends on *two* variables such that the notation must include which variables the Hessian matrix is calculated with respect to. If the Hessian is calculated with respect to the variables $\boldsymbol{Z}_0$ and $\boldsymbol{Z}_1$, the Hessian will be denoted by $\mathcal{H}_{\boldsymbol{Z}_0,\boldsymbol{Z}_1} f$. Later in this section, the exact definition of the complex Hessian matrix of a scalar function f will be given.

Because it is assumed in this section that there exist two input matrix variables $\boldsymbol{Z}$ and $\boldsymbol{Z}^*$, there exist four different complex Hessian matrices of the function f with respect to all ordered combinations of these two matrix variables. It will be shown how these four Hessian matrices of the scalar complex function f can be identified from the second-order complex differential $(d^2 f)$ of the scalar function. These Hessians are the four parts of a bigger matrix, which must be checked to identify whether a stationary point is a local minimum, maximum, or saddle point. This bigger matrix can also be used in deciding convexity or concavity of a scalar objective function f.

When dealing with the Hessian matrix, it is the second-order differential that has to be calculated to identify the Hessian matrix. If $f \in \mathbb{C}$, then,

$$\left(d^2 f\right)^T = d\,(df)^T = d^2 f^T = d^2 f, \tag{5.9}$$

and if $f \in \mathbb{R}$, then,

$$\left(d^2 f\right)^H = d\,(df)^H = d^2 f^H = d^2 f. \tag{5.10}$$

The following proposition will be used to show various symmetry conditions of Hessian matrices in this chapter.

Proposition 5.1 *Let $f : \mathbb{C}^{N\times Q} \times \mathbb{C}^{N\times Q} \rightarrow \mathbb{C}$. It is assumed that $f(\boldsymbol{Z}, \boldsymbol{Z}^*)$ is twice differentiable with respect to all of the variables inside $\boldsymbol{Z} \in \mathbb{C}^{N\times Q}$ and $\boldsymbol{Z}^* \in \mathbb{C}^{N\times Q}$, when these variables are treated as independent variables. Then, by generalizing*

Magnus and Neudecker (1988, Theorem 4, pp. 105–106) to the complex-valued case

$$\frac{\partial^2}{\partial z_{k,l}\partial z_{m,n}} f = \frac{\partial^2}{\partial z_{m,n}\partial z_{k,l}} f, \tag{5.11}$$

$$\frac{\partial^2}{\partial z^*_{k,l}\partial z^*_{m,n}} f = \frac{\partial^2}{\partial z^*_{m,n}\partial z^*_{k,l}} f, \tag{5.12}$$

$$\frac{\partial^2}{\partial z^*_{k,l}\partial z_{m,n}} f = \frac{\partial^2}{\partial z_{m,n}\partial z^*_{k,l}} f, \tag{5.13}$$

where $m, k \in \{0, 1, \ldots, N-1\}$ *and* $n, l \in \{0, 1, \ldots, Q-1\}$.

The following definition is used for the complex Hessian matrix of a scalar function f; it is an extension of the definition given in Magnus and Neudecker (1988, p. 189) to complex scalar functions.

Definition 5.1 (Complex Hessian Matrix of Scalar Function) *Let* $\boldsymbol{Z}_i \in \mathbb{C}^{N_i \times Q_i}$, *where* $i \in \{0, 1\}$, *and let* $f : \mathbb{C}^{N_0 \times Q_0} \times \mathbb{C}^{N_1 \times Q_0} \to \mathbb{C}$. *The* complex Hessian matrix *is denoted by* $\mathcal{H}_{\boldsymbol{Z}_0,\boldsymbol{Z}_1} f$, *and it has size* $N_1 Q_1 \times N_0 Q_0$, *and is defined as*

$$\mathcal{H}_{\boldsymbol{Z}_0,\boldsymbol{Z}_1} f = \mathcal{D}_{\boldsymbol{Z}_0} \left(\mathcal{D}_{\boldsymbol{Z}_1} f\right)^T. \tag{5.14}$$

Remark *Let* $p_i = N_i k_i + l_i$ *where* $i \in \{0, 1\}$, $k_i \in \{0, 1, \ldots, Q_i - 1\}$, *and* $l_i \in \{0, 1, \ldots, N_i - 1\}$. *As a consequence of Definition 5.1 and* (3.82), *it follows that element number* (p_0, p_1) *of* $\mathcal{H}_{\boldsymbol{Z}_0,\boldsymbol{Z}_1} f$ *is given by*

$$\begin{aligned}
\left(\mathcal{H}_{\boldsymbol{Z}_0,\boldsymbol{Z}_1} f\right)_{p_0,p_1} &= \left(\mathcal{D}_{\boldsymbol{Z}_0} \left(\mathcal{D}_{\boldsymbol{Z}_1} f\right)^T\right)_{p_0,p_1} = \left[\frac{\partial}{\partial \operatorname{vec}^T(\boldsymbol{Z}_0)} \left(\frac{\partial}{\partial \operatorname{vec}^T(\boldsymbol{Z}_1)} f\right)^T\right]_{p_0,p_1} \\
&= \frac{\partial}{\partial (\operatorname{vec}(\boldsymbol{Z}_0))_{p_1}} \frac{\partial}{\partial (\operatorname{vec}(\boldsymbol{Z}_1))_{p_0}} f = \frac{\partial}{\partial (\operatorname{vec}(\boldsymbol{Z}_0))_{N_1 k_1 + l_1}} \frac{\partial}{\partial (\operatorname{vec}(\boldsymbol{Z}_1))_{N_0 k_0 + l_0}} f \\
&= \frac{\partial^2 f}{\partial (\boldsymbol{Z}_0)_{l_1,k_1} \partial (\boldsymbol{Z}_1)_{l_0,k_0}}.
\end{aligned} \tag{5.15}$$

And as an immediate consequence of (5.15) *and Proposition 5.1, it follows that, for twice differentiable functions* f,

$$(\mathcal{H}_{\boldsymbol{Z},\boldsymbol{Z}} f)^T = \mathcal{H}_{\boldsymbol{Z},\boldsymbol{Z}} f, \tag{5.16}$$

$$(\mathcal{H}_{\boldsymbol{Z}^*,\boldsymbol{Z}^*} f)^T = \mathcal{H}_{\boldsymbol{Z}^*,\boldsymbol{Z}^*} f, \tag{5.17}$$

$$(\mathcal{H}_{\boldsymbol{Z},\boldsymbol{Z}^*} f)^T = \mathcal{H}_{\boldsymbol{Z}^*,\boldsymbol{Z}} f. \tag{5.18}$$

These properties will also be used later in this chapter for the scalar component functions *of vector and matrix functions.*

To find an identification equation for the complex Hessians of the scalar function f with respect to all four possible combinations of the complex matrix variables $\boldsymbol{Z}$ and $\boldsymbol{Z}^*$, an appropriate form of the expression $d^2 f$ is required. This expression is derived next.

By using the definition of complex-valued matrix derivatives in Definition 3.1 on the scalar function f, the first-order differential of the function $f : \mathbb{C}^{N\times Q} \times \mathbb{C}^{N\times Q} \to \mathbb{C}$, denoted by $f(\boldsymbol{Z}, \boldsymbol{Z}^*)$, can be found from (3.78) as

$$df = (\mathcal{D}_{\boldsymbol{Z}} f) d\,\mathrm{vec}(\boldsymbol{Z}) + (\mathcal{D}_{\boldsymbol{Z}^*} f) d\,\mathrm{vec}(\boldsymbol{Z}^*), \tag{5.19}$$

where $\mathcal{D}_{\boldsymbol{Z}} f \in \mathbb{C}^{1\times NQ}$ and $\mathcal{D}_{\boldsymbol{Z}^*} f \in \mathbb{C}^{1\times NQ}$. When finding the second-order differential of the complex-valued scalar function f, the differential of the two derivatives $\mathcal{D}_{\boldsymbol{Z}} f$ and $\mathcal{D}_{\boldsymbol{Z}^*} f$ is needed. By using Definition 3.1 on the two derivatives $(\mathcal{D}_{\boldsymbol{Z}} f)^T$ and $(\mathcal{D}_{\boldsymbol{Z}^*} f)^T$, the following two expressions are found from (3.78):

$$(d\mathcal{D}_{\boldsymbol{Z}} f)^T = \left[\mathcal{D}_{\boldsymbol{Z}} (\mathcal{D}_{\boldsymbol{Z}} f)^T\right] d\,\mathrm{vec}(\boldsymbol{Z}) + \left[\mathcal{D}_{\boldsymbol{Z}^*} (\mathcal{D}_{\boldsymbol{Z}} f)^T\right] d\,\mathrm{vec}(\boldsymbol{Z}^*), \tag{5.20}$$

and

$$(d\mathcal{D}_{\boldsymbol{Z}^*} f)^T = \left[\mathcal{D}_{\boldsymbol{Z}} (\mathcal{D}_{\boldsymbol{Z}^*} f)^T\right] d\,\mathrm{vec}(\boldsymbol{Z}) + \left[\mathcal{D}_{\boldsymbol{Z}^*} (\mathcal{D}_{\boldsymbol{Z}^*} f)^T\right] d\,\mathrm{vec}(\boldsymbol{Z}^*). \tag{5.21}$$

By taking the transposed expressions on both sides of (5.20) and (5.21), it follows that

$$d\mathcal{D}_{\boldsymbol{Z}} f = \left[d\,\mathrm{vec}^T(\boldsymbol{Z})\right] \left[\mathcal{D}_{\boldsymbol{Z}} (\mathcal{D}_{\boldsymbol{Z}} f)^T\right]^T + \left[d\,\mathrm{vec}^T(\boldsymbol{Z}^*)\right] \left[\mathcal{D}_{\boldsymbol{Z}^*} (\mathcal{D}_{\boldsymbol{Z}} f)^T\right]^T, \tag{5.22}$$

and

$$d\mathcal{D}_{\boldsymbol{Z}^*} f = \left[d\,\mathrm{vec}^T(\boldsymbol{Z})\right] \left[\mathcal{D}_{\boldsymbol{Z}} (\mathcal{D}_{\boldsymbol{Z}^*} f)^T\right]^T + \left[d\,\mathrm{vec}^T(\boldsymbol{Z}^*)\right] \left[\mathcal{D}_{\boldsymbol{Z}^*} (\mathcal{D}_{\boldsymbol{Z}^*} f)^T\right]^T. \tag{5.23}$$

The second-order differential of f can be found by applying the differential operator to both sides of (5.19), and then utilizing the results from (5.1), (5.22), and (5.23) as follows:

$$\begin{aligned}
d^2 f &= (d\mathcal{D}_{\boldsymbol{Z}} f)\, d\,\mathrm{vec}(\boldsymbol{Z}) + (d\mathcal{D}_{\boldsymbol{Z}^*} f) d\,\mathrm{vec}(\boldsymbol{Z}^*) \\
&= \left[d\,\mathrm{vec}^T(\boldsymbol{Z})\right] \left[\mathcal{D}_{\boldsymbol{Z}} (\mathcal{D}_{\boldsymbol{Z}} f)^T\right]^T d\,\mathrm{vec}(\boldsymbol{Z}) \\
&\quad + \left[d\,\mathrm{vec}^T(\boldsymbol{Z}^*)\right] \left[\mathcal{D}_{\boldsymbol{Z}^*} (\mathcal{D}_{\boldsymbol{Z}} f)^T\right]^T d\,\mathrm{vec}(\boldsymbol{Z}) \\
&\quad + \left[d\,\mathrm{vec}^T(\boldsymbol{Z})\right] \left[\mathcal{D}_{\boldsymbol{Z}} (\mathcal{D}_{\boldsymbol{Z}^*} f)^T\right]^T d\,\mathrm{vec}(\boldsymbol{Z}^*) \\
&\quad + \left[d\,\mathrm{vec}^T(\boldsymbol{Z}^*)\right] \left[\mathcal{D}_{\boldsymbol{Z}^*} (\mathcal{D}_{\boldsymbol{Z}^*} f)^T\right]^T d\,\mathrm{vec}(\boldsymbol{Z}^*) \\
&= \left[d\,\mathrm{vec}^T(\boldsymbol{Z})\right] \left[\mathcal{D}_{\boldsymbol{Z}} (\mathcal{D}_{\boldsymbol{Z}} f)^T\right] d\,\mathrm{vec}(\boldsymbol{Z}) \\
&\quad + \left[d\,\mathrm{vec}^T(\boldsymbol{Z})\right] \left[\mathcal{D}_{\boldsymbol{Z}^*} (\mathcal{D}_{\boldsymbol{Z}} f)^T\right] d\,\mathrm{vec}(\boldsymbol{Z}^*) \\
&\quad + \left[d\,\mathrm{vec}^T(\boldsymbol{Z}^*)\right] \left[\mathcal{D}_{\boldsymbol{Z}} (\mathcal{D}_{\boldsymbol{Z}^*} f)^T\right] d\,\mathrm{vec}(\boldsymbol{Z}) \\
&\quad + \left[d\,\mathrm{vec}^T(\boldsymbol{Z}^*)\right] \left[\mathcal{D}_{\boldsymbol{Z}^*} (\mathcal{D}_{\boldsymbol{Z}^*} f)^T\right] d\,\mathrm{vec}(\boldsymbol{Z}^*) \\
&= \left[d\,\mathrm{vec}^T(\boldsymbol{Z}^*),\ d\,\mathrm{vec}^T(\boldsymbol{Z})\right] \begin{bmatrix} \mathcal{D}_{\boldsymbol{Z}} (\mathcal{D}_{\boldsymbol{Z}^*} f)^T & \mathcal{D}_{\boldsymbol{Z}^*} (\mathcal{D}_{\boldsymbol{Z}^*} f)^T \\ \mathcal{D}_{\boldsymbol{Z}} (\mathcal{D}_{\boldsymbol{Z}} f)^T & \mathcal{D}_{\boldsymbol{Z}^*} (\mathcal{D}_{\boldsymbol{Z}} f)^T \end{bmatrix} \begin{bmatrix} d\,\mathrm{vec}(\boldsymbol{Z}) \\ d\,\mathrm{vec}(\boldsymbol{Z}^*) \end{bmatrix} \\
&= \left[d\,\mathrm{vec}^T(\boldsymbol{Z}),\ d\,\mathrm{vec}^T(\boldsymbol{Z}^*)\right] \begin{bmatrix} \mathcal{D}_{\boldsymbol{Z}} (\mathcal{D}_{\boldsymbol{Z}} f)^T & \mathcal{D}_{\boldsymbol{Z}^*} (\mathcal{D}_{\boldsymbol{Z}} f)^T \\ \mathcal{D}_{\boldsymbol{Z}} (\mathcal{D}_{\boldsymbol{Z}^*} f)^T & \mathcal{D}_{\boldsymbol{Z}^*} (\mathcal{D}_{\boldsymbol{Z}^*} f)^T \end{bmatrix} \begin{bmatrix} d\,\mathrm{vec}(\boldsymbol{Z}) \\ d\,\mathrm{vec}(\boldsymbol{Z}^*) \end{bmatrix}.
\end{aligned} \tag{5.24}$$

By using the definition of the complex Hessian of a scalar function (see Definition 5.1), in the last two lines of (5.24), it follows that $d^2 f$ can be rewritten as

$$d^2 f = \left[d\,\text{vec}^T(Z^*)\; d\,\text{vec}^T(Z)\right] \begin{bmatrix} \mathcal{H}_{Z,Z^*} f & \mathcal{H}_{Z^*,Z^*} f \\ \mathcal{H}_{Z,Z} f & \mathcal{H}_{Z^*,Z} f \end{bmatrix} \begin{bmatrix} d\,\text{vec}(Z) \\ d\,\text{vec}(Z^*) \end{bmatrix} \tag{5.25}$$

$$= \left[d\,\text{vec}^T(Z)\; d\,\text{vec}^T(Z^*)\right] \begin{bmatrix} \mathcal{H}_{Z,Z} f & \mathcal{H}_{Z^*,Z} f \\ \mathcal{H}_{Z,Z^*} f & \mathcal{H}_{Z^*,Z^*} f \end{bmatrix} \begin{bmatrix} d\,\text{vec}(Z) \\ d\,\text{vec}(Z^*) \end{bmatrix}. \tag{5.26}$$

Assume that it is possible to find an expression of $d^2 f$ in the following form:

$$\begin{aligned} d^2 f &= \left[d\,\text{vec}^T(Z^*)\right] A_{0,0} d\,\text{vec}(Z) + \left[d\,\text{vec}^T(Z^*)\right] A_{0,1} d\,\text{vec}(Z^*) \\ &\quad + \left[d\,\text{vec}^T(Z)\right] A_{1,0} d\,\text{vec}(Z) + \left[d\,\text{vec}^T(Z)\right] A_{1,1} d\,\text{vec}(Z^*) \\ &= \left[d\,\text{vec}^T(Z^*)\; d\,\text{vec}^T(Z)\right] \begin{bmatrix} A_{0,0} & A_{0,1} \\ A_{1,0} & A_{1,1} \end{bmatrix} \begin{bmatrix} d\,\text{vec}(Z) \\ d\,\text{vec}(Z^*) \end{bmatrix}, \end{aligned} \tag{5.27}$$

where $A_{k,l}$ with $k, l \in \{0, 1\}$ has size $NQ \times NQ$ and can possibly be dependent on Z and Z^*, but not on $d\,\text{vec}(Z)$ or $d\,\text{vec}(Z^*)$. The four complex Hessian matrices in (5.25) can now be identified from the matrices $A_{k,l}$ given in (5.27) in the following way: Subtracting the second-order differentials in (5.25) from (5.27) yields

$$\begin{aligned} &\left[d\,\text{vec}^T(Z)\right] \left(A_{1,0} - \mathcal{H}_{Z,Z} f\right) d\,\text{vec}(Z) \\ &\quad + \left[d\,\text{vec}^T(Z^*)\right] \left(A_{0,0} + A_{1,1}^T - \mathcal{H}_{Z,Z^*} f - (\mathcal{H}_{Z^*,Z} f)^T\right) d\,\text{vec}(Z) \\ &\quad + \left[d\,\text{vec}^T(Z^*)\right] \left(A_{0,1} - \mathcal{H}_{Z^*,Z^*} f\right) d\,\text{vec}(Z^*) = 0, \end{aligned} \tag{5.28}$$

and this is valid for all $dZ \in \mathbb{C}^{N\times Q}$. The expression in (5.28) is now of the same type as the equation used in Lemma 3.2. Recall the symmetry properties in (5.16), (5.17), and (5.18), which will be useful in the following. Lemma 3.2 will now be used, and it is seen that the matrix B_0 in Lemma 3.2 can be identified from (5.28) as

$$B_0 = A_{1,0} - \mathcal{H}_{Z,Z} f. \tag{5.29}$$

From Lemma 3.2, it follows that $B_0 = -B_0^T$, and this can be expressed as

$$A_{1,0} - \mathcal{H}_{Z,Z} f = -\left(A_{1,0} - \mathcal{H}_{Z,Z} f\right)^T. \tag{5.30}$$

By using the fact that the Hessian matrix $\mathcal{H}_{Z,Z} f$ is symmetric, the Hessian $\mathcal{H}_{Z,Z} f$ can be solved from (5.30):

$$\mathcal{H}_{Z,Z} f = \frac{1}{2}\left(A_{1,0} + A_{1,0}^T\right). \tag{5.31}$$

By using Lemma 3.2 on (5.28), the matrix B_2 is identified as

$$B_2 = A_{0,1} - \mathcal{H}_{Z^*,Z^*} f. \tag{5.32}$$

Lemma 3.2 says that $B_2 = -B_2^T$, and by inserting B_2 from (5.32), it is found that

$$A_{0,1} - \mathcal{H}_{Z^*,Z^*} f = -\left(A_{0,1} - \mathcal{H}_{Z^*,Z^*} f\right)^T. \tag{5.33}$$

Table 5.2 Procedure for identifying the complex Hessians of a scalar function $f \in \mathbb{C}$ with respect to complex-valued matrix variables $\mathbf{Z} \in \mathbb{C}^{N\times Q}$ and $\mathbf{Z}^* \in \mathbb{C}^{N\times Q}$.

Step 1:	Compute the second-order differential $d^2 f$.
Step 2:	Manipulate $d^2 f$ into the form given in (5.27) to identify the four $NQ \times NQ$ matrices $\mathbf{A}_{0,0}$, $\mathbf{A}_{0,1}$, $\mathbf{A}_{1,0}$, and $\mathbf{A}_{1,1}$.
Step 3:	Use (5.31), (5.34), (5.36), and (5.37) to identify the four Hessian matrices $\mathcal{H}_{\mathbf{Z},\mathbf{Z}} f$, $\mathcal{H}_{\mathbf{Z}^*,\mathbf{Z}^*} f$, $\mathcal{H}_{\mathbf{Z},\mathbf{Z}^*} f$, and $\mathcal{H}_{\mathbf{Z}^*,\mathbf{Z}} f$.

By using the fact that the Hessian matrix $\mathcal{H}_{\mathbf{Z}^*,\mathbf{Z}^*} f$ is symmetric (see (5.17)), the Hessian $\mathcal{H}_{\mathbf{Z}^*,\mathbf{Z}^*} f$ can be solved from (5.33) to get

$$\mathcal{H}_{\mathbf{Z}^*,\mathbf{Z}^*} f = \frac{1}{2}\left(\mathbf{A}_{0,1} + \mathbf{A}_{0,1}^T\right). \tag{5.34}$$

The matrix $\mathbf{B}_1$ in Lemma 3.2 is identified from (5.28) as

$$\mathbf{B}_1 = \mathbf{A}_{0,0} + \mathbf{A}_{1,1}^T - \mathcal{H}_{\mathbf{Z},\mathbf{Z}^*} f - \left(\mathcal{H}_{\mathbf{Z}^*,\mathbf{Z}} f\right)^T. \tag{5.35}$$

Lemma 3.2 states that $\mathbf{B}_1 = \mathbf{0}_{NQ\times NQ}$, and by using that $\left(\mathcal{H}_{\mathbf{Z}^*,\mathbf{Z}} f\right)^T = \mathcal{H}_{\mathbf{Z},\mathbf{Z}^*} f$, it follows from (5.35) that

$$\mathcal{H}_{\mathbf{Z},\mathbf{Z}^*} f = \frac{1}{2}\left(\mathbf{A}_{0,0} + \mathbf{A}_{1,1}^T\right). \tag{5.36}$$

The last remaining Hessian $\mathcal{H}_{\mathbf{Z}^*,\mathbf{Z}} f$ is given by $\mathcal{H}_{\mathbf{Z}^*,\mathbf{Z}} f = \left(\mathcal{H}_{\mathbf{Z},\mathbf{Z}^*} f\right)^T$; hence, it follows from (5.36) that

$$\mathcal{H}_{\mathbf{Z}^*,\mathbf{Z}} f = \frac{1}{2}\left(\mathbf{A}_{0,0}^T + \mathbf{A}_{1,1}\right). \tag{5.37}$$

The complex Hessian matrices of the scalar function $f \in \mathbb{C}$ can be computed using a three-step procedure given in Table 5.2.

As an application, to check, for instance, convexity and concavity of f, the middle block matrix of size $2NQ \times 2NQ$ on the right-hand side of (5.25) must be positive or negative definite, respectively. In the next lemma, it is shown that this matrix is Hermitian for real-valued scalar functions.

Lemma 5.1 *Let* $f : \mathbb{C}^{N\times Q} \times \mathbb{C}^{N\times Q} \to \mathbb{R}$, *then,*

$$\begin{bmatrix} \mathcal{H}_{\mathbf{Z},\mathbf{Z}^*} f & \mathcal{H}_{\mathbf{Z}^*,\mathbf{Z}^*} f \\ \mathcal{H}_{\mathbf{Z},\mathbf{Z}} f & \mathcal{H}_{\mathbf{Z}^*,\mathbf{Z}} f \end{bmatrix}^H = \begin{bmatrix} \mathcal{H}_{\mathbf{Z},\mathbf{Z}^*} f & \mathcal{H}_{\mathbf{Z}^*,\mathbf{Z}^*} f \\ \mathcal{H}_{\mathbf{Z},\mathbf{Z}} f & \mathcal{H}_{\mathbf{Z}^*,\mathbf{Z}} f \end{bmatrix}. \tag{5.38}$$

Proof By using Definition 5.1, (5.16), (5.17), (5.18), in addition to Lemma 3.3, it is found that

$$\begin{bmatrix} \mathcal{H}_{Z,Z^*} f & \mathcal{H}_{Z^*,Z^*} f \\ \mathcal{H}_{Z,Z} f & \mathcal{H}_{Z^*,Z} f \end{bmatrix}^H = \begin{bmatrix} (\mathcal{H}_{Z,Z^*} f)^H & (\mathcal{H}_{Z,Z} f)^H \\ (\mathcal{H}_{Z^*,Z^*} f)^H & (\mathcal{H}_{Z^*,Z} f)^H \end{bmatrix}$$

$$= \begin{bmatrix} (\mathcal{H}_{Z^*,Z} f)^* & (\mathcal{H}_{Z,Z} f)^* \\ (\mathcal{H}_{Z^*,Z^*} f)^* & (\mathcal{H}_{Z,Z^*} f)^* \end{bmatrix} = \begin{bmatrix} \left(\mathcal{D}_{Z^*} (\mathcal{D}_Z f)^T\right)^* & \left(\mathcal{D}_Z (\mathcal{D}_Z f)^T\right)^* \\ \left(\mathcal{D}_{Z^*} (\mathcal{D}_{Z^*} f)^T\right)^* & \left(\mathcal{D}_Z (\mathcal{D}_{Z^*} f)^T\right)^* \end{bmatrix}$$

$$= \begin{bmatrix} \mathcal{D}_Z (\mathcal{D}_{Z^*} f)^T & \mathcal{D}_{Z^*} (\mathcal{D}_{Z^*} f)^T \\ \mathcal{D}_Z (\mathcal{D}_Z f)^T & \mathcal{D}_{Z^*} (\mathcal{D}_Z f)^T \end{bmatrix} = \begin{bmatrix} \mathcal{H}_{Z,Z^*} f & \mathcal{H}_{Z^*,Z^*} f \\ \mathcal{H}_{Z,Z} f & \mathcal{H}_{Z^*,Z} f \end{bmatrix}, \tag{5.39}$$

which concludes the proof. ∎

The Taylor series for scalar functions and variables can be found in Eriksson, Ollila, and Koivunen (2009). By generalizing Abatzoglou et al. (1991, Eq. (A.1)) to complex-valued *matrix variables*, it is possible to find the second-order Taylor series, and this is stated in the next lemma.

Lemma 5.2 *Let* $f : \mathbb{C}^{N\times Q} \times \mathbb{C}^{N\times Q} \to \mathbb{R}$. *The* second-order Taylor series *of* f *in the point* $\boldsymbol{Z}$ *can be expressed as*

$$\begin{aligned} f(\boldsymbol{Z}+d\boldsymbol{Z}, \boldsymbol{Z}^*+d\boldsymbol{Z}^*) &= f(\boldsymbol{Z}, \boldsymbol{Z}^*) \\ &+ (\mathcal{D}_Z f(\boldsymbol{Z}, \boldsymbol{Z}^*))\, d\operatorname{vec}(\boldsymbol{Z}) + (\mathcal{D}_{Z^*} f(\boldsymbol{Z}, \boldsymbol{Z}^*))\, d\operatorname{vec}(\boldsymbol{Z}^*) \\ &+ \frac{1}{2}\left[d\operatorname{vec}^T(\boldsymbol{Z}^*) \;\; d\operatorname{vec}^T(\boldsymbol{Z})\right] \begin{bmatrix} \mathcal{H}_{Z,Z^*} f & \mathcal{H}_{Z^*,Z^*} f \\ \mathcal{H}_{Z,Z} f & \mathcal{H}_{Z^*,Z} f \end{bmatrix} \begin{bmatrix} d\operatorname{vec}(\boldsymbol{Z}) \\ d\operatorname{vec}(\boldsymbol{Z}^*) \end{bmatrix} + r(d\boldsymbol{Z}, d\boldsymbol{Z}^*), \end{aligned} \tag{5.40}$$

where the function $r : \mathbb{C}^{N\times Q} \times \mathbb{C}^{N\times Q} \to \mathbb{R}$ *satisfies*

$$\lim_{(d\boldsymbol{Z}_0, d\boldsymbol{Z}_1)\to \boldsymbol{0}_{N\times 2Q}} \frac{r(d\boldsymbol{Z}_0, d\boldsymbol{Z}_1)}{\|(d\boldsymbol{Z}_0, d\boldsymbol{Z}_1)\|_F^2} = 0. \tag{5.41}$$

The second-order Taylor series might be very useful to check the nature of a stationary point of a real-valued function $f(\boldsymbol{Z}, \boldsymbol{Z}^*)$. Assume that the function $f(\boldsymbol{Z}, \boldsymbol{Z}^*)$ has a stationary point in $\boldsymbol{Z} = \boldsymbol{C} \in \mathbb{C}^{N\times Q}$. Then, it follows from Theorem 3.2 that

$$\mathcal{D}_Z f(\boldsymbol{C}, \boldsymbol{C}^*) = \boldsymbol{0}_{1\times NQ}, \tag{5.42}$$

$$\mathcal{D}_{Z^*} f(\boldsymbol{C}, \boldsymbol{C}^*) = \boldsymbol{0}_{1\times NQ}. \tag{5.43}$$

If the second-order Taylor series (5.40) is evaluated at $(\boldsymbol{Z}_0, \boldsymbol{Z}_1) = (\boldsymbol{C}, \boldsymbol{C}^*)$, it is found that

$$\begin{aligned} f(\boldsymbol{C}+d\boldsymbol{Z}_0, \boldsymbol{C}^*+d\boldsymbol{Z}_1) &= f(\boldsymbol{C}, \boldsymbol{C}^*) \\ &+ \frac{1}{2}\left[d\operatorname{vec}^T(\boldsymbol{Z}^*) \;\; d\operatorname{vec}^T(\boldsymbol{Z})\right] \begin{bmatrix} \mathcal{H}_{Z,Z^*} f & \mathcal{H}_{Z^*,Z^*} f \\ \mathcal{H}_{Z,Z} f & \mathcal{H}_{Z^*,Z} f \end{bmatrix} \begin{bmatrix} d\operatorname{vec}(\boldsymbol{Z}) \\ d\operatorname{vec}(\boldsymbol{Z}^*) \end{bmatrix} + r(d\boldsymbol{Z}_0, d\boldsymbol{Z}_1). \end{aligned} \tag{5.44}$$

Near the point $\boldsymbol{Z} = \boldsymbol{C}$, it is seen from (5.44) that the function is behaving as a quadratic function in $\text{vec}([d\boldsymbol{Z},\ d\boldsymbol{Z}^*])$. Notice that the second-order term in this variable is $\text{vec}^H([d\boldsymbol{Z},\ d\boldsymbol{Z}^*]) \begin{bmatrix} \mathcal{H}_{\boldsymbol{Z},\boldsymbol{Z}^*} f & \mathcal{H}_{\boldsymbol{Z}^*,\boldsymbol{Z}^*} f \\ \mathcal{H}_{\boldsymbol{Z},\boldsymbol{Z}} f & \mathcal{H}_{\boldsymbol{Z}^*,\boldsymbol{Z}} f \end{bmatrix} \text{vec}([d\boldsymbol{Z},\ d\boldsymbol{Z}^*])$. Hence, to study the nature of a stationary point, it is enough to study if $\begin{bmatrix} \mathcal{H}_{\boldsymbol{Z},\boldsymbol{Z}^*} f & \mathcal{H}_{\boldsymbol{Z}^*,\boldsymbol{Z}^*} f \\ \mathcal{H}_{\boldsymbol{Z},\boldsymbol{Z}} f & \mathcal{H}_{\boldsymbol{Z}^*,\boldsymbol{Z}} f \end{bmatrix}$ is positive definite, negative definite, or indefinite in the stationary point $\boldsymbol{Z} = \boldsymbol{C}$.

In the next section, the theory for finding the complex Hessian of a scalar function will be presented when the input variable to the function is the augmented matrix variable $\mathcal{Z}$.

5.3.2 Complex Hessian Matrices of Scalar Functions Using $\mathcal{Z}$

In this subsection, the matrix-valued function $\boldsymbol{F} : \mathbb{C}^{N\times Q} \times \mathbb{C}^{N\times Q} \to \mathbb{C}^{M\times P}$ is considered. By using the augmented matrix variable $\mathcal{Z}$, the definition of the matrix derivative in (3.78) can be written as

$$
\begin{aligned}
d\,\text{vec}(\boldsymbol{F}) &= (\mathcal{D}_{\boldsymbol{Z}}\boldsymbol{F})\,d\,\text{vec}(\boldsymbol{Z}) + (\mathcal{D}_{\boldsymbol{Z}^*}\boldsymbol{F})\,d\,\text{vec}(\boldsymbol{Z}^*) \\
&= [\mathcal{D}_{\boldsymbol{Z}}\boldsymbol{F}, \mathcal{D}_{\boldsymbol{Z}^*}\boldsymbol{F}] \begin{bmatrix} d\,\text{vec}(\boldsymbol{Z}) \\ d\,\text{vec}(\boldsymbol{Z}^*) \end{bmatrix} \triangleq (\mathcal{D}_{\mathcal{Z}}\boldsymbol{F})\,d\,\text{vec}(\mathcal{Z}),
\end{aligned} \tag{5.45}
$$

where the derivative of the matrix function $\boldsymbol{F}$ with respect to the augmented matrix variable $\mathcal{Z}$ has been defined as

$$
\mathcal{D}_{\mathcal{Z}}\boldsymbol{F} \triangleq [\mathcal{D}_{\boldsymbol{Z}}\boldsymbol{F}\ \mathcal{D}_{\boldsymbol{Z}^*}\boldsymbol{F}] \in \mathbb{C}^{MP\times 2NQ}. \tag{5.46}
$$

The matrix derivative of $\boldsymbol{F}$ with respect to the augmented matrix variable $\mathcal{Z}$ can be identified from the first-order differential in (5.45).

A scalar complex-valued function $f : \mathbb{C}^{N\times 2Q} \to \mathbb{C}$, which depends on $\mathcal{Z} \in \mathbb{C}^{N\times 2Q}$, is denoted by $f(\mathcal{Z})$, and its derivative can be identified by substituting $\boldsymbol{F}$ by f in (5.45) to obtain

$$
\begin{aligned}
df &= (\mathcal{D}_{\boldsymbol{Z}}f)\,d\,\text{vec}(\boldsymbol{Z}) + (\mathcal{D}_{\boldsymbol{Z}^*}f)\,d\,\text{vec}(\boldsymbol{Z}^*) = [\mathcal{D}_{\boldsymbol{Z}}f\ \mathcal{D}_{\boldsymbol{Z}^*}f] \begin{bmatrix} d\,\text{vec}(\boldsymbol{Z}) \\ d\,\text{vec}(\boldsymbol{Z}^*) \end{bmatrix} \\
&= (\mathcal{D}_{\mathcal{Z}}f)\,d\,\text{vec}(\mathcal{Z}),
\end{aligned} \tag{5.47}
$$

where

$$
\mathcal{D}_{\mathcal{Z}}f = [\mathcal{D}_{\boldsymbol{Z}}f\ \mathcal{D}_{\boldsymbol{Z}^*}f], \tag{5.48}
$$

lies in $\mathbb{C}^{1\times 2NQ}$.

The second-order differential is used to identify the Hessian also when finding the complex Hessian with respect to the augmented matrix variable $\mathcal{Z}$. The second-order differential is found by applying the differential operator on both sides of (5.47), and then an expression of the differential of $\mathcal{D}_{\mathcal{Z}}f \in \mathbb{C}^{1\times 2NQ}$ is needed. An expression for the *differential* of the row vector $\mathcal{D}_{\mathcal{Z}}f$ can be found by using (5.45), where $\boldsymbol{F}$ is substituted by $\mathcal{D}_{\mathcal{Z}} f$ to obtain

$$d \operatorname{vec}(\mathcal{D}_{\mathcal{Z}} f) = d (\mathcal{D}_{\mathcal{Z}} f)^T = \left(\mathcal{D}_{\mathcal{Z}} (\mathcal{D}_{\mathcal{Z}} f)^T\right) d \operatorname{vec}(\mathcal{Z}). \tag{5.49}$$

Taking the transposed of both sides of the above equations yields

$$d\mathcal{D}_{\mathcal{Z}} f = \left(d \operatorname{vec}^T(\mathcal{Z})\right) \left(\mathcal{D}_{\mathcal{Z}} (\mathcal{D}_{\mathcal{Z}} f)^T\right)^T. \tag{5.50}$$

The complex Hessian of a scalar function f, which depends on the augmented matrix $\mathcal{Z}$, is defined in a similar way as described previously for complex Hessians in Definition 5.1. The complex Hessian of the scalar function f with respect to $\mathcal{Z}$, and $\mathcal{Z}$ is a symmetric matrix (see Definition 5.1 and the following remark). It is denoted by $\mathcal{H}_{\mathcal{Z},\mathcal{Z}} f \in \mathbb{C}^{2NQ \times 2NQ}$ and is given by

$$\mathcal{H}_{\mathcal{Z},\mathcal{Z}} f = \mathcal{D}_{\mathcal{Z}} (\mathcal{D}_{\mathcal{Z}} f)^T. \tag{5.51}$$

Here, it is assumed that f is twice differentiable with respect to all matrix components of $\mathcal{Z}$. Because there exists only one input matrix variable of the function $f(\mathcal{Z})$, the only Hessian matrix that will be considered is $\mathcal{H}_{\mathcal{Z},\mathcal{Z}} f$. The complex Hessian of f can be identified from the second-order differential of f. The second-order differential of f can be expressed as

$$\begin{aligned} d^2 f &= d(df) = (d\mathcal{D}_{\mathcal{Z}} f) d \operatorname{vec}(\mathcal{Z}) = \left(d \operatorname{vec}^T(\mathcal{Z})\right) \left(\mathcal{D}_{\mathcal{Z}} (\mathcal{D}_{\mathcal{Z}} f)^T\right)^T d \operatorname{vec}(\mathcal{Z}) \\ &= \left(d \operatorname{vec}^T(\mathcal{Z})\right) \left[\mathcal{D}_{\mathcal{Z}} (\mathcal{D}_{\mathcal{Z}} f)^T\right] d \operatorname{vec}(\mathcal{Z}) = \left(d \operatorname{vec}^T(\mathcal{Z})\right) \left[\mathcal{H}_{\mathcal{Z},\mathcal{Z}} f\right] d \operatorname{vec}(\mathcal{Z}), \end{aligned} \tag{5.52}$$

where (5.50) and (5.51) have been used.

Assume that the second-order differential of f can be written in the following way:

$$d^2 f = \left(d \operatorname{vec}^T(\mathcal{Z})\right) \boldsymbol{A}\, d \operatorname{vec}(\mathcal{Z}), \tag{5.53}$$

where $\boldsymbol{A} \in \mathbb{C}^{2NQ \times 2NQ}$ does *not* depend on the differential operator d; however, it might depend on the matrix variables $\mathcal{Z}$ or $\mathcal{Z}^*$. By setting the two expressions of $d^2 f$ in (5.52) and (5.53) as equal, it follows from Lemma 2.15[2] that the Hessian $\mathcal{H}_{\mathcal{Z},\mathcal{Z}} f$ must satisfy

$$\mathcal{H}_{\mathcal{Z},\mathcal{Z}} f + (\mathcal{H}_{\mathcal{Z},\mathcal{Z}} f)^T = 2\mathcal{H}_{\mathcal{Z},\mathcal{Z}} f = \boldsymbol{A} + \boldsymbol{A}^T, \tag{5.54}$$

where it follows from Proposition 5.1 that the Hessian matrix $\mathcal{H}_{\mathcal{Z},\mathcal{Z}} f$ is symmetric. Solving the Hessian $\mathcal{H}_{\mathcal{Z},\mathcal{Z}} f$ from (5.54) leads to

$$\mathcal{H}_{\mathcal{Z},\mathcal{Z}} f = \frac{1}{2}\left[\boldsymbol{A} + \boldsymbol{A}^T\right]. \tag{5.55}$$

This equation suggests a way of identifying the Hessian of a scalar complex-valued function when the augmented matrix variable $\mathcal{Z} \in \mathbb{C}^{N \times 2Q}$ is used. The procedure for finding the complex Hessian of a scalar when $\mathcal{Z}$ is used as a matrix variable is summarized in Table 5.3. Examples of how to calculate the complex Hessian of scalar functions will be given in Subsection 5.6.1.

[2] When using Lemma 2.15 here, the vector variable z in Lemma 2.15 is substituted with the differential vector $d \operatorname{vec}(\mathcal{Z})$, and the middle square matrices $\boldsymbol{A}$ and $\boldsymbol{B}$ in Lemma 2.15 are replaced by $\mathcal{H}_{\mathcal{Z},\mathcal{Z}} f$ (from (5.52)) and $\boldsymbol{A}$ (from (5.53)), respectively.

Table 5.3 Procedure for identifying the complex Hessians of a scalar function $f \in \mathbb{C}$ with respect to the augmented complex-valued matrix variable $\mathcal{Z} \in \mathbb{C}^{N\times 2Q}$.

Step 1:	Compute the second-order differential $d^2 f$.
Step 2:	Manipulate $d^2 f$ into the form given in (5.53) to identify the matrix $\boldsymbol{A} \in \mathbb{C}^{2NQ\times 2NQ}$.
Step 3:	Use (5.55) to find the complex Hessian $\mathcal{H}_{\mathcal{Z},\mathcal{Z}} f$.

5.3.3 Connections between Hessians When Using Two-Matrix Variable Representations

In this subsection, the connection between the two methods presented in Tables 5.2 and 5.3 will be studied.

Lemma 5.3 *The following connections exist between the four Hessians $\mathcal{H}_{Z,Z} f$, $\mathcal{H}_{Z^*,Z} f$, $\mathcal{H}_{Z,Z^*} f$, and $\mathcal{H}_{Z^*,Z^*} f$ and the Hessian with respect to the augmented matrix variable $\mathcal{H}_{\mathcal{Z},\mathcal{Z}} f$:*

$$\mathcal{H}_{\mathcal{Z},\mathcal{Z}} f = \begin{bmatrix} \mathcal{H}_{Z,Z} f & \mathcal{H}_{Z^*,Z} f \\ \mathcal{H}_{Z,Z^*} f & \mathcal{H}_{Z^*,Z^*} f \end{bmatrix} = \begin{bmatrix} \boldsymbol{0}_{NQ\times NQ} & \boldsymbol{I}_{NQ} \\ \boldsymbol{I}_{NQ} & \boldsymbol{0}_{NQ\times NQ} \end{bmatrix} \begin{bmatrix} \mathcal{H}_{Z,Z^*} f & \mathcal{H}_{Z^*,Z^*} f \\ \mathcal{H}_{Z,Z} f & \mathcal{H}_{Z^*,Z} f \end{bmatrix}. \tag{5.56}$$

Proof From (5.48), it follows that

$$(\mathcal{D}_{\mathcal{Z}} f)^T = \begin{bmatrix} (\mathcal{D}_Z f)^T \\ (\mathcal{D}_{Z^*} f)^T \end{bmatrix}. \tag{5.57}$$

Using this result in the definition of $\mathcal{H}_{\mathcal{Z},\mathcal{Z}} f$ leads to

$$\begin{aligned} \mathcal{H}_{\mathcal{Z},\mathcal{Z}} f &= \mathcal{D}_{\mathcal{Z}} (\mathcal{D}_{\mathcal{Z}} f)^T = \mathcal{D}_{\mathcal{Z}} \begin{bmatrix} (\mathcal{D}_Z f)^T \\ (\mathcal{D}_{Z^*} f)^T \end{bmatrix} \\ &= \left[\mathcal{D}_Z \begin{bmatrix} (\mathcal{D}_Z f)^T \\ (\mathcal{D}_{Z^*} f)^T \end{bmatrix}, \mathcal{D}_{Z^*} \begin{bmatrix} (\mathcal{D}_Z f)^T \\ (\mathcal{D}_{Z^*} f)^T \end{bmatrix} \right], \end{aligned} \tag{5.58}$$

where (5.46) was used in the last equality. Before proceeding, an auxiliary result will be needed, and this is presented next.

For vector functions $\boldsymbol{f}_0 : \mathbb{C}^{N\times Q} \times \mathbb{C}^{N\times Q} \to \mathbb{C}^{M\times 1}$ and $\boldsymbol{f}_1 : \mathbb{C}^{N\times Q} \times \mathbb{C}^{N\times Q} \to \mathbb{C}^{M\times 1}$, the following relations are valid:

$$\mathcal{D}_Z \begin{bmatrix} \boldsymbol{f}_0 \\ \boldsymbol{f}_1 \end{bmatrix} = \begin{bmatrix} \mathcal{D}_Z \boldsymbol{f}_0 \\ \mathcal{D}_Z \boldsymbol{f}_1 \end{bmatrix}, \tag{5.59}$$

$$\mathcal{D}_{Z^*} \begin{bmatrix} \boldsymbol{f}_0 \\ \boldsymbol{f}_1 \end{bmatrix} = \begin{bmatrix} \mathcal{D}_{Z^*} \boldsymbol{f}_0 \\ \mathcal{D}_{Z^*} \boldsymbol{f}_1 \end{bmatrix}. \tag{5.60}$$

These can be shown to be valid by using Definition 3.1 repeatedly as follows:

$$d\,\text{vec}\left(\begin{bmatrix} f_0 \\ f_1 \end{bmatrix}\right) = d\begin{bmatrix} f_0 \\ f_1 \end{bmatrix} = \begin{bmatrix} d f_0 \\ d f_1 \end{bmatrix} = \begin{bmatrix} [\mathcal{D}_Z f_0]\, d\,\text{vec}(Z) + [\mathcal{D}_{Z^*} f_0]\, d\,\text{vec}(Z^*) \\ [\mathcal{D}_Z f_1]\, d\,\text{vec}(Z) + [\mathcal{D}_{Z^*} f_1]\, d\,\text{vec}(Z^*) \end{bmatrix}$$

$$= \begin{bmatrix} \mathcal{D}_Z f_0 \\ \mathcal{D}_Z f_1 \end{bmatrix} d\,\text{vec}(Z) + \begin{bmatrix} \mathcal{D}_{Z^*} f_0 \\ \mathcal{D}_{Z^*} f_1 \end{bmatrix} d\,\text{vec}(Z^*). \tag{5.61}$$

By using Definition 3.1 on the above expression, (5.59) and (5.60) follow.

If (5.59) and (5.60) are utilized in (5.58), it is found that the complex Hessian with respect to the augmented matrix variable can be written as

$$\mathcal{H}_{\mathcal{Z},\mathcal{Z}} f = \begin{bmatrix} \mathcal{D}_Z (\mathcal{D}_Z f)^T & \mathcal{D}_{Z^*} (\mathcal{D}_Z f)^T \\ \mathcal{D}_Z (\mathcal{D}_{Z^*} f)^T & \mathcal{D}_{Z^*} (\mathcal{D}_{Z^*} f)^T \end{bmatrix} = \begin{bmatrix} \mathcal{H}_{Z,Z} f & \mathcal{H}_{Z^*,Z} f \\ \mathcal{H}_{Z,Z^*} f & \mathcal{H}_{Z^*,Z^*} f \end{bmatrix}, \tag{5.62}$$

which proves the first equality in the lemma. The second equality in (5.56) follows from block matrix multiplication. ■

Lemma 5.3 gives the connection between the Hessian $\mathcal{H}_{\mathcal{Z},\mathcal{Z}} f$, which was identified in Subsection 5.3.2, and the four Hessians $\mathcal{H}_{Z,Z^*} f$, $\mathcal{H}_{Z^*,Z^*} f$, $\mathcal{H}_{Z,Z} f$, and $\mathcal{H}_{Z^*,Z} f$, which were studied in Subsection 5.3.1. Through the relations in (5.56), the connection between these complex Hessian matrices is found.

Assume that the second-order differential of f can be written as

$$d^2 f = \left[d\,\text{vec}^T(Z)\; d\,\text{vec}^T(Z^*)\right] \begin{bmatrix} A_{1,0} & A_{1,1} \\ A_{0,0} & A_{0,1} \end{bmatrix} \begin{bmatrix} d\,\text{vec}(Z) \\ d\,\text{vec}(Z^*) \end{bmatrix}. \tag{5.63}$$

The middle matrix on the right-hand side of the above equation is identified as A in (5.53) when the procedure in Table 5.3 is used because the first and last factors on the right-hand side of (5.63) are equal to $d\,\text{vec}^T(\mathcal{Z})$ and $d\,\text{vec}(\mathcal{Z})$, respectively. By completing the procedure in Table 5.3, it is seen that the Hessian with respect to the augmented matrix variable $\mathcal{Z}$ can be written as

$$\mathcal{H}_{\mathcal{Z},\mathcal{Z}} f = \frac{1}{2}\left\{ \begin{bmatrix} A_{1,0} & A_{1,1} \\ A_{0,0} & A_{0,1} \end{bmatrix} + \begin{bmatrix} A_{1,0}^T & A_{0,0}^T \\ A_{1,1}^T & A_{0,1}^T \end{bmatrix} \right\} = \frac{1}{2} \begin{bmatrix} A_{1,0} + A_{1,0}^T & A_{1,1} + A_{0,0}^T \\ A_{0,0} + A_{1,1}^T & A_{0,1} + A_{0,1}^T \end{bmatrix}. \tag{5.64}$$

By comparing (5.56) and (5.64), it is seen that the four identification equations for the four Hessians $\mathcal{H}_{Z,Z} f$, $\mathcal{H}_{Z^*,Z} f$, $\mathcal{H}_{Z,Z^*} f$, and $\mathcal{H}_{Z^*,Z^*} f$ in (5.31), (5.37), (5.36), and (5.34) are in agreement with the results found here. Hence, the two methods in Tables 5.2 and 5.3 are in agreement with each other.

Let $f \in \mathbb{R}$ be a real-valued function. The second-order differential from Subsections 5.3.1 and 5.3.2 can be put together in the following manner:

$$d^2 f = \left(d\,\text{vec}^H(\mathcal{Z})\right) \begin{bmatrix} \mathcal{H}_{Z,Z^*} f & \mathcal{H}_{Z^*,Z^*} f \\ \mathcal{H}_{Z,Z} f & \mathcal{H}_{Z^*,Z} f \end{bmatrix} d\,\text{vec}(\mathcal{Z}) \tag{5.65}$$

$$= \left(d\,\text{vec}^T(\mathcal{Z})\right) \left[\mathcal{H}_{\mathcal{Z},\mathcal{Z}} f\right] d\,\text{vec}(\mathcal{Z}). \tag{5.66}$$

From (5.65) and (5.66), it is seen that $d^2 f$ can be expressed in two equivalent ways (i.e., $(dz^H)Adz$ in (5.65) or $(dz^T)Bdz$ in (5.66)). Note that when studying the nature of stationary points of real-valued scalar functions, it is quadratic forms of the type $(dz^H)Adz$ that are considered, and *not* quadratic forms of the shape $(dz^T)Bdz$. From (5.65), it is seen that for a stationary point to be minimum or maximum, the matrix $\begin{bmatrix} \mathcal{H}_{Z,Z^*} f & \mathcal{H}_{Z^*,Z^*} f \\ \mathcal{H}_{Z,Z} f & \mathcal{H}_{Z^*,Z} f \end{bmatrix}$ should be positive or negative definite in the stationary point for a minimum or maximum, respectively. Checking the definiteness of the matrix $\mathcal{H}_{\mathcal{Z},\mathcal{Z}} f$ is *not* relevant for determining the nature of a stationary point. Lemma 5.3 gives the connection between the two middle matrices on the right-hand side of (5.65) and (5.66).

5.4 Complex Hessian Matrices of Vector Functions

In this section, the augmented matrix variable $\mathcal{Z} \in \mathbb{C}^{N \times 2Q}$ is used, and a theory is developed for how to find the complex Hessian of vector functions. Consider the twice differentiable complex-valued *vector function* $\boldsymbol{f}$ defined by $\boldsymbol{f} : \mathbb{C}^{N \times 2Q} \rightarrow \mathbb{C}^{M \times 1}$, which depends only on the matrix $\mathcal{Z}$ and is denoted by $\boldsymbol{f}(\mathcal{Z})$. In Chapters 2, 3, and 4, the vector function that was studied was $\boldsymbol{f} : \mathbb{C}^{N \times Q} \times \mathbb{C}^{N \times Q} \rightarrow \mathbb{C}^{M \times 1}$, and it was denoted by $\boldsymbol{f}(\boldsymbol{Z}, \boldsymbol{Z}^*)$, where the input matrix variables were $\boldsymbol{Z} \in \mathbb{C}^{N \times Q}$ and $\boldsymbol{Z}^* \in \mathbb{C}^{N \times Q}$. To simplify the presentation for finding the complex Hessian of vector functions, the augmented matrix $\mathcal{Z}$ is used in this section.

Let the i-th component of the vector $\boldsymbol{f}$ be denoted f_i. Because all the functions f_i are scalar complex-valued functions $f_i : \mathbb{C}^{N \times 2Q} \rightarrow \mathbb{C}$, we know from Subsection 5.3.2 how to identify the Hessians of the functions $f_i(\mathcal{Z})$ for each $i \in \{0, 1, \ldots, M-1\}$. This can now be used to find the complex Hessian matrix of complex-valued vector functions. It will be shown how the complex Hessian matrix of the vector function $\boldsymbol{f}$ can be identified from the second-order differential of the whole vector function (i.e., $d^2 \boldsymbol{f}$).

Definition 5.2 (Hessian of Complex Vector Functions) *The Hessian matrix of the vector function* $\boldsymbol{f} : \mathbb{C}^{N \times 2Q} \rightarrow \mathbb{C}^{M \times 1}$ *is denoted by* $\mathcal{H}_{\mathcal{Z},\mathcal{Z}} \boldsymbol{f}$ *and has a size of* $2NQM \times 2NQ$. *It is defined as*

$$\mathcal{H}_{\mathcal{Z},\mathcal{Z}} \boldsymbol{f} \triangleq \begin{bmatrix} \mathcal{H}_{\mathcal{Z},\mathcal{Z}} f_0 \\ \mathcal{H}_{\mathcal{Z},\mathcal{Z}} f_1 \\ \vdots \\ \mathcal{H}_{\mathcal{Z},\mathcal{Z}} f_{M-1} \end{bmatrix}, \tag{5.67}$$

where the Hessian matrix of the i-th component function f_i *has size* $2NQ \times 2NQ$ *for all* $i \in \{0, 1, \ldots, M-1\}$ *and is denoted by* $\mathcal{H}_{\mathcal{Z},\mathcal{Z}} f_i$. *The complex Hessian of a scalar function was defined in Definition 5.1. An alternative identical expression of the complex*

Hessian matrix of the vector function $\boldsymbol{f}$ *is*

$$\mathcal{H}_{\mathcal{Z},\mathcal{Z}}\boldsymbol{f} = \mathcal{D}_{\mathcal{Z}}\left(\mathcal{D}_{\mathcal{Z}}\boldsymbol{f}\right)^T, \tag{5.68}$$

which is a natural extension of Definition 5.1.

The second-order differential $d^2\boldsymbol{f} \in \mathbb{C}^{M\times 1}$ can be expressed as

$$d^2\boldsymbol{f} = d\left(d\boldsymbol{f}\right) = d\begin{bmatrix} df_0 \\ df_1 \\ \vdots \\ df_{M-1} \end{bmatrix} = \begin{bmatrix} d^2 f_0 \\ d^2 f_1 \\ \vdots \\ d^2 f_{M-1} \end{bmatrix}. \tag{5.69}$$

Because it was shown in Subsection 5.3.2 how to identify the Hessian of the scalar component function $f_i : \mathbb{C}^{N\times 2Q} \to \mathbb{C}$, this can now be used in the following developments. From (5.52), it follows that $d^2 f_i = \left(d\operatorname{vec}^T\left(\mathcal{Z}\right)\right)\left[\mathcal{H}_{\mathcal{Z},\mathcal{Z}} f_i\right] d\operatorname{vec}\left(\mathcal{Z}\right)$. Using this result in (5.69) leads to

$$\begin{aligned}
d^2\boldsymbol{f} &= \begin{bmatrix} d^2 f_0 \\ d^2 f_1 \\ \vdots \\ d^2 f_{M-1} \end{bmatrix} = \begin{bmatrix} \left(d\operatorname{vec}^T\left(\mathcal{Z}\right)\right)\left[\mathcal{H}_{\mathcal{Z},\mathcal{Z}} f_0\right] d\operatorname{vec}\left(\mathcal{Z}\right) \\ \left(d\operatorname{vec}^T\left(\mathcal{Z}\right)\right)\left[\mathcal{H}_{\mathcal{Z},\mathcal{Z}} f_1\right] d\operatorname{vec}\left(\mathcal{Z}\right) \\ \vdots \\ \left(d\operatorname{vec}^T\left(\mathcal{Z}\right)\right)\left[\mathcal{H}_{\mathcal{Z},\mathcal{Z}} f_{M-1}\right] d\operatorname{vec}\left(\mathcal{Z}\right) \end{bmatrix} \\
&= \begin{bmatrix} \left(d\operatorname{vec}^T\left(\mathcal{Z}\right)\right)\mathcal{H}_{\mathcal{Z},\mathcal{Z}} f_0 \\ \left(d\operatorname{vec}^T\left(\mathcal{Z}\right)\right)\mathcal{H}_{\mathcal{Z},\mathcal{Z}} f_1 \\ \vdots \\ \left(d\operatorname{vec}^T\left(\mathcal{Z}\right)\right)\mathcal{H}_{\mathcal{Z},\mathcal{Z}} f_{M-1} \end{bmatrix} d\operatorname{vec}\left(\mathcal{Z}\right) \\
&= \begin{bmatrix} d\operatorname{vec}^T\left(\mathcal{Z}\right) & \mathbf{0}_{1\times 2NQ} & \cdots & \mathbf{0}_{1\times 2NQ} \\ \mathbf{0}_{1\times 2NQ} & d\operatorname{vec}^T\left(\mathcal{Z}\right) & \cdots & \mathbf{0}_{1\times 2NQ} \\ \vdots & \vdots & \ddots & \vdots \\ \mathbf{0}_{1\times 2NQ} & \mathbf{0}_{1\times 2NQ} & \cdots & d\operatorname{vec}^T\left(\mathcal{Z}\right) \end{bmatrix} \begin{bmatrix} \mathcal{H}_{\mathcal{Z},\mathcal{Z}} f_0 \\ \mathcal{H}_{\mathcal{Z},\mathcal{Z}} f_1 \\ \vdots \\ \mathcal{H}_{\mathcal{Z},\mathcal{Z}} f_{M-1} \end{bmatrix} d\operatorname{vec}\left(\mathcal{Z}\right) \\
&= \left[\boldsymbol{I}_M \otimes d\operatorname{vec}^T\left(\mathcal{Z}\right)\right] \begin{bmatrix} \mathcal{H}_{\mathcal{Z},\mathcal{Z}} f_0 \\ \mathcal{H}_{\mathcal{Z},\mathcal{Z}} f_1 \\ \vdots \\ \mathcal{H}_{\mathcal{Z},\mathcal{Z}} f_{M-1} \end{bmatrix} d\operatorname{vec}\left(\mathcal{Z}\right) \\
&= \left[\boldsymbol{I}_M \otimes d\operatorname{vec}^T\left(\mathcal{Z}\right)\right]\left[\mathcal{H}_{\mathcal{Z},\mathcal{Z}}\boldsymbol{f}\right] d\operatorname{vec}\left(\mathcal{Z}\right),
\end{aligned} \tag{5.70}$$

where Definition 5.2 of the complex Hessian matrix of a vector function (see (5.67)) has been used in the last equality.

If the complex-valued vector function $\boldsymbol{f}$ is twice differentiable of all the components within $\mathcal{Z}$, then all the Hessian matrices $\mathcal{H}_{\mathcal{Z},\mathcal{Z}} f_i$ are symmetric and the Hessian matrix $\mathcal{H}_{\mathcal{Z},\mathcal{Z}} \boldsymbol{f}$ is said to be *column symmetric* (see Definition 2.13).

Assume that the second-order differential expression $d^2 \boldsymbol{f}$ can be written as follows:

$$d^2 \boldsymbol{f} = \left[\boldsymbol{I}_M \otimes d\, \mathrm{vec}^T(\mathcal{Z})\right] \boldsymbol{B}\, d\, \mathrm{vec}(\mathcal{Z}), \tag{5.71}$$

where the matrix $\boldsymbol{B} \in \mathbb{C}^{2NQM \times 2NQ}$ may depend on $\mathcal{Z}$ and $\mathcal{Z}^*$; however, it does *not* depend on the differential operator d. The matrix $\boldsymbol{B}$ can be expressed as

$$\boldsymbol{B} = \begin{bmatrix} \boldsymbol{B}_0 \\ \boldsymbol{B}_1 \\ \vdots \\ \boldsymbol{B}_{M-1} \end{bmatrix}, \tag{5.72}$$

where $\boldsymbol{B}_i \in \mathbb{C}^{2NQ \times 2NQ}$ is a complex square matrix for all $i \in \{0, 1, \ldots, M-1\}$. The transposed of the matrix $\boldsymbol{B}$ can be written as follows:

$$\boldsymbol{B}^T = \left[\boldsymbol{B}_0^T \; \boldsymbol{B}_1^T \; \cdots \; \boldsymbol{B}_{M-1}^T\right]. \tag{5.73}$$

To identify the Hessian matrix of $\boldsymbol{f}$, the following matrix is needed:

$$\mathrm{vecb}\left(\boldsymbol{B}^T\right) = \begin{bmatrix} \boldsymbol{B}_0^T \\ \boldsymbol{B}_1^T \\ \vdots \\ \boldsymbol{B}_{M-1}^T \end{bmatrix}, \tag{5.74}$$

where the block vectorization operator $\mathrm{vecb}(\cdot)$ from Definition 2.13 is used.

Because $d^2 \boldsymbol{f}$ is on the left-hand side of both (5.70) and (5.71), the right-hand side expressions of these equations have to be equal as well. Using Lemma 2.19[3] on the right-hand-side expressions in (5.70) and (5.71), it follows that

$$\mathcal{H}_{\mathcal{Z},\mathcal{Z}} \boldsymbol{f} + \mathrm{vecb}([\mathcal{H}_{\mathcal{Z},\mathcal{Z}} \boldsymbol{f}]^T) = \boldsymbol{B} + \mathrm{vecb}\left(\boldsymbol{B}^T\right). \tag{5.75}$$

For a twice differentiable vector function, the Hessian $\mathcal{H}_{\mathcal{Z},\mathcal{Z}} \boldsymbol{f}$ must be column symmetric; hence, the relation $\mathrm{vecb}([\mathcal{H}_{\mathcal{Z},\mathcal{Z}} \boldsymbol{f}]^T) = \mathcal{H}_{\mathcal{Z},\mathcal{Z}} \boldsymbol{f}$ is valid. By using the column symmetry in (5.75), it follows that

$$\mathcal{H}_{\mathcal{Z},\mathcal{Z}} \boldsymbol{f} = \frac{1}{2}\left[\boldsymbol{B} + \mathrm{vecb}\left(\boldsymbol{B}^T\right)\right]. \tag{5.76}$$

The identification equation (5.76) for complex-valued Hessian matrices of vector functions is a generalization of identification in Magnus and Neudecker (1988, p. 108) to

[3] Notice that when using Lemma 2.19 here, the vector variable z in Lemma 2.19 is substituted with the differential vector $d\, \mathrm{vec}(\mathcal{Z})$ and the matrices $\boldsymbol{A}$ and $\boldsymbol{B}$ in Lemma 2.19 are replaced by $\mathcal{H}_{\mathcal{Z},\mathcal{Z}} \boldsymbol{f}$ (from (5.67)) and $\boldsymbol{B}$ (from (5.71)), respectively.

Table 5.4 Procedure for identifying the complex Hessians of a vector function $\boldsymbol{f} \in \mathbb{C}^{M\times 1}$ with respect to the augmented complex-valued matrix variable $\boldsymbol{\mathcal{Z}} \in \mathbb{C}^{N\times 2Q}$.

Step 1:	Compute the second-order differential $d^2\boldsymbol{f}$.
Step 2:	Manipulate $d^2\boldsymbol{f}$ into the form given in (5.71) in order to identify the matrix $\boldsymbol{B} \in \mathbb{C}^{2NQM\times 2NQ}$.
Step 3:	Use (5.76) to find the complex Hessian $\mathcal{H}_{\boldsymbol{\mathcal{Z}},\boldsymbol{\mathcal{Z}}}\boldsymbol{f}$.

the case of *complex-valued* vector functions. The procedure for finding the complex Hessian matrix of a vector function is summarized in Table 5.4.

5.5 Complex Hessian Matrices of Matrix Functions

Let $\boldsymbol{F} : \mathbb{C}^{N\times Q} \times \mathbb{C}^{N\times Q} \to \mathbb{C}^{M\times P}$ be a matrix function that depends on the two matrices $\boldsymbol{Z} \in \mathbb{C}^{N\times Q}$ and $\boldsymbol{Z}^* \in \mathbb{C}^{N\times Q}$. An alternative equivalent representation of this function is $\boldsymbol{F} : \mathbb{C}^{N\times 2Q} \to \mathbb{C}^{M\times P}$ and is denoted by $\boldsymbol{F}(\boldsymbol{\mathcal{Z}})$, where the augmented matrix variable $\boldsymbol{\mathcal{Z}} \in \mathbb{C}^{N\times 2Q}$ is used. The last representation will be used in this section.

To identify the Hessian of a complex-valued matrix function, the second-order differential expression $d^2 \operatorname{vec}(\boldsymbol{F})$ will be used. This is a natural generalization of the second-order differential expressions used for scalar- and vector-valued functions presented earlier in this chapter; it can also be remembered as the differential of the differential expression that is used to identify the first-order derivatives of a matrix function in Definition 3.1 (i.e., $d\,(d \operatorname{vec}(\boldsymbol{F}))$).

Let the (k, l)-th component function of $\boldsymbol{F}$ be denoted by $f_{k,l}$, such that $f_{k,l} : \mathbb{C}^{N\times 2Q} \to \mathbb{C}$ is the scalar component function where $k \in \{0, 1, \ldots, M-1\}$ and $l \in \{0, 1, \ldots, P-1\}$ are the row and column numbers of the matrix function $\boldsymbol{F}$. Second-order differential $d^2 \operatorname{vec}(\boldsymbol{F}) \in \mathbb{C}^{MP\times 1}$ can be expressed as follows:

$$d^2 \operatorname{vec}(\boldsymbol{F}) = d\,(d \operatorname{vec}(\boldsymbol{F})) = d \begin{bmatrix} df_{0,0} \\ df_{1,0} \\ \vdots \\ df_{M-1,0} \\ df_{0,1} \\ \vdots \\ df_{0,P-1} \\ \vdots \\ df_{M-1,P-1} \end{bmatrix} = \begin{bmatrix} d^2 f_{0,0} \\ d^2 f_{1,0} \\ \vdots \\ d^2 f_{M-1,0} \\ d^2 f_{0,1} \\ \vdots \\ d^2 f_{0,P-1} \\ \vdots \\ d^2 f_{M-1,P-1} \end{bmatrix}. \quad (5.77)$$

Next, the definition of the complex Hessian matrix of a matrix function of the type $\boldsymbol{F} : \mathbb{C}^{N\times 2Q} \to \mathbb{C}^{M\times P}$ is stated.

Definition 5.3 (Hessian Matrix of Complex Matrix Function) *The Hessian of the matrix function* $\boldsymbol{F} : \mathbb{C}^{N\times 2Q} \to \mathbb{C}^{M\times P}$ *is a matrix of size* $2NQMP \times 2NQ$ *and is defined by the* MP *scalar component functions within* $\boldsymbol{F}$ *in the following way:*

$$\mathcal{H}_{\mathcal{Z},\mathcal{Z}}\boldsymbol{F} \triangleq \begin{bmatrix} \mathcal{H}_{\mathcal{Z},\mathcal{Z}} f_{0,0} \\ \mathcal{H}_{\mathcal{Z},\mathcal{Z}} f_{1,0} \\ \vdots \\ \mathcal{H}_{\mathcal{Z},\mathcal{Z}} f_{M-1,0} \\ \mathcal{H}_{\mathcal{Z},\mathcal{Z}} f_{0,1} \\ \vdots \\ \mathcal{H}_{\mathcal{Z},\mathcal{Z}} f_{0,P-1} \\ \vdots \\ \mathcal{H}_{\mathcal{Z},\mathcal{Z}} f_{M-1,P-1} \end{bmatrix}, \tag{5.78}$$

where the matrix $\mathcal{H}_{\mathcal{Z},\mathcal{Z}} f_{i,j}$ *of size* $2NQ \times 2NQ$ *is the complex Hessian of the component function* $f_{i,j}$ *given in Definition 5.1. The Hessian matrix of* $\boldsymbol{F}$ *can equivalently be expressed as*

$$\mathcal{H}_{\mathcal{Z},\mathcal{Z}}\boldsymbol{F} = \mathcal{D}_{\mathcal{Z}} \left(\mathcal{D}_{\mathcal{Z}}\boldsymbol{F}\right)^T. \tag{5.79}$$

By comparing (5.14) and (5.79), it is seen that Definition 5.3 is a natural extension of Definition 5.1. The two expressions (3.82) and (5.79) are used to find the following alternative expression of the complex Hessian of a matrix function:

$$\mathcal{H}_{\mathcal{Z},\mathcal{Z}}\boldsymbol{F} = \mathcal{D}_{\mathcal{Z}} \left[\frac{\partial \operatorname{vec}(\boldsymbol{F})}{\partial \operatorname{vec}^T(\mathcal{Z})}\right]^T = \frac{\partial}{\partial \operatorname{vec}^T(\mathcal{Z})} \operatorname{vec}\left(\left[\frac{\partial \operatorname{vec}(\boldsymbol{F})}{\partial \operatorname{vec}^T(\mathcal{Z})}\right]^T\right). \tag{5.80}$$

In this chapter, it is assumed that all component functions of $\boldsymbol{F}(\mathcal{Z})$, which are defined as $f_{i,j} : \mathbb{C}^{N\times 2Q} \to \mathbb{C}$, are twice differentiable; hence, the complex Hessian matrix $\mathcal{H}_{\mathcal{Z},\mathcal{Z}} f_{i,j}$ is symmetric such that the Hessian matrix $\mathcal{H}_{\mathcal{Z},\mathcal{Z}}\boldsymbol{F}$ is column symmetric. The column symmetry of the complex Hessian matrix $\mathcal{H}_{\mathcal{Z},\mathcal{Z}}\boldsymbol{F}$ can be expressed as

$$\operatorname{vecb}\left(\left[\mathcal{H}_{\mathcal{Z},\mathcal{Z}}\boldsymbol{F}\right]^T\right) = \mathcal{H}_{\mathcal{Z},\mathcal{Z}}\boldsymbol{F}. \tag{5.81}$$

To find an expression of the complex Hessian matrix of the matrix function $\boldsymbol{F}$, the following calculations are used:

$$d^2 \operatorname{vec}(\boldsymbol{F}) = \begin{bmatrix} d^2 f_{0,0} \\ d^2 f_{1,0} \\ \vdots \\ d^2 f_{M-1,0} \\ d^2 f_{0,1} \\ \vdots \\ d^2 f_{0,P-1} \\ \vdots \\ d^2 f_{M-1,P-1} \end{bmatrix} = \begin{bmatrix} \left(d \operatorname{vec}^T(\mathcal{Z})\right) \left[\mathcal{H}_{\mathcal{Z},\mathcal{Z}} f_{0,0}\right] d \operatorname{vec}(\mathcal{Z}) \\ \left(d \operatorname{vec}^T(\mathcal{Z})\right) \left[\mathcal{H}_{\mathcal{Z},\mathcal{Z}} f_{1,0}\right] d \operatorname{vec}(\mathcal{Z}) \\ \vdots \\ \left(d \operatorname{vec}^T(\mathcal{Z})\right) \left[\mathcal{H}_{\mathcal{Z},\mathcal{Z}} f_{M-1,0}\right] d \operatorname{vec}(\mathcal{Z}) \\ \left(d \operatorname{vec}^T(\mathcal{Z})\right) \left[\mathcal{H}_{\mathcal{Z},\mathcal{Z}} f_{0,1}\right] d \operatorname{vec}(\mathcal{Z}) \\ \vdots \\ \left(d \operatorname{vec}^T(\mathcal{Z})\right) \left[\mathcal{H}_{\mathcal{Z},\mathcal{Z}} f_{0,P-1}\right] d \operatorname{vec}(\mathcal{Z}) \\ \vdots \\ \left(d \operatorname{vec}^T(\mathcal{Z})\right) \left[\mathcal{H}_{\mathcal{Z},\mathcal{Z}} f_{M-1,P-1}\right] d \operatorname{vec}(\mathcal{Z}) \end{bmatrix}$$

$$= \begin{bmatrix} \left(d \operatorname{vec}^T(\mathcal{Z})\right) \mathcal{H}_{\mathcal{Z},\mathcal{Z}} f_{0,0} \\ \vdots \\ \left(d \operatorname{vec}^T(\mathcal{Z})\right) \mathcal{H}_{\mathcal{Z},\mathcal{Z}} f_{M-1,0} \\ \vdots \\ \left(d \operatorname{vec}^T(\mathcal{Z})\right) \mathcal{H}_{\mathcal{Z},\mathcal{Z}} f_{M-1,P-1} \end{bmatrix} d \operatorname{vec}(\mathcal{Z})$$

$$= \begin{bmatrix} d \operatorname{vec}^T(\mathcal{Z}) & \cdots & \boldsymbol{0}_{1\times 2NQ} & \cdots & \boldsymbol{0}_{1\times 2NQ} \\ & \ddots & & \ddots & \\ \boldsymbol{0}_{1\times 2NQ} & \cdots & d \operatorname{vec}^T(\mathcal{Z}) & \cdots & \boldsymbol{0}_{1\times 2NQ} \\ & \ddots & & \ddots & \\ \boldsymbol{0}_{1\times 2NQ} & \cdots & \boldsymbol{0}_{1\times 2NQ} & \cdots & d \operatorname{vec}^T(\mathcal{Z}) \end{bmatrix} \begin{bmatrix} \mathcal{H}_{\mathcal{Z},\mathcal{Z}} f_{0,0} \\ \vdots \\ \mathcal{H}_{\mathcal{Z},\mathcal{Z}} f_{M-1,0} \\ \vdots \\ \mathcal{H}_{\mathcal{Z},\mathcal{Z}} f_{M-1,P-1} \end{bmatrix} d \operatorname{vec}(\mathcal{Z})$$

$$= \left(\boldsymbol{I}_{MP} \otimes d \operatorname{vec}^T(\mathcal{Z})\right) \left[\mathcal{H}_{\mathcal{Z},\mathcal{Z}} \boldsymbol{F}\right] d \operatorname{vec}(\mathcal{Z}), \tag{5.82}$$

where the definition in (5.78) has been used in the last equality.

To identify the Hessian of a complex-valued matrix function, assume that the following expression can be found:

$$d^2 \operatorname{vec}(\boldsymbol{F}) = \left(\boldsymbol{I}_{MP} \otimes d \operatorname{vec}^T(\mathcal{Z})\right) \boldsymbol{C} d \operatorname{vec}(\mathcal{Z}), \tag{5.83}$$

where the matrix $\boldsymbol{C} \in \mathbb{C}^{2NQMP \times 2NQ}$ may depend on $\mathcal{Z}$ and $\mathcal{Z}^*$; however, it may *not* depend on the differential operator d. The matrix $\boldsymbol{C}$ is given by

$$\boldsymbol{C} = \begin{bmatrix} \boldsymbol{C}_{0,0} \\ \vdots \\ \boldsymbol{C}_{M-1,0} \\ \vdots \\ \boldsymbol{C}_{M-1,P-1} \end{bmatrix}, \tag{5.84}$$

Table 5.5 Procedure for identifying the complex Hessian matrix of the matrix function $\boldsymbol{F} \in \mathbb{C}^{M\times P}$ with respect to the augmented complex-valued matrix variable $\boldsymbol{\mathcal{Z}} \in \mathbb{C}^{N\times 2Q}$.

Step 1:	Compute the second-order differential $d^2 \operatorname{vec}(\boldsymbol{F})$.
Step 2:	Manipulate $d^2 \operatorname{vec}(\boldsymbol{F})$ into the form given in (5.83) to identify the matrix $\boldsymbol{C} \in \mathbb{C}^{2NQMP\times 2NQ}$.
Step 3:	Use (5.87) to find the complex Hessian $\mathcal{H}_{\boldsymbol{\mathcal{Z}},\boldsymbol{\mathcal{Z}}}\boldsymbol{F}$.

where $\boldsymbol{C}_{k,l} \in \mathbb{C}^{2NQ\times 2NQ}$ is square complex-valued matrices. The transposed of the matrix $\boldsymbol{C}$ can be expressed as

$$\boldsymbol{C}^T = \left[\boldsymbol{C}_{0,0}^T \cdots \boldsymbol{C}_{M-1,0}^T \cdots \boldsymbol{C}_{M-1,P-1}^T\right]. \tag{5.85}$$

The block vectorization applied on $\boldsymbol{C}^T$ is given by

$$\operatorname{vecb}\left(\boldsymbol{C}^T\right) = \begin{bmatrix} \boldsymbol{C}_{0,0}^T \\ \vdots \\ \boldsymbol{C}_{M-1,0}^T \\ \vdots \\ \boldsymbol{C}_{M-1,P-1}^T \end{bmatrix}. \tag{5.86}$$

The expression $\operatorname{vecb}\left(\boldsymbol{C}^T\right)$ will be used as part of the expression that finds the complex Hessian of matrix functions.

For a twice differentiable matrix function $\boldsymbol{F}$, the Hessian matrix $\mathcal{H}_{\boldsymbol{\mathcal{Z}},\boldsymbol{\mathcal{Z}}}\boldsymbol{F}$ is column symmetric such that it satisfies (5.81). Because the left-hand-side expressions of (5.82) and (5.83) are identical for all $d\boldsymbol{\mathcal{Z}}$, Lemma 2.19 can be used on the right-hand-side expressions of (5.82) and (5.83). When using Lemma 2.19, the matrices $\boldsymbol{A}$ and $\boldsymbol{B}$ of this lemma are substituted by $\mathcal{H}_{\boldsymbol{\mathcal{Z}},\boldsymbol{\mathcal{Z}}}\boldsymbol{F}$ and $\boldsymbol{C}$, respectively, and the vector $\boldsymbol{z}$ of Lemma 2.19 is replaced by $d\operatorname{vec}(\boldsymbol{\mathcal{Z}})$. Making these substitutions in (2.125) and solving the equation for $\mathcal{H}_{\boldsymbol{\mathcal{Z}},\boldsymbol{\mathcal{Z}}}\boldsymbol{F}$ gives us the following identification equation for the complex Hessian of a matrix function:

$$\mathcal{H}_{\boldsymbol{\mathcal{Z}},\boldsymbol{\mathcal{Z}}}\boldsymbol{F} = \frac{1}{2}\left[\boldsymbol{C} + \operatorname{vecb}\left(\boldsymbol{C}^T\right)\right]. \tag{5.87}$$

Based on the above result, the procedure of finding the complex Hessian of a matrix function is summarized in Table 5.5.

A theory has now been developed for finding the complex Hessian matrix of all three types of scalar, vector, and matrix functions given in Table 5.1. For these three function types that depend on the augmented matrix variable $\boldsymbol{\mathcal{Z}} \in \mathbb{C}^{N\times 2Q}$, and treated in Subsection 5.3.2, Sections 5.4, and 5.5, respectively, the identifying relations for finding the complex Hessian matrix are summarized in Table 5.6. From this table, it can be seen that the vector case is a special case of the matrix case by setting $P = 1$. Furthermore, the scalar case is a special case of the vector case when $M = 1$.

Table 5.6 Identification table for complex Hessian matrices of scalar, vector, and matrix functions, which depend on the augmented matrix variable $\mathcal{Z} \in \mathbb{C}^{N\times 2Q}$.

Function type	Second-order differential	Hessian wrt. $\mathcal{Z}$	Size of Hessian
$f : \mathbb{C}^{N\times 2Q} \to \mathbb{C}$	$d^2 f = \left(d\,\mathrm{vec}^T(\boldsymbol{Z})\right) \boldsymbol{A} d\,\mathrm{vec}(\boldsymbol{Z})$	$\mathcal{H}_{\mathcal{Z},\mathcal{Z}} f = \frac{1}{2}\left[\boldsymbol{A} + \boldsymbol{A}^T\right]$	$2NQ \times 2NQ$
$\boldsymbol{f} : \mathbb{C}^{N\times 2Q} \to \mathbb{C}^{M\times 1}$	$d^2 \boldsymbol{f} = \left[\boldsymbol{I}_M \otimes d\,\mathrm{vec}^T(\mathcal{Z})\right] \boldsymbol{B} d\,\mathrm{vec}(\mathcal{Z})$	$\mathcal{H}_{\mathcal{Z},\mathcal{Z}} \boldsymbol{f} = \frac{1}{2}\left[\boldsymbol{B} + \mathrm{vecb}\left(\boldsymbol{B}^T\right)\right]$	$2NQM \times 2NQ$
$\boldsymbol{F} : \mathbb{C}^{N\times 2Q} \to \mathbb{C}^{M\times P}$	$d^2\,\mathrm{vec}(\boldsymbol{F}) = \left(\boldsymbol{I}_{MP} \otimes d\,\mathrm{vec}^T(\mathcal{Z})\right) \boldsymbol{C} d\,\mathrm{vec}(\mathcal{Z})$	$\mathcal{H}_{\mathcal{Z},\mathcal{Z}} \boldsymbol{F} = \frac{1}{2}\left[\boldsymbol{C} + \mathrm{vecb}\left(\boldsymbol{C}^T\right)\right]$	$2NQMP \times 2NQ$

5.5.1 Alternative Expression of Hessian Matrix of Matrix Function

In this subsection, an alternative explicit formula will be developed for finding the complex Hessian matrix of the matrix function $\boldsymbol{F} : \mathbb{C}^{N \times 2Q} \to \mathbb{C}^{M \times P}$. By using (3.85), the derivative of $\boldsymbol{F} \in \mathbb{C}^{M \times P}$ with respect to $\boldsymbol{\mathcal{Z}} \in \mathbb{C}^{N \times 2Q}$ is given by

$$\mathcal{D}_{\boldsymbol{\mathcal{Z}}} \boldsymbol{F} = \sum_{n=0}^{N-1} \sum_{q=0}^{2Q-1} \operatorname{vec}\left(\frac{\partial \boldsymbol{F}}{\partial w_{n,q}} \right) \operatorname{vec}^T \left(\boldsymbol{E}_{n,q} \right), \tag{5.88}$$

where $\boldsymbol{E}_{n,q}$ is an $N \times 2Q$ matrix with zeros everywhere and $+1$ at position number (n, q), and the (n, q)-th element of $\boldsymbol{\mathcal{Z}}$ is denoted by $w_{n,q}$ because the symbol $z_{n,q}$ is used earlier to denote the (n, q)-th element of $\boldsymbol{Z}$, which is a submatrix of $\boldsymbol{\mathcal{Z}}$, see (5.2). By using (5.79) and (5.88), the following calculations are done to find an explicit formula for the complex Hessian of the matrix function $\boldsymbol{F}$:

$$\begin{aligned}
\mathcal{H}_{\boldsymbol{\mathcal{Z}},\boldsymbol{\mathcal{Z}}} \boldsymbol{F} &= \mathcal{D}_{\boldsymbol{\mathcal{Z}}} \left(\mathcal{D}_{\boldsymbol{\mathcal{Z}}} \boldsymbol{F} \right)^T = \mathcal{D}_{\boldsymbol{\mathcal{Z}}} \left(\sum_{i=0}^{N-1} \sum_{j=0}^{2Q-1} \operatorname{vec}\left(\boldsymbol{E}_{i,j} \right) \operatorname{vec}^T \left(\frac{\partial \boldsymbol{F}}{\partial w_{i,j}} \right) \right) \\
&= \sum_{i=0}^{N-1} \sum_{j=0}^{2Q-1} \mathcal{D}_{\boldsymbol{\mathcal{Z}}} \left(\operatorname{vec}\left(\boldsymbol{E}_{i,j} \right) \operatorname{vec}^T \left(\frac{\partial \boldsymbol{F}}{\partial w_{i,j}} \right) \right) \\
&= \sum_{n=0}^{N-1} \sum_{q=0}^{2Q-1} \sum_{i=0}^{N-1} \sum_{j=0}^{2Q-1} \operatorname{vec}\left(\frac{\partial \left[\operatorname{vec}\left(\boldsymbol{E}_{i,j} \right) \operatorname{vec}^T \left(\frac{\partial \boldsymbol{F}}{\partial w_{i,j}} \right) \right]}{\partial w_{n,q}} \right) \operatorname{vec}^T \left(\boldsymbol{E}_{n,q} \right) \\
&= \sum_{n=0}^{N-1} \sum_{q=0}^{2Q-1} \sum_{i=0}^{N-1} \sum_{j=0}^{2Q-1} \operatorname{vec}\left(\operatorname{vec}\left(\boldsymbol{E}_{i,j} \right) \operatorname{vec}^T \left(\frac{\partial^2 \boldsymbol{F}}{\partial w_{n,q} \partial w_{i,j}} \right) \right) \operatorname{vec}^T \left(\boldsymbol{E}_{n,q} \right) \\
&= \sum_{n=0}^{N-1} \sum_{q=0}^{2Q-1} \sum_{i=0}^{N-1} \sum_{j=0}^{2Q-1} \left[\operatorname{vec}\left(\frac{\partial^2 \boldsymbol{F}}{\partial w_{n,q} \partial w_{i,j}} \right) \otimes \operatorname{vec}\left(\boldsymbol{E}_{i,j} \right) \right] \left[1 \otimes \operatorname{vec}^T \left(\boldsymbol{E}_{n,q} \right) \right] \\
&= \sum_{n=0}^{N-1} \sum_{q=0}^{2Q-1} \sum_{i=0}^{N-1} \sum_{j=0}^{2Q-1} \operatorname{vec}\left(\frac{\partial^2 \boldsymbol{F}}{\partial w_{n,q} \partial w_{i,j}} \right) \left[\operatorname{vec}\left(\boldsymbol{E}_{i,j} \right) \operatorname{vec}^T \left(\boldsymbol{E}_{n,q} \right) \right],
\end{aligned} \tag{5.89}$$

where (2.101) was used in the second to last equality above. The expression in (5.89) can be used to derive the complex Hessian matrix $\mathcal{H}_{\boldsymbol{\mathcal{Z}},\boldsymbol{\mathcal{Z}}} \boldsymbol{F}$ directly without going all the way through the second-order differential as mentioned earlier.

5.5.2 Chain Rule for Complex Hessian Matrices

In this subsection, the chain rule for finding the complex Hessian matrix is derived.

Theorem 5.1 (Chain Rule of Complex Hessian) *Let $\mathcal{S} \subseteq \mathbb{C}^{N \times 2Q}$, and let $\boldsymbol{F} : \mathcal{S} \to \mathbb{C}^{M \times P}$ be differentiable at an interior point $\boldsymbol{\mathcal{Z}}$ of the set $\mathcal{S}$. Let $\mathcal{T} \subseteq \mathbb{C}^{M \times P}$ be such that $\boldsymbol{F}(\boldsymbol{\mathcal{Z}}) \in \mathcal{T}$ for all $\boldsymbol{\mathcal{Z}} \in \mathcal{S}$. Assume that $\boldsymbol{G} : \mathcal{T} \to \mathbb{C}^{R \times S}$ is differentiable at an inner* *point $\boldsymbol{F}(\mathcal{Z}) \in \mathcal{T}$. Define the composite function $\boldsymbol{H} : \mathcal{S} \to \mathbb{C}^{R\times S}$ by*

$$\boldsymbol{H}(\mathcal{Z}) = \boldsymbol{G}(\boldsymbol{F}(\mathcal{Z})). \tag{5.90}$$

The complex Hessian $\mathcal{H}_{\mathcal{Z},\mathcal{Z}}\boldsymbol{H}$ is given by

$$\mathcal{H}_{\mathcal{Z},\mathcal{Z}}\boldsymbol{H} = \left[\mathcal{D}_{\boldsymbol{F}}\boldsymbol{G} \otimes \boldsymbol{I}_{2NQ}\right] \mathcal{H}_{\mathcal{Z},\mathcal{Z}}\boldsymbol{F} + \left[\boldsymbol{I}_{RS} \otimes (\mathcal{D}_{\mathcal{Z}}\boldsymbol{F})^T\right] \left[\mathcal{H}_{\boldsymbol{F},\boldsymbol{F}}\boldsymbol{G}\right] \mathcal{D}_{\mathcal{Z}}\boldsymbol{F}. \tag{5.91}$$

Proof By Theorem 3.1, it follows that $\mathcal{D}_{\mathcal{Z}}\boldsymbol{H} = (\mathcal{D}_{\boldsymbol{F}}\boldsymbol{G})\,\mathcal{D}_{\mathcal{Z}}\boldsymbol{F}$; hence,

$$(\mathcal{D}_{\mathcal{Z}}\boldsymbol{H})^T = (\mathcal{D}_{\mathcal{Z}}\boldsymbol{F})^T (\mathcal{D}_{\boldsymbol{F}}\boldsymbol{G})^T. \tag{5.92}$$

By the definition of the complex Hessian matrix (see Definition 5.3), the complex Hessian matrix of $\boldsymbol{H}$ can be found by taking the derivative with respect to $\mathcal{Z}$ of both sides of (5.92):

$$\begin{aligned}
\mathcal{H}_{\mathcal{Z},\mathcal{Z}}\boldsymbol{H} &= \mathcal{D}_{\mathcal{Z}}(\mathcal{D}_{\mathcal{Z}}\boldsymbol{H})^T = \mathcal{D}_{\mathcal{Z}}\left[(\mathcal{D}_{\mathcal{Z}}\boldsymbol{F})^T (\mathcal{D}_{\boldsymbol{F}}\boldsymbol{G})^T\right] \\
&= \left[\mathcal{D}_{\boldsymbol{F}}\boldsymbol{G} \otimes \boldsymbol{I}_{2NQ}\right] \mathcal{D}_{\mathcal{Z}}(\mathcal{D}_{\mathcal{Z}}\boldsymbol{F})^T + \left[\boldsymbol{I}_{RS} \otimes (\mathcal{D}_{\mathcal{Z}}\boldsymbol{F})^T\right] \mathcal{D}_{\mathcal{Z}}(\mathcal{D}_{\boldsymbol{F}}\boldsymbol{G})^T \\
&= \left[\mathcal{D}_{\boldsymbol{F}}\boldsymbol{G} \otimes \boldsymbol{I}_{2NQ}\right] \mathcal{H}_{\mathcal{Z},\mathcal{Z}}\boldsymbol{F} + \left[\boldsymbol{I}_{RS} \otimes (\mathcal{D}_{\mathcal{Z}}\boldsymbol{F})^T\right] \left[\mathcal{D}_{\boldsymbol{F}}(\mathcal{D}_{\boldsymbol{F}}\boldsymbol{G})^T\right] \mathcal{D}_{\mathcal{Z}}\boldsymbol{F},
\end{aligned} \tag{5.93}$$

where the derivative of a product from Lemma 3.4 has been used. In the last equality above, the chain rule was used because $\mathcal{D}_{\mathcal{Z}}(\mathcal{D}_{\boldsymbol{F}}\boldsymbol{G})^T = \left[\mathcal{D}_{\boldsymbol{F}}(\mathcal{D}_{\boldsymbol{F}}\boldsymbol{G})^T\right] \mathcal{D}_{\mathcal{Z}}\boldsymbol{F}$. By using $\mathcal{D}_{\boldsymbol{F}}(\mathcal{D}_{\boldsymbol{F}}\boldsymbol{G})^T = \mathcal{H}_{\boldsymbol{F},\boldsymbol{F}}\boldsymbol{G}$, the expression in (5.91) is obtained. ■

5.6 Examples of Finding Complex Hessian Matrices

This section contains three subsections. Subsection 5.6.1 shows several examples of how to find the complex Hessian matrices of scalar functions. Examples for how to find the Hessians of complex vector and matrix functions are shown in Subsections 5.6.2 and 5.6.3, respectively.

5.6.1 Examples of Finding Complex Hessian Matrices of Scalar Functions

---

Example 5.1 Let $f : \mathbb{C}^{N\times 1} \times \mathbb{C}^{N\times 1} \to \mathbb{C}$ be defined as

$$f(\boldsymbol{z}, \boldsymbol{z}^*) = \boldsymbol{z}^H \boldsymbol{\Phi} \boldsymbol{z}, \tag{5.94}$$

where $\boldsymbol{\Phi} \in \mathbb{C}^{N\times N}$ is independent of $\boldsymbol{z}$ and $\boldsymbol{z}^*$. The second-order differential of f is given by

$$d^2 f = 2\left(d\boldsymbol{z}^H\right) \boldsymbol{\Phi} d\boldsymbol{z} = \left[d\boldsymbol{z}^H \; d\boldsymbol{z}^T\right] \begin{bmatrix} 2\boldsymbol{\Phi} & \boldsymbol{0}_{N\times N} \\ \boldsymbol{0}_{N\times N} & \boldsymbol{0}_{N\times N} \end{bmatrix} \begin{bmatrix} d\boldsymbol{z} \\ d\boldsymbol{z}^* \end{bmatrix}. \tag{5.95}$$

From the above expression, the $\boldsymbol{A}_{k,l}$ matrices in (5.27) can be identified as

$$\boldsymbol{A}_{0,0} = 2\boldsymbol{\Phi}, \quad \boldsymbol{A}_{0,1} = \boldsymbol{A}_{1,0} = \boldsymbol{A}_{1,1} = \boldsymbol{0}_{N\times N}. \tag{5.96}$$

And from (5.31), (5.34), (5.36), and (5.37), the four complex Hessian matrices of f are found

$$\mathcal{H}_{Z,Z} f = \mathbf{0}_{N \times N}, \; \mathcal{H}_{Z^*,Z^*} f = \mathbf{0}_{N \times N}, \; \mathcal{H}_{Z^*,Z} f = \boldsymbol{\Phi}^T, \; \mathcal{H}_{Z,Z^*} f = \boldsymbol{\Phi}. \tag{5.97}$$

The function f is often used in array signal processing (Jonhson & Dudgeon 1993) and adaptive filtering (Diniz 2008). To check the convexity of the function f, use the $2NQ \times 2NQ$ middle matrix on the right-hand side of (5.25). Here, this matrix is given by

$$\begin{bmatrix} \mathcal{H}_{Z,Z^*} f & \mathcal{H}_{Z^*,Z^*} f \\ \mathcal{H}_{Z,Z} f & \mathcal{H}_{Z^*,Z} f \end{bmatrix} = \begin{bmatrix} \boldsymbol{\Phi} & \mathbf{0}_{N \times N} \\ \mathbf{0}_{N \times N} & \boldsymbol{\Phi}^T \end{bmatrix}. \tag{5.98}$$

If this is positive semidefinite, then the problem is convex.

Example 5.2 Reconsider the function in Example 5.1 given in (5.94); however, now the augmented matrix variable given in (5.2) will be used. Here, the input variables z and z^* are vectors in $\mathbb{C}^{N\times 1}$; hence, the augmented matrix variable $\mathcal{Z}$ is given by

$$\mathcal{Z} \triangleq \left[z \; z^* \right] \in \mathbb{C}^{N \times 2}. \tag{5.99}$$

The connection between the input vector variables $z \in \mathbb{C}^{N\times 1}$ and $z^* \in \mathbb{C}^{N\times 1}$ is given by the following two relations:

$$z = \mathcal{Z} \boldsymbol{e}_0, \tag{5.100}$$

$$z^* = \mathcal{Z} \boldsymbol{e}_1, \tag{5.101}$$

where the two unit vectors $\boldsymbol{e}_0 = [1 \; 0]^T$ and $\boldsymbol{e}_1 = [0 \; 1]^T$ have size 2×1.

Let us now express the function f, given in (5.94), in terms of the augmented matrix variable $\mathcal{Z}$. The function f is defined as $f : \mathbb{C}^{N\times 2} \to \mathbb{C}$, and it can be expressed as

$$f(\mathcal{Z}) = \boldsymbol{e}_1^T \mathcal{Z}^T \boldsymbol{\Phi} \mathcal{Z} \boldsymbol{e}_0, \tag{5.102}$$

where (5.100) and (5.101) are used to find expressions for z and z^H, respectively. The first-order differential of f is given by

$$df = \boldsymbol{e}_1^T \left(d\mathcal{Z}^T \right) \boldsymbol{\Phi} \mathcal{Z} \boldsymbol{e}_0 + \boldsymbol{e}_1^T \mathcal{Z}^T \boldsymbol{\Phi} \left(d\mathcal{Z} \right) \boldsymbol{e}_0. \tag{5.103}$$

Using the fact that $d^2 \mathcal{Z} = \mathbf{0}_{N\times 2Q}$, the second-order differential can be found as follows:

$$\begin{aligned} d^2 f &= 2 \boldsymbol{e}_1^T \left(d\mathcal{Z}^T \right) \boldsymbol{\Phi} \left(d\mathcal{Z} \right) \boldsymbol{e}_0 = 2 \operatorname{Tr} \left\{ \boldsymbol{e}_1^T \left(d\mathcal{Z}^T \right) \boldsymbol{\Phi} \left(d\mathcal{Z} \right) \boldsymbol{e}_0 \right\} \\ &= 2 \operatorname{Tr} \left\{ \boldsymbol{e}_0 \boldsymbol{e}_1^T \left(d\mathcal{Z}^T \right) \boldsymbol{\Phi} d\mathcal{Z} \right\} = 2 \operatorname{vec}^T \left(\boldsymbol{\Phi}^T \left(d\mathcal{Z} \right) \boldsymbol{e}_1 \boldsymbol{e}_0^T \right) d \operatorname{vec}\left(\mathcal{Z} \right) \\ &= 2 \left\{ \left[\left[\boldsymbol{e}_0 \boldsymbol{e}_1^T \right] \otimes \boldsymbol{\Phi}^T \right] d \operatorname{vec}\left(\mathcal{Z} \right) \right\}^T d \operatorname{vec}\left(\mathcal{Z} \right) \\ &= 2 \left(d \operatorname{vec}^T \left(\mathcal{Z} \right) \right) \left[\left[\boldsymbol{e}_1 \boldsymbol{e}_0^T \right] \otimes \boldsymbol{\Phi} \right] d \operatorname{vec}\left(\mathcal{Z} \right). \end{aligned} \tag{5.104}$$

The second-order differential expression $d^2 f$ is now of the form given in (5.53), such that the matrix $\boldsymbol{A}$, used in the method presented in Subsection 5.3.2, is identified as

$$\boldsymbol{A} = 2\left[\boldsymbol{e}_1\boldsymbol{e}_0^T\right] \otimes \boldsymbol{\Phi}. \tag{5.105}$$

The following two matrices are needed later in this example:

$$\boldsymbol{e}_1\boldsymbol{e}_0^T = \begin{bmatrix} 0 \\ 1 \end{bmatrix} [1\ 0] = \begin{bmatrix} 0 & 0 \\ 1 & 0 \end{bmatrix}, \tag{5.106}$$

$$\boldsymbol{e}_0\boldsymbol{e}_1^T = \begin{bmatrix} 1 \\ 0 \end{bmatrix} [0\ 1] = \begin{bmatrix} 0 & 1 \\ 0 & 0 \end{bmatrix}. \tag{5.107}$$

Using (5.55) to identify the Hessian matrix $\mathcal{H}_{\mathcal{Z},\mathcal{Z}} f$ with $\boldsymbol{A}$ given in (5.105)

$$\begin{aligned}\mathcal{H}_{\mathcal{Z},\mathcal{Z}} f &= \frac{1}{2}\left(\boldsymbol{A} + \boldsymbol{A}^T\right) = \frac{1}{2}\left(2\left[\boldsymbol{e}_1\boldsymbol{e}_0^T\right] \otimes \boldsymbol{\Phi} + 2\left[\boldsymbol{e}_0\boldsymbol{e}_1^T\right] \otimes \boldsymbol{\Phi}^T\right) \\ &= \begin{bmatrix} 0 & 0 \\ 1 & 0 \end{bmatrix} \otimes \boldsymbol{\Phi} + \begin{bmatrix} 0 & 1 \\ 0 & 0 \end{bmatrix} \otimes \boldsymbol{\Phi}^T = \begin{bmatrix} \boldsymbol{0}_{N\times N} & \boldsymbol{\Phi}^T \\ \boldsymbol{\Phi} & \boldsymbol{0}_{N\times N} \end{bmatrix},\end{aligned} \tag{5.108}$$

which is in line with the right-hand side of (5.98), having in mind the relations between the $\mathcal{H}_{\mathcal{Z},\mathcal{Z}} f$ and the four matrices $\mathcal{H}_{Z,Z} f$, $\mathcal{H}_{Z^*,Z} f$, $\mathcal{H}_{Z,Z^*} f$, and $\mathcal{H}_{Z^*,Z^*} f$ is given in Lemma 5.3. Remember that it is the matrix $\begin{bmatrix} \mathcal{H}_{Z,Z^*} f & \mathcal{H}_{Z^*,Z^*} f \\ \mathcal{H}_{Z,Z} f & \mathcal{H}_{Z^*,Z} f \end{bmatrix}$ that has to be checked to find out about the nature of the stationary point, *not* the matrix $\mathcal{H}_{\mathcal{Z},\mathcal{Z}} f$.

Example 5.3 The second-order differential of the eigenvalue function $\lambda(\boldsymbol{Z})$ is now found at $\boldsymbol{Z} = \boldsymbol{Z}_0$. This derivation is similar to the one in Magnus and Neudecker (1988, p. 166), where the same result for $d^2\lambda$ was found. See the discussion in (4.71) to (4.74) for an introduction to the eigenvalue and eigenvector notations.

Applying the differential operator to both sides of (4.75) results in

$$2\,(d\boldsymbol{Z})\,(d\boldsymbol{u}) + \boldsymbol{Z}_0 d^2\boldsymbol{u} = \left(d^2\lambda\right)\boldsymbol{u}_0 + 2\,(d\lambda)\,d\boldsymbol{u} + \lambda_0 d^2\boldsymbol{u}. \tag{5.109}$$

According to Horn and Johnson (1985, Lemma 6.3.10), the following inner product is nonzero: $\boldsymbol{v}_0^H\boldsymbol{u}_0 \neq 0$; hence, it is possible to divide by $\boldsymbol{v}_0^H\boldsymbol{u}_0$. Left-multiplying this equation by the vector $\boldsymbol{v}_0^H$ and solving for $d^2\lambda$ gives

$$\begin{aligned}d^2\lambda &= \frac{2\boldsymbol{v}_0^H\left(d\boldsymbol{Z} - \boldsymbol{I}_N d\lambda\right)d\boldsymbol{u}}{\boldsymbol{v}_0^H\boldsymbol{u}_0} = \frac{2\boldsymbol{v}_0^H\left(d\boldsymbol{Z} - \boldsymbol{I}_N\frac{\boldsymbol{v}_0^H(d\boldsymbol{Z})\boldsymbol{u}_0}{\boldsymbol{v}_0^H\boldsymbol{u}_0}\right)d\boldsymbol{u}}{\boldsymbol{v}_0^H\boldsymbol{u}_0} \\ &= \frac{2\left(\boldsymbol{v}_0^H d\boldsymbol{Z} - \frac{\boldsymbol{v}_0^H(d\boldsymbol{Z})\boldsymbol{u}_0\boldsymbol{v}_0^H}{\boldsymbol{v}_0^H\boldsymbol{u}_0}\right)d\boldsymbol{u}}{\boldsymbol{v}_0^H\boldsymbol{u}_0} \\ &= \frac{2\boldsymbol{v}_0^H\left(d\boldsymbol{Z}\right)\left(\boldsymbol{I}_N - \frac{\boldsymbol{u}_0\boldsymbol{v}_0^H}{\boldsymbol{v}_0^H\boldsymbol{u}_0}\right)\left(\lambda_0\boldsymbol{I}_N - \boldsymbol{Z}_0\right)^+\left(\boldsymbol{I}_N - \frac{\boldsymbol{u}_0\boldsymbol{v}_0^H}{\boldsymbol{v}_0^H\boldsymbol{u}_0}\right)\left(d\boldsymbol{Z}\right)\boldsymbol{u}_0}{\boldsymbol{v}_0^H\boldsymbol{u}_0},\end{aligned} \tag{5.110}$$

where (4.77) and (4.87) were utilized.

The second-order differential $d^2\lambda$, in (5.110), can be reformulated by means of (2.100) and (2.116) in the following way:

$$
\begin{aligned}
d^2\lambda &= \frac{2}{\boldsymbol{v}_0^H \boldsymbol{u}_0} \operatorname{Tr}\left\{ \boldsymbol{u}_0 \boldsymbol{v}_0^H (d\boldsymbol{Z}) \left(\boldsymbol{I}_N - \frac{\boldsymbol{u}_0 \boldsymbol{v}_0^H}{\boldsymbol{v}_0^H \boldsymbol{u}_0} \right) (\lambda_0 \boldsymbol{I}_N - \boldsymbol{Z}_0)^+ \left(\boldsymbol{I}_N - \frac{\boldsymbol{u}_0 \boldsymbol{v}_0^H}{\boldsymbol{v}_0^H \boldsymbol{u}_0} \right) d\boldsymbol{Z} \right\} \\
&= \frac{2}{\boldsymbol{v}_0^H \boldsymbol{u}_0} d \operatorname{vec}^T(\boldsymbol{Z}) \left[\left\{ \left(\boldsymbol{I}_N - \frac{\boldsymbol{u}_0 \boldsymbol{v}_0^H}{\boldsymbol{v}_0^H \boldsymbol{u}_0} \right) (\lambda_0 \boldsymbol{I}_N - \boldsymbol{Z}_0)^+ \left(\boldsymbol{I}_N - \frac{\boldsymbol{u}_0 \boldsymbol{v}_0^H}{\boldsymbol{v}_0^H \boldsymbol{u}_0} \right) \right\} \right. \\
&\qquad \left. \otimes \boldsymbol{v}_0^* \boldsymbol{u}_0^T \right] d \operatorname{vec}\left(\boldsymbol{Z}^T\right) \\
&= \frac{2}{\boldsymbol{v}_0^H \boldsymbol{u}_0} d \operatorname{vec}^T(\boldsymbol{Z}) \left[\left\{ \left(\boldsymbol{I}_N - \frac{\boldsymbol{u}_0 \boldsymbol{v}_0^H}{\boldsymbol{v}_0^H \boldsymbol{u}_0} \right) (\lambda_0 \boldsymbol{I}_N - \boldsymbol{Z}_0)^+ \left(\boldsymbol{I}_N - \frac{\boldsymbol{u}_0 \boldsymbol{v}_0^H}{\boldsymbol{v}_0^H \boldsymbol{u}_0} \right) \right\} \right. \\
&\qquad \left. \otimes \boldsymbol{v}_0^* \boldsymbol{u}_0^T \right] \boldsymbol{K}_{N,N} d \operatorname{vec}(\boldsymbol{Z}),
\end{aligned}
\tag{5.111}
$$

where Lemma 2.14 was used in the second equality. From (5.111), it is possible to identify the four complex Hessian matrices by means of (5.27), (5.31), (5.34), (5.36), and (5.37).

Example 5.4 Let $f : \mathbb{C}^{N\times Q} \times \mathbb{C}^{N\times Q} \to \mathbb{C}$ be given by

$$f(\boldsymbol{Z}, \boldsymbol{Z}^*) = \operatorname{Tr}\left\{\boldsymbol{Z}\boldsymbol{A}\boldsymbol{Z}^H\right\}, \tag{5.112}$$

where $\boldsymbol{Z}$ and $\boldsymbol{A}$ have sizes $N \times Q$ and $Q \times Q$, respectively. The matrix $\boldsymbol{A}$ is independent of the two matrix variables $\boldsymbol{Z}$ and $\boldsymbol{Z}^*$. By using (2.116), the function f can be rewritten as

$$f = \operatorname{vec}^T(\boldsymbol{Z}^*) \left(\boldsymbol{A}^T \otimes \boldsymbol{I}_N\right) \operatorname{vec}(\boldsymbol{Z}). \tag{5.113}$$

By applying the differential operator twice to (5.113), it follows that the second-order differential of f can be expressed as

$$d^2 f = 2 \left(d \operatorname{vec}^T(\boldsymbol{Z}^*)\right) \left(\boldsymbol{A}^T \otimes \boldsymbol{I}_N\right) d \operatorname{vec}(\boldsymbol{Z}). \tag{5.114}$$

From this expression, it is possible to identify the four complex Hessian matrices by means of (5.27), (5.31), (5.34), (5.36), and (5.37).

The following example is a slightly modified version of Hjørungnes and Gesbert (2007*b*, Example 3).

Example 5.5 Define the *Frobenius norm* of the matrix $\boldsymbol{Z} \in \mathbb{C}^{N\times Q}$ as $\|\boldsymbol{Z}\|_F^2 \triangleq \operatorname{Tr}\left\{\boldsymbol{Z}^H\boldsymbol{Z}\right\}$. Let $f : \mathbb{C}^{N\times Q} \times \mathbb{C}^{N\times Q} \to \mathbb{R}$ be defined as

$$f(\boldsymbol{Z}, \boldsymbol{Z}^*) = \|\boldsymbol{Z}\|_F^2 - \operatorname{Tr}\left\{\boldsymbol{Z}^T\boldsymbol{Z} + \boldsymbol{Z}^H\boldsymbol{Z}^*\right\}, \tag{5.115}$$

where the Frobenius norm is used. The first-order differential is given by

$$\begin{aligned} df &= \operatorname{Tr}\left\{\left(d\boldsymbol{Z}^{H}\right)\boldsymbol{Z}+\boldsymbol{Z}^{H}d\boldsymbol{Z}-\left(d\boldsymbol{Z}^{T}\right)\boldsymbol{Z}-\boldsymbol{Z}^{T}d\boldsymbol{Z}-\left(d\boldsymbol{Z}^{H}\right)\boldsymbol{Z}^{*}-\boldsymbol{Z}^{H}d\boldsymbol{Z}^{*}\right\} \\ &= \operatorname{Tr}\left\{\left(\boldsymbol{Z}^{H}-2\boldsymbol{Z}^{T}\right)d\boldsymbol{Z}+\left(\boldsymbol{Z}^{T}-2\boldsymbol{Z}^{H}\right)d\boldsymbol{Z}^{*}\right\} \\ &= \left(\operatorname{vec}^{T}(\boldsymbol{Z}^{*})-2\operatorname{vec}^{T}(\boldsymbol{Z})\right)d\operatorname{vec}(\boldsymbol{Z})+\left(\operatorname{vec}^{T}(\boldsymbol{Z})-2\operatorname{vec}^{T}(\boldsymbol{Z}^{*})\right)d\operatorname{vec}(\boldsymbol{Z}^{*}). \end{aligned} \tag{5.116}$$

Therefore, the derivatives of f with respect to $\boldsymbol{Z}$ and $\boldsymbol{Z}^*$ are given by

$$\mathcal{D}_{\boldsymbol{Z}} f = \operatorname{vec}^{T}(\boldsymbol{Z}^{*}-2\boldsymbol{Z}), \tag{5.117}$$

and

$$\mathcal{D}_{\boldsymbol{Z}^*} f = \operatorname{vec}^{T}(\boldsymbol{Z}-2\boldsymbol{Z}^{*}). \tag{5.118}$$

By solving the necessary conditions for optimality from Theorem 3.2, it is seen from the equation $\mathcal{D}_{\boldsymbol{Z}^*} f = \boldsymbol{0}_{1\times NQ}$ that $\boldsymbol{Z} = \boldsymbol{0}_{N\times Q}$ is a stationary point of f, and now the nature of the stationary point $\boldsymbol{Z} = \boldsymbol{0}_{N\times Q}$ is checked by studying four complex Hessian matrices. The second-order differential is given by

$$\begin{aligned} d^2 f &= \left(d\operatorname{vec}^{T}(\boldsymbol{Z})-2d\operatorname{vec}^{T}(\boldsymbol{Z}^{*})\right)d\operatorname{vec}(\boldsymbol{Z}^{*}) \\ &\quad+\left(d\operatorname{vec}^{T}(\boldsymbol{Z}^{*})-2d\operatorname{vec}^{T}(\boldsymbol{Z})\right)d\operatorname{vec}(\boldsymbol{Z}) \\ &= \left[d\operatorname{vec}^{T}(\boldsymbol{Z}^{*})\;\; d\operatorname{vec}^{T}(\boldsymbol{Z})\right]\begin{bmatrix} \boldsymbol{I}_{NQ} & -2\boldsymbol{I}_{NQ} \\ -2\boldsymbol{I}_{NQ} & \boldsymbol{I}_{NQ} \end{bmatrix}\begin{bmatrix} d\operatorname{vec}(\boldsymbol{Z}) \\ d\operatorname{vec}(\boldsymbol{Z}^{*}) \end{bmatrix}. \end{aligned} \tag{5.119}$$

From $d^2 f$, the four Hessians in (5.31), (5.34), (5.36), and (5.37) are identified as

$$\mathcal{H}_{\boldsymbol{Z}^*,\boldsymbol{Z}} f = \boldsymbol{I}_{NQ}, \tag{5.120}$$

$$\mathcal{H}_{\boldsymbol{Z},\boldsymbol{Z}^*} f = \boldsymbol{I}_{NQ}, \tag{5.121}$$

$$\mathcal{H}_{\boldsymbol{Z},\boldsymbol{Z}} f = -2\boldsymbol{I}_{NQ}, \tag{5.122}$$

$$\mathcal{H}_{\boldsymbol{Z}^*,\boldsymbol{Z}^*} f = -2\boldsymbol{I}_{NQ}. \tag{5.123}$$

This shows that the two matrices $\mathcal{H}_{\boldsymbol{Z}^*,\boldsymbol{Z}} f$ and $\mathcal{H}_{\boldsymbol{Z},\boldsymbol{Z}^*} f$ are positive definite; however, the bigger matrix

$$\begin{bmatrix} \mathcal{H}_{\boldsymbol{Z},\boldsymbol{Z}^*} f & \mathcal{H}_{\boldsymbol{Z}^*,\boldsymbol{Z}^*} f \\ \mathcal{H}_{\boldsymbol{Z},\boldsymbol{Z}} f & \mathcal{H}_{\boldsymbol{Z}^*,\boldsymbol{Z}} f \end{bmatrix} = \begin{bmatrix} \boldsymbol{I}_{NQ} & -2\boldsymbol{I}_{NQ} \\ -2\boldsymbol{I}_{NQ} & \boldsymbol{I}_{NQ} \end{bmatrix}, \tag{5.124}$$

is indefinite (Horn & Johnson 1985, p. 397) because

$$\boldsymbol{e}_0^H \begin{bmatrix} \mathcal{H}_{\boldsymbol{Z},\boldsymbol{Z}^*} f & \mathcal{H}_{\boldsymbol{Z}^*,\boldsymbol{Z}^*} f \\ \mathcal{H}_{\boldsymbol{Z},\boldsymbol{Z}} f & \mathcal{H}_{\boldsymbol{Z}^*,\boldsymbol{Z}} f \end{bmatrix} \boldsymbol{e}_0 > 0 > \boldsymbol{1}_{2NQ\times 1}^H \begin{bmatrix} \mathcal{H}_{\boldsymbol{Z},\boldsymbol{Z}^*} f & \mathcal{H}_{\boldsymbol{Z}^*,\boldsymbol{Z}^*} f \\ \mathcal{H}_{\boldsymbol{Z},\boldsymbol{Z}} f & \mathcal{H}_{\boldsymbol{Z}^*,\boldsymbol{Z}} f \end{bmatrix} \boldsymbol{1}_{2NQ\times 1}, \tag{5.125}$$

meaning that f has a saddle point at the origin. This shows the importance of checking the *whole* matrix $\begin{bmatrix} \mathcal{H}_{\boldsymbol{Z},\boldsymbol{Z}^*} f & \mathcal{H}_{\boldsymbol{Z}^*,\boldsymbol{Z}^*} f \\ \mathcal{H}_{\boldsymbol{Z},\boldsymbol{Z}} f & \mathcal{H}_{\boldsymbol{Z}^*,\boldsymbol{Z}} f \end{bmatrix}$ when deciding whether or not a stationary

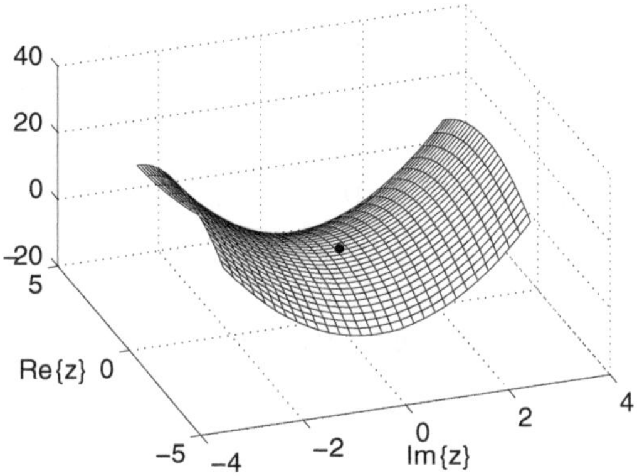

Figure 5.1 Function from Example 5.5 with $N = Q = 1$.

point is a local minimum, local maximum, or saddle point. Figure 5.1 shows f for $N = Q = 1$, and it is seen that the origin (marked as $\bullet$) is indeed a saddle point.

5.6.2 Examples of Finding Complex Hessian Matrices of Vector Functions

---

Example 5.6 Let $\boldsymbol{f} : \mathbb{C}^{N\times 2Q} \to \mathbb{C}^{M\times 1}$ be given by

$$\boldsymbol{f}(\boldsymbol{\mathcal{Z}}) = \boldsymbol{A}\boldsymbol{\mathcal{Z}}\boldsymbol{\mathcal{Z}}^T\boldsymbol{b}, \tag{5.126}$$

where $\boldsymbol{A} \in \mathbb{C}^{M\times N}$ and $\boldsymbol{b} \in \mathbb{C}^{N\times 1}$ are independent of all matrix components within $\boldsymbol{\mathcal{Z}} \in \mathbb{C}^{N\times 2Q}$. The first-order differential of $\boldsymbol{f}$ can be expressed as

$$d\boldsymbol{f} = \boldsymbol{A}(d\boldsymbol{\mathcal{Z}})\,\boldsymbol{\mathcal{Z}}^T\boldsymbol{b} + \boldsymbol{A}\boldsymbol{\mathcal{Z}}\left(d\boldsymbol{\mathcal{Z}}^T\right)\boldsymbol{b}. \tag{5.127}$$

The second-order differential of $\boldsymbol{f}$ can be found as

$$d^2\boldsymbol{f} = \boldsymbol{A}(d\boldsymbol{\mathcal{Z}})\left(d\boldsymbol{\mathcal{Z}}^T\right)\boldsymbol{b} + \boldsymbol{A}(d\boldsymbol{\mathcal{Z}})\left(d\boldsymbol{\mathcal{Z}}^T\right)\boldsymbol{b} = 2\boldsymbol{A}(d\boldsymbol{\mathcal{Z}})\left(d\boldsymbol{\mathcal{Z}}^T\right)\boldsymbol{b}. \tag{5.128}$$

Following the procedure in Table 5.4, it is seen that the next step is to try to put the second-order differential expression $d^2\boldsymbol{f}$ into the same form as given in (5.71). This task can be accomplished by first trying to find the complex Hessian of the component functions $f_i : \mathbb{C}^{N\times 2Q} \to \mathbb{C}$ of the vector function $\boldsymbol{f}$, where $i \in \{0, 1, \ldots, M-1\}$. The second-order differential of component function number i, (i.e., $d^2 f_i$) can be written as

$$\begin{aligned}
d^2 f_i &= \left(2\boldsymbol{A}(d\boldsymbol{\mathcal{Z}})\left(d\boldsymbol{\mathcal{Z}}^T\right)\boldsymbol{b}\right)_i = 2\boldsymbol{A}_{i,:}(d\boldsymbol{\mathcal{Z}})\left(d\boldsymbol{\mathcal{Z}}^T\right)\boldsymbol{b} \\
&= 2\,\mathrm{Tr}\left\{\left(d\boldsymbol{\mathcal{Z}}^T\right)\boldsymbol{b}\boldsymbol{A}_{i,:}d\boldsymbol{\mathcal{Z}}\right\} = 2\,\mathrm{vec}^T\left(\boldsymbol{A}_{i,:}^T\boldsymbol{b}^T d\boldsymbol{\mathcal{Z}}\right)d\,\mathrm{vec}(\boldsymbol{\mathcal{Z}}) \\
&= 2\left[\left\{\boldsymbol{I}_{2Q}\otimes\left(\boldsymbol{A}_{i,:}^T\boldsymbol{b}^T\right)\right\}d\,\mathrm{vec}(\boldsymbol{\mathcal{Z}})\right]^T d\,\mathrm{vec}(\boldsymbol{\mathcal{Z}}) \\
&= \left(d\,\mathrm{vec}^T(\boldsymbol{\mathcal{Z}})\right)2\left[\boldsymbol{I}_{2Q}\otimes(\boldsymbol{b}\boldsymbol{A}_{i,:})\right]d\,\mathrm{vec}(\boldsymbol{\mathcal{Z}}).
\end{aligned} \tag{5.129}$$

From this expression, it is possible to identify the complex Hessian matrix of the component function f_i. However, the main task is to find the Hessian of the whole vector function $\boldsymbol{f}$, and the next step is to study in greater detail the expression $d^2 \boldsymbol{f}$. By using the expression found in (5.129) inside $d^2 \boldsymbol{f}$, it is seen that

$$d^2 \boldsymbol{f} = \begin{bmatrix} d^2 f_0 \\ d^2 f_1 \\ \vdots \\ d^2 f_{M-1} \end{bmatrix} = 2 \begin{bmatrix} \left(d \operatorname{vec}^T(\mathcal{Z})\right) \left[\boldsymbol{I}_{2Q} \otimes (\boldsymbol{b}\boldsymbol{A}_{0,:})\right] d \operatorname{vec}(\mathcal{Z}) \\ \left(d \operatorname{vec}^T(\mathcal{Z})\right) \left[\boldsymbol{I}_{2Q} \otimes (\boldsymbol{b}\boldsymbol{A}_{1,:})\right] d \operatorname{vec}(\mathcal{Z}) \\ \vdots \\ \left(d \operatorname{vec}^T(\mathcal{Z})\right) \left[\boldsymbol{I}_{2Q} \otimes (\boldsymbol{b}\boldsymbol{A}_{M-1,:})\right] d \operatorname{vec}(\mathcal{Z}) \end{bmatrix}$$

$$= \left[\boldsymbol{I}_M \otimes d \operatorname{vec}^T(\mathcal{Z})\right] 2 \begin{bmatrix} \boldsymbol{I}_{2Q} \otimes (\boldsymbol{b}\boldsymbol{A}_{0,:}) \\ \boldsymbol{I}_{2Q} \otimes (\boldsymbol{b}\boldsymbol{A}_{1,:}) \\ \vdots \\ \boldsymbol{I}_{2Q} \otimes (\boldsymbol{b}\boldsymbol{A}_{M-1,:}) \end{bmatrix} d \operatorname{vec}(\mathcal{Z}). \tag{5.130}$$

Now, $d^2 \boldsymbol{f}$ has been developed into the form given in (5.71), and the matrix $\boldsymbol{B} \in \mathbb{C}^{2NQM \times 2NQ}$ can be identified as

$$\boldsymbol{B} = 2 \begin{bmatrix} \boldsymbol{I}_{2Q} \otimes (\boldsymbol{b}\boldsymbol{A}_{0,:}) \\ \boldsymbol{I}_{2Q} \otimes (\boldsymbol{b}\boldsymbol{A}_{1,:}) \\ \vdots \\ \boldsymbol{I}_{2Q} \otimes (\boldsymbol{b}\boldsymbol{A}_{M-1,:}) \end{bmatrix}. \tag{5.131}$$

The complex Hessian matrix of a vector function is given by (5.76), and the last matrix that is needed in (5.76) is $\operatorname{vecb}\left(\boldsymbol{B}^T\right)$, which can be found from (5.131) as

$$\operatorname{vecb}\left(\boldsymbol{B}^T\right) = 2 \begin{bmatrix} \boldsymbol{I}_{2Q} \otimes \left(\boldsymbol{A}_{0,:}^T \boldsymbol{b}^T\right) \\ \boldsymbol{I}_{2Q} \otimes \left(\boldsymbol{A}_{1,:}^T \boldsymbol{b}^T\right) \\ \vdots \\ \boldsymbol{I}_{2Q} \otimes \left(\boldsymbol{A}_{M-1,:}^T \boldsymbol{b}^T\right) \end{bmatrix}. \tag{5.132}$$

By using (5.76) and the above expressions for $\boldsymbol{B}$ and $\operatorname{vecb}\left(\boldsymbol{B}^T\right)$, the complex Hessian matrix of $\boldsymbol{f}$ is found as

$$\mathcal{H}_{\mathcal{Z},\mathcal{Z}} \boldsymbol{f} = \begin{bmatrix} \boldsymbol{I}_{2Q} \otimes \left(\boldsymbol{b}\boldsymbol{A}_{0,:} + \boldsymbol{A}_{0,:}^T \boldsymbol{b}^T\right) \\ \boldsymbol{I}_{2Q} \otimes \left(\boldsymbol{b}\boldsymbol{A}_{1,:} + \boldsymbol{A}_{1,:}^T \boldsymbol{b}^T\right) \\ \vdots \\ \boldsymbol{I}_{2Q} \otimes \left(\boldsymbol{b}\boldsymbol{A}_{M-1,:} + \boldsymbol{A}_{M-1,:}^T \boldsymbol{b}^T\right) \end{bmatrix}. \tag{5.133}$$

From (5.133), it is observed that the complex Hessian $\mathcal{H}_{\mathcal{Z},\mathcal{Z}} \boldsymbol{f}$ is column symmetric, which is always the case for the Hessian of twice differentiable functions in all the matrix variables inside $\mathcal{Z}$.

This example shows that one useful strategy for finding the complex Hessian matrix of a complex vector function is to first find the Hessian of the component functions, and then use them in the expression of $d^2 \boldsymbol{f}$, making the expression into the appropriate form given by (5.71).

Example 5.7 The complex Hessian matrix of a vector function $\boldsymbol{f} : \mathbb{C}^{N\times 2Q} \rightarrow \mathbb{C}^{M\times 1}$ will be found when the function is given by

$$\boldsymbol{f}(\boldsymbol{\mathcal{Z}}) = \boldsymbol{a} f(\boldsymbol{\mathcal{Z}}), \tag{5.134}$$

where the scalar function $f(\boldsymbol{\mathcal{Z}})$ has a symmetric complex Hessian given by $\mathcal{H}_{\boldsymbol{\mathcal{Z}},\boldsymbol{\mathcal{Z}}} f$, which is assumed to be known. Hence, the goal is to derive the complex Hessian of a vector function $\boldsymbol{f}$ when the complex Hessian matrix of the scalar function f is already known. The vector $\boldsymbol{a} \in \mathbb{C}^{M\times 1}$ is independent of $\boldsymbol{\mathcal{Z}}$.

The first-order differential of the vector function $\boldsymbol{f}$ is given by

$$d\boldsymbol{f} = \boldsymbol{a} df. \tag{5.135}$$

It is assumed that the scalar function f is twice differentiable, such that its Hessian matrix is symmetric, that is,

$$(\mathcal{H}_{\boldsymbol{\mathcal{Z}},\boldsymbol{\mathcal{Z}}} f)^T = \mathcal{H}_{\boldsymbol{\mathcal{Z}},\boldsymbol{\mathcal{Z}}} f. \tag{5.136}$$

The second-order differential of $\boldsymbol{f}$ can be calculated as

$$\begin{aligned} d^2\boldsymbol{f} &= \boldsymbol{a} d^2 f = \boldsymbol{a} \left(d \operatorname{vec}^T(\boldsymbol{\mathcal{Z}})\right) \left[\mathcal{H}_{\boldsymbol{\mathcal{Z}},\boldsymbol{\mathcal{Z}}} f\right] d \operatorname{vec}(\boldsymbol{\mathcal{Z}}) \\ &= \left[\boldsymbol{a} \otimes \left(d \operatorname{vec}^T(\boldsymbol{\mathcal{Z}})\right)\right] \left[\mathcal{H}_{\boldsymbol{\mathcal{Z}},\boldsymbol{\mathcal{Z}}} f\right] d \operatorname{vec}(\boldsymbol{\mathcal{Z}}) \\ &= \left[\{\boldsymbol{I}_M \boldsymbol{a}\} \otimes \left\{\left(d \operatorname{vec}^T(\boldsymbol{\mathcal{Z}})\right) \boldsymbol{I}_{2NQ}\right\}\right] \left[\mathcal{H}_{\boldsymbol{\mathcal{Z}},\boldsymbol{\mathcal{Z}}} f\right] d \operatorname{vec}(\boldsymbol{\mathcal{Z}}) \\ &= \left[\boldsymbol{I}_M \otimes d \operatorname{vec}^T(\boldsymbol{\mathcal{Z}})\right] \left[\boldsymbol{a} \otimes \boldsymbol{I}_{2NQ}\right] \left[\mathcal{H}_{\boldsymbol{\mathcal{Z}},\boldsymbol{\mathcal{Z}}} f\right] d \operatorname{vec}(\boldsymbol{\mathcal{Z}}). \end{aligned} \tag{5.137}$$

From this expression, the matrix $\boldsymbol{B} \in \mathbb{C}^{2NQM\times 2NQ}$ can be identified as

$$\boldsymbol{B} = \left[\boldsymbol{a} \otimes \boldsymbol{I}_{2NQ}\right] \mathcal{H}_{\boldsymbol{\mathcal{Z}},\boldsymbol{\mathcal{Z}}} f. \tag{5.138}$$

The following expression is needed when finding the complex Hessian:

$$\begin{aligned} \operatorname{vecb}\left(\boldsymbol{B}^T\right) &= \operatorname{vecb}\left((\mathcal{H}_{\boldsymbol{\mathcal{Z}},\boldsymbol{\mathcal{Z}}} f)^T \left[\boldsymbol{a}^T \otimes \boldsymbol{I}_{2NQ}\right]\right) = \operatorname{vecb}\left(\mathcal{H}_{\boldsymbol{\mathcal{Z}},\boldsymbol{\mathcal{Z}}} f \left[\boldsymbol{a}^T \otimes \boldsymbol{I}_{2NQ}\right]\right) \\ &= \begin{bmatrix} a_0 \mathcal{H}_{\boldsymbol{\mathcal{Z}},\boldsymbol{\mathcal{Z}}} f \\ a_1 \mathcal{H}_{\boldsymbol{\mathcal{Z}},\boldsymbol{\mathcal{Z}}} f \\ \vdots \\ a_{M-1} \mathcal{H}_{\boldsymbol{\mathcal{Z}},\boldsymbol{\mathcal{Z}}} f \end{bmatrix} = \left[\boldsymbol{a} \otimes \boldsymbol{I}_{2NQ}\right] \mathcal{H}_{\boldsymbol{\mathcal{Z}},\boldsymbol{\mathcal{Z}}} f = \boldsymbol{B}, \end{aligned} \tag{5.139}$$

where $\boldsymbol{a} = [a_0, a_1, \ldots, a_{M-1}]^T$. Hence, the complex Hessian matrix of $\boldsymbol{f}$ can now be found as

$$\mathcal{H}_{\boldsymbol{\mathcal{Z}},\boldsymbol{\mathcal{Z}}} \boldsymbol{f} = \frac{1}{2}\left(\boldsymbol{B} + \operatorname{vecb}\left(\boldsymbol{B}^T\right)\right) = \boldsymbol{B} = \left[\boldsymbol{a} \otimes \boldsymbol{I}_{2NQ}\right] \mathcal{H}_{\boldsymbol{\mathcal{Z}},\boldsymbol{\mathcal{Z}}} f. \tag{5.140}$$

This example is an extension of the real-valued case where the input variable was a *vector* (see Magnus & Neudecker 1988, p. 194, Section 7).

5.6.3 Examples of Finding Complex Hessian Matrices of Matrix Functions

Example 5.8 Let the function $\boldsymbol{F} : \mathbb{C}^{N\times 2Q} \to \mathbb{C}^{M\times P}$ be given by

$$\boldsymbol{F}(\boldsymbol{\mathcal{Z}}) = \boldsymbol{U}\boldsymbol{\mathcal{Z}}\boldsymbol{D}\boldsymbol{\mathcal{Z}}^T\boldsymbol{E}, \tag{5.141}$$

where the three matrices $\boldsymbol{U} \in \mathbb{C}^{M\times N}$, $\boldsymbol{D} \in \mathbb{C}^{2Q\times 2Q}$, and $\boldsymbol{E} \in \mathbb{C}^{N\times P}$ are independent of all elements within the matrix variable $\boldsymbol{\mathcal{Z}} \in \mathbb{C}^{N\times 2Q}$. The first-order differential of the matrix function $\boldsymbol{F}$ is given by

$$d\boldsymbol{F} = \boldsymbol{U}\left(d\boldsymbol{\mathcal{Z}}\right)\boldsymbol{D}\boldsymbol{\mathcal{Z}}^T\boldsymbol{E} + \boldsymbol{U}\boldsymbol{\mathcal{Z}}\boldsymbol{D}\left(d\boldsymbol{\mathcal{Z}}^T\right)\boldsymbol{E}. \tag{5.142}$$

The second-order differential of $\boldsymbol{F}$ can be found by applying the differential operator on both sides of (5.142); this results in

$$d^2\boldsymbol{F} = 2\boldsymbol{U}\left(d\boldsymbol{\mathcal{Z}}\right)\boldsymbol{D}\left(d\boldsymbol{\mathcal{Z}}^T\right)\boldsymbol{E}, \tag{5.143}$$

because $d^2\boldsymbol{\mathcal{Z}} = \boldsymbol{0}_{N\times 2Q}$.

Let the (i, j)-th component function of the matrix function $\boldsymbol{F}$ be denoted by $f_{i,j}$, where $i \in \{0, 1, \ldots, M-1\}$ and $j \in \{0, 1, \ldots, P-1\}$, hence, $f_{i,j} : \mathbb{C}^{N\times 2Q} \to \mathbb{C}$ is a scalar function. The second-order differential of $f_{i,j}$ can be found from (5.143) and is given by

$$d^2 f_{i,j} = 2\boldsymbol{U}_{i,:}\left(d\boldsymbol{\mathcal{Z}}\right)\boldsymbol{D}\left(d\boldsymbol{\mathcal{Z}}^T\right)\boldsymbol{E}_{:,j}. \tag{5.144}$$

The expression of $d^2 f_{i,j}$ is first manipulated in such a manner that it is expressed as in (5.53)

$$\begin{aligned}
d^2 f_{i,j} &= 2\boldsymbol{U}_{i,:}\left(d\boldsymbol{\mathcal{Z}}\right)\boldsymbol{D}\left(d\boldsymbol{\mathcal{Z}}^T\right)\boldsymbol{E}_{:,j} = 2\operatorname{Tr}\left\{\boldsymbol{U}_{i,:}\left(d\boldsymbol{\mathcal{Z}}\right)\boldsymbol{D}\left(d\boldsymbol{\mathcal{Z}}^T\right)\boldsymbol{E}_{:,j}\right\} \\
&= 2\operatorname{Tr}\left\{\boldsymbol{D}\left(d\boldsymbol{\mathcal{Z}}^T\right)\boldsymbol{E}_{:,j}\boldsymbol{U}_{i,:}d\boldsymbol{\mathcal{Z}}\right\} = 2\operatorname{vec}^T\left(\boldsymbol{U}_{i,:}^T\boldsymbol{E}_{:,j}^T\left(d\boldsymbol{\mathcal{Z}}\right)\boldsymbol{D}^T\right)d\operatorname{vec}\left(\boldsymbol{\mathcal{Z}}\right) \\
&= 2\left[\left\{\boldsymbol{D}\otimes\left(\boldsymbol{U}_{i,:}^T\boldsymbol{E}_{:,j}^T\right)\right\}d\operatorname{vec}\left(\boldsymbol{\mathcal{Z}}\right)\right]^T d\operatorname{vec}\left(\boldsymbol{\mathcal{Z}}\right) \\
&= 2\left(d\operatorname{vec}^T\left(\boldsymbol{\mathcal{Z}}\right)\right)\left[\boldsymbol{D}^T\otimes\left(\boldsymbol{E}_{:,j}\boldsymbol{U}_{i,:}\right)\right]d\operatorname{vec}\left(\boldsymbol{\mathcal{Z}}\right).
\end{aligned} \tag{5.145}$$

Now, the expression of $d^2 \operatorname{vec}(\boldsymbol{F})$ in (5.77) can be studied by inserting the results from (5.145) into (5.77):

$$d^2\operatorname{vec}(\boldsymbol{F}) = \begin{bmatrix} d^2 f_{0,0} \\ d^2 f_{1,0} \\ \vdots \\ d^2 f_{M-1,0} \\ d^2 f_{0,1} \\ \vdots \\ d^2 f_{0,P-1} \\ \vdots \\ d^2 f_{M-1,P-1} \end{bmatrix} = 2\begin{bmatrix} \left(d\operatorname{vec}^T(\boldsymbol{\mathcal{Z}})\right)\left[\boldsymbol{D}^T \otimes (\boldsymbol{E}_{:,0}\boldsymbol{U}_{0,:})\right] d\operatorname{vec}(\boldsymbol{\mathcal{Z}}) \\ \left(d\operatorname{vec}^T(\boldsymbol{\mathcal{Z}})\right)\left[\boldsymbol{D}^T \otimes (\boldsymbol{E}_{:,0}\boldsymbol{U}_{1,:})\right] d\operatorname{vec}(\boldsymbol{\mathcal{Z}}) \\ \vdots \\ \left(d\operatorname{vec}^T(\boldsymbol{\mathcal{Z}})\right)\left[\boldsymbol{D}^T \otimes (\boldsymbol{E}_{:,0}\boldsymbol{U}_{M-1,:})\right] d\operatorname{vec}(\boldsymbol{\mathcal{Z}}) \\ \left(d\operatorname{vec}^T(\boldsymbol{\mathcal{Z}})\right)\left[\boldsymbol{D}^T \otimes (\boldsymbol{E}_{:,1}\boldsymbol{U}_{0,:})\right] d\operatorname{vec}(\boldsymbol{\mathcal{Z}}) \\ \vdots \\ \left(d\operatorname{vec}^T(\boldsymbol{\mathcal{Z}})\right)\left[\boldsymbol{D}^T \otimes (\boldsymbol{E}_{:,P-1}\boldsymbol{U}_{0,:})\right] d\operatorname{vec}(\boldsymbol{\mathcal{Z}}) \\ \vdots \\ \left(d\operatorname{vec}^T(\boldsymbol{\mathcal{Z}})\right)\left[\boldsymbol{D}^T \otimes (\boldsymbol{E}_{:,P-1}\boldsymbol{U}_{M-1,:})\right] d\operatorname{vec}(\boldsymbol{\mathcal{Z}}) \end{bmatrix}$$

$$= \left[\boldsymbol{I}_{MP} \otimes d\operatorname{vec}^T(\boldsymbol{\mathcal{Z}})\right] 2 \begin{bmatrix} \boldsymbol{D}^T \otimes (\boldsymbol{E}_{:,0}\boldsymbol{U}_{0,:}) \\ \boldsymbol{D}^T \otimes (\boldsymbol{E}_{:,0}\boldsymbol{U}_{1,:}) \\ \vdots \\ \boldsymbol{D}^T \otimes (\boldsymbol{E}_{:,0}\boldsymbol{U}_{M-1,:}) \\ \boldsymbol{D}^T \otimes (\boldsymbol{E}_{:,1}\boldsymbol{U}_{0,:}) \\ \vdots \\ \boldsymbol{D}^T \otimes (\boldsymbol{E}_{:,P-1}\boldsymbol{U}_{0,:}) \\ \vdots \\ \boldsymbol{D}^T \otimes (\boldsymbol{E}_{:,P-1}\boldsymbol{U}_{M-1,:}) \end{bmatrix} d\operatorname{vec}(\boldsymbol{\mathcal{Z}}). \tag{5.146}$$

The expression $d^2\operatorname{vec}(\boldsymbol{F})$ is now put into the form given by (5.83), and the middle matrix $\boldsymbol{C} \in \mathbb{C}^{2NQMP \times 2NQ}$ of (5.83) can be identified as

$$\boldsymbol{C} = 2 \begin{bmatrix} \boldsymbol{D}^T \otimes (\boldsymbol{E}_{:,0}\boldsymbol{U}_{0,:}) \\ \boldsymbol{D}^T \otimes (\boldsymbol{E}_{:,0}\boldsymbol{U}_{1,:}) \\ \vdots \\ \boldsymbol{D}^T \otimes (\boldsymbol{E}_{:,0}\boldsymbol{U}_{M-1,:}) \\ \boldsymbol{D}^T \otimes (\boldsymbol{E}_{:,1}\boldsymbol{U}_{0,:}) \\ \vdots \\ \boldsymbol{D}^T \otimes (\boldsymbol{E}_{:,P-1}\boldsymbol{U}_{0,:}) \\ \vdots \\ \boldsymbol{D}^T \otimes (\boldsymbol{E}_{:,P-1}\boldsymbol{U}_{M-1,:}) \end{bmatrix}. \tag{5.147}$$

From the above equation, the matrix vecb $\left(\boldsymbol{C}^T\right)$ can be found, and if that expression is inserted into (5.87), the complex Hessian of the matrix function $\boldsymbol{F}$ is

$$\mathcal{H}_{\mathcal{Z},\mathcal{Z}}\boldsymbol{F} = \begin{bmatrix} \boldsymbol{D}^T \otimes (\boldsymbol{E}_{:,0}\boldsymbol{U}_{0,:}) + \boldsymbol{D} \otimes \left(\boldsymbol{U}_{0,:}^T \boldsymbol{E}_{:,0}^T\right) \\ \boldsymbol{D}^T \otimes (\boldsymbol{E}_{:,0}\boldsymbol{U}_{1,:}) + \boldsymbol{D} \otimes \left(\boldsymbol{U}_{1,:}^T \boldsymbol{E}_{:,0}^T\right) \\ \vdots \\ \boldsymbol{D}^T \otimes (\boldsymbol{E}_{:,0}\boldsymbol{U}_{M-1,:}) + \boldsymbol{D} \otimes \left(\boldsymbol{U}_{M-1,:}^T \boldsymbol{E}_{:,0}^T\right) \\ \boldsymbol{D}^T \otimes (\boldsymbol{E}_{:,1}\boldsymbol{U}_{0,:}) + \boldsymbol{D} \otimes \left(\boldsymbol{U}_{0,:}^T \boldsymbol{E}_{:,1}^T\right) \\ \vdots \\ \boldsymbol{D}^T \otimes (\boldsymbol{E}_{:,P-1}\boldsymbol{U}_{0,:}) + \boldsymbol{D} \otimes \left(\boldsymbol{U}_{0,:}^T \boldsymbol{E}_{:,P-1}^T\right) \\ \vdots \\ \boldsymbol{D}^T \otimes (\boldsymbol{E}_{:,P-1}\boldsymbol{U}_{M-1,:}) + \boldsymbol{D} \otimes \left(\boldsymbol{U}_{M-1,:}^T \boldsymbol{E}_{:,P-1}^T\right) \end{bmatrix}. \tag{5.148}$$

From the above expression of the complex Hessian matrix of $\boldsymbol{F}$, it is seen that it is column symmetric.

As in Example 5.6, it is seen that the procedure for finding the complex Hessian matrix in Example 5.8 was first to consider each of the component functions of $\boldsymbol{F}$ individually, and then to collect the second-order differential of these functions into the expression $d^2 \operatorname{vec}(\boldsymbol{F})$. In Exercises 5.9 and 5.10, we will study how the results of Examples 5.6 and 5.8 can be found in a more direct manner without considering each component function individually.

Example 5.9 Let the complex Hessian matrix of the vector function $\boldsymbol{f} : \mathbb{C}^{N\times 2NP} \rightarrow \mathbb{C}^{M\times 1}$ be known and denoted as $\mathcal{H}_{\mathcal{Z},\mathcal{Z}}\boldsymbol{f}$. Let the matrix function $\boldsymbol{F} : \mathbb{C}^{N\times 2NQ} \rightarrow \mathbb{C}^{M\times P}$ be given by

$$\boldsymbol{F}(\mathcal{Z}) = \boldsymbol{A}\boldsymbol{f}(\mathcal{Z})\boldsymbol{b}, \tag{5.149}$$

where the matrix $\boldsymbol{A} \in \mathbb{C}^{M\times M}$ and the vector $\boldsymbol{b} \in \mathbb{C}^{1\times P}$ are independent of all components of $\mathcal{Z}$. In this example, an expression for the complex Hessian matrix of $\boldsymbol{F}$ will be found as a function of the complex Hessian matrix of $\boldsymbol{f}$.

Because the complex Hessian of $\boldsymbol{f}$ is known, it can be expressed as

$$d^2\boldsymbol{f} = \left[\boldsymbol{I}_M \otimes d\operatorname{vec}^T(\mathcal{Z})\right]\left[\mathcal{H}_{\mathcal{Z},\mathcal{Z}}\boldsymbol{f}\right] d\operatorname{vec}(\mathcal{Z}). \tag{5.150}$$

The first-order differential of $\boldsymbol{F}$ can be written as

$$d\boldsymbol{F} = \boldsymbol{A}(d\boldsymbol{f})\boldsymbol{b}. \tag{5.151}$$

To find the Hessian of $\boldsymbol{F}$, the following expression must be studied:

$$\begin{aligned} d^2\operatorname{vec}(\boldsymbol{F}) &= \operatorname{vec}(d(d\boldsymbol{F})) = \operatorname{vec}\left(\boldsymbol{A}\left(d^2\boldsymbol{f}\right)\boldsymbol{b}\right) = \left[\boldsymbol{b}^T \otimes \boldsymbol{A}\right] d^2\boldsymbol{f} \\ &= \left[\boldsymbol{b}^T \otimes \boldsymbol{A}\right]\left[\boldsymbol{I}_M \otimes d\operatorname{vec}^T(\mathcal{Z})\right]\left[\mathcal{H}_{\mathcal{Z},\mathcal{Z}}\boldsymbol{f}\right] d\operatorname{vec}(\mathcal{Z}). \end{aligned} \tag{5.152}$$

The next task is to manipulate the last expression such that it has the form given in (5.83). To achieve this rewriting, (2.101), (2.103), and (2.110) will be used several times. The product of the two first factors in (5.152) can be rewritten using the *commutation matrix*

$$
\begin{aligned}
\left[\boldsymbol{b}^T \otimes \boldsymbol{A}\right]\left[\boldsymbol{I}_M \otimes d\,\mathrm{vec}^T(\mathcal{Z})\right] &= \boldsymbol{K}_{P,M}\left[\boldsymbol{A}\otimes\boldsymbol{b}^T\right]\left[\boldsymbol{I}_M \otimes d\,\mathrm{vec}^T(\mathcal{Z})\right] \\
&= \boldsymbol{K}_{P,M}\left[\boldsymbol{A}\otimes\left(\boldsymbol{b}^T d\,\mathrm{vec}^T(\mathcal{Z})\right)\right] = \left[\left(\boldsymbol{b}^T d\,\mathrm{vec}^T(\mathcal{Z})\right)\otimes\boldsymbol{A}\right]\boldsymbol{K}_{2NQ,M} \\
&= \left[\boldsymbol{b}^T \otimes \left(d\,\mathrm{vec}^T(\mathcal{Z})\right)\otimes\boldsymbol{A}\right]\boldsymbol{K}_{2NQ,M} = \left[\left(d\,\mathrm{vec}^T(\mathcal{Z})\right)\otimes\boldsymbol{b}^T\otimes\boldsymbol{A}\right]\boldsymbol{K}_{2NQ,M} \\
&= \left[\left(d\,\mathrm{vec}^T(\mathcal{Z})\,\boldsymbol{I}_{2NQ}\right)\otimes\left(\boldsymbol{I}_{MP}\left\{\boldsymbol{b}^T\otimes\boldsymbol{A}\right\}\right)\right]\boldsymbol{K}_{2NQ,M} \\
&= \left[\left(d\,\mathrm{vec}^T(\mathcal{Z})\right)\otimes\boldsymbol{I}_{MP}\right]\left[\boldsymbol{I}_{2NQ}\otimes\left\{\boldsymbol{b}^T\otimes\boldsymbol{A}\right\}\right]\boldsymbol{K}_{2NQ,M} \\
&= \left[\left(d\,\mathrm{vec}^T(\mathcal{Z})\right)\otimes\boldsymbol{I}_{MP}\right]\boldsymbol{K}_{2NQ,MP}\left[\left\{\boldsymbol{b}^T\otimes\boldsymbol{A}\right\}\otimes\boldsymbol{I}_{2NQ}\right] \\
&= \boldsymbol{K}_{1,MP}\left[\boldsymbol{I}_{MP}\otimes d\,\mathrm{vec}^T(\mathcal{Z})\right]\left[\boldsymbol{b}^T\otimes\boldsymbol{A}\otimes\boldsymbol{I}_{2NQ}\right] \\
&= \left[\boldsymbol{I}_{MP}\otimes d\,\mathrm{vec}^T(\mathcal{Z})\right]\left[\boldsymbol{b}^T\otimes\boldsymbol{A}\otimes\boldsymbol{I}_{2NQ}\right],
\end{aligned}
\tag{5.153}
$$

where it was used such that $\boldsymbol{K}_{1,MP} = \boldsymbol{I}_{MP} = \boldsymbol{K}_{MP,1}$ which follows from Exercise 2.6. By substituting the product of the first two factors of (5.152) with the result in (5.153), it is found that

$$
d^2\,\mathrm{vec}(\boldsymbol{F}) = \left[\boldsymbol{I}_{MP}\otimes d\,\mathrm{vec}^T(\mathcal{Z})\right]\left[\boldsymbol{b}^T\otimes\boldsymbol{A}\otimes\boldsymbol{I}_{2NQ}\right]\left[\mathcal{H}_{\mathcal{Z},\mathcal{Z}}\boldsymbol{f}\right]d\,\mathrm{vec}(\mathcal{Z}). \tag{5.154}
$$

The expression for $d^2\,\mathrm{vec}(\boldsymbol{F})$ now has the same form as in (5.83), such that the matrix $\boldsymbol{C}$ is identified as

$$
\boldsymbol{C} = \left[\boldsymbol{b}^T\otimes\boldsymbol{A}\otimes\boldsymbol{I}_{2NQ}\right]\mathcal{H}_{\mathcal{Z},\mathcal{Z}}\boldsymbol{f}. \tag{5.155}
$$

By using the result from Exercise 2.19, it is seen that $\boldsymbol{C}$ is column symmetric, that is, $\mathrm{vecb}\left(\boldsymbol{C}^T\right) = \boldsymbol{C}$. The complex Hessian of $\boldsymbol{F}$ can be found as

$$
\mathcal{H}_{\mathcal{Z},\mathcal{Z}}\boldsymbol{F} = \frac{1}{2}\left[\boldsymbol{C} + \mathrm{vecb}\left(\boldsymbol{C}^T\right)\right] = \boldsymbol{C} = \left[\boldsymbol{b}^T\otimes\boldsymbol{A}\otimes\boldsymbol{I}_{2NQ}\right]\mathcal{H}_{\mathcal{Z},\mathcal{Z}}\boldsymbol{f}. \tag{5.156}
$$

5.7 Exercises

5.1 Let $f : \mathbb{C}^{N\times 2Q} \to \mathbb{C}$ be given by

$$
f(\mathcal{Z}) = \mathrm{Tr}\left\{\boldsymbol{A}\mathcal{Z}\right\}, \tag{5.157}
$$

where $\mathcal{Z} \in \mathbb{C}^{N\times 2Q}$ is the augmented matrix variable, and $\boldsymbol{A} \in \mathbb{C}^{2Q\times N}$ is independent of $\mathcal{Z}$. Show that the complex Hessian matrix of the function f is given by

$$
\mathcal{H}_{\mathcal{Z},\mathcal{Z}}f = \boldsymbol{0}_{2NQ\times 2NQ}. \tag{5.158}
$$

5.2 Let the function $f : \mathbb{C}^{N\times N} \to \mathbb{C}$ be given by

$$
f(\boldsymbol{Z}) = \ln\left(\det(\boldsymbol{Z})\right), \tag{5.159}
$$

where the domain of f is the set of matrices in $\mathbb{C}^{N\times N}$ that have a determinant that is not both real and nonpositive. Calculate the second-order differential of f, and show that it can be expressed as

$$d^2 f = -\left(d\,\mathrm{vec}^T(\boldsymbol{Z})\right)\left\{\boldsymbol{K}_{N,N}\left[\boldsymbol{Z}^{-T}\otimes\boldsymbol{Z}^{-1}\right]\right\}d\,\mathrm{vec}(\boldsymbol{Z}). \tag{5.160}$$

Show from the above expression of $d^2 f$ that the complex Hessian of f with respect to $\boldsymbol{Z}$ and $\boldsymbol{Z}$ is given by

$$\mathcal{H}_{\boldsymbol{Z},\boldsymbol{Z}}f = -\boldsymbol{K}_{N,N}\left(\boldsymbol{Z}^{-T}\otimes\boldsymbol{Z}^{-1}\right). \tag{5.161}$$

5.3 Show that the complex Hessian matrix found in Example 5.8 reduces to the complex Hessian matrix found in Example 5.6 for an appropriate choice of the constant matrices and vectors involved in these examples.

5.4 This exercise is a continuation of Exercise 4.8. Assume that the conditions in Exercise 4.8 are valid, then show that

$$d^2\lambda = 2\boldsymbol{u}_0^H(d\boldsymbol{Z})(\lambda_0\boldsymbol{I}_N-\boldsymbol{Z}_0)^+(d\boldsymbol{Z})\,\boldsymbol{u}_0. \tag{5.162}$$

This is a generalization of Magnus and Neudecker (1988, Theorem 10, p. 166), which treats the real symmetric case to the complex-valued Hermitian matrices.

5.5 Assume that the conditions in Exercise 4.8 are fulfilled. Show that the second-order differential of the eigenvector function $\boldsymbol{u}$ evaluated at $\boldsymbol{Z}_0$ can be written as

$$d^2\boldsymbol{u} = 2(\lambda_0\boldsymbol{I}_N-\boldsymbol{Z}_0)^+\left[d\boldsymbol{Z}-\boldsymbol{u}_0^H(d\boldsymbol{Z}_0)\,\boldsymbol{u}_0\boldsymbol{I}_N\right](\lambda_0\boldsymbol{I}_N-\boldsymbol{Z}_0)^+(d\boldsymbol{Z})\,\boldsymbol{u}_0. \tag{5.163}$$

5.6 Let the complex Hessian matrix of the vector function $\boldsymbol{g}:\mathbb{C}^{N\times 2Q}\to\mathbb{C}^{P\times 1}$ be known and denoted by $\mathcal{H}_{\boldsymbol{\mathcal{Z}},\boldsymbol{\mathcal{Z}}}\boldsymbol{g}$. Let the matrix $\boldsymbol{A}\in\mathbb{C}^{M\times P}$ be independent of $\boldsymbol{\mathcal{Z}}$. Define the vector function $\boldsymbol{f}:\mathbb{C}^{N\times 2Q}\to\mathbb{C}^{M\times 1}$ by

$$\boldsymbol{f}(\boldsymbol{\mathcal{Z}}) = \boldsymbol{A}\boldsymbol{g}(\boldsymbol{\mathcal{Z}}). \tag{5.164}$$

Show that the second-order differential of $\boldsymbol{f}$ can be written as

$$d^2\boldsymbol{f} = \left[\boldsymbol{I}_M\otimes d\,\mathrm{vec}^T(\boldsymbol{\mathcal{Z}})\right]\left[\boldsymbol{A}\otimes\boldsymbol{I}_{2NQ}\right]\left[\mathcal{H}_{\boldsymbol{\mathcal{Z}},\boldsymbol{\mathcal{Z}}}\boldsymbol{g}\right]d\,\mathrm{vec}(\boldsymbol{\mathcal{Z}}). \tag{5.165}$$

From the above expression of $d^2\boldsymbol{f}$, show that the Hessian of $\boldsymbol{f}$ is given by

$$\mathcal{H}_{\boldsymbol{\mathcal{Z}},\boldsymbol{\mathcal{Z}}}\boldsymbol{f} = \left[\boldsymbol{A}\otimes\boldsymbol{I}_{2NQ}\right]\mathcal{H}_{\boldsymbol{\mathcal{Z}},\boldsymbol{\mathcal{Z}}}\boldsymbol{g}. \tag{5.166}$$

5.7 Let the complex Hessian matrix of the scalar function $f:\mathbb{C}^{N\times 2Q}\to\mathbb{C}$ be known and given by $\mathcal{H}_{\boldsymbol{\mathcal{Z}},\boldsymbol{\mathcal{Z}}}f$. The matrix function $\boldsymbol{F}:\mathbb{C}^{N\times 2Q}\to\mathbb{C}^{M\times P}$ is given by

$$\boldsymbol{F}(\boldsymbol{\mathcal{Z}}) = \boldsymbol{A}f(\boldsymbol{\mathcal{Z}}), \tag{5.167}$$

where the matrix $\boldsymbol{A}\in\mathbb{C}^{M\times P}$ is independent of all components of $\boldsymbol{\mathcal{Z}}$. Show that the second-order differential of $\mathrm{vec}(\boldsymbol{F})$ can be expressed as

$$d^2\,\mathrm{vec}(\boldsymbol{F}) = \left[\boldsymbol{I}_{MP}\otimes d\,\mathrm{vec}^T(\boldsymbol{\mathcal{Z}})\right]\left[\mathrm{vec}(\boldsymbol{A})\otimes\boldsymbol{I}_{2NQ}\right]\left[\mathcal{H}_{\boldsymbol{\mathcal{Z}},\boldsymbol{\mathcal{Z}}}f\right]d\,\mathrm{vec}(\boldsymbol{\mathcal{Z}}). \tag{5.168}$$

Show from the above second-order differential that the complex Hessian matrix of $\boldsymbol{F}$ is given by

$$\mathcal{H}_{\mathcal{Z},\mathcal{Z}}\boldsymbol{F} = \left[\operatorname{vec}(\boldsymbol{A}) \otimes \boldsymbol{I}_{2NQ}\right] \mathcal{H}_{\mathcal{Z},\mathcal{Z}} f. \tag{5.169}$$

5.8 This exercise is a continuation of Exercise 3.10, where the function $f : \mathbb{C}^{N\times 1} \times \mathbb{C}^{N\times 1} \rightarrow \mathbb{R}$ is defined in (3.142). Let the augmented matrix variable be given by

$$\mathcal{Z} \triangleq \left[z \; z^*\right] \in \mathbb{C}^{N\times 2}. \tag{5.170}$$

Show that the second-order differential of f can be expressed as

$$d^2 f = 2\left(d\operatorname{vec}^T(\mathcal{Z})\right)\left[\begin{bmatrix} 0 & 0 \\ 1 & 0 \end{bmatrix} \otimes \boldsymbol{R}\right] d\operatorname{vec}(\mathcal{Z}). \tag{5.171}$$

From the expression of $d^2 f$, show that the complex Hessian of f is given by

$$\mathcal{H}_{\mathcal{Z},\mathcal{Z}} f = \begin{bmatrix} \boldsymbol{0}_{N\times N} & \boldsymbol{R}^* \\ \boldsymbol{R} & \boldsymbol{0}_{N\times N} \end{bmatrix}. \tag{5.172}$$

5.9 In this exercise, an alternative derivation of the results in Example 5.6 is made where the results are *not* found in a component-wise manner, but in a more direct approach. Show that the second-order differential of $\boldsymbol{f}$ can be written as

$$d^2 \boldsymbol{f} = 2\left[\boldsymbol{I}_M \otimes d\operatorname{vec}^T(\mathcal{Z})\right] \boldsymbol{K}_{M,2NQ}\left[\boldsymbol{I}_{2Q} \otimes \boldsymbol{b} \otimes \boldsymbol{A}\right] d\operatorname{vec}(\mathcal{Z}). \tag{5.173}$$

Show that the Hessian of $\boldsymbol{f}$ can be written as

$$\mathcal{H}_{\mathcal{Z},\mathcal{Z}} \boldsymbol{f} = \boldsymbol{K}_{M,2NQ}\left[\boldsymbol{I}_{2Q} \otimes \boldsymbol{b} \otimes \boldsymbol{A}\right] + \operatorname{vecb}\left(\left[\boldsymbol{I}_{2Q} \otimes \boldsymbol{b}^T \otimes \boldsymbol{A}^T\right] \boldsymbol{K}_{2NQ,M}\right). \tag{5.174}$$

Make a MATLAB implementation of the function vecb$(\cdot)$. By writing a MATLAB program, verify that the expressions in (5.133) and (5.174) give the same numerical values for $\mathcal{H}_{\mathcal{Z},\mathcal{Z}} \boldsymbol{f}$.

5.10 In this exercise, the complex Hessian matrix of Example 5.8 is derived in an alternative way. Use the results from Exercise 2.11 to show that the $d^2\operatorname{vec}(\boldsymbol{F})$ can be written as

$$d^2\operatorname{vec}(\boldsymbol{F}) = 2\left[\boldsymbol{I}_{MP} \otimes d\operatorname{vec}^T(\mathcal{Z})\right] \boldsymbol{K}_{MP,2NQ}\left[\boldsymbol{D} \otimes \left\{\boldsymbol{G}\boldsymbol{E}^T\right\}\right] d\operatorname{vec}(\mathcal{Z}), \tag{5.175}$$

where the matrix $\boldsymbol{G} \in \mathbb{C}^{MPN\times P}$ is given by

$$\boldsymbol{G} = (\boldsymbol{K}_{N,P} \otimes \boldsymbol{I}_M)(\boldsymbol{I}_P \otimes \operatorname{vec}(\boldsymbol{U})). \tag{5.176}$$

Show from the above expression that the complex Hessian can be expressed as

$$\mathcal{H}_{\mathcal{Z},\mathcal{Z}}\boldsymbol{F} = \boldsymbol{K}_{MP,2NQ}\left[\boldsymbol{D} \otimes \left\{\boldsymbol{G}\boldsymbol{E}^T\right\}\right] + \operatorname{vecb}\left(\left[\boldsymbol{D}^T \otimes \left\{\boldsymbol{E}\boldsymbol{G}^T\right\}\right] \boldsymbol{K}_{2NQ,MP}\right). \tag{5.177}$$

Write a MATLAB program that verifies numerically that the results in (5.148) and (5.177) give the same result for the complex Hessian matrix $\mathcal{H}_{\mathcal{Z},\mathcal{Z}}\boldsymbol{F}$.

5.11 Assume that the complex Hessian matrix of $\boldsymbol{F} : \mathbb{C}^{N\times 2Q} \rightarrow \mathbb{C}^{M\times P}$ is known and given by $\mathcal{H}_{\mathcal{Z},\mathcal{Z}}\boldsymbol{F}$. Show that the second-order differential $d^2\operatorname{vec}(\boldsymbol{F}^*)$ can be expressed as

$$d^2 \operatorname{vec}(\boldsymbol{F}^*) = \left[\boldsymbol{I}_{MP} \otimes d \operatorname{vec}^T(\mathcal{Z})\right] \left(\boldsymbol{I}_{MP} \otimes \begin{bmatrix} \boldsymbol{0}_{NQ\times NQ} & \boldsymbol{I}_{NQ} \\ \boldsymbol{I}_{NQ} & \boldsymbol{0}_{NQ\times NQ} \end{bmatrix}\right) (\mathcal{H}_{\mathcal{Z},\mathcal{Z}}\boldsymbol{F})^* \begin{bmatrix} \boldsymbol{0}_{NQ\times NQ} & \boldsymbol{I}_{NQ} \\ \boldsymbol{I}_{NQ} & \boldsymbol{0}_{NQ\times NQ} \end{bmatrix} d \operatorname{vec}(\mathcal{Z}). \tag{5.178}$$

Show that the complex Hessian matrix of $\boldsymbol{F}^*$ is given by

$$\mathcal{H}_{\mathcal{Z},\mathcal{Z}}\boldsymbol{F}^* = \left(\boldsymbol{I}_{MP} \otimes \begin{bmatrix} \boldsymbol{0}_{NQ\times NQ} & \boldsymbol{I}_{NQ} \\ \boldsymbol{I}_{NQ} & \boldsymbol{0}_{NQ\times NQ} \end{bmatrix}\right) (\mathcal{H}_{\mathcal{Z},\mathcal{Z}}\boldsymbol{F})^* \begin{bmatrix} \boldsymbol{0}_{NQ\times NQ} & \boldsymbol{I}_{NQ} \\ \boldsymbol{I}_{NQ} & \boldsymbol{0}_{NQ\times NQ} \end{bmatrix}. \tag{5.179}$$

6 Generalized Complex-Valued Matrix Derivatives

6.1 Introduction

Often in signal processing and communications, problems appear for which we have to find a complex-valued matrix that minimizes or maximizes a real-valued objective function under the constraint that the matrix belongs to a set of matrices with a structure or pattern (i.e., where there exist some functional dependencies among the matrix elements). The theory presented in previous chapters is not suited for the case of functional dependencies among elements of the matrix. In this chapter, a systematic method is presented for finding the generalized derivative of complex-valued matrix functions, which depend on matrix arguments that have a certain structure. In Chapters 2 through 5, theory has been presented for how to find derivatives and Hessians of complex-valued functions $\boldsymbol{F} : \mathbb{C}^{N\times Q} \times \mathbb{C}^{N\times Q} \rightarrow \mathbb{C}^{M\times P}$ with respect to the complex-valued matrix $\boldsymbol{Z} \in \mathbb{C}^{N\times Q}$ and its complex conjugate $\boldsymbol{Z}^* \in \mathbb{C}^{N\times Q}$. As seen from Lemma 3.1, the differential variables $d\,\text{vec}(\boldsymbol{Z})$ and $d\,\text{vec}(\boldsymbol{Z}^*)$ should be treated as independent when finding derivatives. This is the main reason why the function $\boldsymbol{F} : \mathbb{C}^{N\times Q} \times \mathbb{C}^{N\times Q} \rightarrow \mathbb{C}^{M\times P}$ is denoted by two complex-valued input arguments $\boldsymbol{F}(\boldsymbol{Z}, \boldsymbol{Z}^*)$ because $\boldsymbol{Z} \in \mathbb{C}^{N\times Q}$ and $\boldsymbol{Z}^* \in \mathbb{C}^{N\times Q}$ should be treated independently when finding complex-valued matrix derivatives (see Lemma 3.1). Based on the presented theory, up to this point, it has been assumed that all elements of the input matrix variable $\boldsymbol{Z}$ contain *independent elements*. The type of derivative studied in Chapters 3, 4, and 5 is called a complex-valued matrix derivative or *unpatterned* complex-valued matrix derivative. A matrix that contains *independent elements* will be called *unpatterned*. Hence, any matrix variable that is not unpatterned is called a *patterned matrix*. An example of a patterned vector is $[z, \; z^*]$, where $z \in \mathbb{C}$.

In Tracy and Jinadasa (1988), a method was proposed for finding generalized derivatives when the matrices contain *real-valued* components. The method proposed in Tracy and Jinadasa (1988) is not adequate for the complex-valued matrix case; however, the method presented in the current chapter can be used. As in Tracy and Jinadasa (1988), a method is presented for finding the derivative of a function that depends on structured matrices; however, in this chapter, the matrices can be *complex-valued*. In Palomar and Verdú (2006), some results on derivatives of scalar functions with respect to complex-valued matrices were provided, as were results for derivatives of complex-valued scalar functions with respect to matrices with structure. Some of the results in Palomar and Verdú (2006) are studied in detail in the exercises. It is shown in Palomar and Verdú

(2007) how to link estimation theory and information theory through derivatives for arbitrary channels. Gradients were used in Feiten, Hanly, and Mathar (2007) to show the connection between the minimum MSE estimator and its error matrix. In Hjørungnes and Palomar (2008*a*–2008*b*), a theory was proposed for finding derivatives of functions that depend on complex-valued matrices with structure. Central to this theory is the so-called *pattern producing function*. In this chapter, this function will be called the *parameterization function*, because in general, its only requirement is that it is a diffeomorphism. Because the identity map is also a diffeomorphism, generalized derivatives include the complex-valued matrix derivatives from Chapters 3 to 5, where all matrix components are independent. Hence, the parameterization function does not necessarily introduce any structure, and the name *parameterization function* will be used instead of *pattern producing function*. Some of the material presented in this chapter is contained in Hjørungnes and Palomar (2008*a*–2008*b*). The traditional chain rule for unpatterned complex-valued matrix derivatives will be used to find the derivative of functions by indicating the *independent variables* that build up the function. The reason why the functions must have independent input variable matrices is that in the traditional chain rule for unpatterned matrix derivatives, it is a requirement that the functions within the chain rule must have freely chosen input arguments.

In this chapter, theory will be developed for finding complex-valued derivatives with respect to matrices belonging to a certain set of matrices. This theory will be called *generalized complex-valued matrix derivatives*. This is a natural extension of complex-valued matrix derivatives and contains the theory of the previous chapters as a special case. However, the theory presented in this chapter will show that it is *not* possible to find generalized complex-valued derivatives with respect to an *arbitrary* set of complex-valued patterned matrices. It will be shown that it is possible to find complex-valued matrices only for certain sets of matrices. A Venn diagram for the relation between patterned and unpatterned matrices, in addition to the connection between complex-valued matrix derivatives and generalized complex-valued matrices, is shown in Figure 6.1. The rectangle on the left side of the figure is the set of all unpatterned matrices, and the rectangle on the right side is the set of all patterned matrices. The sets of unpatterned and patterned matrices are disjoint, and their union is the set of all complex-valued matrix variables. Complex-valued matrix derivatives, presented in Chapters 2 through 5, are defined when the input variables are unpatterned; hence, in the Venn diagram, the set of unpatterned matrices and the complex-valued matrix derivatives are the same (left rectangle). Generalized complex-valued matrix derivatives are defined for unpatterned matrices, in addition to a *proper subset* of patterned matrices. Thus, the set of matrices for which the generalized complex-valued matrix derivatives is defined represents the union of the left rectangle and the half circle in Figure 6.1. The theory developed in this chapter presents the conditions and the sets of matrices for which the generalized complex-valued matrix derivatives are defined.

Central to the theory of generalized complex-valued matrix derivatives is the so-called *parameterization function*, which will be defined in this chapter. By means of the chain rule, the derivatives with respect to the *input variables* of the parameterization function will first be calculated. By introducing terms from the theory of manifolds (Spivak 2005)

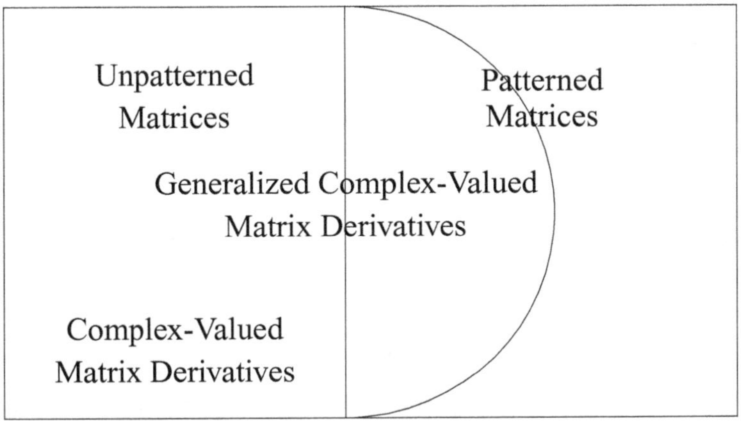

Figure 6.1 Venn diagram of the relationships between unpatterned and patterned matrices, in addition to complex-valued matrix derivatives and generalized complex-valued matrix derivatives.

and combining them with formal derivatives (see Definition 2.2), it is possible to find explicit expressions for the generalized matrix derivatives with respect to the matrices belonging to certain sets of matrices called manifolds. The presented theory is general and can be applied to find derivatives of functions that depend on matrices of linear and nonlinear parameterization functions when the matrix with structure lies in a so-called *complex-valued manifold*. If the parameterization function is linear in its input matrix variables, then the manifold is called linear. Illustrative examples with relevance to problems in signal processing and communications are presented.

In the theory of manifolds, the mathematicians (see, for example, Guillemin & Pollack 1974) have defined what is meant by derivatives with respect to objects within a manifold. The derivative with respect to elements within the manifold must follow several main requirements (Guillemin & Pollack 1974): (1) The chain rule should be valid for the generalized derivative. (2) The generalized derivative contains unpatterned derivatives as a special case. (3) The generalized derivatives are mappings between tangent spaces. (4) In the theory of manifolds, commutative diagrams for functions are used such that functions that start and end at the same set produce the same composite function. A book about optimization algorithms on manifolds was written by Absil, Mahony, and Sepulchre (2008).

The parameterization function should be a so-called diffeomorphism; this means that the function should have several properties. One such property is that the parameterization function depends on variables with independent differentials, and its domain must have the same dimension as the dimension of the manifold. In addition, this function must produce all the matrices within the manifold of interest. Another property of a diffeomorphism is that it is a one-to-one smooth mapping of variables with independent differentials to a set of matrices that the derivative will be found with respect to (i.e., the manifold). Instead of working on the complicated set of matrices with a certain structure, the diffeomorphism makes it possible to map the problem into a problem where the variables with independent differentials are used.

In the following example, it will be shown that the method presented in earlier chapters (i.e., using the differential of the function under consideration) is *not* applicable when *linear dependencies* exist among elements of the input matrix variable.

Example 6.1 Let the complex-valued function $g : \mathbb{C}^{2\times 2} \to \mathbb{C}$ be denoted by $g(\boldsymbol{W})$, where $\boldsymbol{W} \in \mathbb{C}^{2\times 2}$ is given by

$$g(\boldsymbol{W}) = \operatorname{Tr}\{\boldsymbol{A}\boldsymbol{W}\}, \tag{6.1}$$

and assume that the matrix $\boldsymbol{W}$ is symmetric such that $\boldsymbol{W}^T = \boldsymbol{W}$, that is, $(\boldsymbol{W})_{0,1} = w_{0,1} = (\boldsymbol{W})_{1,0} = w_{1,0} \triangleq w$. The arbitrary matrix $\boldsymbol{A} \in \mathbb{C}^{2\times 2}$ is assumed to be independent of the symmetric matrix $\boldsymbol{W}$ and $(\boldsymbol{A})_{k,l} = a_{k,l}$. The function g can be written is several ways. Here are some identical ways to express the function g

$$\begin{aligned} g(\boldsymbol{W}) = \operatorname{vec}^T\left(\boldsymbol{A}^T\right)\operatorname{vec}(\boldsymbol{W}) &= \left[a_{0,0}, a_{0,1}, a_{1,0}, a_{1,1}\right]\begin{bmatrix} w_{0,0} \\ w \\ w \\ w_{1,1}\end{bmatrix} \\ &= \left[a_{0,0}, a_{1,0}, a_{0,1}, a_{1,1}\right]\begin{bmatrix} w_{0,0} \\ w \\ w \\ w_{1,1}\end{bmatrix} \\ &= \left[a_{0,0}, \alpha(a_{0,1}+a_{1,0}), (1-\alpha)(a_{0,1}+a_{1,0}), a_{1,1}\right]\begin{bmatrix} w_{0,0} \\ w \\ w \\ w_{1,1}\end{bmatrix} \\ &= \left[a_{0,0}, a_{0,1}+a_{1,0}, a_{1,1}\right]\begin{bmatrix} w_{0,0} \\ w \\ w_{1,1}\end{bmatrix}, \end{aligned} \tag{6.2}$$

where $\alpha \in \mathbb{C}$ may be chosen arbitrarily. The alternative representations in (6.2) show that it does *not make sense* to try to identify the derivative by using the differential of g when there is a dependency between the elements within $\boldsymbol{W}$, because the differential of g can be expressed in multiple ways. In this chapter, we present a method for finding the derivative of a function that depends on a matrix that belongs to a manifold; hence, it may contain structure. In Subsection 6.5.4, it will be shown how to find the derivative of the function presented in this example.

In this chapter, the two operators $\dim_{\mathbb{R}}\{\cdot\}$ and $\dim_{\mathbb{C}}\{\cdot\}$ return, respectively, the real and complex dimensions of the space they are applied to.

The rest of this chapter is organized as follows: In Section 6.2, the procedure for finding *unpatterned* complex-valued derivatives is modified to include the case where one of the unpatterned input matrices is real-valued, in addition to another complex-valued matrix and its complex conjugate. The chain rule and the steepest descent method are also derived for the mixture of real- and complex-valued matrix variables in Section 6.2. Background material from the theory of manifolds is presented in Section 6.3. The method for finding generalized derivatives of functions that depend on complex-valued matrices belonging to a manifold is presented in Section 6.4. Several examples are given in Section 6.5 for how to find generalized complex-valued matrices derivatives for different types of manifolds that are relevant for problems in signal processing and communications. Finally, exercises related to the theory presented in this chapter are presented in Section 6.6.

6.2 Derivatives of Mixture of Real- and Complex-Valued Matrix Variables

In this chapter, it is assumed that all elements of the matrices $\boldsymbol{X} \in \mathbb{R}^{K\times L}$ and $\boldsymbol{Z} \in \mathbb{C}^{N\times Q}$ are independent; in addition, $\boldsymbol{X}$ and $\boldsymbol{Z}$ are *independent* of each other. Note that $\boldsymbol{X} \in \mathbb{R}^{K\times L}$ and $\boldsymbol{Z} \in \mathbb{C}^{N\times Q}$ have different sizes in general. For an introduction to complex differentials, see Sections 3.2 and 3.3.[1]

Because the real variables $\boldsymbol{X}$, $\mathrm{Re}\{\boldsymbol{Z}\}$, and $\mathrm{Im}\{\boldsymbol{Z}\}$ are independent of each other, so are their differentials. Although the complex variables $\boldsymbol{Z}$ and $\boldsymbol{Z}^*$ are related, the differentials of $\boldsymbol{X}$, $\boldsymbol{Z}$, and $\boldsymbol{Z}^*$ are linearly independent in the following way:

Lemma 6.1 *Let $\boldsymbol{A}_i \in \mathbb{C}^{M\times NQ}$, $\boldsymbol{B} \in \mathbb{C}^{M\times KL}$, $\boldsymbol{X} \in \mathbb{R}^{K\times L}$, and $\boldsymbol{Z} \in \mathbb{C}^{N\times Q}$, where the matrices $\boldsymbol{B}$, $\boldsymbol{A}_0$, and $\boldsymbol{A}_1$ might depend on $\boldsymbol{X}$, $\boldsymbol{Z}$, and $\boldsymbol{Z}^*$, but* not *on the differential operator. If*

$$\boldsymbol{B} d\,\mathrm{vec}(\boldsymbol{X}) + \boldsymbol{A}_0 d\,\mathrm{vec}(\boldsymbol{Z}) + \boldsymbol{A}_1 d\,\mathrm{vec}(\boldsymbol{Z}^*) = \boldsymbol{0}_{M\times 1}, \tag{6.3}$$

for all $d\boldsymbol{X} \in \mathbb{R}^{K\times L}$ and $d\boldsymbol{Z} \in \mathbb{C}^{N\times Q}$, then $\boldsymbol{B} = \boldsymbol{0}_{M\times KL}$ and $\boldsymbol{A}_i = \boldsymbol{0}_{M\times NQ}$ for $i \in \{0, 1\}$.

Proof If $d\,\mathrm{vec}(\boldsymbol{Z}) = d\,\mathrm{vec}(\mathrm{Re}\{\boldsymbol{Z}\}) + \jmath d\,\mathrm{vec}(\mathrm{Im}\{\boldsymbol{Z}\})$, then it follows by complex conjugation that $d\,\mathrm{vec}(\boldsymbol{Z}^*) = d\,\mathrm{vec}(\mathrm{Re}\{\boldsymbol{Z}\}) - \jmath d\,\mathrm{vec}(\mathrm{Im}\{\boldsymbol{Z}\})$. If $d\,\mathrm{vec}(\boldsymbol{Z})$ and $d\,\mathrm{vec}(\boldsymbol{Z}^*)$ are substituted into (6.3), then it follows that

$$\begin{aligned}\boldsymbol{B} d\,\mathrm{vec}(\boldsymbol{X}) &+ \boldsymbol{A}_0(d\,\mathrm{vec}(\mathrm{Re}\{\boldsymbol{Z}\}) + \jmath d\,\mathrm{vec}(\mathrm{Im}\{\boldsymbol{Z}\})) \\ &+ \boldsymbol{A}_1(d\,\mathrm{vec}(\mathrm{Re}\{\boldsymbol{Z}\}) - \jmath d\,\mathrm{vec}(\mathrm{Im}\{\boldsymbol{Z}\})) = \boldsymbol{0}_{M\times 1}.\end{aligned} \tag{6.4}$$

The last expression is equivalent to

$$\boldsymbol{B} d\,\mathrm{vec}(\boldsymbol{X}) + (\boldsymbol{A}_0 + \boldsymbol{A}_1) d\,\mathrm{vec}(\mathrm{Re}\{\boldsymbol{Z}\}) + \jmath(\boldsymbol{A}_0 - \boldsymbol{A}_1) d\,\mathrm{vec}(\mathrm{Im}\{\boldsymbol{Z}\}) = \boldsymbol{0}_{M\times 1}. \tag{6.5}$$

[1] The mixture of real- and complex-valued Gaussian distribution was treated in van den Bos (1998), and this is a generalization of the general *complex* Gaussian distribution studied in van den Bos (1995*b*). In van den Bos (1995*a*), complex-valued derivatives were used to solve a Fourier coefficient estimation problem for which real- and complex-valued parameters were estimated jointly.

Table 6.1 Procedure for finding the unpatterned derivatives for a mixture of real- and complex-valued matrix variables, $\boldsymbol{X} \in \mathbb{R}^{K \times L}$, $\boldsymbol{Z} \in \mathbb{C}^{N \times Q}$, and $\boldsymbol{Z}^* \in \mathbb{C}^{N \times Q}$.

Step 1:	Compute the differential $d\,\text{vec}(\boldsymbol{F})$.
Step 2:	Manipulate the expression into the form given in (6.6).
Step 3:	The matrix $\mathcal{D}_{\boldsymbol{X}}\boldsymbol{F}(\boldsymbol{X}, \boldsymbol{Z}, \boldsymbol{Z}^*)$, $\mathcal{D}_{\boldsymbol{Z}}\boldsymbol{F}(\boldsymbol{X}, \boldsymbol{Z}, \boldsymbol{Z}^*)$, and $\mathcal{D}_{\boldsymbol{Z}^*}\boldsymbol{F}(\boldsymbol{X}, \boldsymbol{Z}, \boldsymbol{Z}^*)$ can now be read out by using Definition 6.1.

Because the three differentials $d\boldsymbol{X}$, $d\,\text{Re}\{\boldsymbol{Z}\}$, and $d\,\text{Im}\{\boldsymbol{Z}\}$ are independent, so are $d\,\text{vec}(\boldsymbol{X})$, $d\,\text{vec}(\text{Re}\{\boldsymbol{Z}\})$, and $d\,\text{vec}(\text{Im}\{\boldsymbol{Z}\})$. Thus, (6.5) leads to $\boldsymbol{B} = \boldsymbol{0}_{M \times KL}$, $\boldsymbol{A}_0 + \boldsymbol{A}_1 = \boldsymbol{0}_{M \times NQ}$, and $\boldsymbol{A}_0 - \boldsymbol{A}_1 = \boldsymbol{0}_{M \times NQ}$. From these relations, it follows that $\boldsymbol{A}_0 = \boldsymbol{A}_1 = \boldsymbol{0}_{M \times NQ}$, which concludes the proof. ■

Next, Definition 3.1 is extended to fit the case of handling generalized complex-valued matrix derivatives. This is done by including a real-valued matrix in the domain of the function under consideration, in addition to the complex-valued matrix variable and its complex conjugate.

Definition 6.1 (Unpatterned Derivatives wrt. Real- and Complex-Valued Matrices) *Let $\boldsymbol{F} : \mathbb{R}^{K \times L} \times \mathbb{C}^{N \times Q} \times \mathbb{C}^{N \times Q} \to \mathbb{C}^{M \times P}$. Then, the derivative of the matrix function $\boldsymbol{F}(\boldsymbol{X}, \boldsymbol{Z}, \boldsymbol{Z}^*) \in \mathbb{C}^{M \times P}$ with respect to $\boldsymbol{X} \in \mathbb{R}^{K \times L}$ is denoted by $\mathcal{D}_{\boldsymbol{X}}\boldsymbol{F}(\boldsymbol{X}, \boldsymbol{Z}, \boldsymbol{Z}^*)$ and has size $MP \times KL$; the derivative with respect to $\boldsymbol{Z} \in \mathbb{C}^{N \times Q}$ is denoted by $\mathcal{D}_{\boldsymbol{Z}}\boldsymbol{F}(\boldsymbol{X}, \boldsymbol{Z}, \boldsymbol{Z}^*)$, and the derivative of $\boldsymbol{F}(\boldsymbol{X}, \boldsymbol{Z}, \boldsymbol{Z}^*) \in \mathbb{C}^{M \times P}$ with respect to $\boldsymbol{Z}^* \in \mathbb{C}^{N \times Q}$ is denoted by $\mathcal{D}_{\boldsymbol{Z}^*}\boldsymbol{F}(\boldsymbol{X}, \boldsymbol{Z}, \boldsymbol{Z}^*)$. The size of both $\mathcal{D}_{\boldsymbol{Z}}\boldsymbol{F}(\boldsymbol{X}, \boldsymbol{Z}, \boldsymbol{Z}^*)$ and $\mathcal{D}_{\boldsymbol{Z}^*}\boldsymbol{F}(\boldsymbol{X}, \boldsymbol{Z}, \boldsymbol{Z}^*)$ is $MP \times NQ$. The three matrix derivatives $\mathcal{D}_{\boldsymbol{X}}\boldsymbol{F}(\boldsymbol{X}, \boldsymbol{Z}, \boldsymbol{Z}^*)$, $\mathcal{D}_{\boldsymbol{Z}}\boldsymbol{F}(\boldsymbol{X}, \boldsymbol{Z}, \boldsymbol{Z}^*)$, and $\mathcal{D}_{\boldsymbol{Z}^*}\boldsymbol{F}(\boldsymbol{X}, \boldsymbol{Z}, \boldsymbol{Z}^*)$ are defined by the following differential expression:*

$$
\begin{aligned}
d\,\text{vec}(\boldsymbol{F}) = {} & (\mathcal{D}_{\boldsymbol{X}}\boldsymbol{F}(\boldsymbol{X}, \boldsymbol{Z}, \boldsymbol{Z}^*))\, d\,\text{vec}(\boldsymbol{X}) \\
& + (\mathcal{D}_{\boldsymbol{Z}}\boldsymbol{F}(\boldsymbol{X}, \boldsymbol{Z}, \boldsymbol{Z}^*))\, d\,\text{vec}(\boldsymbol{Z}) + (\mathcal{D}_{\boldsymbol{Z}^*}\boldsymbol{F}(\boldsymbol{X}, \boldsymbol{Z}, \boldsymbol{Z}^*))\, d\,\text{vec}(\boldsymbol{Z}^*). \qquad (6.6)
\end{aligned}
$$

The three matrices $\mathcal{D}_{\boldsymbol{X}}\boldsymbol{F}(\boldsymbol{X}, \boldsymbol{Z}, \boldsymbol{Z}^*)$, $\mathcal{D}_{\boldsymbol{Z}}\boldsymbol{F}(\boldsymbol{X}, \boldsymbol{Z}, \boldsymbol{Z}^*)$, and $\mathcal{D}_{\boldsymbol{Z}^*}\boldsymbol{F}(\boldsymbol{X}, \boldsymbol{Z}, \boldsymbol{Z}^*)$ are called the Jacobian matrices *of $\boldsymbol{F}$ with respect to $\boldsymbol{X}$, $\boldsymbol{Z}$, and $\boldsymbol{Z}^*$, respectively.*

When finding the derivatives with respect to $\boldsymbol{X}$, $\boldsymbol{Z}$, and $\boldsymbol{Z}^*$, these matrix variables should be treated as *independent matrix variables*. The reason for this is Lemma 6.1, which shows that $\boldsymbol{X}$, $\boldsymbol{Z}$, and $\boldsymbol{Z}^*$ have independent differentials. Finding the derivative of the complex-valued matrix function $\boldsymbol{F}$ with respect to the real- and complex-valued matrices $\boldsymbol{X}$, $\boldsymbol{Z}$, and $\boldsymbol{Z}^*$ can be achieved using the three-step procedure shown in Table 6.1. An example for how this can be done is shown next.

Example 6.2 Let the complex-valued function $\boldsymbol{F} : \mathbb{R}^{K \times L} \times \mathbb{C}^{N \times Q} \times \mathbb{C}^{N \times Q} \to \mathbb{C}^{M \times P}$ be given by

$$
\boldsymbol{F}(\boldsymbol{X}, \boldsymbol{Z}, \boldsymbol{Z}^*) = \boldsymbol{A}\boldsymbol{X}\boldsymbol{B}\boldsymbol{Z}\boldsymbol{C}\boldsymbol{Z}^H\boldsymbol{D}, \qquad (6.7)
$$

where the four matrices $A \in \mathbb{C}^{M\times K}$, $B \in \mathbb{C}^{L\times N}$, $C \in \mathbb{C}^{Q\times Q}$, and $D \in \mathbb{C}^{N\times P}$ are independent of the matrix variables $X \in \mathbb{R}^{K\times L}$, $Z \in \mathbb{C}^{N\times Q}$, and $Z^* \in \mathbb{C}^{N\times Q}$. By following the procedure in Table 6.1, we get

$$d\,\text{vec}\,(F) = d\,\text{vec}\left(AXBZCZ^H D\right) = \left[\left(D^T Z^* C^T Z^T B^T\right) \otimes A\right] d\,\text{vec}\,(X)$$
$$+ \left(D^T Z^* C^T\right) \otimes (AXB)\, d\,\text{vec}\,(Z) + \left[D^T \otimes (AXBZC)\right] K_{N,Q} d\,\text{vec}\,(Z^*). \quad (6.8)$$

It is observed that $d\,\text{vec}\,(F)$ has been manipulated into the form given by (6.6); hence, the derivatives of F with respect to X, Z, and Z^* are identified as

$$\mathcal{D}_X F = \left(D^T Z^* C^T Z^T B^T\right) \otimes A, \quad (6.9)$$
$$\mathcal{D}_Z F = \left(D^T Z^* C^T\right) \otimes (AXB), \quad (6.10)$$
$$\mathcal{D}_{Z^*} F = \left[D^T \otimes (AXBZC)\right] K_{N,Q}. \quad (6.11)$$

The rest of this section contains two subsections with results for the case of a mixture of real- and complex-valued matrix variables. In Subsection 6.2.1, the chain rule for mixed real- and complex-variable matrix variables is presented. The steepest descent method is derived in Subsection 6.2.2 for mixed real- and complex-valued input matrix variables for real-valued scalar functions.

6.2.1 Chain Rule for Mixture of Real- and Complex-Valued Matrix Variables

The chain rule is now stated for the mixture of real- and complex-valued matrices. This is a generalization of Theorem 3.1 that is better suited for handling the case of finding generalized complex-valued derivatives.

Theorem 6.1 (Chain Rule) *Let $(\mathcal{S}_0, \mathcal{S}_1, \mathcal{S}_2) \subseteq \mathbb{R}^{K\times L} \times \mathbb{C}^{N\times Q} \times \mathbb{C}^{N\times Q}$, and let $F : \mathcal{S}_0 \times \mathcal{S}_1 \times \mathcal{S}_2 \to \mathbb{C}^{M\times P}$ be differentiable with respect to its first, second, and third arguments at an interior point (X, Z, Z^*) in the set $\mathcal{S}_0 \times \mathcal{S}_1 \times \mathcal{S}_2$. Let $\mathcal{T}_0 \times \mathcal{T}_1 \subseteq \mathbb{C}^{M\times P} \times \mathbb{C}^{M\times P}$ be such that $(F(X, Z, Z^*), F^*(X, Z, Z^*)) \in \mathcal{T}_0 \times \mathcal{T}_1$ for all $(X, Z, Z^*) \in \mathcal{S}_0 \times \mathcal{S}_1 \times \mathcal{S}_2$. Assume that $G : \mathcal{T}_0 \times \mathcal{T}_1 \to \mathbb{C}^{R\times S}$ is differentiable at an interior point $(F(X, Z, Z^*), F^*(X, Z, Z^*)) \in \mathcal{T}_0 \times \mathcal{T}_1$. Define the composite function $H : \mathcal{S}_0 \times \mathcal{S}_1 \times \mathcal{S}_2 \to \mathbb{C}^{R\times S}$ by*

$$H(X, Z, Z^*) \triangleq G\left(F(X, Z, Z^*), F^*(X, Z, Z^*)\right). \quad (6.12)$$

The derivatives $\mathcal{D}_X H$, $\mathcal{D}_Z H$, and $\mathcal{D}_{Z^*} H$ are obtained through

$$\mathcal{D}_X H = (\mathcal{D}_F G)\,\mathcal{D}_X F + (\mathcal{D}_{F^*} G)\,\mathcal{D}_X F^*, \quad (6.13)$$
$$\mathcal{D}_Z H = (\mathcal{D}_F G)\,\mathcal{D}_Z F + (\mathcal{D}_{F^*} G)\,\mathcal{D}_Z F^*, \quad (6.14)$$
$$\mathcal{D}_{Z^*} H = (\mathcal{D}_F G)\,\mathcal{D}_{Z^*} F + (\mathcal{D}_{F^*} G)\,\mathcal{D}_{Z^*} F^*. \quad (6.15)$$

Proof The differential of the function vec(H) can be written as

$$d\,\text{vec}\,(H) = d\,\text{vec}\,(G) = (\mathcal{D}_F G)\, d\,\text{vec}(F) + (\mathcal{D}_{F^*} G)\, d\,\text{vec}(F^*). \quad (6.16)$$

The differentials $d\,\text{vec}(\boldsymbol{F})$ and $d\,\text{vec}(\boldsymbol{F}^*)$ are given by

$$d\,\text{vec}(\boldsymbol{F}) = (\mathcal{D}_{\boldsymbol{X}}\boldsymbol{F})\,d\,\text{vec}(\boldsymbol{X}) + (\mathcal{D}_{\boldsymbol{Z}}\boldsymbol{F})\,d\,\text{vec}(\boldsymbol{Z}) + (\mathcal{D}_{\boldsymbol{Z}^*}\boldsymbol{F})\,d\,\text{vec}(\boldsymbol{Z}^*), \tag{6.17}$$

$$d\,\text{vec}(\boldsymbol{F}^*) = (\mathcal{D}_{\boldsymbol{X}}\boldsymbol{F}^*)\,d\,\text{vec}(\boldsymbol{X}) + (\mathcal{D}_{\boldsymbol{Z}}\boldsymbol{F}^*)\,d\,\text{vec}(\boldsymbol{Z}) + (\mathcal{D}_{\boldsymbol{Z}^*}\boldsymbol{F}^*)\,d\,\text{vec}(\boldsymbol{Z}^*), \tag{6.18}$$

respectively. Inserting the two differentials $d\,\text{vec}(\boldsymbol{F})$ and $d\,\text{vec}(\boldsymbol{F}^*)$ from (6.17) and (6.18) into (6.16) gives the following expression:

$$\begin{aligned} d\,\text{vec}(\boldsymbol{H}) = &\left[(\mathcal{D}_{\boldsymbol{F}}\boldsymbol{G})\,\mathcal{D}_{\boldsymbol{X}}\boldsymbol{F} + (\mathcal{D}_{\boldsymbol{F}^*}\boldsymbol{G})\,\mathcal{D}_{\boldsymbol{X}}\boldsymbol{F}^*\right] d\,\text{vec}(\boldsymbol{X}) \\ &+ \left[(\mathcal{D}_{\boldsymbol{F}}\boldsymbol{G})\,\mathcal{D}_{\boldsymbol{Z}}\boldsymbol{F} + (\mathcal{D}_{\boldsymbol{F}^*}\boldsymbol{G})\,\mathcal{D}_{\boldsymbol{Z}}\boldsymbol{F}^*\right] d\,\text{vec}(\boldsymbol{Z}) \\ &+ \left[(\mathcal{D}_{\boldsymbol{F}}\boldsymbol{G})\,\mathcal{D}_{\boldsymbol{Z}^*}\boldsymbol{F} + (\mathcal{D}_{\boldsymbol{F}^*}\boldsymbol{G})\,\mathcal{D}_{\boldsymbol{Z}^*}\boldsymbol{F}^*\right] d\,\text{vec}(\boldsymbol{Z}^*). \end{aligned} \tag{6.19}$$

By using Definition 6.1 on (6.19), the derivatives of $\boldsymbol{H}$ with respect to $\boldsymbol{X}$, $\boldsymbol{Z}$, and $\boldsymbol{Z}^*$ can be identified as (6.13), (6.14), and (6.15), respectively. ■

By using Theorem 6.1 on the complex conjugate of $\boldsymbol{H}$, it is found that the derivatives of $\boldsymbol{H}^*$ with respect to $\boldsymbol{X}$, $\boldsymbol{Z}$, and $\boldsymbol{Z}^*$ are given by

$$\mathcal{D}_{\boldsymbol{X}}\boldsymbol{H}^* = (\mathcal{D}_{\boldsymbol{F}}\boldsymbol{G}^*)\,\mathcal{D}_{\boldsymbol{X}}\boldsymbol{F} + (\mathcal{D}_{\boldsymbol{F}^*}\boldsymbol{G}^*)\,\mathcal{D}_{\boldsymbol{X}}\boldsymbol{F}^*, \tag{6.20}$$

$$\mathcal{D}_{\boldsymbol{Z}}\boldsymbol{H}^* = (\mathcal{D}_{\boldsymbol{F}}\boldsymbol{G}^*)\,\mathcal{D}_{\boldsymbol{Z}}\boldsymbol{F} + (\mathcal{D}_{\boldsymbol{F}^*}\boldsymbol{G}^*)\,\mathcal{D}_{\boldsymbol{Z}}\boldsymbol{F}^*, \tag{6.21}$$

$$\mathcal{D}_{\boldsymbol{Z}^*}\boldsymbol{H}^* = (\mathcal{D}_{\boldsymbol{F}}\boldsymbol{G}^*)\,\mathcal{D}_{\boldsymbol{Z}^*}\boldsymbol{F} + (\mathcal{D}_{\boldsymbol{F}^*}\boldsymbol{G}^*)\,\mathcal{D}_{\boldsymbol{Z}^*}\boldsymbol{F}^*, \tag{6.22}$$

respectively. A diagram for how to find the derivatives of the functions $\boldsymbol{H}$ and $\boldsymbol{H}^*$ with respect to $\boldsymbol{X} \in \mathbb{R}^{K\times L}$, $\boldsymbol{Z} \in \mathbb{C}^{N\times Q}$, and $\boldsymbol{Z}^* \in \mathbb{C}^{N\times Q}$ is shown in Figure 6.2. This diagram helps to remember the chain rule.[2]

Remark *To use the chain rule in Theorem 6.1, all variables within $\boldsymbol{X} \in \mathbb{R}^{K\times L}$ and $\boldsymbol{Z} \in \mathbb{C}^{N\times Q}$ must be independent, and these matrices must be independent of each other. In addition, in the definition of the function $\boldsymbol{G}$, the* arguments *of this function should be chosen with independent matrix components.*

Example 6.3 (Use of Chain Rule) The chain rule will be used to derive two well-known results, which were found in Example 2.2. Let $\mathcal{W} = \{\boldsymbol{w} \in \mathbb{C}^{2\times 1} \mid \boldsymbol{w} = [z,\ z^*]^T, z \in \mathbb{C}\}$. The two functions $\boldsymbol{f}$ and g from the chain rule are defined first. These play the same role as in the chain rule. Define the function $\boldsymbol{f} : \mathbb{C} \times \mathbb{C} \to \mathcal{W}$ by

$$\boldsymbol{f}(z, z^*) = \begin{bmatrix} z \\ z^* \end{bmatrix}, \tag{6.23}$$

and let the function $g : \mathbb{C}^{2\times 1} \to \mathbb{C}$ be given by

$$g\left(\begin{bmatrix} z_0 \\ z_1 \end{bmatrix}\right) = z_0 z_1. \tag{6.24}$$

[2] Similar diagrams were developed in Edwards and Penney (1986, Section 15–7).

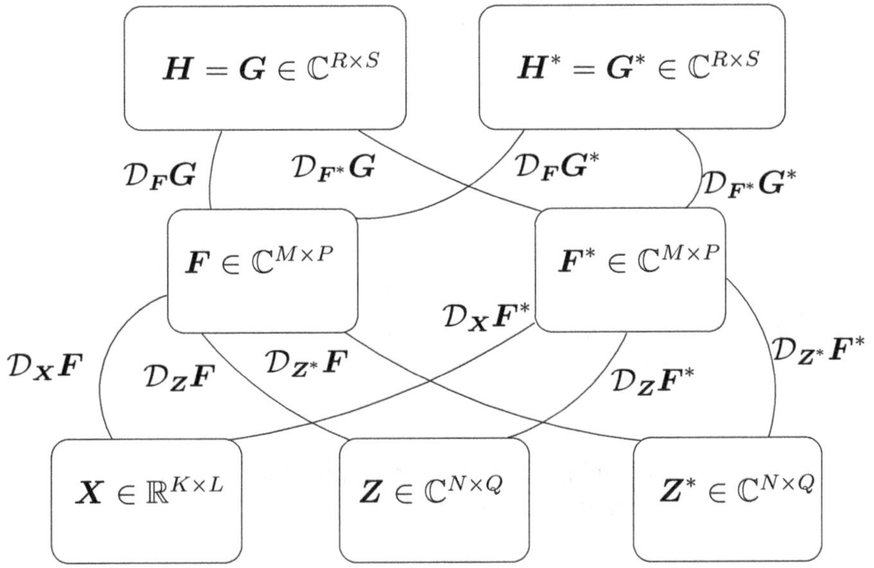

Figure 6.2 Diagram showing how to find the derivatives of $\boldsymbol{H}(\boldsymbol{X}, \boldsymbol{Z}, \boldsymbol{Z}^*) = \boldsymbol{G}(\boldsymbol{F}, \boldsymbol{F}^*)$, with respect to $\boldsymbol{X} \in \mathbb{R}^{K \times L}$, $\boldsymbol{Z} \in \mathbb{C}^{N \times Q}$, and $\boldsymbol{Z}^* \in \mathbb{C}^{N \times Q}$, where $\boldsymbol{F}$ and $\boldsymbol{F}^*$ are functions of the three matrix variables $\boldsymbol{X} \in \mathbb{R}^{K \times L}$, $\boldsymbol{Z} \in \mathbb{C}^{N \times Q}$, and $\boldsymbol{Z}^* \in \mathbb{C}^{N \times Q}$. The derivatives are shown along the curves connecting the boxes.

In the chain rule, the following derivatives are needed:

$$\mathcal{D}_z \boldsymbol{f} = \begin{bmatrix} 1 \\ 0 \end{bmatrix}, \tag{6.25}$$

$$\mathcal{D}_{z^*} \boldsymbol{f} = \begin{bmatrix} 0 \\ 1 \end{bmatrix}, \tag{6.26}$$

$$\mathcal{D}_{[z_0, z_1]} g = [z_1, \; z_0]. \tag{6.27}$$

Let $h : \mathbb{C} \times \mathbb{C} \to \mathbb{C}$ be defined as

$$h(z, z^*) = g\left(\boldsymbol{f}\left(\begin{bmatrix} z \\ z^* \end{bmatrix} \right) \right) = z z^* = |z|^2, \tag{6.28}$$

that is, the functions represent the squared Euclidean distance from the origin to Z. This is the same function that is plotted in Figure 3.1. From the definition of formal partial derivative (see Definition 2.2), it is seen that the following results are valid:

$$\mathcal{D}_z h(z, z^*) = z^*, \tag{6.29}$$

$$\mathcal{D}_{z^*} h(z, z^*) = z. \tag{6.30}$$

These results are in agreement with the derivatives derived in Example 3.1.

Now, these two results are derived by the use of the chain rule. From the chain rule in Theorem 6.1, it follows that

$$\mathcal{D}_z h(z) = \mathcal{D}_z g \left(\begin{array}{c} z_0 \\ z_1 \end{array} \right) \Bigg|_{f(z)} \mathcal{D}_z \boldsymbol{f}(z) = [z^*, \ z] \begin{bmatrix} 1 \\ 0 \end{bmatrix} = z^*, \tag{6.31}$$

and

$$\mathcal{D}_{z^*} h(z) = \mathcal{D}_z g \left(\begin{array}{c} z_0 \\ z_1 \end{array} \right) \Bigg|_{f(z)} \mathcal{D}_{z^*} \boldsymbol{f}(z) = [z^*, \ z] \begin{bmatrix} 0 \\ 1 \end{bmatrix} = z, \tag{6.32}$$

and these are the same as in (6.29) and (6.30), respectively.

6.2.2 Steepest Ascent and Descent Methods for Mixture of Real- and Complex-Valued Matrix Variables

When a real-valued scalar function is dependent on a mixture of real- and complex-valued matrix variables, the steepest descent method has to be modified. This is detailed in the following theorem:

Theorem 6.2 *Let $f : \mathbb{R}^{K \times L} \times \mathbb{C}^{N \times Q} \times \mathbb{C}^{N \times Q} \to \mathbb{R}$. The directions where the function f has the maximum and minimum rate of change with respect to the vector $[\text{vec}^T(\boldsymbol{X}), \ \text{vec}^T(\boldsymbol{Z})]$ are given by $[\mathcal{D}_{\boldsymbol{X}} f(\boldsymbol{X}, \boldsymbol{Z}, \boldsymbol{Z}^*), \ 2\mathcal{D}_{\boldsymbol{Z}^*} f(\boldsymbol{X}, \boldsymbol{Z}, \boldsymbol{Z}^*)]$ and $-[\mathcal{D}_{\boldsymbol{X}} f(\boldsymbol{X}, \boldsymbol{Z}, \boldsymbol{Z}^*), \ 2\mathcal{D}_{\boldsymbol{Z}^*} f(\boldsymbol{X}, \boldsymbol{Z}, \boldsymbol{Z}^*)]$, respectively.*

Proof Because $f \in \mathbb{R}$, it is possible to write df in two ways:

$$df = (\mathcal{D}_{\boldsymbol{X}} f)\, d\,\text{vec}(\boldsymbol{X}) + (\mathcal{D}_{\boldsymbol{Z}} f)\, d\,\text{vec}(\boldsymbol{Z}) + (\mathcal{D}_{\boldsymbol{Z}^*} f)\, d\,\text{vec}(\boldsymbol{Z}^*), \tag{6.33}$$

$$df^* = (\mathcal{D}_{\boldsymbol{X}} f)\, d\,\text{vec}(\boldsymbol{X}) + (\mathcal{D}_{\boldsymbol{Z}} f)^*\, d\,\text{vec}(\boldsymbol{Z}^*) + (\mathcal{D}_{\boldsymbol{Z}^*} f)^*\, d\,\text{vec}(\boldsymbol{Z}), \tag{6.34}$$

where $df = df^*$ since $f \in \mathbb{R}$. By subtracting (6.33) from (6.34) and then applying Lemma 6.1, it follows[3] that $\mathcal{D}_{\boldsymbol{Z}^*} f = (\mathcal{D}_{\boldsymbol{Z}} f)^*$. By using this result, (6.33) can be rewritten as follows:

$$\begin{aligned} df &= (\mathcal{D}_{\boldsymbol{X}} f)\, d\,\text{vec}(\boldsymbol{X}) + 2\,\text{Re}\,\{(\mathcal{D}_{\boldsymbol{Z}} f)\, d\,\text{vec}(\boldsymbol{Z})\} \\ &= (\mathcal{D}_{\boldsymbol{X}} f)\, d\,\text{vec}(\boldsymbol{X}) + 2\,\text{Re}\,\left\{(\mathcal{D}_{\boldsymbol{Z}^*} f)^*\, d\,\text{vec}(\boldsymbol{Z})\right\}. \end{aligned} \tag{6.35}$$

If $\boldsymbol{a}_i \in \mathbb{C}^{K \times 1}$, where $i \in \{0, 1\}$, then,

$$\text{Re}\left\{\boldsymbol{a}_0^H \boldsymbol{a}_1\right\} = \left\langle \begin{bmatrix} \text{Re}\,\{\boldsymbol{a}_0\} \\ \text{Im}\,\{\boldsymbol{a}_0\} \end{bmatrix}, \begin{bmatrix} \text{Re}\,\{\boldsymbol{a}_1\} \\ \text{Im}\,\{\boldsymbol{a}_1\} \end{bmatrix} \right\rangle, \tag{6.36}$$

where $\langle \cdot, \cdot \rangle$ is the ordinary Euclidean inner product (Young 1990) between real vectors in $\mathbb{R}^{2K \times 1}$. By using the inner product between real-valued vectors and rewriting the

[3] A similar result was obtained earlier in Lemma 3.3 for functions of the type $\boldsymbol{F} : \mathbb{C}^{N \times Q} \times \mathbb{C}^{N \times Q} \to \mathbb{C}^{M \times P}$ (i.e., not a mixture of real- and complex-valued variables).

right-hand side of (6.35) using (6.36), the differential of f can be written as

$$df = \left\langle \begin{bmatrix} (\mathcal{D}_{\boldsymbol{X}} f)^T \\ 2\,\mathrm{Re}\left\{(\mathcal{D}_{\boldsymbol{Z}^*} f)^T\right\} \\ 2\,\mathrm{Im}\left\{(\mathcal{D}_{\boldsymbol{Z}^*} f)^T\right\} \end{bmatrix}, \begin{bmatrix} d\,\mathrm{vec}(\boldsymbol{X}) \\ d\,\mathrm{Re}\left\{\mathrm{vec}(\boldsymbol{Z})\right\} \\ d\,\mathrm{Im}\left\{\mathrm{vec}(\boldsymbol{Z})\right\} \end{bmatrix} \right\rangle. \tag{6.37}$$

By applying Cauchy-Schwartz inequality (Young 1990) for inner products, it can be shown that the maximum value of df occurs when the two vectors in the inner product are parallel. This can be rewritten as

$$[d\,\mathrm{vec}^T(\boldsymbol{X}),\ d\,\mathrm{vec}^T(\boldsymbol{Z})] = \alpha[\mathcal{D}_{\boldsymbol{X}} f,\ 2\mathcal{D}_{\boldsymbol{Z}^*} f], \tag{6.38}$$

for $\alpha > 0$. And the minimum rate of change occurs when

$$[d\,\mathrm{vec}^T(\boldsymbol{X}),\ d\,\mathrm{vec}^T(\boldsymbol{Z})] = -\beta[\mathcal{D}_{\boldsymbol{X}} f,\ 2\mathcal{D}_{\boldsymbol{Z}^*} f], \tag{6.39}$$

where $\beta > 0$. ■

If a real-valued function f is being optimized with respect to the parameter matrices $\boldsymbol{X}$ and $\boldsymbol{Z}$ by means of the steepest descent method, it follows from Theorem 6.2 that the updating term must be proportional to $[\mathcal{D}_{\boldsymbol{X}} f,\ 2\mathcal{D}_{\boldsymbol{Z}^*} f]$. The update equation for optimizing the real-valued function in Theorem 6.2 by means of the steepest ascent or descent method can be expressed as

$$\begin{bmatrix} \mathrm{vec}(\boldsymbol{X}_{k+1}) \\ \mathrm{vec}(\boldsymbol{Z}_{k+1}) \end{bmatrix} = \begin{bmatrix} \mathrm{vec}(\boldsymbol{X}_{k}) \\ \mathrm{vec}(\boldsymbol{Z}_{k}) \end{bmatrix} + \mu \begin{bmatrix} \left(\mathcal{D}_{\boldsymbol{X}} f(\boldsymbol{X}_k, \boldsymbol{Z}_k, \boldsymbol{Z}_k^*)\right)^T \\ 2\left(\mathcal{D}_{\boldsymbol{Z}^*} f(\boldsymbol{X}_k, \boldsymbol{Z}_k, \boldsymbol{Z}_k^*)\right)^T \end{bmatrix}, \tag{6.40}$$

where μ is a *real positive constant if it is a maximization problem* or a *real negative constant if it is a minimization problem*, and where $\boldsymbol{X}_k \in \mathbb{R}^{K \times L}$ and $\boldsymbol{Z}_k \in \mathbb{C}^{N \times Q}$ are the values of the unknown parameter matrices after k iterations.

The next example illustrates the importance of factor 2 in front of $\mathcal{D}_{\boldsymbol{Z}^*} f$ in Theorem 6.2.

Example 6.4 Consider the following non-negative real-valued function: $h : \mathbb{R} \times \mathbb{C} \times \mathbb{C} \to \mathbb{R}$, given by

$$h(x, z, z^*) = x^2 + zz^* = x^2 + |z|^2, \tag{6.41}$$

where $x \in \mathbb{R}$ and $z \in \mathbb{C}$. The minimum value of the non-negative function h is in the origin where its value is 0, that is, $h(0, 0, 0) = 0$. The derivatives of this function with respect to x, z, and z^* are given by

$$\mathcal{D}_x h = 2x, \tag{6.42}$$

$$\mathcal{D}_z h = z^*, \tag{6.43}$$

$$\mathcal{D}_{z^*} h = z, \tag{6.44}$$

respectively. To test the validity of factor 2 in the steepest descent method of this function, let us replace 2 with a factor called β. The modified steepest descent equations can be expressed as

$$\begin{bmatrix} x_{k+1} \\ z_{k+1} \end{bmatrix} = \begin{bmatrix} x_k \\ z_k \end{bmatrix} - \mu \begin{bmatrix} \mathcal{D}_x h \\ \beta \mathcal{D}_{z^*} h \end{bmatrix} \Bigg|_{[x,z]=[x_k,z_k]} = \begin{bmatrix} x_k \\ z_k \end{bmatrix} - \mu \begin{bmatrix} 2x_k \\ \beta z_k \end{bmatrix}$$
$$= \begin{bmatrix} (1-2\mu)x_k \\ (1-\beta\mu)z_k \end{bmatrix}, \tag{6.45}$$

where k is the iteration index. By studying the function h carefully, it is seen that the three real-valued variables x, $\mathrm{Re}\{z\}$, and $\mathrm{Im}\{z\}$ should have the *same* rate of change when going toward the minimum. Hence, from the final expression in (6.45), it is seen that $\beta = 2$ corresponds to this choice. That $\beta = 2$ is the best choice in general is shown in Theorem 6.2.

6.3 Definitions from the Theory of Manifolds

A rich mathematical literature exists on manifolds and complex manifolds (see, e.g., Guillemin & Pollack 1974; Remmert 1991; Fritzsche & Grauert 2002; Spivak 2005, and Wells, Jr. 2008). Interested readers are encouraged to go deeper into the mathematical literature on manifolds.

To use the theory of manifolds for finding the derivatives with respect to matrices within a certain manifold, some basic definitions are given in this section. Some of these are taken from Guillemin and Pollack (1974).

Definition 6.2 (Smooth Function) *A function is called* smooth *if it has continuous partial derivatives of all orders with respect to all its input variables.*

Definition 6.3 (Diffeomorphism) *A smooth bijective*[4] *function is called a* diffeomorphism *if the inverse function is also smooth.*

Because all diffeomorphisms are one-to-one and onto, their domains and image set have the same real- or complex-valued dimensions.

Example 6.5 Notice that the function $f : \mathbb{R} \to \mathbb{R}$ given by $f(x) = x^3$ is both bijective and smooth; however, its inverse function $f^{-1}(x) = x^{\frac{1}{3}}$ is *not* smooth because it is not differentiable at $x = 0$. Therefore, this function is *not* a diffeomorphism.

Definition 6.4 (Real-Valued Manifold) *Let $\mathcal{X}$ be a subset of a big ambient*[5] *Euclidean space $\mathbb{R}^{N\times 1}$. Then, $\mathcal{X}$, is a k-dimensional manifold if it is locally diffeomorphic to $\mathbb{R}^{k\times 1}$, where $k \leq N$.*

[4] A function is bijective if it is both one-to-one (injective) and onto (surjective).

[5] An ambient space is a space surrounding a mathematical object.

For an introduction to manifolds, see Guillemin and Pollack (1974). Manifolds are a very general framework; however, we will use this theory to find generalized complex-valued matrix derivatives with respect to complex-valued matrices that belong to a manifold. This means that the matrices might contain a certain structure (see Figure 6.1).

When working with generalized complex-valued matrix derivatives, there often exists a mixture of independent real- and complex-valued matrix variables. The formal partial derivatives are used when finding the derivatives with respect to the complex-valued matrix variables with independent differentials. Hence, when a complex-valued matrix variable is present (e.g., $\boldsymbol{Z} \in \mathbb{C}^{N \times Q}$), this matrix variable has to be treated as independent of its complex conjugate $\boldsymbol{Z}^* \in \mathbb{C}^{N \times Q}$ when finding the derivatives. The phrase "treated as independent" can be handled with a procedure where the complex-valued matrix variable $\boldsymbol{Z} \in \mathbb{C}^{N \times Q}$ is replaced with the matrix variable $\boldsymbol{Z}_0 \in \mathbb{C}^{N \times Q}$, and the complex-valued matrix variable $\boldsymbol{Z}^* \in \mathbb{C}^{N \times Q}$ is replaced with the matrix variable $\boldsymbol{Z}_1 \in \mathbb{C}^{N \times Q}$, where the two matrix variables $\boldsymbol{Z}_0$ and $\boldsymbol{Z}_1$ are treated as independent.[6]

Definition 6.5 (Mixed Real- and Complex-Valued Manifold) *Let $\mathcal{W}$ be a subset of the complex space $\mathbb{C}^{M \times P}$. Then $\mathcal{W}$ is a $(KL + 2NQ)$-real-dimensional manifold*[7] *if it is locally diffeomorphic to $\mathbb{R}^{K \times L} \times \mathbb{C}^{N \times Q} \times \mathbb{C}^{N \times Q}$, where $KL + 2NQ \leq 2MP$, and where the diffeomorphism $\boldsymbol{F} : \mathbb{R}^{K \times L} \times \mathbb{C}^{N \times Q} \times \mathbb{C}^{N \times Q} \to \mathcal{W} \subseteq \mathbb{C}^{M \times P}$ is denoted by $\boldsymbol{F}(\boldsymbol{X}, \boldsymbol{Z}, \boldsymbol{Z}^*)$, and where $\boldsymbol{X} \in \mathbb{R}^{K \times L}$ and $\boldsymbol{Z} \in \mathbb{C}^{N \times Q}$ have independent components. The components of $\boldsymbol{Z}$ and $\boldsymbol{Z}^*$ should be treated as independent when finding complex-valued matrix derivatives.*

It is possible to find the derivative with respect to matrices that belong to a certain manifold, and this includes some special types of patterned matrices. Hence, with the theory presented in this chapter, it is *not* possible to find derivatives with respect to an *arbitrary* pattern, but only with respect to matrices that belong to a manifold.

Definition 6.6 (Tangent Space) *Assume that $\mathcal{W}$ is a mixed real- and complex-valued manifold given by Definition 6.5; hence, $\mathcal{W}$ is the image of the parameterization function $\boldsymbol{F} : \mathbb{R}^{K \times L} \times \mathbb{C}^{N \times Q} \times \mathbb{C}^{N \times Q} \to \mathcal{W}$. Let $\boldsymbol{\Phi} : (-\epsilon, \epsilon) \to \mathcal{W}$ be a smooth curve on the manifold $\mathcal{W}$, where $\epsilon > 0$. Let $\boldsymbol{U} \in \mathbb{C}^{M \times P}$ be the direction of the tangent of the curve $\boldsymbol{\Phi}(t)$ at the point $\boldsymbol{\Phi}(0) = \boldsymbol{W} \in \mathcal{W}$. The matrix function $\boldsymbol{\Phi}$ can be expressed as*

$$\operatorname{vec}(\boldsymbol{\Phi}(t)) = \operatorname{vec}(\boldsymbol{W}) + t \operatorname{vec}(\boldsymbol{U}), \tag{6.46}$$

where $t \in (-\epsilon, \epsilon)$. The tangent space *is denoted by $\mathcal{T}_{\boldsymbol{W}}$ and is defined as the set of all pairs of the form $(\boldsymbol{W}, \boldsymbol{U})$ called tangent vectors at the point $\boldsymbol{W} \in \mathcal{W}$.*

Example 6.6 (Hermitian Matrix) Let $\mathcal{W}$ be the set of Hermitian $N \times N$ matrices, that is, $\mathcal{W} = \{\boldsymbol{W} \in \mathbb{C}^{N \times N} \mid \boldsymbol{W}^H = \boldsymbol{W}\}$. Then $\mathcal{W}$ is a subset of the ambient complex Euclidean space $\mathbb{C}^{N \times N}$. Clearly, the actual matrix components of the Hermitian matrix

[6] In Brandwood (1983, pp. 11–12), a similar procedure was used when replacing the two scalars z and z^* with the two independent variables z_1 and z_2, respectively.

[7] The complex dimension of $\mathcal{W}$ is $\frac{KL}{2} + NQ$, that is, $\dim_{\mathbb{C}}\{\mathcal{W}\} = \frac{KL}{2} + NQ = \frac{\dim_{\mathbb{R}}\{\mathcal{W}\}}{2}$.

are dependent of each other because the elements *strictly below* the main diagonal are a function of the elements *strictly above* the main diagonal. Hence, matrices in $\mathcal{W}$ are patterned. The differentials of all matrix elements of $\boldsymbol{W} \in \mathcal{W}$ are independent.

The function $\boldsymbol{F} : \mathbb{R}^{N\times 1} \times \mathbb{C}^{\frac{(N-1)N}{2}} \times \mathbb{C}^{\frac{(N-1)N}{2}} \to \mathcal{W}$ denoted by $\boldsymbol{F}(\boldsymbol{x}, \boldsymbol{z}, \boldsymbol{z}^*)$ and given by

$$\text{vec}(\boldsymbol{F}(\boldsymbol{x}, \boldsymbol{z}, \boldsymbol{z}^*)) = \boldsymbol{L}_d \boldsymbol{x} + \boldsymbol{L}_l \boldsymbol{z} + \boldsymbol{L}_u \boldsymbol{z}^*, \tag{6.47}$$

is one-to-one and onto $\mathcal{W}$. It is also smooth, and its inverse is also smooth; hence, it is a diffeomorphism. Therefore, $\mathcal{W}$ is a manifold with real dimension given by $\dim_{\mathbb{R}}\{\mathcal{W}\} = N + 2\frac{(N-1)N}{2} = N^2$.

It is very important that the parameterization function *cannot* have too many input variables. For example, for producing Hermitian matrices, the function $\boldsymbol{H} : \mathbb{C}^{N\times N} \times \mathbb{C}^{N\times N} \to \mathcal{W}$ given by

$$\boldsymbol{H}(\boldsymbol{Z}, \boldsymbol{Z}^*) = \frac{1}{2}\left(\boldsymbol{Z} + \boldsymbol{Z}^H\right), \tag{6.48}$$

will produce all Hermitian matrices in $\mathcal{W}$ when $\boldsymbol{Z} \in \mathbb{C}^{N\times N}$ is an unpatterned matrix. The function $\boldsymbol{H}$ is *not* one-to-one. There are too many input variables, so this function is not a bijection, which is one of the requirements for a diffeomorphism.

Example 6.7 (Symmetric Matrix) Let $\mathbb{K}$ denote the field of real $\mathbb{R}$ or complex $\mathbb{C}$ numbers, and let $\mathcal{W} = \{\boldsymbol{Z} \in \mathbb{K}^{N\times N} | \boldsymbol{Z}^T = \boldsymbol{Z}\}$ be the set of all symmetric $N \times N$ matrices with elements in $\mathbb{K}$. Then $\mathcal{W}$ is a linear manifold studied in Magnus (1988) for real-valued components. If $k \neq l$, then $(\boldsymbol{Z})_{k,l} = (\boldsymbol{Z})_{l,k}$; hence, $\boldsymbol{Z}$ is patterned.

Example 6.8 (Matrices Containing One Constant) Let $\mathcal{W}$ be the set of all complex-valued matrices of size $N \times Q$ containing the constant c in row number k and column number l, where $k, l \in \{0, 1, \ldots, N-1\}$. The set $\mathcal{W}$ is then a manifold because a diffeomorphism can be found for generating $\mathcal{W}$ from $NQ - 1$ independent complex-valued parameters.

Magnus (1988) and Abadir and Magnus (2005) studied real-valued *linear structures*. A linear structure is equivalent to a linear manifold, meaning that the parameterization function is linear. Methods for how to find derivatives with respect to the real-valued independent Euclidean input parameters to the parameterization function are presented in Magnus (1988) and Wiens (1985). For linear manifolds; one set of basis vectors can be chosen for the whole manifold; however, this is *not* the case for a nonlinear manifold. Because the requirement for a manifold is that it is *locally* diffeomorphic, the choice of basis vectors might be different for each point for nonlinear manifolds. Hence, when working with nonlinear manifolds, it might be best to try to optimize the function with respect to the parameters in the space of variables with independent differentials (i.e.,

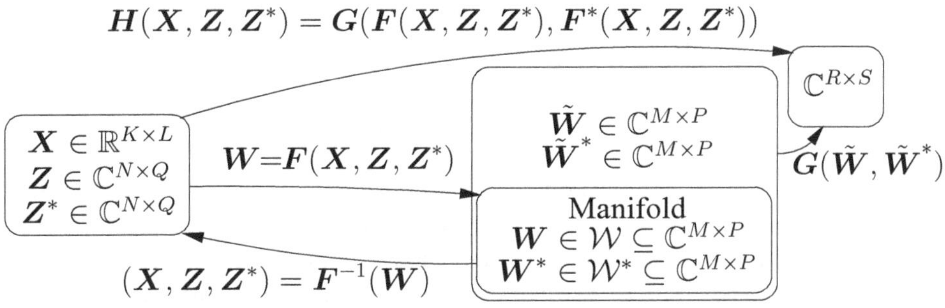

Figure 6.3 Functions used to find generalized matrix derivatives with respect to matrices $\boldsymbol{W}$ in the manifold $\mathcal{W}$ and matrices $\boldsymbol{W}^*$ in $\mathcal{W}^*$. Adapted from Hjørungnes and Palomar (2008a), © 2008 IEEE.

the input variables to the parameterization function), since one set of basis vectors is enough for the space of input variables of the parameterization function.

6.4 Finding Generalized Complex-Valued Matrix Derivatives

In this section, a method is presented for how to find generalized matrix derivatives with respect to matrices that belong to a manifold. The method will be derived by means of the chain rule, the theory of manifolds, and formal derivatives.

6.4.1 Manifolds and Parameterization Function

We now want to develop a theory by which we can find the derivative also with respect to a matrix belonging to a manifold $\boldsymbol{W} = \boldsymbol{F}(\boldsymbol{X}, \boldsymbol{Z}, \boldsymbol{Z}^*)$; to achieve this, we need the existence of $\mathcal{D}_{\boldsymbol{W}}\boldsymbol{F}^{-1}$ and $\mathcal{D}_{\boldsymbol{W}^*}\boldsymbol{F}^{-1}$, which exist only when $\boldsymbol{F}$ is a *diffeomorphism* (see Definition 6.5).

Figure 6.3 shows the situation we are working under by depicting how several functions are defined. As indicated by this figure, let the three matrices $\boldsymbol{X} \in \mathbb{R}^{K\times L}$, $\boldsymbol{Z} \in \mathbb{C}^{N\times Q}$, and $\boldsymbol{Z}^* \in \mathbb{C}^{N\times Q}$ be matrices that contain *independent* differentials, such that they can be treated as independent when finding derivatives. It is assumed that the three input matrices $\boldsymbol{X}$, $\boldsymbol{Z}$, and $\boldsymbol{Z}^*$ are used to produce all matrices in a considered manifold $\mathcal{W}$, as they are the input variables of the parameterization function $\boldsymbol{F}(\boldsymbol{X}, \boldsymbol{Z}, \boldsymbol{Z}^*)$, which is a *diffeomorphic* function (see Definition 6.5). The range and the image set of $\boldsymbol{F}$ are equal, and they are given by $\mathcal{W}$, which is a subset of $\mathbb{C}^{M\times P}$. One arbitrary member of $\mathcal{W}$ is denoted by $\boldsymbol{W}$ (see the middle part of Figure 6.3). Hence, the matrix $\boldsymbol{W}$ represents a potentially[8] patterned matrix that belongs to $\mathcal{W}$. Let $\tilde{\boldsymbol{W}} \in \mathbb{C}^{M\times P}$ be a matrix of independent components. Hence, the matrices $\tilde{\boldsymbol{W}} \in \mathbb{C}^{M\times P}$ and $\tilde{\boldsymbol{W}}^* \in \mathbb{C}^{M\times P}$

[8] Notice that generalized complex-valued matrices exist for unpatterned matrices and a subset of all patterned matrices (see Figure 6.1).

are unpatterned versions of the matrices $\boldsymbol{W} \in \mathcal{W}$ and $\boldsymbol{W}^* \in \mathcal{W}^* \triangleq \{\boldsymbol{W}^* \in \mathbb{C}^{M\times P} \mid \boldsymbol{W} \in \mathcal{W}\}$, respectively. It is assumed that all matrices $\boldsymbol{W}$ within $\mathcal{W}$ can be produced by the parameterization function $\boldsymbol{F} : \mathbb{R}^{K\times L} \times \mathbb{C}^{N\times Q} \times \mathbb{C}^{N\times Q} \to \mathcal{W} \subseteq \mathbb{C}^{M\times P}$, given by

$$\boldsymbol{W} = \boldsymbol{F}(\boldsymbol{X}, \boldsymbol{Z}, \boldsymbol{Z}^*). \tag{6.49}$$

In Figure 6.3, the function $\boldsymbol{F}$ is shown as a diffeomorphism from the matrices $\boldsymbol{X} \in \mathbb{R}^{K\times L}$, $\boldsymbol{Z} \in \mathbb{C}^{N\times Q}$, and $\boldsymbol{Z}^* \in \mathbb{C}^{N\times Q}$ onto the manifold $\mathcal{W} \subseteq \mathbb{C}^{M\times P}$. Figure 6.3 is a *commutative diagram*, such that maps that start and end at the same set in this figure return the same functions. In order to use the theory of complex-valued matrix derivatives (see Chapter 3), both $\boldsymbol{Z}$ and $\boldsymbol{Z}^*$ are used explicitly as variables, and they are treated as independent when finding derivatives.

The parameterization function $\boldsymbol{F} : \mathbb{R}^{K\times L} \times \mathbb{C}^{N\times Q} \times \mathbb{C}^{N\times Q} \to \mathcal{W}$ is onto and one-to-one; hence, the range and image set of $\boldsymbol{F}$ are both equal to $\mathcal{W}$. Because $\boldsymbol{F}$ is a bijection, the inverse function $\boldsymbol{F}^{-1} : \mathcal{W} \to \mathbb{R}^{K\times L} \times \mathbb{C}^{N\times Q} \times \mathbb{C}^{N\times Q}$ exists and is denoted by

$$\boldsymbol{F}^{-1}(\boldsymbol{W}) = (\boldsymbol{X}, \boldsymbol{Z}, \boldsymbol{Z}^*) = \left(\boldsymbol{F}_{\boldsymbol{X}}^{-1}(\boldsymbol{W})\ \boldsymbol{F}_{\boldsymbol{Z}}^{-1}(\boldsymbol{W})\ \boldsymbol{F}_{\boldsymbol{Z}^*}^{-1}(\boldsymbol{W})\right), \tag{6.50}$$

where the three functions $\boldsymbol{F}_{\boldsymbol{X}}^{-1} : \mathcal{W} \to \mathbb{R}^{K\times L}$, $\boldsymbol{F}_{\boldsymbol{Z}}^{-1} : \mathcal{W} \to \mathbb{C}^{N\times Q}$, and $\boldsymbol{F}_{\boldsymbol{Z}^*}^{-1} : \mathcal{W} \to \mathbb{C}^{N\times Q}$ are introduced in such a way that $\boldsymbol{X} = \boldsymbol{F}_{\boldsymbol{X}}^{-1}(\boldsymbol{W})$, $\boldsymbol{Z} = \boldsymbol{F}_{\boldsymbol{Z}}^{-1}(\boldsymbol{W})$, and $\boldsymbol{Z}^* = \boldsymbol{F}_{\boldsymbol{Z}^*}^{-1}(\boldsymbol{W})$. It is required that the function $\boldsymbol{F} : \mathbb{R}^{K\times L} \times \mathbb{C}^{N\times Q} \times \mathbb{C}^{N\times Q} \to \mathcal{W}$ is differentiable when using formal derivatives. Another requirement is that the function $\boldsymbol{F}^{-1}(\boldsymbol{W})$ should be differentiable when using formal derivatives.

The image set of the diffeomorphic function $\boldsymbol{F}$ is $\mathcal{W}$, and it has the same dimension as its domain, such that $\dim_{\mathbb{R}}\{\mathcal{W}\} = \dim_{\mathbb{R}}\{\mathbb{R}^{K\times L} \times \mathbb{C}^{N\times Q}\} = KL + 2NQ \leq \dim_{\mathbb{R}}\{\mathbb{C}^{M\times P}\} = 2MP$. This means that the set of matrices $\mathcal{W}$ can be parameterized with KL independent *real* variables collected within $\boldsymbol{X} \in \mathbb{R}^{K\times L}$ and NQ *independent* complex-valued variables inside the matrix $\boldsymbol{Z} \in \mathbb{C}^{N\times Q}$ and its complex conjugate $\boldsymbol{Z}^* \in \mathbb{C}^{N\times Q}$ through (6.49). The two complex-valued matrix variables $\boldsymbol{Z}$ and $\boldsymbol{Z}^*$ cannot be varied independently because they are the complex conjugate of each other; however, they should be treated as independent when finding generalized complex-valued matrix derivatives.

The inverse of the parameterization function $\boldsymbol{F}$ is denoted $\boldsymbol{F}^{-1}$, and it must satisfy

$$\boldsymbol{F}^{-1}(\boldsymbol{F}(\boldsymbol{X}, \boldsymbol{Z}, \boldsymbol{Z}^*)) = (\boldsymbol{X}, \boldsymbol{Z}, \boldsymbol{Z}^*), \tag{6.51}$$

$$\boldsymbol{F}(\boldsymbol{F}^{-1}(\boldsymbol{W})) = \boldsymbol{W}, \tag{6.52}$$

for all $\boldsymbol{X} \in \mathbb{R}^{K\times L}$, $\boldsymbol{Z} \in \mathbb{C}^{N\times Q}$, $\boldsymbol{Z}^* \in \mathbb{C}^{N\times Q}$, and $\boldsymbol{W} \in \mathcal{W}$. Here, the space $\mathbb{R}^{K\times L} \times \mathbb{C}^{N\times Q} \times \mathbb{C}^{N\times Q}$ contains variables that should be treated as independent when finding derivatives, and $\mathcal{W}$ is the set that contains matrices in the manifold. The real dimension of the tangent space of $\mathcal{W}$, produced by the parameterization function $\boldsymbol{F}$ defined in (6.49), is $KL + 2NQ$. To find the generalized complex-valued matrix derivatives, a basis for expressing vectors of the form $\text{vec}(\boldsymbol{W})$ should be chosen, where $\boldsymbol{W} \in \mathcal{W}$. Because the manifold is expressed with $KL + 2NQ$ real- and complex-valued parameters inside $\boldsymbol{X} \in \mathbb{R}^{K\times L}$, $\boldsymbol{Z} \in \mathbb{C}^{N\times Q}$, and $\boldsymbol{Z}^* \in \mathbb{C}^{N\times Q}$ with independent differentials, the number of basis vectors used to express the vector $\text{vec}(\boldsymbol{W}) \in \mathbb{C}^{MP\times 1}$ is $KL + 2NQ$. This will serve as a basis for the tangent space of $\mathcal{W}$. When using these $KL + 2NQ$ basis vectors to express vectors of the form vec ($\boldsymbol{W}$), the size of the generalized complex-valued matrix derivative $\mathcal{D}_{\boldsymbol{W}}\boldsymbol{F}^{-1}$ is $(KL + 2NQ) \times (KL + 2NQ)$; hence, this is a *square* matrix. When using the same basis vectors of the tangent space of $\mathcal{W}$ to express the three derivatives $\mathcal{D}_{\boldsymbol{X}}\boldsymbol{F}$, $\mathcal{D}_{\boldsymbol{Z}}\boldsymbol{F}$, and $\mathcal{D}_{\boldsymbol{Z}^*}\boldsymbol{F}$, the sizes of these three derivatives are $(KL + 2NQ) \times KL$, $(KL + 2NQ) \times NQ$, and $(KL + 2NQ) \times NQ$, respectively.

Taking the derivative with respect to $\boldsymbol{X}$ on both sides of (6.51) leads to

$$\left(\mathcal{D}_{\boldsymbol{W}}\boldsymbol{F}^{-1}\right)\mathcal{D}_{\boldsymbol{X}}\boldsymbol{F} = \mathcal{D}_{\boldsymbol{X}}(\boldsymbol{X}, \boldsymbol{Z}, \boldsymbol{Z}^*) = \begin{pmatrix} \boldsymbol{I}_{KL} \\ \boldsymbol{0}_{NQ\times KL} \\ \boldsymbol{0}_{NQ\times KL} \end{pmatrix}, \tag{6.53}$$

where $\mathcal{D}_{\boldsymbol{W}}\boldsymbol{F}^{-1}$ and $\mathcal{D}_{\boldsymbol{X}}\boldsymbol{F}$ are expressed in terms of the basis for the tangent space of $\mathcal{W}$, and they have size $(KL + 2NQ) \times (KL + 2NQ)$, and $(KL + 2NQ) \times KL$, respectively. In a similar manner, taking the derivatives of both sides of (6.51) with respect to $\boldsymbol{Z}$ gives

$$\left(\mathcal{D}_{\boldsymbol{W}}\boldsymbol{F}^{-1}\right)\mathcal{D}_{\boldsymbol{Z}}\boldsymbol{F} = \mathcal{D}_{\boldsymbol{Z}}(\boldsymbol{X}, \boldsymbol{Z}, \boldsymbol{Z}^*) = \begin{pmatrix} \boldsymbol{0}_{KL\times NQ} \\ \boldsymbol{I}_{NQ} \\ \boldsymbol{0}_{NQ\times NQ} \end{pmatrix}, \tag{6.54}$$

where the size of $\mathcal{D}_{\boldsymbol{W}}\boldsymbol{F}^{-1}$ and $\mathcal{D}_{\boldsymbol{Z}}\boldsymbol{F}$ are $(KL + 2NQ) \times (KL + 2NQ)$ and $(KL + 2NQ) \times NQ$ when expressed in terms of the basis of the tangent space of $\mathcal{W}$. Calculating the derivatives with respect to $\boldsymbol{Z}^*$ of both sides of (6.51) yields

$$\left(\mathcal{D}_{\boldsymbol{W}}\boldsymbol{F}^{-1}\right)\mathcal{D}_{\boldsymbol{Z}^*}\boldsymbol{F} = \mathcal{D}_{\boldsymbol{Z}^*}(\boldsymbol{X}, \boldsymbol{Z}, \boldsymbol{Z}^*) = \begin{pmatrix} \boldsymbol{0}_{KL\times NQ} \\ \boldsymbol{0}_{NQ\times NQ} \\ \boldsymbol{I}_{NQ} \end{pmatrix}, \tag{6.55}$$

where $\mathcal{D}_{\boldsymbol{W}}\boldsymbol{F}^{-1}$ has size $(KL + 2NQ) \times (KL + 2NQ)$ and $\mathcal{D}_{\boldsymbol{Z}^*}\boldsymbol{F}$ has size $(KL + 2NQ) \times NQ$ when expressed in terms of the basis of the tangent space of $\mathcal{W}$.

The three results in (6.53), (6.54), and (6.55) can be put together into a single equation as follows:

$$\left(\mathcal{D}_{\boldsymbol{W}}\boldsymbol{F}^{-1}\right)\left[\mathcal{D}_{\boldsymbol{X}}\boldsymbol{F}, \mathcal{D}_{\boldsymbol{Z}}\boldsymbol{F}, \mathcal{D}_{\boldsymbol{Z}^*}\boldsymbol{F}\right] = \boldsymbol{I}_{KL+2NQ}, \tag{6.56}$$

where the size of $\left[\mathcal{D}_{\boldsymbol{X}}\boldsymbol{F}, \mathcal{D}_{\boldsymbol{Z}}\boldsymbol{F}, \mathcal{D}_{\boldsymbol{Z}^*}\boldsymbol{F}\right]$ is $(KL + 2NQ) \times (KL + 2NQ)$ when expressed in terms of the basis of the tangent space of $\mathcal{W}$.

By calculating the derivatives of both sides of (6.52) with respect to $\boldsymbol{W}$ and expressing the derivatives with respect to the basis of the tangent space of $\mathcal{W}$, it is found that

$$\left[\mathcal{D}_{\boldsymbol{X}}\boldsymbol{F}, \mathcal{D}_{\boldsymbol{Z}}\boldsymbol{F}, \mathcal{D}_{\boldsymbol{Z}^*}\boldsymbol{F}\right]\mathcal{D}_{\boldsymbol{W}}\boldsymbol{F}^{-1} = \mathcal{D}_{\boldsymbol{W}}\boldsymbol{W} = \boldsymbol{I}_{KL+2NQ}, \tag{6.57}$$

where the size of both $\left[\mathcal{D}_{\boldsymbol{X}}\boldsymbol{F}, \mathcal{D}_{\boldsymbol{Z}}\boldsymbol{F}, \mathcal{D}_{\boldsymbol{Z}^*}\boldsymbol{F}\right]$ and $\mathcal{D}_{\boldsymbol{W}}\boldsymbol{F}^{-1}$ are $(KL + 2NQ) \times (KL + 2NQ)$, when the basis for the tangent space of $\mathcal{W}$ is used.

In various examples, it will be shown how the basis of $\mathcal{W}$ can be chosen such that $\mathcal{D}_{\boldsymbol{W}}\boldsymbol{F}^{-1} = \boldsymbol{I}_{KL+2NQ}$ for *linear manifolds*. For linear manifolds, one global choice of basis vectors for the tangent space is sufficient. In principle, we have the freedom to choose the $KL + 2NQ$ basis vectors as we like, and the derivative $\mathcal{D}_{\boldsymbol{W}} \boldsymbol{F}^{-1}$ depends on this choice.

Here is a list of some of the most important requirements that the parameterization function $\boldsymbol{F} : \mathbb{R}^{K \times L} \times \mathbb{C}^{N \times Q} \times \mathbb{C}^{N \times Q} \rightarrow \mathcal{W}$ must satisfy

- For any matrix $\boldsymbol{W} \in \mathcal{W}$, there should exist variables $(\boldsymbol{X}, \boldsymbol{Z}, \boldsymbol{Z}^*)$ such that $\boldsymbol{F}(\boldsymbol{X}, \boldsymbol{Z}, \boldsymbol{Z}^*) = \boldsymbol{W}$. This means that the parameterization function can produce *all* matrices within the manifold $\mathcal{W}$ of interest.
- The parameterization function should produce matrices only *within* the manifold. This means that for all allowable values of the independent parameters $(\boldsymbol{X}, \boldsymbol{Z}, \boldsymbol{Z}^*)$, the value of $\boldsymbol{F}(\boldsymbol{X}, \boldsymbol{Z}, \boldsymbol{Z}^*)$ should always give a matrix within the manifold $\mathcal{W}$ of interest. It should never give any matrix outside this manifold.
- The number of input variables of the parameterization function should be kept to a minimum. This means that no redundant variables should be introduced in the domain of the parameterization function. The parameterization function should be bijective; hence, the dimension of the domain of the parameterization function and the dimension of the manifold of interest will be identical.
- Even though the input variables $\boldsymbol{Z}$ and $\boldsymbol{Z}^*$ of the parameterization function should be treated as independent variables when finding the derivatives of this function, they are a mathematical function of each other. Formal derivatives (Wirtinger derivatives) should be used.
- When finding a candidate for the parameterization function, this function should satisfy the two relations in (6.56) and (6.57).

Some candidate functions for parameterization functions (diffeomorphisms) are presented next; some can be parameterization functions and others cannot.

Example 6.9 Let $f : \mathbb{C} \rightarrow \mathbb{C}$ be given by

$$f(z) = z. \tag{6.58}$$

It is observed that this function does not have z^* as an input variable, and f is the identity function. The function f satisfies equivalent versions when only z is the input variable of (6.56) and (6.57). Hence, the function f is a diffeomorphism because the function is one-to-one and smooth, and its inverse function is also smooth.

Example 6.10 Let $f : \mathbb{C} \rightarrow \mathbb{C}$ be given by

$$f(z^*) = z^*. \tag{6.59}$$

This function is equivalent to the function in Example 6.9; hence, the function f is a diffeomorphism because the function is one-to-one and smooth, and its inverse function is also smooth.

Example 6.11 Let $f : \mathbb{C} \to \mathbb{C}$ be the function given by

$$f(z) = z^*. \tag{6.60}$$

For this function, $\mathcal{D}_z f = 0$; hence, it is impossible that equivalent versions of (6.56) and (6.57), for one input variable z, are satisfied. Therefore, the function f is not a parameterization function.

Example 6.12 Let $f : \mathbb{C} \to \mathbb{C}$ be the function given by

$$f(z^*) = z. \tag{6.61}$$

This function is equivalent to the function in Example 6.11. Thus, the function f is not a parameterization function.

Example 6.13 Let $f : \mathbb{C} \times \mathbb{C} \to \mathbb{C}$ be given by

$$f(z, z^*) = z. \tag{6.62}$$

In this example. $\mathcal{D}_w f^{-1} = \begin{bmatrix} 1 \\ 0 \end{bmatrix}$, and $[\mathcal{D}_z f, \mathcal{D}_{z^*} f] = [1, 0]$. Hence, (6.57) is satisfied; however, (6.56) is *not* satisfied. The function f is not a diffeomorphism.

Example 6.14 Let $\boldsymbol{f} : \mathbb{C} \to \mathbb{C} \times \mathbb{C}$ be given by

$$\boldsymbol{f}(z) = \begin{bmatrix} z \\ z^* \end{bmatrix}. \tag{6.63}$$

In this example, it is found that $\mathcal{D}_z \boldsymbol{f} = \begin{bmatrix} 1 \\ 0 \end{bmatrix}$ and $\mathcal{D}_w \boldsymbol{f}^{-1} = [1, 0]$. It is then observed that (6.56) is satisfied, but (6.57) is not satisfied. Hence, the function $\boldsymbol{f}$ is *not* a diffeomorphism.

Example 6.15 Let $\mathcal{W} = \left\{ \boldsymbol{w} \in \mathbb{C}^{2\times 1} \;\middle|\; \boldsymbol{w} = \begin{bmatrix} z \\ z^* \end{bmatrix}, z \in \mathbb{C} \right\}$. Let $\boldsymbol{f} : \mathbb{C} \times \mathbb{C} \to \mathcal{W}$ be given by

$$\boldsymbol{f}(z, z^*) = \begin{bmatrix} z \\ z^* \end{bmatrix} = \boldsymbol{w}. \tag{6.64}$$

In this case, $\mathcal{D}_{[z,z^*]}\boldsymbol{f} = \boldsymbol{I}_2$, and $\mathcal{D}_{w}\boldsymbol{f}^{-1} = \boldsymbol{I}_2$ and both (6.56) and (6.57) are satisfied. The function $\boldsymbol{f}$ satisfies all requirements for a diffeomorphism; hence, the function $\boldsymbol{f}$ is a parameterization function.

6.4.2 Finding the Derivative of $\boldsymbol{H}(\boldsymbol{X}, \boldsymbol{Z}, \boldsymbol{Z}^*)$

In this subsection, we find the derivative of the function $\boldsymbol{H}(\boldsymbol{X}, \boldsymbol{Z}, \boldsymbol{Z}^*)$ by using the chain rule stated in Theorem 6.1. As seen from Figure 6.3, the composed function $\boldsymbol{H} : \mathbb{R}^{K\times L} \times \mathbb{C}^{N\times Q} \times \mathbb{C}^{N\times Q} \to \mathbb{C}^{R\times S}$ denoted by $\boldsymbol{H}(\boldsymbol{X}, \boldsymbol{Z}, \boldsymbol{Z}^*)$ is given by

$$\begin{aligned}\boldsymbol{H}(\boldsymbol{X}, \boldsymbol{Z}, \boldsymbol{Z}^*) &\triangleq \boldsymbol{G}(\tilde{\boldsymbol{W}}, \tilde{\boldsymbol{W}}^*)|_{\tilde{\boldsymbol{W}}=\boldsymbol{W}=\boldsymbol{F}(\boldsymbol{X},\boldsymbol{Z},\boldsymbol{Z}^*)} \\ &= \boldsymbol{G}(\boldsymbol{F}(\boldsymbol{X}, \boldsymbol{Z}, \boldsymbol{Z}^*), \boldsymbol{F}^*(\boldsymbol{X}, \boldsymbol{Z}, \boldsymbol{Z}^*)) = \boldsymbol{G}(\boldsymbol{W}, \boldsymbol{W}^*). \qquad (6.65)\end{aligned}$$

One of the requirements for using the chain rule in Theorem 6.1 is that the matrix functions $\boldsymbol{F}$ and $\boldsymbol{G}$ must be differentiable; this requires that these functions depend on matrix variables that *do not* contain any patterns. The unpatterned matrix input variables of the function $\boldsymbol{G}$ are $\tilde{\boldsymbol{W}}$ and $\tilde{\boldsymbol{W}}^*$, and they should be treated as independent when finding complex-valued matrix derivatives. Let the matrix function $\boldsymbol{G} : \mathbb{C}^{M\times P} \times \mathbb{C}^{M\times P} \to \mathbb{C}^{R\times S}$ be defined such that the domain of this function is the set of unpatterned matrices $(\tilde{\boldsymbol{W}}, \tilde{\boldsymbol{W}}^*) \in \mathbb{C}^{M\times P} \times \mathbb{C}^{M\times P}$. We want to calculate the generalized derivative of $\boldsymbol{G}(\boldsymbol{W}, \boldsymbol{W}^*) = \boldsymbol{G}(\tilde{\boldsymbol{W}}, \tilde{\boldsymbol{W}}^*)\big|_{\tilde{\boldsymbol{W}}=\boldsymbol{W}}$ with respect to $\boldsymbol{W} \in \mathcal{W}$ and $\boldsymbol{W}^* \in \mathcal{W}^*$. The chain rule can now be used for finding the derivative of the matrix function $\boldsymbol{H}(\boldsymbol{X}, \boldsymbol{Z}, \boldsymbol{Z}^*)$ because in both function definitions,

$$\boldsymbol{G} : \mathbb{C}^{M\times P} \times \mathbb{C}^{M\times P} \to \mathbb{C}^{R\times S}, \qquad (6.66)$$

and

$$\boldsymbol{F} : \mathbb{R}^{K\times L} \times \mathbb{C}^{N\times Q} \times \mathbb{C}^{N\times Q} \to \mathcal{W} \subseteq \mathbb{C}^{M\times P}, \qquad (6.67)$$

the input arguments of $\boldsymbol{G}$ and $\boldsymbol{F}$ can be independently chosen because all input variables of $\boldsymbol{G}(\tilde{\boldsymbol{W}}, \tilde{\boldsymbol{W}}^*)$ and $\boldsymbol{F}(\boldsymbol{X}, \boldsymbol{Z}, \boldsymbol{Z}^*)$ should be treated as independent when finding complex-valued matrix derivatives. In addition, both of these functions are assumed to be differentiable with respect to its matrix arguments. By using the chain rule, we find the derivative of $\boldsymbol{H}(\boldsymbol{X}, \boldsymbol{Z}, \boldsymbol{Z}^*)$, in (6.65), with respect to $\boldsymbol{X}$, $\boldsymbol{Z}$, and $\boldsymbol{Z}^*$, respectively, as

$$\begin{aligned}\mathcal{D}_{\boldsymbol{X}}\boldsymbol{H}(\boldsymbol{X}, \boldsymbol{Z}, \boldsymbol{Z}^*) &= \left(\mathcal{D}_{\tilde{\boldsymbol{W}}}\boldsymbol{G}(\tilde{\boldsymbol{W}}, \tilde{\boldsymbol{W}}^*)|_{\tilde{\boldsymbol{W}}=\boldsymbol{F}(\boldsymbol{X},\boldsymbol{Z},\boldsymbol{Z}^*)}\right)\mathcal{D}_{\boldsymbol{X}}\boldsymbol{F}(\boldsymbol{X}, \boldsymbol{Z}, \boldsymbol{Z}^*) \\ &+ \left(\mathcal{D}_{\tilde{\boldsymbol{W}}^*}\boldsymbol{G}(\tilde{\boldsymbol{W}}, \tilde{\boldsymbol{W}}^*)|_{\tilde{\boldsymbol{W}}=\boldsymbol{F}(\boldsymbol{X},\boldsymbol{Z},\boldsymbol{Z}^*)}\right)\mathcal{D}_{\boldsymbol{X}}\boldsymbol{F}^*(\boldsymbol{X}, \boldsymbol{Z}, \boldsymbol{Z}^*), \qquad (6.68)\end{aligned}$$

$$\begin{aligned}\mathcal{D}_{\boldsymbol{Z}}\boldsymbol{H}(\boldsymbol{X}, \boldsymbol{Z}, \boldsymbol{Z}^*) &= \left(\mathcal{D}_{\tilde{\boldsymbol{W}}}\boldsymbol{G}(\tilde{\boldsymbol{W}}, \tilde{\boldsymbol{W}}^*)|_{\tilde{\boldsymbol{W}}=\boldsymbol{F}(\boldsymbol{X},\boldsymbol{Z},\boldsymbol{Z}^*)}\right)\mathcal{D}_{\boldsymbol{Z}}\boldsymbol{F}(\boldsymbol{X}, \boldsymbol{Z}, \boldsymbol{Z}^*) \\ &+ \left(\mathcal{D}_{\tilde{\boldsymbol{W}}^*}\boldsymbol{G}(\tilde{\boldsymbol{W}}, \tilde{\boldsymbol{W}}^*)|_{\tilde{\boldsymbol{W}}=\boldsymbol{F}(\boldsymbol{X},\boldsymbol{Z},\boldsymbol{Z}^*)}\right)\mathcal{D}_{\boldsymbol{Z}}\boldsymbol{F}^*(\boldsymbol{X}, \boldsymbol{Z}, \boldsymbol{Z}^*), \qquad (6.69)\end{aligned}$$

and

$$\mathcal{D}_{Z^*} H(X, Z, Z^*) = \left(\mathcal{D}_{\tilde{W}} G(\tilde{W}, \tilde{W}^*)|_{\tilde{W}=F(X,Z,Z^*)}\right) \mathcal{D}_{Z^*} F(X, Z, Z^*)$$
$$+ \left(\mathcal{D}_{\tilde{W}^*} G(\tilde{W}, \tilde{W}^*)|_{\tilde{W}=F(X,Z,Z^*)}\right) \mathcal{D}_{Z^*} F^*(X, Z, Z^*). \quad (6.70)$$

From (6.68), (6.69), and (6.70), it is seen that the derivatives of the function H can be found from several different derivatives of ordinary unpatterned derivatives; then the theory of unpatterned derivatives from Section 6.2 can be applied. Hence, a method has been found for calculating the derivative of the function $H(X, Z, Z^*)$ with respect to the three matrices X, Z, and Z^*.

6.4.3 Finding the Derivative of $G(W, W^*)$

We want to find a way of finding the derivative of the complex-valued matrix function $G : \mathcal{W} \times \mathcal{W}^* \to \mathbb{C}^{R\times S}$ with respect to $W \in \mathcal{W}$. This function is written $G(W, W^*)$, where it is assumed that it depends on the matrix $W \in \mathcal{W}$ and its complex conjugate $W^* \in \mathcal{W}^*$. Generalized derivatives of the function G with respect to elements within the manifold $\mathcal{W}$ represent a mapping between the tangent space of $\mathcal{W}$ onto the tangent space of the function G (Guillemin & Pollack 1974). The derivatives $\mathcal{D}_W G$ and $\mathcal{D}_{W^*} G$ exist exactly when there exists a diffeomorphism, as stated in Definition 6.5. From (6.49), it follows that $W^* = F^*(X, Z, Z^*)$. Because the diagram in Figure 6.3 is commutative, it follows that $\mathcal{D}_W G$ and $\mathcal{D}_{W^*} G$ can be found as

$$\mathcal{D}_W G = [\mathcal{D}_X H, \mathcal{D}_Z H, \mathcal{D}_{Z^*} H]\, \mathcal{D}_W F^{-1}, \quad (6.71)$$

$$\mathcal{D}_{W^*} G = [\mathcal{D}_X H, \mathcal{D}_Z H, \mathcal{D}_{Z^*} H]\, \mathcal{D}_{W^*} F^{-1}, \quad (6.72)$$

where $\mathcal{D}_X H, \mathcal{D}_Z H,$ and $\mathcal{D}_{Z^*} H$ can be found from (6.68), (6.69), and (6.70), respectively, while $\mathcal{D}_W F^{-1}$ and $\mathcal{D}_{W^*} F^{-1}$ can be identified after a basis is chosen for the tangent space of $\mathcal{W}$, and the sizes of both these derivatives are $(KL + 2NQ) \times (KL + 2NQ)$. The dimension of the tangent space of $\mathcal{W}$ is $(KL + 2NQ)$, such that the sizes of both $\mathcal{D}_W G$ and $\mathcal{D}_{W^*} G$ are $RS \times (KL + 2NQ)$.

6.4.4 Specialization to Unpatterned Derivatives

If the matrix W is *unpatterned* and complex-valued, then we can choose $(K, L) = (0, 0)$ (the real parameter matrix X is not needed), $(N, Q) = (M, P)$, and $W = F(Z, Z^*) = Z$, and this leads to $\mathcal{D}_{Z^*} F(Z, Z^*) = 0_{NQ\times NQ} = \mathcal{D}_Z F^*(Z, Z^*)$ and $\mathcal{D}_Z F(Z, Z^*) = I_{NQ} = \mathcal{D}_{Z^*} F^*(Z, Z^*)$. Therefore, the derived method in (6.69) and (6.70) reduces to the method of finding unpatterned complex-valued matrix derivatives as presented in Chapters 3 and 4.

6.4.5 Specialization to Real-Valued Derivatives

If we try to apply the presented method to the real-valued derivatives, no functions depend on $\boldsymbol{Z}$ and $\boldsymbol{Z}^*$; hence, $N = Q = 0$. The method can then be modified by using the function $\boldsymbol{F} : \mathbb{R}^{K\times L} \to \mathbb{R}^{M\times P}$ to produce all matrices $\boldsymbol{W} = \boldsymbol{F}(\boldsymbol{X}) \in \mathcal{W} \subseteq \mathbb{R}^{M\times P}$ in the real-valued manifold $\mathcal{W}$, where the unpatterned real-valued parameter matrix is denoted $\boldsymbol{X} \in \mathbb{R}^{K\times L}$. The function of interest is $\boldsymbol{G} : \mathbb{R}^{M\times P} \to \mathbb{R}^{R\times S}$ and is denoted by $\boldsymbol{G}(\boldsymbol{W})$. The unpatterned real-valued variables in $\mathbb{R}^{M\times P}$ are collected in $\tilde{\boldsymbol{W}}$; the composite function is defined as $\boldsymbol{H} : \mathbb{R}^{K\times L} \to \mathbb{R}^{R\times S}$ and is given by $\boldsymbol{H}(\boldsymbol{X}) = \boldsymbol{G}(\boldsymbol{W}) = \boldsymbol{G}(\boldsymbol{F}(\boldsymbol{X}))$. By using the chain rule, the derivative of $\boldsymbol{H}$ with respect to $\boldsymbol{X}$ is given by

$$\mathcal{D}_{\boldsymbol{X}}\boldsymbol{H}(\boldsymbol{X}) = \left(\mathcal{D}_{\tilde{\boldsymbol{W}}}\boldsymbol{G}(\tilde{\boldsymbol{W}})|_{\tilde{\boldsymbol{W}}=\boldsymbol{F}(\boldsymbol{X})}\right) \mathcal{D}_{\boldsymbol{X}}\boldsymbol{F}(\boldsymbol{X}). \tag{6.73}$$

This result is consistent with the real-valued case given in Tracy and Jinadasa (1988); hence, the presented method is a natural extension of Tracy and Jinadasa (1988) to the complex-valued case. In Tracy and Jinadasa (1988), investigators did not use manifolds to find generalized derivatives, but they used the chain rule to find the derivative with respect to the input variables to the function, which produces matrices belonging to a specific set. The presented theory in this chapter can also be used to find real-valued generalized matrix derivatives. One important condition is that the parameterization function $\boldsymbol{F} : \mathbb{R}^{K\times L} \to \mathcal{W}$ should be a diffeomorphism.

6.4.6 Specialization to Scalar Function of Square Complex-Valued Matrices

One situation that appears frequently in signal processing and communication problems involves functions of the type $g : \mathbb{C}^{N\times N} \times \mathbb{C}^{N\times N} \to \mathbb{R}$ denoted by $g(\tilde{\boldsymbol{W}}, \tilde{\boldsymbol{W}}^*)$, which should be optimized when $\boldsymbol{W} \in \mathcal{W} \subseteq \mathbb{C}^{N\times N}$, where $\mathcal{W}$ is a manifold. One way of solving these types of problems is by using generalized complex-valued matrix derivatives. Several ways in which this can be done are shown for various manifolds in Exercise 6.15. A natural definition of partial derivatives with respect to matrices that belong to a manifold $\mathcal{W} \subseteq \mathbb{C}^{N\times N}$ follows.

Definition 6.7 *Assume that $\mathcal{W}$ is a manifold, and that $g : \mathbb{C}^{N\times N} \times \mathbb{C}^{N\times N} \to \mathbb{C}$. Let $\boldsymbol{W} \in \mathcal{W} \subseteq \mathbb{C}^{N\times N}$ and $\boldsymbol{W}^* \in \mathcal{W}^* \subseteq \mathbb{C}^{N\times N}$. The derivatives of the scalar function g with respect to $\boldsymbol{W}$ and $\boldsymbol{W}^*$ are defined as*

$$\frac{\partial g}{\partial \boldsymbol{W}} = \sum_{k=0}^{N-1}\sum_{l=0}^{N-1} \frac{\partial g}{\partial\left(\boldsymbol{W}\right)_{k,l}} \boldsymbol{E}_{k,l}, \tag{6.74}$$

$$\frac{\partial g}{\partial \boldsymbol{W}^*} = \sum_{k=0}^{N-1}\sum_{l=0}^{N-1} \frac{\partial g}{\partial\left(\boldsymbol{W}^*\right)_{k,l}} \boldsymbol{E}_{k,l}, \tag{6.75}$$

where $\boldsymbol{E}_{k,l}$ is an $N \times N$ matrix with zeros everywhere and $+1$ at position number (k, l) (see Definition 2.16). If the function g is independent *of the component $(\boldsymbol{W})_{k,l}$ of the matrix $\boldsymbol{W} \in \mathcal{W}$, then the corresponding component of $\frac{\partial g}{\partial \boldsymbol{W}}$ is equal to 0, that is,*

$\left(\frac{\partial g}{\partial \boldsymbol{W}}\right)_{k,l} = 0$. *Hence, if the function g is independent of all components of the matrix* $\boldsymbol{W}$, *then* $\frac{\partial g}{\partial \boldsymbol{W}} = \boldsymbol{0}_{N \times N}$.

This definition leads to results for complex-valued derivatives with structure that are in accordance with results in Palomar and Verdú (2006) and Vaidyanathan et al. (2010, Chapter 20). By using the operator vec(·) on both sides of (6.74), it is seen that

$$\operatorname{vec}\left(\frac{\partial g}{\partial \boldsymbol{W}}\right) = \sum_{k=l} \operatorname{vec}(\boldsymbol{E}_{k,l}) \frac{\partial g}{\partial (\boldsymbol{W})_{k,l}} + \sum_{k<l} \operatorname{vec}(\boldsymbol{E}_{k,l}) \frac{\partial g}{\partial (\boldsymbol{W})_{k,l}}$$

$$+ \sum_{k>l} \operatorname{vec}(\boldsymbol{E}_{k,l}) \frac{\partial g}{\partial (\boldsymbol{W})_{k,l}} = \boldsymbol{L}_d \frac{\partial g}{\partial \operatorname{vec}_d(\boldsymbol{W})} + \boldsymbol{L}_l \frac{\partial g}{\partial \operatorname{vec}_l(\boldsymbol{W})} + \boldsymbol{L}_u \frac{\partial g}{\partial \operatorname{vec}_u(\boldsymbol{W})}, \quad (6.76)$$

where (2.157), (2.163), and (2.170) were used. Assume that the parameterization function $\boldsymbol{F} : \mathbb{R}^{N \times 1} \times \mathbb{C}^{\frac{(N-1)N}{2} \times 1} \times \mathbb{C}^{\frac{(N-1)N}{2} \times 1} \rightarrow \mathcal{W}$ of the manifold $\mathcal{W}$ is denoted by $\boldsymbol{W} = \boldsymbol{F}(\boldsymbol{x}, \boldsymbol{z}, \boldsymbol{z}^*)$. Let the composed function $h : \mathbb{R}^{N \times 1} \times \mathbb{C}^{\frac{(N-1)N}{2} \times 1} \times \mathbb{C}^{\frac{(N-1)N}{2} \times 1} \rightarrow \mathbb{C}$ be defined as

$$h(\boldsymbol{x}, \boldsymbol{z}, \boldsymbol{z}^*) = g(\tilde{\boldsymbol{W}}, \tilde{\boldsymbol{W}}^*)\Big|_{\tilde{\boldsymbol{W}} = \boldsymbol{W} = \boldsymbol{F}(\boldsymbol{x}, \boldsymbol{z}, \boldsymbol{z}^*)} = g(\boldsymbol{W}, \boldsymbol{W}^*). \quad (6.77)$$

This relation shows that the functions $h(\boldsymbol{x}, \boldsymbol{z}, \boldsymbol{z}^*)$ and $g(\boldsymbol{W}, \boldsymbol{W}^*)$ are *identical*. If this fact is used in (6.76), it is found that

$$\operatorname{vec}\left(\frac{\partial g}{\partial \boldsymbol{W}}\right) = \boldsymbol{L}_d \frac{\partial h}{\partial \operatorname{vec}_d(\boldsymbol{W})} + \boldsymbol{L}_l \frac{\partial h}{\partial \operatorname{vec}_l(\boldsymbol{W})} + \boldsymbol{L}_u \frac{\partial h}{\partial \operatorname{vec}_u(\boldsymbol{W})}. \quad (6.78)$$

This expression will be used later in this chapter to find partial derivatives of the form $\frac{\partial g}{\partial \boldsymbol{W}}$ when the matrix $\boldsymbol{W}$ belongs to complex-valued diagonal, symmetric, skew-symmetric, Hermitian, or skew-Hermitian sets of matrices.

If some of the elements within $\boldsymbol{W}$ have *dependent differentials* when finding complex-valued derivatives, then a nonstandard basis for the space $\mathcal{W}$ is often necessary. The manifold $\mathcal{W}$ is a proper subset of $\mathbb{C}^{N \times N}$ in these cases, and the number of the tangent space of the manifold $\mathcal{W}$ is *strictly less* than the number of basis vectors that span $\mathbb{C}^{N \times N}$. The number of basis vectors that spans $\mathbb{C}^{N \times N}$ is N^2. In cases with dependent differentials within $\boldsymbol{W}$

$$\mathcal{D}_{\boldsymbol{W}} g \neq \operatorname{vec}^T\left(\frac{\partial g}{\partial \boldsymbol{W}}\right), \quad (6.79)$$

because the sizes of each side are different. The size of $\mathcal{D}_{\boldsymbol{W}} g$ is a row vector of length equal to the number of elements in $\boldsymbol{W} \in \mathcal{W}$ that have independent differentials. The size of $\frac{\partial g}{\partial \boldsymbol{W}}$ is $N \times N$, such that the size of $\operatorname{vec}^T\left(\frac{\partial g}{\partial \boldsymbol{W}}\right)$ is $1 \times N^2$. For example, if $\boldsymbol{W}$ belongs to the set of symmetric $N \times N$ matrices, then the size of $\mathcal{D}_{\boldsymbol{W}} g$ is $1 \times \frac{N(N+1)}{2}$, and $\operatorname{vec}^T\left(\frac{\partial g}{\partial \boldsymbol{W}}\right)$ has size $1 \times N^2$. This will be shown in Example 6.22.

If all the elements in $\boldsymbol{W}$ have *independent differentials*, then,

$$\mathcal{D}_{\boldsymbol{W}} g = \operatorname{vec}^T\left(\frac{\partial g}{\partial \boldsymbol{W}}\right). \quad (6.80)$$

For example, when $\boldsymbol{W}$ belongs to the set of Hermitian matrices, all its components have independent differentials; this will be shown in Example 6.25, and (6.80) will hold.

Now let $g : \mathbb{C}^{M\times P} \times \mathbb{C}^{M\times P} \rightarrow \mathbb{R}$, that is, g is a real-valued function denoted by $g(\tilde{\boldsymbol{W}}, \tilde{\boldsymbol{W}}^*)$, where $\tilde{\boldsymbol{W}}$ is unpatterned. Assume that $\mathcal{W}$ can be produced by the parameterization function $\boldsymbol{F} : \mathbb{R}^{K\times L} \times \mathbb{C}^{N\times Q} \times \mathbb{C}^{N\times Q} \rightarrow \mathcal{W}$, and let $h(\boldsymbol{X}, \boldsymbol{Z}, \boldsymbol{Z}^*)$ be given by

$$\begin{aligned} h(\boldsymbol{X}, \boldsymbol{Z}, \boldsymbol{Z}^*) &\triangleq g(\tilde{\boldsymbol{W}}, \tilde{\boldsymbol{W}}^*)|_{\tilde{\boldsymbol{W}}=\boldsymbol{W}=\boldsymbol{F}(\boldsymbol{X},\boldsymbol{Z},\boldsymbol{Z}^*)} \\ &= g(\boldsymbol{F}(\boldsymbol{X}, \boldsymbol{Z}, \boldsymbol{Z}^*), \boldsymbol{F}^*(\boldsymbol{X}, \boldsymbol{Z}, \boldsymbol{Z}^*)) = g(\boldsymbol{W}, \boldsymbol{W}^*), \end{aligned} \tag{6.81}$$

To solve

$$\min_{\boldsymbol{W}\in\mathcal{W}} g, \tag{6.82}$$

the following two (among others) solution procedures can be used:

(1) Solve the two equations:

$$\mathcal{D}_{\boldsymbol{X}} h(\boldsymbol{X}, \boldsymbol{Z}, \boldsymbol{Z}^*) = \boldsymbol{0}_{K\times L}, \tag{6.83}$$

$$\mathcal{D}_{\boldsymbol{Z}^*} h(\boldsymbol{X}, \boldsymbol{Z}, \boldsymbol{Z}^*) = \boldsymbol{0}_{N\times Q}. \tag{6.84}$$

The total number of equations here is $KL + NQ$. Because $h \in \mathbb{R}$, (6.84) is equivalent to

$$\mathcal{D}_{\boldsymbol{Z}} h(\boldsymbol{X}, \boldsymbol{Z}, \boldsymbol{Z}^*) = \boldsymbol{0}_{N\times Q}. \tag{6.85}$$

There exist examples for which it might be easier to solve (6.83), (6.84), and (6.85) jointly, rather than just (6.83) and (6.84).

(2) Solve the equation:

$$\frac{\partial g}{\partial \boldsymbol{W}} = \boldsymbol{0}_{M\times P}. \tag{6.86}$$

The number of equations here is MP.

If all elements of $\boldsymbol{W} \in \mathcal{W}$ have independent differentials, then $\mathcal{D}_{\boldsymbol{W}} \boldsymbol{F}^{-1}$ is invertible, and solving

$$\mathcal{D}_{\boldsymbol{W}} g = \text{vec}^T\left(\frac{\partial g}{\partial \boldsymbol{W}}\right) = \boldsymbol{0}_{1\times N^2}, \tag{6.87}$$

might be easier than solving (6.83), (6.84), and (6.85). When $\mathcal{D}_{\boldsymbol{W}} \boldsymbol{F}^{-1}$ is invertible, it follows from (6.71) that solving (6.87) is equivalent to finding the solutions of (6.83), (6.84), and (6.85) jointly.

The procedures (1) and (2) above are equivalent, and the way to choose depends on the problem under consideration.

6.5 Examples of Generalized Complex Matrix Derivatives

When working with problems of finding the generalized complex matrix derivatives, the main difficulty is to identify the parameterization function $\boldsymbol{F}$ that produces all matrices in the manifold $\mathcal{W}$, which has a domain of the same dimension as the dimension of $\mathcal{W}$. The parameterization function $\boldsymbol{F}$ should satisfy all requirements stated in Section 6.3. In this section, it will be shown how $\boldsymbol{F}$ can be chosen for several examples with applications in signal processing and communications. Examples include diagonal, symmetric, skew-symmetric, Hermitian, skew-Hermitian, unitary, and positive semidefinite matrices.

This section contains several subsections in which related examples are grouped together. The rest of this section is organized as follows: Subsection 6.5.1 contains some examples that show how generalized matrix derivatives can be found for scalar functions that depend on scalar variables. Subsection 6.5.2 contains an example of how to find the generalized complex-valued derivative with respect to patterned vectors. Diagonal matrices are studied in Subsection 6.5.3, and derivatives with respect to symmetric matrices are found in Subsection 6.5.4. Several examples of generalized matrix derivatives with respect to Hermitian matrices are shown in Subsection 6.5.5, including an example in which the capacity of a MIMO system is studied. Generalized derivatives of matrices that are skew-symmetric and skew-Hermitian are found in Subsections 6.5.6 and 6.5.7, respectively. Optimization with respect to orthogonal and unitary matrices is discussed in Subsections 6.5.8 and 6.5.9, while optimization with respect to positive semidefinite matrices is considered in Subsection 6.5.10.

6.5.1 Generalized Derivative with Respect to Scalar Variables

---

Example 6.16 Let $\mathcal{W} = \{w \in \mathbb{C} \mid w^* = w\}$. Hence, $\mathcal{W}$ consists of all real-valued points within $\mathbb{C}$. One parameterization function[9] would be $f : \mathbb{R} \to \mathcal{W}$ given by

$$w = f(x) = x, \tag{6.88}$$

that is, the identity map; hence, $f^{-1} : \mathcal{W} \to \mathbb{R}$ is also the identity map, and $\mathcal{D}_x f = \mathcal{D}_w f^{-1} = 1$. Consider the function $g : \mathbb{C} \times \mathbb{C} \to \mathbb{R}$, given by $g(\tilde{w}, \tilde{w}^*) = |\tilde{w}|^2$. The unconstrained derivatives of this function are given by $\mathcal{D}_{\tilde{w}} g = \tilde{w}^*$ and $\mathcal{D}_{\tilde{w}^*} g = \tilde{w}$; these agree with the corresponding results found in Example 2.2. Define the composed function $h : \mathbb{R} \to \mathbb{R}$ by

$$h(x) = g(\tilde{w}, \tilde{w}^*)_{\tilde{w}=w=f(x)} = |f(x)|^2 = x^2. \tag{6.89}$$

[9] Another function that could be considered is $t : \mathbb{C} \times \mathbb{C} \to \mathcal{W}$, given by $t(z, z^*) = \frac{1}{2}(z + z^*)$. This function is not a parameterization function because $\dim_{\mathbb{R}}\{\mathcal{W}\} = 1$; the dimension of the domain of this function is $\dim_{\mathbb{R}}\{\mathbb{C} \times \mathbb{C}\} = 2$, where the fact that the first and second arguments of $t(z, z^*)$ *cannot* be chosen freely is used. The first and second input variables of $t(z, z^*)$ are the complex conjugate of each other. Any diffeomorphism has the same dimension as its domain and image set.

The derivative of h with respect to x is found by the chain rule,

$$\mathcal{D}_x h = (\mathcal{D}_{\tilde{w}} g)_{\tilde{w}=w=f(x)} \mathcal{D}_x f + (\mathcal{D}_{\tilde{w}^*} g)_{\tilde{w}=w=f(x)} \mathcal{D}_x f^* = x \cdot 1 + x \cdot 1 = 2x. \quad (6.90)$$

Using the method of generalized derivatives, the derivative of g with respect to w is given by

$$\mathcal{D}_w g = (\mathcal{D}_x h) \mathcal{D}_x f^{-1} = 2x \cdot 1 = 2x, \quad (6.91)$$

which is the expected result.

Example 6.17 Let $g : \mathbb{C} \times \mathbb{C} \to \mathbb{R}$ be given by

$$g(\tilde{w}, \tilde{w}^*) = \tilde{w}\tilde{w}^* = |\tilde{w}|^2. \quad (6.92)$$

This is a very popular function in signal processing and communications, and it is equal to the function shown in Figure 3.1, which is the squared Euclidean distance between the origin and $\tilde{w}$. We want to find the derivative of this function when w lies on a circle with center at the origin and radius $r \in \mathbb{R}^+$, that is, w lies in the set

$$\mathcal{W} \triangleq \left\{ w \in \mathbb{C} \mid w = re^{\jmath\theta}, \theta \in (-\pi, \pi) \right\}. \quad (6.93)$$

This problem can be seen as a generalized derivative because the variables w and w^* are assumed to be constrained to $\mathcal{W}$ and $\mathcal{W}^*$, respectively. It will be shown below how a parameterization function can be found for producing the set $\mathcal{W} \subset \mathbb{C}$. From inspection of (6.92), we know that this function should not vary along $\mathcal{W}$; hence, we expect that the generalized derivative is zero. See also Figure 3.1, which shows the contour plot of the function $g(\tilde{w}, \tilde{w}^*)$. Now, we use the presented method to find the generalized derivative of g with respect to $w \in \mathcal{W}$.

First, we consider the function $g(\tilde{w}, \tilde{w}^*)$, where $\tilde{w} \in \mathbb{C}$ is unconstrained. To find the derivatives of g with respect to $\tilde{w}$ and $\tilde{w}^*$, the differential of g is found as

$$dg = \tilde{w}^* d\tilde{w} + \tilde{w} d\tilde{w}^*. \quad (6.94)$$

This implies that $\mathcal{D}_{\tilde{w}} g = \tilde{w}^*$ and $\mathcal{D}_{\tilde{w}^*} g = \tilde{w}$.

Next, we need to find a function that depends on independent variables and parameterizes $\mathcal{W}$ in (6.93). This can be done by using *one real-valued* parameter because $\mathcal{W}$ can be mapped over to a straight line in the real domain. Let us name the independent variable $x \in (-\pi, \pi)$ (in an open interval[10]) by using the following nonlinear function $f : (-\pi, \pi) \to \mathcal{W} \subset \mathbb{C}$:

$$w = f(x) = re^{\jmath x}, \quad (6.95)$$

[10] A parameterization function must be a diffeomorphism; hence, in particular, it is a *homeomorphism* (Munkres 2000, p. 105). This means that the parameterization function should be continuous. The inverse parameterization function's map of the circle should be open; hence, the domain of the parameterization function should be an open interval. When deciding the domain of the parameterization function, it is also important that the function is one-to-one and onto.

then it follows that

$$w^* = f^*(x) = re^{-\jmath x}. \tag{6.96}$$

In this example, the independent parameter that is defining the function is *real valued*, so $K = L = 1$, and $N = Q = 0$; hence, $\boldsymbol{Z}$ and $\boldsymbol{Z}^*$ are not present (see Definition 6.5). The derivatives of f and f^* with respect to x can be found as

$$\mathcal{D}_x f = \jmath r e^{\jmath x}, \tag{6.97}$$

$$\mathcal{D}_x f^* = -\jmath r e^{-\jmath x}. \tag{6.98}$$

The function $h(x)$, in the presented method for finding generalized derivatives, is given by $h(x) = g(w, w^*)|_{w=f(x)} = g(f(x), f^*(x)) = r^2$. Now, we can use the chain rule to find the derivative of $h(x)$:

$$\begin{aligned}\mathcal{D}_x h(x) &= \left(\mathcal{D}_{\tilde{w}} g(\tilde{w}, \tilde{w}^*)|_{\tilde{w}=f(x)}\right) \mathcal{D}_x f(x) + \left(\mathcal{D}_{\tilde{w}^*} g(\tilde{w}, \tilde{w}^*)|_{\tilde{w}=f(x)}\right) \mathcal{D}_x f^*(x) \\ &= e^{-\jmath x} \jmath r e^{\jmath x} + e^{\jmath x}(-\jmath r e^{-\jmath x}) = 0,\end{aligned} \tag{6.99}$$

as expected because $h(x) = r^2$ is independent of x. The derivative of g with respect to $w \in \mathcal{W}$ can be found by the method in (6.71), and it is seen that

$$\mathcal{D}_w g = (\mathcal{D}_x h) \mathcal{D}_w f^{-1} = 0 \cdot \mathcal{D}_w f^{-1} = 0, \tag{6.100}$$

which is the expected result because the function $g(w, w^*)$ stays constant when moving along a circle with center at the origin and radius r.

Remark *It is possible to imagine the function $t : \mathbb{C} \to \mathcal{W} \subset \mathbb{C}$ given by $t(z) = re^{\jmath \angle z}$ as an alternative function for producing the manifold $\mathcal{W}$ in Example 6.17. The image set of the function t is $\mathcal{W}$; however, the problem with this function is that its domain has one dimension in the complex domain and $\mathcal{W}$ has only one dimension in the* real domain*; this is impossible for a diffeomorphism. It is possible to parameterize $\mathcal{W}$ with a function that depends on only one* real *variable, and this can, for example, be done by means of the function given in* (6.95). *This shows the importance of parameterizing the manifold with a function that is a diffeomorphism.*

Example 6.18 Let $\tilde{w} \in \mathbb{C}$ be an unconstrained complex variable. Three examples of scalar manifolds will now be presented.

(1) The set defined as

$$\mathcal{W} = \{w \in \mathbb{C} \mid w = \text{Re}\{w\}\}, \tag{6.101}$$

is a manifold because the following function $f : \mathbb{R} \to \mathcal{W} \subset \mathbb{C}$, given by

$$w = f(x) = x, \tag{6.102}$$

is a diffeomorphism that corresponds to a parameterization function for $\mathcal{W}$.

(2) Let the set $\mathcal{V}$ be defined as

$$\mathcal{V} = \{w \in \mathbb{C} \mid w = \jmath \operatorname{Im}\{w\}\}. \tag{6.103}$$

This is a manifold because the function $g : \mathbb{R} \to \mathcal{V} \subset \mathbb{C}$, defined as

$$g(x) = \jmath x, \tag{6.104}$$

is a diffeomorphism.

(3) The set $\mathcal{U}$ defined as follows:

$$\mathcal{U} = \{w \in \mathbb{C} \mid \operatorname{Re}\{w\} = w, \operatorname{Re}\{w\} > 0\}, \tag{6.105}$$

is also a manifold. The reason for this is that it can be parameterized by the function $h : \mathbb{R}^+ \to \mathcal{U} \subset \mathbb{C}$, given by

$$w = h(x) = x, \tag{6.106}$$

and h is a diffeomorphism.

6.5.2 Generalized Derivative with Respect to Vector Variables

---

Example 6.19 Let $g : \mathbb{C}^{2\times 1} \times \mathbb{C}^{2\times 1} \to \mathbb{R}$ be given by

$$g(\tilde{\boldsymbol{w}}, \tilde{\boldsymbol{w}}^*) = \|\boldsymbol{A}\tilde{\boldsymbol{w}} - \boldsymbol{b}\|^2 = \tilde{\boldsymbol{w}}^H \boldsymbol{A}^H \boldsymbol{A}\tilde{\boldsymbol{w}} - \tilde{\boldsymbol{w}}^H \boldsymbol{A}^H \boldsymbol{b} - \boldsymbol{b}^H \boldsymbol{A}\tilde{\boldsymbol{w}} + \boldsymbol{b}^H \boldsymbol{b}, \tag{6.107}$$

where $\boldsymbol{A} \in \mathbb{C}^{N\times 2}$ and $\boldsymbol{b} \in \mathbb{C}^{N\times 1}$ contain elements that are independent of $\tilde{\boldsymbol{w}}$ and $\tilde{\boldsymbol{w}}^*$ and $\operatorname{rank}(\boldsymbol{A}) = 2$. The vector $\tilde{\boldsymbol{w}}$ is unpatterned. First, the unconstrained optimization problem of minimizing g over the set $\tilde{\boldsymbol{w}} \in \mathbb{C}^{2\times 1}$ is solved. To find necessary conditions for optimality, the equations $\mathcal{D}_{\tilde{\boldsymbol{w}}} g = \boldsymbol{0}_{1\times 2}$ or $\mathcal{D}_{\tilde{\boldsymbol{w}}^*} g = \boldsymbol{0}_{1\times 2}$ can be used. The derivatives $\mathcal{D}_{\tilde{\boldsymbol{w}}} g$ and $\mathcal{D}_{\tilde{\boldsymbol{w}}^*} g$ can be found from

$$\begin{aligned} dg &= \left(d\tilde{\boldsymbol{w}}^H\right) \boldsymbol{A}^H \boldsymbol{A}\tilde{\boldsymbol{w}} + \tilde{\boldsymbol{w}}^H \boldsymbol{A}^H \boldsymbol{A} d\tilde{\boldsymbol{w}} - \left(d\tilde{\boldsymbol{w}}^H\right) \boldsymbol{A}^H \boldsymbol{b} - \boldsymbol{b}^H \boldsymbol{A} d\tilde{\boldsymbol{w}} \\ &= \left(\tilde{\boldsymbol{w}}^H \boldsymbol{A}^H \boldsymbol{A} - \boldsymbol{b}^H \boldsymbol{A}\right) d\tilde{\boldsymbol{w}} + \left(\tilde{\boldsymbol{w}}^T \boldsymbol{A}^T \boldsymbol{A}^* - \boldsymbol{b}^T \boldsymbol{A}^*\right) d\tilde{\boldsymbol{w}}^*. \end{aligned} \tag{6.108}$$

Hence, the two derivatives $\mathcal{D}_{\tilde{\boldsymbol{w}}} g$ and $\mathcal{D}_{\tilde{\boldsymbol{w}}^*} g$ are given by

$$\mathcal{D}_{\tilde{\boldsymbol{w}}} g = \tilde{\boldsymbol{w}}^H \boldsymbol{A}^H \boldsymbol{A} - \boldsymbol{b}^H \boldsymbol{A}, \tag{6.109}$$

$$\mathcal{D}_{\tilde{\boldsymbol{w}}^*} g = \tilde{\boldsymbol{w}}^T \boldsymbol{A}^T \boldsymbol{A}^* - \boldsymbol{b}^T \boldsymbol{A}^*, \tag{6.110}$$

respectively. Necessary conditions for the unconstrained problem $\min_{\tilde{\boldsymbol{w}} \in \mathbb{C}^{2\times 1}} g(\tilde{\boldsymbol{w}}, \tilde{\boldsymbol{w}}^*)$ can be found by, for example, $\mathcal{D}_{\tilde{\boldsymbol{w}}^*} g = \boldsymbol{0}_{1\times 2}$, and this leads to

$$\tilde{\boldsymbol{w}} = \left(\boldsymbol{A}^H \boldsymbol{A}\right)^{-1} \boldsymbol{A}^H \boldsymbol{b} = \boldsymbol{A}^+ \boldsymbol{b}, \tag{6.111}$$

where $\operatorname{rank}(\boldsymbol{A}) = \operatorname{rank}\left(\boldsymbol{A}^H \boldsymbol{A}\right) = 2$ and (2.80) were used.

Now, a constrained set is introduced such that the function g should be minimized when its argument lies in $\mathcal{W}$. Let $\mathcal{W}$ be given by

$$\mathcal{W} = \left\{ \boldsymbol{w} \in \mathbb{C}^{2\times 1} \,\middle|\, \boldsymbol{w} = \begin{bmatrix} 1 \\ 1 \end{bmatrix} x + \jmath \begin{bmatrix} 1 \\ -1 \end{bmatrix} y,\ x, y \in \mathbb{R} \right\}$$
$$= \left\{ \boldsymbol{w} \in \mathbb{C}^{2\times 1} \,\middle|\, \boldsymbol{w} = \begin{bmatrix} z \\ z^* \end{bmatrix}, z \in \mathbb{C} \right\}. \tag{6.112}$$

Let $\boldsymbol{w} \in \mathcal{W}$, meaning that when it is enforced, the unconstrained vector $\tilde{\boldsymbol{w}}$ should lie inside the set $\mathcal{W}$, then it is named $\boldsymbol{w}$. The dimension of $\mathcal{W}$ is given by $\dim_{\mathbb{R}}\{\mathcal{W}\} = 2$ or, equivalently, $\dim_{\mathbb{C}}\{\mathcal{W}\} = 1$.

In the rest of this example, it will be shown how to solve the constrained complex-valued optimization problem $\min_{\boldsymbol{w} \in \mathcal{W}} g(\boldsymbol{w}, \boldsymbol{w}^*)$ in two alternative ways.

(a) Let the function $\boldsymbol{f} : \mathbb{C} \times \mathbb{C} \to \mathcal{W}$ be defined as

$$\boldsymbol{f}(z, z^*) = \begin{bmatrix} z \\ z^* \end{bmatrix} = \boldsymbol{w}. \tag{6.113}$$

As in all previous chapters, when it is written that $\boldsymbol{f} : \mathbb{C} \times \mathbb{C} \to \mathcal{W}$, this means that the first input argument z of $\boldsymbol{f}$ takes values from $\mathbb{C}$, and simultaneously the second input argument z^* takes values in $\mathbb{C}$; however, the two input arguments are complex conjugates of each other. Hence, they cannot be varied independently of each other. When calculating formal partial derivatives with respect to these two input variables, they are treated independently. The total complex dimension of the space of the input variables z and z^* is 1, and this is the same dimension as the manifold $\mathcal{W}$ in (6.112). It is seen that the function $\boldsymbol{f}$ produces all elements in $\mathcal{W}$; hence, it is onto $\mathcal{W}$. It is also seen that $\boldsymbol{f}$ is one-to-one. Hence, $\boldsymbol{f}$ is invertible. The derivatives of $\boldsymbol{f}$ with respect to z and z^* are given by

$$\mathcal{D}_z \boldsymbol{f} = \begin{bmatrix} 1 \\ 0 \end{bmatrix}, \tag{6.114}$$

$$\mathcal{D}_{z^*} \boldsymbol{f} = \begin{bmatrix} 0 \\ 1 \end{bmatrix}, \tag{6.115}$$

respectively. From the above two derivatives, it follows from Lemma 3.3 that

$$\mathcal{D}_z \boldsymbol{f}^* = \begin{bmatrix} 0 \\ 1 \end{bmatrix}, \tag{6.116}$$

$$\mathcal{D}_{z^*} \boldsymbol{f}^* = \begin{bmatrix} 1 \\ 0 \end{bmatrix}. \tag{6.117}$$

Define the composed function $h : \mathbb{C} \times \mathbb{C} \to \mathbb{R}$ by

$$h(z, z^*) = g(\tilde{\boldsymbol{w}}, \tilde{\boldsymbol{w}}^*)\big|_{\tilde{\boldsymbol{w}}=\boldsymbol{w}=\boldsymbol{f}(z,z^*)} = g(\boldsymbol{w}, \boldsymbol{w}^*) = g(\boldsymbol{f}(z, z^*), \boldsymbol{f}^*(z, z^*)). \tag{6.118}$$

The derivatives of h with respect to z and z^* can be found by the chain rule as follows:

$$\begin{aligned}\mathcal{D}_z h &= \mathcal{D}_{\tilde{w}} g|_{\tilde{w}=w}\, \mathcal{D}_z f + \mathcal{D}_{\tilde{w}^*} g|_{\tilde{w}=w}\, \mathcal{D}_z f^* \\ &= \left[w^H A^H A - b^H A\right] \begin{bmatrix} 1 \\ 0 \end{bmatrix} + \left[w^T A^T A^* - b^T A^*\right] \begin{bmatrix} 0 \\ 1 \end{bmatrix},\end{aligned} \tag{6.119}$$

and

$$\begin{aligned}\mathcal{D}_{z^*} h &= \mathcal{D}_{\tilde{w}} g|_{\tilde{w}=w}\, \mathcal{D}_{z^*} f + \mathcal{D}_{\tilde{w}^*} g|_{\tilde{w}=w}\, \mathcal{D}_{z^*} f^* \\ &= \left[w^H A^H A - b^H A\right] \begin{bmatrix} 0 \\ 1 \end{bmatrix} + \left[w^T A^T A^* - b^T A^*\right] \begin{bmatrix} 1 \\ 0 \end{bmatrix}.\end{aligned} \tag{6.120}$$

Note that (6.119) and (6.120) are the complex conjugates of each other, and this is in agreement with Lemma 3.3. Necessary conditions for optimality can be found by solving $\mathcal{D}_z h = 0$ or, equivalently, $\mathcal{D}_{z^*} h = 0$ (see Theorem 3.2). In Exercise 6.1, we observe that each of these equations has the same shape as (6.269), and it is shown how such equations can be solved.

(b) Alternatively, the constrained optimization problem can be solved by considering the generalized complex-valued matrix derivative $\mathcal{D}_w g$, and this is done next. Let the $N \times N$ reverse identity matrix (Bernstein 2005, p. 20) be denoted by J_N, and it has zeros everywhere except $+1$ on the main reverse diagonal such that, for example, $J_2 = \begin{bmatrix} 0 & 1 \\ 1 & 0 \end{bmatrix}$. The function $f : \mathbb{C} \times \mathbb{C} \to \mathcal{W}$ is one-to-one, onto, and differentiable. The inverse function $f^{-1} : \mathcal{W} \to \mathbb{C} \times \mathbb{C}$, which is unique, is given by

$$f^{-1}(w) = f^{-1}\left(\begin{bmatrix} z \\ z^* \end{bmatrix}\right) = \begin{bmatrix} z \\ z^* \end{bmatrix} = w = J_2 w^*, \tag{6.121}$$

hence, $f^{-1}(w) = w$ and the derivative of f with respect to w is given by

$$\mathcal{D}_w f^{-1} = I_2. \tag{6.122}$$

Because the elements of w^* have independent differentials, we get

$$\mathcal{D}_{w^*} f^{-1} = \mathcal{D}_{w^*}\left(J_2 w^*\right) = J_2. \tag{6.123}$$

Now, the derivatives of g with respect to w and w^* are determined. It follows from (6.71) that

$$\begin{aligned}\mathcal{D}_w g &= [\mathcal{D}_z h,\ \mathcal{D}_{z^*} h]\Big|_{\begin{bmatrix} z \\ z^* \end{bmatrix} = f^{-1}(w)} \mathcal{D}_w f^{-1} = [\mathcal{D}_z h,\ \mathcal{D}_{z^*} h]\Big|_{\begin{bmatrix} z \\ z^* \end{bmatrix} = f^{-1}(w)} \\ &= \Bigg[\left[w^H A^H A - b^H A\right] \begin{bmatrix} 1 \\ 0 \end{bmatrix} + \left[w^T A^T A^* - b^T A^*\right] \begin{bmatrix} 0 \\ 1 \end{bmatrix}, \\ &\qquad \left[w^H A^H A - b^H A\right] \begin{bmatrix} 0 \\ 1 \end{bmatrix} + \left[w^T A^T A^* - b^T A^*\right] \begin{bmatrix} 1 \\ 0 \end{bmatrix}\Bigg].\end{aligned} \tag{6.124}$$

When solving the equation $\mathcal{D}_{w} g = \mathbf{0}_{1\times 2}$ using the above expression, the following equation must be solved:

$$\left[\boldsymbol{w}^H \boldsymbol{A}^H \boldsymbol{A} - \boldsymbol{b}^H \boldsymbol{A}\right] \boldsymbol{I}_2 + \left[\boldsymbol{w}^T \boldsymbol{A}^T \boldsymbol{A}^* - \boldsymbol{b}^T \boldsymbol{A}^*\right] \boldsymbol{J}_2 = \mathbf{0}_{1\times 2}. \tag{6.125}$$

Using that $\boldsymbol{w}^* = \boldsymbol{J}_2 \boldsymbol{w}$, (6.125) is solved as

$$\boldsymbol{w} = \left[\boldsymbol{A}^H \boldsymbol{A} + \boldsymbol{J}_2 \boldsymbol{A}^T \boldsymbol{A}^* \boldsymbol{J}_2\right]^{-1} \left[\boldsymbol{A}^H \boldsymbol{b} + \boldsymbol{J}_2 \boldsymbol{A}^T \boldsymbol{b}^*\right]. \tag{6.126}$$

In Exercise 6.2, it is shown that the solution in (6.126) satisfies $\boldsymbol{J}_2 \boldsymbol{w} = \boldsymbol{w}^*$. From (6.72), it follows that the derivative of g with respect to $\boldsymbol{w}^*$ is given by

$$\mathcal{D}_{w^*} g = \left[\mathcal{D}_z h,\ \mathcal{D}_{z^*} h\right]\Big|_{\begin{bmatrix} z \\ z^* \end{bmatrix} = f^{-1}(w)} \mathcal{D}_{w^*} f^{-1} = (\mathcal{D}_w g)\, \boldsymbol{J}_2. \tag{6.127}$$

From (6.127), it is seen that the solution of $\mathcal{D}_{w^*} g = \mathbf{0}_{1\times 2}$ is also given by (6.126).

In Exercise 6.3, another case of generalized derivatives is studied with respect to a vector where a structure of the time reverse complex conjugate is considered. This is a structure that is related to *linear phase FIR filters*. Let the coefficients of the causal FIR filter $H(z) = \sum_{k=0}^{N-1} h(k) z^{-k}$ form the vector $\boldsymbol{h} \triangleq [h(0), h(1), \ldots, h(N-1)]^T$. Then, this filter has linear phase (Vaidyanathan 1993, p. 37, Eq. (2.4.8)) if and only if

$$\boldsymbol{h} = d \boldsymbol{J}_N \boldsymbol{h}^*, \tag{6.128}$$

where $|d| = 1$. Linearly constrained adaptive filters are studied in de Campos, Werner, and Apolinário Jr. (2004) and Diniz (2008, Section 2.5). In cases where the constraint can be formulated as a manifold, the theory of this chapter can be used to optimize such adaptive filters. The solution of linear equations can be written as the sum of the particular solution and the homogeneous solution (Strang 1988, Chapter 2). This can be used to parameterize all solutions of the set of linear equations, and it is useful, for example, when working with linearly constrained adaptive filters.

6.5.3 Generalized Matrix Derivatives with Respect to Diagonal Matrices

---

Example 6.20 (Complex-Valued Diagonal Matrix) Let $\mathcal{W}$ be the set of diagonal $N \times N$ complex-valued matrices, that is,

$$\mathcal{W} = \left\{\boldsymbol{W} \in \mathbb{C}^{N\times N} \mid \boldsymbol{W} \odot \boldsymbol{I}_N = \boldsymbol{W}\right\}, \tag{6.129}$$

where $\odot$ denotes the Hadamard product (see Definition 2.7). For $\mathcal{W}$, the following parameterization function $\boldsymbol{F} : \mathbb{C}^{N\times 1} \to \mathcal{W} \subseteq \mathbb{C}^{N\times N}$ can be used:

$$\operatorname{vec}(\boldsymbol{W}) = \operatorname{vec}(\boldsymbol{F}(\boldsymbol{z})) = \boldsymbol{L}_d \boldsymbol{z}, \tag{6.130}$$

where $\boldsymbol{z} \in \mathbb{C}^{N\times 1}$ contains the diagonal elements of the matrices $\boldsymbol{W} \in \mathcal{W}$, and where the $N^2 \times N$ matrix $\boldsymbol{L}_d$ is given in Definition 2.12. The function $\boldsymbol{F}$ is a diffeomorphism;

hence, $\mathcal{W}$ is a manifold. From (6.130), it follows that $d\,\text{vec}\,(\boldsymbol{F}(\boldsymbol{z})) = \boldsymbol{L}_d d\boldsymbol{z}$, and from this differential, the derivative of $\boldsymbol{F}$ with respect to $\boldsymbol{z}$ can be identified as

$$\mathcal{D}_{\boldsymbol{z}} \boldsymbol{F} = \boldsymbol{L}_d. \tag{6.131}$$

Let $g : \mathbb{C}^{N\times N} \to \mathbb{C}$ be given by

$$g(\tilde{\boldsymbol{W}}) = \text{Tr}\left\{\boldsymbol{A}\tilde{\boldsymbol{W}}\right\}, \tag{6.132}$$

where $\boldsymbol{A} \in \mathbb{C}^{N\times N}$ is an arbitrary complex-valued matrix, and where $\tilde{\boldsymbol{W}} \in \mathbb{C}^{N\times N}$ contains independent components. The differential of g can be written as $dg = \text{Tr}\left\{\boldsymbol{A}d\tilde{\boldsymbol{W}}\right\} = \text{vec}^T\left(\boldsymbol{A}^T\right) d\,\text{vec}\left(\tilde{\boldsymbol{W}}\right)$. Hence, the derivative of g with respect to $\tilde{\boldsymbol{W}}$ is given by

$$\mathcal{D}_{\tilde{\boldsymbol{W}}} g = \text{vec}^T\left(\boldsymbol{A}^T\right), \tag{6.133}$$

and the size of $\mathcal{D}_{\tilde{\boldsymbol{W}}} g$ is $1 \times N^2$.

Define the composed function $h : \mathbb{C}^{N\times 1} \to \mathbb{C}$ by

$$h(\boldsymbol{z}) = g(\tilde{\boldsymbol{W}})\big|_{\tilde{\boldsymbol{W}}=\boldsymbol{W}=\boldsymbol{F}(\boldsymbol{z})} = g(\boldsymbol{W}) = g(\boldsymbol{F}(\boldsymbol{z})). \tag{6.134}$$

The derivative of h with respect to $\boldsymbol{z}$ can be found by the chain rule

$$\mathcal{D}_{\boldsymbol{z}} h = \left(\mathcal{D}_{\tilde{\boldsymbol{W}}} g\right)\big|_{\tilde{\boldsymbol{W}}=\boldsymbol{W}} \mathcal{D}_{\boldsymbol{z}} \boldsymbol{F} = \text{vec}^T\left(\boldsymbol{A}^T\right) \boldsymbol{L}_d = \text{vec}_d^T\left(\boldsymbol{A}\right), \tag{6.135}$$

where (2.140) was used.

Here, the dimension of the tangent space of $\mathcal{W}$ is N. We need to choose N basis vectors for this space. Let these be the $N \times N$ matrices denoted by $\boldsymbol{E}_{i,i}$, where $\boldsymbol{E}_{i,i}$ contains only 0s except $+1$ at the i-th main diagonal element, where $i \in \{0, 1, \ldots, N-1\}$ (see Definition 2.16). Any element $\boldsymbol{W} \in \mathcal{W}$ can be expressed as

$$\boldsymbol{W} = z_0 \boldsymbol{E}_{0,0} + z_1 \boldsymbol{E}_{1,1} + \cdots + z_{N-1} \boldsymbol{E}_{N-1,N-1} \triangleq [\boldsymbol{z}]_{\{\boldsymbol{E}_{i,i}\}}, \tag{6.136}$$

where $[\boldsymbol{z}]_{\{\boldsymbol{E}_{i,i}\}}$ contains the N coefficients z_i in terms of the basis vectors $\boldsymbol{E}_{i,i}$. If we look at the function $\boldsymbol{F} : \mathbb{C}^{N\times 1} \to \mathcal{W}$ and express the output in terms of the basis $\boldsymbol{E}_{i,i}$, this function is the identity map, that is, $\boldsymbol{F}(\boldsymbol{z}) = [\boldsymbol{z}]_{\{\boldsymbol{E}_{i,i}\}}$; here, it is important to be aware that $\boldsymbol{z}$ inside $\boldsymbol{F}(\boldsymbol{z})$ is expressed in terms of the standard basis $\boldsymbol{e}_i$ in $\mathbb{C}^{N\times 1}$, but inside $[\boldsymbol{z}]_{\{\boldsymbol{E}_{i,i}\}}$ the $\boldsymbol{z}$ is expressed in terms of the basis $\boldsymbol{E}_{i,i}$. Definition 2.16 and (2.153) lead to the following relation:

$$\left[\text{vec}\left(\boldsymbol{E}_{0,0}\right), \text{vec}\left(\boldsymbol{E}_{1,1}\right), \ldots, \text{vec}\left(\boldsymbol{E}_{N-1,N-1}\right)\right] = \boldsymbol{L}_d. \tag{6.137}$$

The inverse function $\boldsymbol{F}^{-1} : \mathcal{W} \to \mathbb{C}^{N\times 1}$ can be expressed as

$$\boldsymbol{F}^{-1}(\boldsymbol{W}) = \boldsymbol{F}^{-1}([\boldsymbol{z}]_{\{\boldsymbol{E}_{i,i}\}}) = \boldsymbol{z}. \tag{6.138}$$

Therefore, $\mathcal{D}_{\boldsymbol{W}} \boldsymbol{F}^{-1} = \mathcal{D}_{[\boldsymbol{z}]_{\{\boldsymbol{E}_{i,i}\}}} \boldsymbol{F}^{-1} = \boldsymbol{I}_N$. We can now use the theory of manifolds to find $\mathcal{D}_{\boldsymbol{W}} g$. The derivative of g with respect to $\boldsymbol{W}$ can be found by the method in (6.71):

$$\mathcal{D}_{\boldsymbol{W}} g = \left(\mathcal{D}_{\boldsymbol{z}} h\right) \mathcal{D}_{\boldsymbol{W}} \boldsymbol{F}^{-1} = \text{vec}_d^T\left(\boldsymbol{A}\right) \boldsymbol{I}_N = \text{vec}_d^T\left(\boldsymbol{A}\right). \tag{6.139}$$

Note that the size of $\mathcal{D}_{\boldsymbol{W}} g$ is $1 \times N$, and it is expressed in terms of the basis chosen for $\mathcal{W}$. Hence, even though the size of both $\boldsymbol{W}$ and $\tilde{\boldsymbol{W}}$ is $N \times N$, the sizes of $\mathcal{D}_{\boldsymbol{W}} g$ and

$\mathcal{D}_{\tilde{W}} g$ are *different*, and they are given by $1 \times N$ and $1 \times N^2$, respectively. This illustrates that the sizes of the generalized and unpatterned derivatives are different in general.

If $\text{vec}_d(\boldsymbol{W}) = \boldsymbol{z}$ and $\text{vec}_l(\boldsymbol{W}) = \text{vec}_u(\boldsymbol{W}) = \boldsymbol{0}_{\frac{(N-1)N}{2} \times 1}$ are used in (6.78), an expression for $\frac{\partial g}{\partial \boldsymbol{W}}$ can be found as follows:

$$\begin{aligned}\text{vec}\left(\frac{\partial g}{\partial \boldsymbol{W}}\right) &= \text{vec}\left(\frac{\partial h}{\partial \boldsymbol{W}}\right) = \boldsymbol{L}_d \frac{\partial h}{\partial \, \text{vec}_d(\boldsymbol{W})} + \boldsymbol{L}_l \frac{\partial h}{\partial \, \text{vec}_l(\boldsymbol{W})} + \boldsymbol{L}_u \frac{\partial h}{\partial \, \text{vec}_u(\boldsymbol{W})} \\ &= \boldsymbol{L}_d \frac{\partial h}{\partial \boldsymbol{z}} = \boldsymbol{L}_d \left(\mathcal{D}_{\boldsymbol{z}} h\right)^T = \boldsymbol{L}_d \, \text{vec}_d(\boldsymbol{A}) = \text{vec}\left(\boldsymbol{A} \odot \boldsymbol{I}_N\right), \end{aligned} \tag{6.140}$$

where Definition 6.7 and Lemma 2.23 were utilized. Hence, it is observed that $\frac{\partial g}{\partial \boldsymbol{W}}$ is diagonal and given by

$$\frac{\partial g}{\partial \boldsymbol{W}} = \boldsymbol{A} \odot \boldsymbol{I}_N. \tag{6.141}$$

Example 6.21 Assume that diagonal matrices are considered such that $\boldsymbol{W} \in \mathcal{W}$, where $\mathcal{W}$ is given in (6.129). Assume that the function $g : \mathbb{C}^{N\times N} \times \mathbb{C}^{N\times N} \to \mathbb{C}$ is given, and that expressions for $\mathcal{D}_{\tilde{W}} g = \text{vec}^T\left(\frac{\partial g}{\partial \tilde{\boldsymbol{W}}}\right)$ and $\mathcal{D}_{\tilde{W}^*} g = \text{vec}^T\left(\frac{\partial g}{\partial \tilde{\boldsymbol{W}}^*}\right)$ are available. The parameterization function for $\mathcal{W}$ is given in (6.130), and from this equation, it is deduced that

$$\mathcal{D}_{\boldsymbol{z}^*} \boldsymbol{F} = \boldsymbol{0}_{N^2 \times N}, \tag{6.142}$$

$$\mathcal{D}_{\boldsymbol{z}} \boldsymbol{F}^* = \boldsymbol{0}_{N^2 \times N}, \tag{6.143}$$

$$\mathcal{D}_{\boldsymbol{z}^*} \boldsymbol{F}^* = \boldsymbol{L}_d. \tag{6.144}$$

Define the composed function $h : \mathbb{C}^{N\times 1} \to \mathbb{C}$ by

$$h(\boldsymbol{z}) = g(\tilde{\boldsymbol{W}}, \tilde{\boldsymbol{W}}^*)\Big|_{\tilde{\boldsymbol{W}}=\boldsymbol{W}} = g(\boldsymbol{W}, \boldsymbol{W}^*). \tag{6.145}$$

The derivative of h with respect to $\boldsymbol{z}$ is found by the chain rule as

$$\begin{aligned}\mathcal{D}_{\boldsymbol{z}} h &= \mathcal{D}_{\tilde{W}} g\big|_{\tilde{\boldsymbol{W}}=\boldsymbol{W}} \mathcal{D}_{\boldsymbol{z}} \boldsymbol{F} + \mathcal{D}_{\tilde{W}^*} g\big|_{\tilde{\boldsymbol{W}}=\boldsymbol{W}} \mathcal{D}_{\boldsymbol{z}} \boldsymbol{F}^* = \text{vec}^T\left(\frac{\partial g}{\partial \tilde{\boldsymbol{W}}}\right)\Bigg|_{\tilde{\boldsymbol{W}}=\boldsymbol{W}} \boldsymbol{L}_d \\ &= \text{vec}_d^T\left(\frac{\partial g}{\partial \tilde{\boldsymbol{W}}}\right)\Bigg|_{\tilde{\boldsymbol{W}}=\boldsymbol{W}}. \end{aligned} \tag{6.146}$$

When $\boldsymbol{W} \in \mathcal{W}$, it follows from (6.130) that the following three relations hold:

$$\text{vec}_d(\boldsymbol{W}) = \boldsymbol{z}, \tag{6.147}$$

$$\text{vec}_l(\boldsymbol{W}) = \boldsymbol{0}_{\frac{(N-1)N}{2} \times 1}, \tag{6.148}$$

$$\text{vec}_u(\boldsymbol{W}) = \boldsymbol{0}_{\frac{(N-1)N}{2} \times 1}. \tag{6.149}$$

From Definition 6.7, it follows that because all off-diagonal elements of $\boldsymbol{W}$ are zero, the derivative of g with respect to off-diagonal elements of $\boldsymbol{W}$ is zero. Hence, it follows that

$$\frac{\partial g}{\partial \operatorname{vec}_l(\boldsymbol{W})} = \frac{\partial h}{\partial \operatorname{vec}_l(\boldsymbol{W})} = \boldsymbol{0}_{\frac{(N-1)N}{2}\times 1}, \tag{6.150}$$

$$\frac{\partial g}{\partial \operatorname{vec}_u(\boldsymbol{W})} = \frac{\partial h}{\partial \operatorname{vec}_u(\boldsymbol{W})} = \boldsymbol{0}_{\frac{(N-1)N}{2}\times 1}. \tag{6.151}$$

Since $\operatorname{vec}_d(\boldsymbol{W}) = \boldsymbol{z}$ contains components with independent differentials, it is found from (6.78) that

$$\begin{aligned}\operatorname{vec}\left(\frac{\partial g}{\partial \boldsymbol{W}}\right) &= \boldsymbol{L}_d \frac{\partial h}{\partial \operatorname{vec}_d(\boldsymbol{W})} + \boldsymbol{L}_l \frac{\partial h}{\partial \operatorname{vec}_l(\boldsymbol{W})} + \boldsymbol{L}_u \frac{\partial h}{\partial \operatorname{vec}_u(\boldsymbol{W})} \\ &= \boldsymbol{L}_d (\mathcal{D}_{\boldsymbol{z}} h)^T = \boldsymbol{L}_d \operatorname{vec}_d\left(\frac{\partial g}{\partial \tilde{\boldsymbol{W}}}\right)\bigg|_{\tilde{\boldsymbol{W}}=\boldsymbol{W}} = \operatorname{vec}\left(\boldsymbol{I}_N \odot \frac{\partial g}{\partial \tilde{\boldsymbol{W}}}\right)\bigg|_{\tilde{\boldsymbol{W}}=\boldsymbol{W}}.\end{aligned} \tag{6.152}$$

From this, it follows that $\frac{\partial g}{\partial \boldsymbol{W}}$ is diagonal and is given by

$$\frac{\partial g}{\partial \boldsymbol{W}} = \boldsymbol{I}_N \odot \frac{\partial g}{\partial \tilde{\boldsymbol{W}}}\bigg|_{\tilde{\boldsymbol{W}}=\boldsymbol{W}}. \tag{6.153}$$

It is observed that the result in (6.141) is in agreement with the result found in (6.153).

6.5.4 Generalized Matrix Derivative with Respect to Symmetric Matrices

Derivatives with respect to symmetric real-valued matrices are mentioned in Payaró and Palomar (2009, Appendix B)[11].

Example 6.22 (Symmetric Complex Matrices) Consider symmetric matrices such that the set of matrices studied is

$$\mathcal{W} = \{\boldsymbol{W} \in \mathbb{C}^{N\times N} \mid \boldsymbol{W}^T = \boldsymbol{W}\} \subseteq \mathbb{C}^{N\times N}. \tag{6.154}$$

A parameterization function $\boldsymbol{F} : \mathbb{C}^{N\times 1} \times \mathbb{C}^{\frac{(N-1)N}{2}\times 1} \to \mathcal{W}$ denoted by $\boldsymbol{F}(\boldsymbol{x}, \boldsymbol{y}) = \boldsymbol{W}$ is given by

$$\operatorname{vec}(\boldsymbol{W}) = \operatorname{vec}(\boldsymbol{F}(\boldsymbol{x}, \boldsymbol{y})) = \boldsymbol{L}_d \boldsymbol{x} + (\boldsymbol{L}_l + \boldsymbol{L}_u)\boldsymbol{y}, \tag{6.155}$$

where $\boldsymbol{x} = \operatorname{vec}_d(\boldsymbol{W}) \in \mathbb{C}^{N\times 1}$ contains the main diagonal elements of $\boldsymbol{W}$, and $\boldsymbol{y} = \operatorname{vec}_l(\boldsymbol{W}) = \operatorname{vec}_u(\boldsymbol{W}) \in \mathbb{C}^{\frac{(N-1)N}{2}\times 1}$ contains the elements strictly below and also strictly above the main diagonal. From (6.155), it is seen that the derivatives of $\boldsymbol{F}$ with respect

[11] It is suggested in Wiens (1985) (see also Payaró and Palomar 2009, Appendix B) to replace the $\operatorname{vec}(\cdot)$ operator with the $v(\cdot)$ operator for finding derivatives with respect to symmetric matrices.

to $\boldsymbol{x}$ and $\boldsymbol{y}$ are, respectively,

$$\mathcal{D}_{\boldsymbol{x}} \boldsymbol{F} = \boldsymbol{L}_d, \tag{6.156}$$

$$\mathcal{D}_{\boldsymbol{y}} \boldsymbol{F} = \boldsymbol{L}_l + \boldsymbol{L}_u. \tag{6.157}$$

Let us consider the same function g as in Example 6.20, such that $g(\tilde{\boldsymbol{W}})$ is defined in (6.132), and its derivative with respect to the unpatterned matrix $\tilde{\boldsymbol{W}}$ is given by (6.133). To apply the method presented in this chapter for finding generalized matrix derivatives of functions, define the composed function $h : \mathbb{C}^{N\times 1} \times \mathbb{C}^{\frac{(N-1)N}{2}\times 1} \to \mathbb{C}$ as

$$h(\boldsymbol{x}, \boldsymbol{y}) = g\left(\tilde{\boldsymbol{W}}\right)\big|_{\tilde{\boldsymbol{W}}=\boldsymbol{W}} = g(\boldsymbol{W}) = g(\boldsymbol{F}(\boldsymbol{x}, \boldsymbol{y})). \tag{6.158}$$

The derivatives of h with respect to $\boldsymbol{x}$ and $\boldsymbol{y}$ can now be found by the chain rule as

$$\mathcal{D}_{\boldsymbol{x}} h = \left(\mathcal{D}_{\tilde{\boldsymbol{W}}} g\right)\big|_{\tilde{\boldsymbol{W}}=\boldsymbol{W}} \mathcal{D}_{\boldsymbol{x}} \boldsymbol{F} = \operatorname{vec}^T\left(\boldsymbol{A}^T\right) \boldsymbol{L}_d = \operatorname{vec}_d^T(\boldsymbol{A}), \tag{6.159}$$

$$\mathcal{D}_{\boldsymbol{y}} h = \left(\mathcal{D}_{\tilde{\boldsymbol{W}}} g\right)\big|_{\tilde{\boldsymbol{W}}=\boldsymbol{W}} \mathcal{D}_{\boldsymbol{y}} \boldsymbol{F} = \operatorname{vec}^T\left(\boldsymbol{A}^T\right) (\boldsymbol{L}_l + \boldsymbol{L}_u) = \operatorname{vec}_l^T\left(\boldsymbol{A} + \boldsymbol{A}^T\right), \tag{6.160}$$

respectively.

The dimension of the tangent space of $\mathcal{W}$ is $\frac{N(N+1)}{2}$. To use the theory of manifolds, we need to find a basis for this space. Let $\boldsymbol{E}_{i,i}$, defined in Definition 2.16, be a basis for the diagonal elements, where $i \in \{0, 1, \ldots, N-1\}$. Furthermore, let $\boldsymbol{G}_i$ be the *symmetric* $N \times N$ matrix with zeros on the main diagonal and given by the following relations:

$$\operatorname{vec}_l(\boldsymbol{G}_i) = \operatorname{vec}_u(\boldsymbol{G}_i) = (\boldsymbol{L}_l)_{:,i}, \tag{6.161}$$

where $i \in \{0, 1, \ldots, \frac{(N-1)N}{2} - 1\}$. This means that the matrix $\boldsymbol{G}_i$ is symmetric and contains two components that are $+1$ in accordance with (6.161), and all other components are zeros. As two examples, $\boldsymbol{G}_i$ for $i \in \{0, 1\}$ is given by

$$\boldsymbol{G}_0 = \begin{bmatrix} 0 & 1 & 0 & \cdots & 0 \\ 1 & 0 & 0 & \cdots & 0 \\ 0 & 0 & 0 & \cdots & 0 \\ \vdots & \vdots & & \cdots & \vdots \\ 0 & 0 & 0 & \cdots & 0 \end{bmatrix}, \quad \boldsymbol{G}_1 = \begin{bmatrix} 0 & 0 & 1 & \cdots & 0 \\ 0 & 0 & 0 & \cdots & 0 \\ 1 & 0 & 0 & \cdots & 0 \\ \vdots & \vdots & & \cdots & \vdots \\ 0 & 0 & 0 & \cdots & 0 \end{bmatrix}. \tag{6.162}$$

Define $\boldsymbol{x} = [x_0, x_1, \ldots, x_{N-1}]^T$ and $\boldsymbol{y} = \left[y_0, y_1, \ldots, y_{\frac{(N-1)N}{2}-1}\right]^T$, then any $\boldsymbol{W} \in \mathcal{W}$ can be expressed as

$$\begin{aligned} \boldsymbol{W} = {} & x_0 \boldsymbol{E}_{0,0} + x_1 \boldsymbol{E}_{1,1} + \cdots + x_{N-1} \boldsymbol{E}_{N-1,N-1} + y_0 \boldsymbol{G}_0 + y_1 \boldsymbol{G}_1 + \cdots \\ & + y_{\frac{(N-1)N}{2}-1} \boldsymbol{G}_{\frac{(N-1)N}{2}-1} \triangleq \left[[\boldsymbol{x}]_{\{\boldsymbol{E}_{i,i}\}}, [\boldsymbol{y}]_{\{\boldsymbol{G}_i\}}\right], \end{aligned} \tag{6.163}$$

where the notation $\left[[\boldsymbol{x}]_{\{\boldsymbol{E}_{i,i}\}}, [\boldsymbol{y}]_{\{\boldsymbol{G}_i\}}\right]$ means that the basis matrices defined above in $\{\boldsymbol{E}_{i,i}\}_{i=0}^{N-1}$ and $\{\boldsymbol{G}_i\}_{i=0}^{\frac{(N-1)N}{2}-1}$ are used. Notice that (6.137) and the following relation are valid:

$$\left[\operatorname{vec}(\boldsymbol{G}_0), \operatorname{vec}(\boldsymbol{G}_1), \ldots, \operatorname{vec}\left(\boldsymbol{G}_{\frac{(N-1)N}{2}-1}\right)\right] = \boldsymbol{L}_l + \boldsymbol{L}_u. \tag{6.164}$$

The result in (6.164) follows from (2.159) and (2.166).

When studying the parameterization function $\boldsymbol{F}$ in terms of the basis for the tangent space of $\mathcal{W}$, we have $\boldsymbol{W} = \boldsymbol{F}(\boldsymbol{x}, \boldsymbol{y}) = \left[[\boldsymbol{x}]_{\{\boldsymbol{E}_{i,i}\}}, [\boldsymbol{y}]_{\{\boldsymbol{G}_i\}} \right]$; hence, this is the identity map. Its inverse is also the identity map:

$$\boldsymbol{F}^{-1}(\boldsymbol{W}) = \boldsymbol{F}^{-1}\left(\left[[\boldsymbol{x}]_{\{\boldsymbol{E}_{i,i}\}}, [\boldsymbol{y}]_{\{\boldsymbol{G}_i\}} \right] \right) = (\boldsymbol{x}, \boldsymbol{y}). \tag{6.165}$$

The derivative of $\boldsymbol{F}^{-1}(\boldsymbol{W})$ with respect to $\boldsymbol{W}$ is given by

$$\mathcal{D}_{\boldsymbol{W}} \boldsymbol{F}^{-1}(\boldsymbol{W}) = \mathcal{D}_{\left[[\boldsymbol{x}]_{\{\boldsymbol{E}_{i,i}\}}, [\boldsymbol{y}]_{\{\boldsymbol{G}_i\}} \right]} \boldsymbol{F}^{-1}\left(\left[[\boldsymbol{x}]_{\{\boldsymbol{E}_{i,i}\}}, [\boldsymbol{y}]_{\{\boldsymbol{G}_i\}} \right] \right) = \boldsymbol{I}_{\frac{N(N+1)}{2}}. \tag{6.166}$$

Now, we are ready to find $\mathcal{D}_{\boldsymbol{W}} g$ by the method presented in (6.71):

$$\mathcal{D}_{\boldsymbol{W}} g = \left[\mathcal{D}_{\boldsymbol{x}} h, \mathcal{D}_{\boldsymbol{y}} h \right] \mathcal{D}_{\boldsymbol{W}} \boldsymbol{F}^{-1} = \left[\operatorname{vec}_d^T(\boldsymbol{A}), \operatorname{vec}_l^T\left(\boldsymbol{A} + \boldsymbol{A}^T\right) \right]_{[\boldsymbol{L}_d, \boldsymbol{L}_l + \boldsymbol{L}_u]}. \tag{6.167}$$

Here, $\mathcal{D}_{\boldsymbol{W}} g$ is expressed in terms of the basis chosen for $\mathcal{W}$, which is indicated by the subscript $[\boldsymbol{L}_d, \boldsymbol{L}_l + \boldsymbol{L}_u]$. The size of $\mathcal{D}_{\boldsymbol{W}} g$ expressed in terms of the basis $[\boldsymbol{L}_d, \boldsymbol{L}_l + \boldsymbol{L}_u]$ is $1 \times \frac{N(N+1)}{2}$, and this is different from the size of $\mathcal{D}_{\tilde{\boldsymbol{W}}} g$, which is $1 \times N^2$ (see (6.133)). Hence, this shows that, in general, $\mathcal{D}_{\boldsymbol{W}} g \neq \mathcal{D}_{\tilde{\boldsymbol{W}}} g$.

If $\operatorname{vec}_d(\boldsymbol{W}) = \boldsymbol{x}$ and $\operatorname{vec}_l(\boldsymbol{W}) = \operatorname{vec}_u(\boldsymbol{W}) = \boldsymbol{y}$ are used in (6.78), it is found that

$$\begin{aligned}
\operatorname{vec}\left(\frac{\partial g}{\partial \boldsymbol{W}} \right) &= \boldsymbol{L}_d \frac{\partial h}{\partial \operatorname{vec}_d(\boldsymbol{W})} + \boldsymbol{L}_l \frac{\partial h}{\partial \operatorname{vec}_l(\boldsymbol{W})} + \boldsymbol{L}_u \frac{\partial h}{\partial \operatorname{vec}_u(\boldsymbol{W})} \\
&= \boldsymbol{L}_d (\mathcal{D}_{\boldsymbol{x}} h)^T + \boldsymbol{L}_l \left(\mathcal{D}_{\boldsymbol{y}} h\right)^T + \boldsymbol{L}_u \left(\mathcal{D}_{\boldsymbol{y}} h\right)^T \\
&= \boldsymbol{L}_d \operatorname{vec}_d(\boldsymbol{A}) + (\boldsymbol{L}_l + \boldsymbol{L}_u)(\operatorname{vec}_l(\boldsymbol{A}) + \operatorname{vec}_u(\boldsymbol{A})) \\
&= \boldsymbol{L}_d \operatorname{vec}_d(\boldsymbol{A}) + \boldsymbol{L}_l \operatorname{vec}_l(\boldsymbol{A}) + \boldsymbol{L}_u \operatorname{vec}_u(\boldsymbol{A}) + \boldsymbol{L}_l \operatorname{vec}_u(\boldsymbol{A}) + \boldsymbol{L}_u \operatorname{vec}_l(\boldsymbol{A}) \\
&= \operatorname{vec}(\boldsymbol{A}) + \boldsymbol{L}_l \operatorname{vec}_l\left(\boldsymbol{A}^T\right) + \boldsymbol{L}_u \operatorname{vec}_u\left(\boldsymbol{A}^T\right) + \boldsymbol{L}_d \operatorname{vec}_d\left(\boldsymbol{A}^T\right) - \boldsymbol{L}_d \operatorname{vec}_d\left(\boldsymbol{A}^T\right) \\
&= \operatorname{vec}(\boldsymbol{A}) + \operatorname{vec}\left(\boldsymbol{A}^T\right) - \operatorname{vec}(\boldsymbol{A} \odot \boldsymbol{I}_N) = \operatorname{vec}\left(\boldsymbol{A} + \boldsymbol{A}^T - \boldsymbol{A} \odot \boldsymbol{I}_N\right),
\end{aligned} \tag{6.168}$$

where Definition 2.12 and Lemmas 2.20 and 2.23 were utilized. Hence, $\frac{\partial g}{\partial \boldsymbol{W}}$ is symmetric and given by

$$\frac{\partial g}{\partial \boldsymbol{W}} = \boldsymbol{A} + \boldsymbol{A}^T - \boldsymbol{A} \odot \boldsymbol{I}_N. \tag{6.169}$$

In the next example, an alternative way of defining a parameterization function for symmetric complex-valued matrices will be given by means of the *duplication matrix* (see Definition 2.14).

Example 6.23 (Symmetric Complex Matrix by the Duplication Matrix) Let $\boldsymbol{W} \in \mathbb{C}^{N \times N}$ be symmetric such that $\boldsymbol{W} \in \mathcal{W}$, where $\mathcal{W}$ is defined in (6.154). Then, $\mathcal{W}$ can be parameterized by the parameterization function $\boldsymbol{F} : \mathbb{C}^{\frac{N(N+1)}{2} \times 1} \to \mathcal{W} \subseteq \mathbb{C}^{N \times N}$, given by

$$\operatorname{vec}(\boldsymbol{W}) = \operatorname{vec}(\boldsymbol{F}(\boldsymbol{z})) = \boldsymbol{D}_N \boldsymbol{z}, \tag{6.170}$$

where $\boldsymbol{D}_N$ is the *duplication matrix* (see Definition 2.14) of size $N^2 \times \frac{N(N+1)}{2}$, and $\boldsymbol{z} \in \mathbb{C}^{\frac{N(N+1)}{2} \times 1}$ contains all the independent complex variables necessary for producing all matrices $\boldsymbol{W}$ in the manifold $\mathcal{W}$. Some connections between the duplication matrix $\boldsymbol{D}_N$ and the three matrices $\boldsymbol{L}_d$, $\boldsymbol{L}_l$, and $\boldsymbol{L}_u$ are given in Lemma 2.32. Using the differential operator on (6.170) results in $d \operatorname{vec}(\boldsymbol{F}) = \boldsymbol{D}_N d\boldsymbol{z}$, and this leads to $\mathcal{D}_{\boldsymbol{z}}\boldsymbol{F} = \boldsymbol{D}_N$.

To find the inverse parameterization function, a basis is needed for the set of matrices within the manifold $\mathcal{W}$. Let this basis be denoted by $\boldsymbol{H}_i \in \mathbb{Z}_2^{N\times N}$, such that

$$\operatorname{vec}(\boldsymbol{H}_i) = (\boldsymbol{D}_N)_{:,i} \in \mathbb{Z}_2^{N^2\times 1}, \tag{6.171}$$

where $i \in \left\{0, 1, \ldots, \frac{N(N+1)}{2} - 1\right\}$. A consequence of this is that

$$\boldsymbol{D}_N = \left[\operatorname{vec}(\boldsymbol{H}_0), \operatorname{vec}(\boldsymbol{H}_1), \ldots, \operatorname{vec}\left(\boldsymbol{H}_{\frac{N(N+1)}{2}-1}\right)\right]. \tag{6.172}$$

If $\boldsymbol{W} \in \mathcal{W}$, then $\boldsymbol{W}$ can be expressed as

$$\operatorname{vec}(\boldsymbol{W}) = \operatorname{vec}(\boldsymbol{F}(\boldsymbol{z})) = \boldsymbol{D}_N\boldsymbol{z} = \sum_{i=0}^{\frac{N(N+1)}{2}-1} (\boldsymbol{D}_N)_{:,i}\, z_i = \sum_{i=0}^{\frac{N(N+1)}{2}-1} \operatorname{vec}(\boldsymbol{H}_i)\, z_i. \tag{6.173}$$

This is equivalent to the following expression:

$$\boldsymbol{W} = \boldsymbol{F}(\boldsymbol{z}) = \sum_{i=0}^{\frac{N(N+1)}{2}-1} \boldsymbol{H}_i z_i \triangleq [\boldsymbol{z}]_{\{\boldsymbol{H}_i\}}, \tag{6.174}$$

where the notation $[\boldsymbol{z}]_{\{\boldsymbol{H}_i\}}$ means that the basis $\{\boldsymbol{H}_i\}_{i=0}^{\frac{N(N+1)}{2}-1}$ is used to express an arbitrary element $\boldsymbol{W} \in \mathcal{W}$. This shows that the parameterization function $\boldsymbol{F}(\boldsymbol{z}) = [\boldsymbol{z}]_{\{\boldsymbol{H}_i\}}$ is the identity function such that its inverse is also given by the identity function, that is, $\boldsymbol{F}^{-1}([\boldsymbol{z}]_{\{\boldsymbol{H}_i\}}) = \boldsymbol{z}$. Hence, the derivative of the inverse of the parameterization function is given by $\mathcal{D}_{\boldsymbol{W}}\boldsymbol{F}^{-1} = \mathcal{D}_{[\boldsymbol{z}]_{\{\boldsymbol{H}_i\}}}\boldsymbol{F}^{-1} = \boldsymbol{I}_{\frac{N(N+1)}{2}}$.

Consider the function g defined in (6.132), and define the composed function $h : \mathbb{C}^{\frac{N(N+1)}{2}\times 1} \to \mathbb{C}$ by

$$h(\boldsymbol{z}) = g(\tilde{\boldsymbol{W}})|_{\tilde{\boldsymbol{W}}=\boldsymbol{W}} = g(\boldsymbol{W}) = g(\boldsymbol{F}(\boldsymbol{z})). \tag{6.175}$$

The derivative of h with respect to $\boldsymbol{z}$ can be found by the chain rule as

$$\mathcal{D}_{\boldsymbol{z}}h = \mathcal{D}_{\tilde{\boldsymbol{W}}}g|_{\tilde{\boldsymbol{W}}=\boldsymbol{W}}\mathcal{D}_{\boldsymbol{z}}\boldsymbol{F} = \operatorname{vec}^T\left(\boldsymbol{A}^T\right)\boldsymbol{D}_N. \tag{6.176}$$

By using the method in (6.71), it is possible to find the generalized derivative of g with respect to $\boldsymbol{W} \in \mathcal{W}$ as follows:

$$\mathcal{D}_{\boldsymbol{W}}g = (\mathcal{D}_{\boldsymbol{z}}h)\mathcal{D}_{\boldsymbol{W}}\boldsymbol{F}^{-1} = \operatorname{vec}^T\left(\boldsymbol{A}^T\right)\boldsymbol{D}_N, \tag{6.177}$$

when the basis $\{\boldsymbol{H}_i\}_{i=0}^{\frac{N(N+1)}{2}-1}$ is used to express the elements in the manifold $\mathcal{W}$.

To show how (6.177) is related to the result found in (6.167), the result presented in (2.175) is used to reformulate (6.177) in the following way:

$$\begin{aligned}\mathrm{vec}^T\left(A^T\right) D_N &= \mathrm{vec}^T\left(A^T\right) L_d V_d^T + \mathrm{vec}^T\left(A^T\right)(L_l + L_u) V_l^T \\ &= \mathrm{vec}_d^T\left(A^T\right) V_d^T + \mathrm{vec}_l^T\left(A^T\right) V_l^T + \mathrm{vec}_u^T\left(A^T\right) V_l^T \\ &= \mathrm{vec}_d^T\left(A^T\right) V_d^T + \mathrm{vec}_l^T\left(A^T\right) V_l^T + \mathrm{vec}_l^T\left(A\right) V_l^T \\ &= \mathrm{vec}_d^T\left(A^T\right) V_d^T + \mathrm{vec}_l^T\left(A + A^T\right) V_l^T \\ &= \left[\mathrm{vec}_d^T\left(A^T\right),\ \mathrm{vec}_l^T\left(A + A^T\right)\right] \begin{bmatrix} V_d^T \\ V_l^T \end{bmatrix}. \end{aligned} \tag{6.178}$$

From (2.182), it follows that

$$[L_d,\ L_l + L_u] \begin{bmatrix} V_d^T \\ V_l^T \end{bmatrix} = D_N. \tag{6.179}$$

From (6.178) and (6.179), it is seen that (6.177) is equivalent to the result found in (6.167) because we have found an invertible matrix $V \in \mathbb{C}^{\frac{N(N+1)}{2} \times \frac{N(N+1)}{2}}$, which transforms one basis to another, that is, $D_N V = [L_d,\ L_l + L_u]$, where $V = [V_d,\ V_l]$ is given in Definition 2.15.

Because the matrix W is symmetric, the matrix $\frac{\partial g}{\partial W}$ will also be symmetric. Let us use the matrices $T_{i,j}$ defined in Exercise 2.13. Using the vec($\cdot$) operator on (6.74) leads to

$$\begin{aligned} \mathrm{vec}\left(\frac{\partial g}{\partial W}\right) &= \sum_{i=0}^{N-1}\sum_{j=0}^{N-1} \mathrm{vec}\left(E_{i,j}\right) \frac{\partial h}{\partial (W)_{i,j}} = \sum_{i \geq j} \mathrm{vec}\left(T_{i,j}\right) \frac{\partial h}{\partial (W)_{i,j}} \\ &= \left[\mathrm{vec}\left(T_{0,0}\right), \mathrm{vec}\left(T_{1,0}\right), \cdots, \mathrm{vec}\left(T_{N-1,0}\right), \mathrm{vec}\left(T_{1,1}\right), \cdots, \mathrm{vec}\left(T_{N-1,N-1}\right)\right] \\ &\times \begin{bmatrix} \frac{\partial h}{\partial (W)_{0,0}} \\ \frac{\partial h}{\partial (W)_{1,0}} \\ \vdots \\ \frac{\partial h}{\partial (W)_{N-1,0}} \\ \frac{\partial h}{\partial (W)_{1,1}} \\ \vdots \\ \frac{\partial h}{\partial (W)_{N-1,N-1}} \end{bmatrix} = D_N \frac{\partial h}{\partial v(W)}, \end{aligned} \tag{6.180}$$

where (6.175) has been used in the first equality by replacing g with h, and (2.212) was used in the last equality. If $v(W) = z$ is inserted into (6.180), it is found from (2.139), (2.150), and (2.214) that

$$\begin{aligned} \mathrm{vec}\left(\frac{\partial g}{\partial W}\right) &= D_N \frac{\partial h}{\partial v(W)} = D_N \left(\mathcal{D}_z h\right)^T = D_N D_N^T \mathrm{vec}\left(A^T\right) \\ &= D_N D_N^T K_{N,N} \mathrm{vec}\left(A\right) = D_N D_N^T \mathrm{vec}\left(A\right) \\ &= \left(I_{N^2} + K_{N,N} - K_{N,N} \odot I_{N^2}\right) \mathrm{vec}\left(A\right) = \mathrm{vec}\left(A + A^T - A \odot I_N\right). \end{aligned} \tag{6.181}$$

This result is in agreement with (6.169).

Examples 6.22 and 6.23 show that different choices of the basis vectors for expanding the elements of the manifold $\boldsymbol{W}$ may lead to different equivalent expressions for the generalized derivative. The results found in Examples 6.22 and 6.23 are in agreement with the more general result derived in Exercise 6.4 (see (6.277)).

6.5.5 Generalized Matrix Derivative with Respect to Hermitian Matrices

---

Example 6.24 (Hermitian) Let us define the following manifold:

$$\mathcal{W} = \{\boldsymbol{W} \in \mathbb{C}^{N\times N} \mid \boldsymbol{W}^H = \boldsymbol{W}\} \subset \mathbb{C}^{N\times N}. \tag{6.182}$$

An arbitrary Hermitian matrix $\boldsymbol{W} \in \mathcal{W}$ can be parameterized by the real-valued vector $\boldsymbol{x} = \text{vec}_d(\boldsymbol{W}) \in \mathbb{R}^{N\times 1}$, which contains the real-valued main diagonal elements, and the complex-valued vector $\boldsymbol{z} = \text{vec}_l(\boldsymbol{W}) = (\text{vec}_u(\boldsymbol{W}))^* \in \mathbb{C}^{\frac{(N-1)N}{2}\times 1}$, which contains the strictly below diagonal elements, and it is also equal to the complex conjugate of the strictly above diagonal elements. One way of generating any Hermitian $N \times N$ matrix $\boldsymbol{W} \in \mathcal{W}$ is by using the parameterization function $\boldsymbol{F} : \mathbb{R}^{N\times 1} \times \mathbb{C}^{\frac{(N-1)N}{2}\times 1} \times \mathbb{C}^{\frac{(N-1)N}{2}\times 1} \to \mathcal{W}$ given by

$$\text{vec}(\boldsymbol{W}) = \text{vec}\left(\boldsymbol{F}(\boldsymbol{x}, \boldsymbol{z}, \boldsymbol{z}^*)\right) = \boldsymbol{L}_d\boldsymbol{x} + \boldsymbol{L}_l\boldsymbol{z} + \boldsymbol{L}_u\boldsymbol{z}^*. \tag{6.183}$$

From (6.183), it is seen that $\mathcal{D}_{\boldsymbol{x}}\boldsymbol{F} = \boldsymbol{L}_d$, $\mathcal{D}_{\boldsymbol{z}}\boldsymbol{F} = \boldsymbol{L}_l$, and $\mathcal{D}_{\boldsymbol{z}^*}\boldsymbol{F} = \boldsymbol{L}_u$. The dimension of the tangent space of $\mathcal{W}$ is N^2, and all elements within $\boldsymbol{W} \in \mathcal{W}$ can be treated as independent when finding derivatives. If we choose as a basis for $\mathcal{W}$ the N^2 matrices of size $N \times N$ found by reshaping each of the *columns* of $\boldsymbol{L}_d$, $\boldsymbol{L}_l$, and $\boldsymbol{L}_u$ by "inverting" the vec-operation, we can represent $\boldsymbol{W} \in \mathcal{W}$ as $\boldsymbol{W} \triangleq [[\boldsymbol{x}], [\boldsymbol{z}], [\boldsymbol{z}^*]]_{[\boldsymbol{L}_d,\boldsymbol{L}_l,\boldsymbol{L}_u]}$. With this representation, the function $\boldsymbol{F}$ is the identity function because $\boldsymbol{W} = \boldsymbol{F}(\boldsymbol{x}, \boldsymbol{z}, \boldsymbol{z}^*) = [[\boldsymbol{x}], [\boldsymbol{z}], [\boldsymbol{z}^*]]_{[\boldsymbol{L}_d,\boldsymbol{L}_l,\boldsymbol{L}_u]}$. Therefore, the inverse function $\boldsymbol{F}^{-1} : \mathcal{W} \to \mathbb{R}^{N\times 1} \times \mathbb{C}^{\frac{(N-1)N}{2}\times 1} \times \mathbb{C}^{\frac{(N-1)N}{2}\times 1}$ can be expressed as

$$\begin{aligned}\boldsymbol{F}^{-1}(\boldsymbol{W}) &= \boldsymbol{F}^{-1}([[\boldsymbol{x}], [\boldsymbol{z}], [\boldsymbol{z}^*]]_{[\boldsymbol{L}_d,\boldsymbol{L}_l,\boldsymbol{L}_u]}) = (\text{vec}_d(\boldsymbol{W}), \text{vec}_l(\boldsymbol{W}), \text{vec}_u(\boldsymbol{W}))\\ &= (\boldsymbol{x}, \boldsymbol{z}, \boldsymbol{z}^*),\end{aligned} \tag{6.184}$$

which is also the identity function. Therefore, it follows that $\mathcal{D}_{\boldsymbol{W}}\boldsymbol{F}^{-1} = \boldsymbol{I}_{N^2}$.

Example 6.25 Let us assume that $\boldsymbol{W} \in \mathcal{W} \subset \mathbb{C}^{N\times N}$, where $\mathcal{W}$ is the manifold defined in (6.182). These matrices can be produced by the parameterization function $\boldsymbol{F} : \mathbb{R}^{N\times 1} \times \mathbb{C}^{\frac{(N-1)N}{2}\times 1} \times \mathbb{C}^{\frac{(N-1)N}{2}\times 1} \to \mathcal{W}$ given in (6.183). The derivatives of $\boldsymbol{F}$ with respect to $\boldsymbol{x} \in \mathbb{R}^{N\times 1}$, $\boldsymbol{z} \in \mathbb{C}^{\frac{(N-1)N}{2}\times 1}$, and $\boldsymbol{z}^* \in \mathbb{C}^{\frac{(N-1)N}{2}\times 1}$ are given in Example 6.24. From (6.183), the differential of the complex conjugate of the parameterization function $\boldsymbol{F}$ is given by

$$d\,\mathrm{vec}(\boldsymbol{W}^*) = d\,\mathrm{vec}(\boldsymbol{F}^*) = \boldsymbol{L}_d d\boldsymbol{x} + \boldsymbol{L}_u d\boldsymbol{z} + \boldsymbol{L}_l d\boldsymbol{z}^*. \tag{6.185}$$

And from (6.185), it follows that the following derivatives can be identified: $\mathcal{D}_{\boldsymbol{x}}\boldsymbol{F}^* = \boldsymbol{L}_d$, $\mathcal{D}_{\boldsymbol{z}}\boldsymbol{F}^* = \boldsymbol{L}_u$, and $\mathcal{D}_{\boldsymbol{z}^*}\boldsymbol{F}^* = \boldsymbol{L}_l$.

Let $g : \mathbb{C}^{N\times N} \times \mathbb{C}^{N\times N} \to \mathbb{C}$ be given by $g(\tilde{\boldsymbol{W}}, \tilde{\boldsymbol{W}}^*)$, where $\tilde{\boldsymbol{W}} \in \mathbb{C}^{N\times N}$ is a matrix containing only independent variables. Assume that the two unconstrained complex-valued matrix derivatives of $g(\tilde{\boldsymbol{W}}, \tilde{\boldsymbol{W}}^*)$ of size $1 \times N^2$ are available, and given by $\mathcal{D}_{\tilde{\boldsymbol{W}}}g = \mathrm{vec}^T\left(\frac{\partial g}{\partial \tilde{\boldsymbol{W}}}\right)$ and $\mathcal{D}_{\tilde{\boldsymbol{W}}^*}g = \mathrm{vec}^T\left(\frac{\partial g}{\partial \tilde{\boldsymbol{W}}^*}\right)$. Define the composed function $h : \mathbb{R}^{N\times 1} \times \mathbb{C}^{\frac{(N-1)N}{2}\times 1} \times \mathbb{C}^{\frac{(N-1)N}{2}\times 1} \to \mathbb{C}$ by

$$h(\boldsymbol{x}, \boldsymbol{z}, \boldsymbol{z}^*) = g(\tilde{\boldsymbol{W}}, \tilde{\boldsymbol{W}}^*)\Big|_{\tilde{\boldsymbol{W}}=\boldsymbol{W}} = g(\boldsymbol{W}, \boldsymbol{W}^*) = g(\boldsymbol{F}(\boldsymbol{x}, \boldsymbol{z}, \boldsymbol{z}^*), \boldsymbol{F}^*(\boldsymbol{x}, \boldsymbol{z}, \boldsymbol{z}^*)). \tag{6.186}$$

The derivatives of the function h with respect to $\boldsymbol{x}$, $\boldsymbol{z}$, and $\boldsymbol{z}^*$ can be found by the chain rule as follows:

$$\begin{aligned}
\mathcal{D}_{\boldsymbol{x}}h &= \mathcal{D}_{\tilde{\boldsymbol{W}}}g\big|_{\tilde{\boldsymbol{W}}=\boldsymbol{W}}\mathcal{D}_{\boldsymbol{x}}\boldsymbol{F} + \mathcal{D}_{\tilde{\boldsymbol{W}}^*}g\big|_{\tilde{\boldsymbol{W}}=\boldsymbol{W}}\mathcal{D}_{\boldsymbol{x}}\boldsymbol{F}^* = \left[\mathcal{D}_{\tilde{\boldsymbol{W}}}g + \mathcal{D}_{\tilde{\boldsymbol{W}}^*}g\right]_{\tilde{\boldsymbol{W}}=\boldsymbol{W}}\boldsymbol{L}_d \\
&= \mathrm{vec}^T\left(\frac{\partial g}{\partial \tilde{\boldsymbol{W}}} + \frac{\partial g}{\partial \tilde{\boldsymbol{W}}^*}\right)\Bigg|_{\tilde{\boldsymbol{W}}=\boldsymbol{W}}\boldsymbol{L}_d = \mathrm{vec}_d^T\left(\frac{\partial g}{\partial \tilde{\boldsymbol{W}}} + \left(\frac{\partial g}{\partial \tilde{\boldsymbol{W}}^*}\right)^T\right)\Bigg|_{\tilde{\boldsymbol{W}}=\boldsymbol{W}},
\end{aligned} \tag{6.187}$$

$$\begin{aligned}
\mathcal{D}_{\boldsymbol{z}}h &= \mathcal{D}_{\tilde{\boldsymbol{W}}}g\big|_{\tilde{\boldsymbol{W}}=\boldsymbol{W}}\mathcal{D}_{\boldsymbol{z}}\boldsymbol{F} + \mathcal{D}_{\tilde{\boldsymbol{W}}^*}g\big|_{\tilde{\boldsymbol{W}}=\boldsymbol{W}}\mathcal{D}_{\boldsymbol{z}}\boldsymbol{F}^* \\
&= \mathrm{vec}^T\left(\frac{\partial g}{\partial \tilde{\boldsymbol{W}}}\right)\Bigg|_{\tilde{\boldsymbol{W}}=\boldsymbol{W}}\boldsymbol{L}_l + \mathrm{vec}^T\left(\frac{\partial g}{\partial \tilde{\boldsymbol{W}}^*}\right)\Bigg|_{\tilde{\boldsymbol{W}}=\boldsymbol{W}}\boldsymbol{L}_u \\
&= \mathrm{vec}_l^T\left(\frac{\partial g}{\partial \tilde{\boldsymbol{W}}} + \left(\frac{\partial g}{\partial \tilde{\boldsymbol{W}}^*}\right)^T\right)\Bigg|_{\tilde{\boldsymbol{W}}=\boldsymbol{W}},
\end{aligned} \tag{6.188}$$

and

$$\begin{aligned}
\mathcal{D}_{\boldsymbol{z}^*}h &= \mathcal{D}_{\tilde{\boldsymbol{W}}}g\big|_{\tilde{\boldsymbol{W}}=\boldsymbol{W}}\mathcal{D}_{\boldsymbol{z}^*}\boldsymbol{F} + \mathcal{D}_{\tilde{\boldsymbol{W}}^*}g\big|_{\tilde{\boldsymbol{W}}=\boldsymbol{W}}\mathcal{D}_{\boldsymbol{z}^*}\boldsymbol{F}^* \\
&= \mathrm{vec}^T\left(\frac{\partial g}{\partial \tilde{\boldsymbol{W}}}\right)\Bigg|_{\tilde{\boldsymbol{W}}=\boldsymbol{W}}\boldsymbol{L}_u + \mathrm{vec}^T\left(\frac{\partial g}{\partial \tilde{\boldsymbol{W}}^*}\right)\Bigg|_{\tilde{\boldsymbol{W}}=\boldsymbol{W}}\boldsymbol{L}_l \\
&= \mathrm{vec}_u^T\left(\frac{\partial g}{\partial \tilde{\boldsymbol{W}}} + \left(\frac{\partial g}{\partial \tilde{\boldsymbol{W}}^*}\right)^T\right)\Bigg|_{\tilde{\boldsymbol{W}}=\boldsymbol{W}}.
\end{aligned} \tag{6.189}$$

The sizes of the three derivatives $\mathcal{D}_{\boldsymbol{x}}h$, $\mathcal{D}_{\boldsymbol{z}}h$, and $\mathcal{D}_{\boldsymbol{z}^*}h$ are $1 \times N$, $1 \times \frac{(N-1)N}{2}$, and $1 \times \frac{(N-1)N}{2}$, respectively. The total number of components within the three derivatives $\mathcal{D}_{\boldsymbol{x}}h$, $\mathcal{D}_{\boldsymbol{z}}h$, and $\mathcal{D}_{\boldsymbol{z}^*}h$ are $N + \frac{(N-1)N}{2} + \frac{(N-1)N}{2} = N^2$. If the method given in (6.71) is used, the derivative $\mathcal{D}_{W} g$ can now be expressed as

$$
\begin{aligned}
[\mathcal{D}_{W} g]_{[L_d, L_l, L_u]} &= [\mathcal{D}_{x} h, \mathcal{D}_{z} h, \mathcal{D}_{z^*} h] \, \mathcal{D}_{W} F^{-1} \\
&= \left[\operatorname{vec}^T \left(\frac{\partial g}{\partial \tilde{W}} + \left(\frac{\partial g}{\partial \tilde{W}^*} \right)^T \right) \Bigg|_{\tilde{W}=W} [L_d, L_l, L_u] \right]_{[L_d, L_l, L_u]}, \quad (6.190)
\end{aligned}
$$

where the results from (6.187), (6.188), and (6.189), in addition to $\mathcal{D}_{W} F^{-1} = I_{N^2}$ from Example 6.24, have been utilized. In (6.190), the derivative of g with respect to the matrix $W \in \mathcal{W}$ is expressed with the basis chosen in Example 6.24, that is, the N first basis vectors of $\mathcal{W}$ are given by the N columns of L_d, then the next $\frac{(N-1)N}{2}$ as the $\frac{(N-1)N}{2}$ columns of L_l, and the last $\frac{(N-1)N}{2}$ basis vectors are given by the $\frac{(N-1)N}{2}$ columns of L_u; this is indicated by the subscript $[L_d, L_l, L_u]$.

Let the matrix $Z \in \mathbb{C}^{N \times N}$ be unpatterned. From Chapter 3, we know that unpatterned derivatives are identified from the differential of the function, that the unpatterned matrix variable should be written in the form $d\operatorname{vec}(Z)$, and that the standard basis used to express these *unpatterned derivatives* is e_i of size $N^2 \times 1$ (see Definition 2.16). To introduce this basis for the example under consideration, observe that (2.42) is equivalent to

$$
\operatorname{vec}^T(A) = \left[\operatorname{vec}_d^T(A), \ \operatorname{vec}_l^T(A), \ \operatorname{vec}_u^T(A) \right] \begin{bmatrix} L_d^T \\ L_l^T \\ L_u^T \end{bmatrix}. \quad (6.191)
$$

From this expression, it is seen that if we want to use the *standard basis* $E_{i,j}$ (see Definition 2.16) to express $\mathcal{D}_{W} g$, this can be done in the following way:

$$
\begin{aligned}
[\mathcal{D}_{W} g]_{[E_{i,j}]} &= [\mathcal{D}_{W} g]_{[L_d, L_l, L_u]} \begin{bmatrix} L_d^T \\ L_l^T \\ L_u^T \end{bmatrix} = [\mathcal{D}_{x} h, \ \mathcal{D}_{z} h, \ \mathcal{D}_{z^*} h] \begin{bmatrix} L_d^T \\ L_l^T \\ L_u^T \end{bmatrix} \\
&= \operatorname{vec}^T \left(\frac{\partial g}{\partial \tilde{W}} + \left(\frac{\partial g}{\partial \tilde{W}^*} \right)^T \right) \Bigg|_{\tilde{W}=W} [L_d, L_l, L_u] \begin{bmatrix} L_d^T \\ L_l^T \\ L_u^T \end{bmatrix} \\
&= \left[\operatorname{vec}^T \left(\frac{\partial g}{\partial \tilde{W}} \right) + \operatorname{vec}^T \left(\left(\frac{\partial g}{\partial \tilde{W}^*} \right)^T \right) \right]_{\tilde{W}=W} \quad (6.192) \\
&= \left[\mathcal{D}_{\tilde{W}} g + \left\{ K_{N,N} \operatorname{vec} \left(\frac{\partial g}{\partial \tilde{W}^*} \right) \right\}^T \right]_{\tilde{W}=W} = \left[\mathcal{D}_{\tilde{W}} g + \left(\mathcal{D}_{\tilde{W}^*} g \right) K_{N,N} \right]_{\tilde{W}=W}, \quad (6.193)
\end{aligned}
$$

where the notation $[\cdot]_{[E_{i,j}]}$ means that the standard basis $E_{i,j}$ is used, the notation $[\cdot]_{[L_d, L_l, L_u]}$ means that the transform is expressed with $[L_d, L_l, L_u]$ as basis, and $L_d L_d^T + L_l L_l^T + L_u L_u^T = I_{N^2}$ has been used (see Lemma 2.21). Notice that, in this example, where the matrices are Hermitian, the sizes of $\mathcal{D}_{\tilde{W}} g$ and $\mathcal{D}_{W} g$ are both $1 \times N^2$; the reason for this is that all components inside the matrix $W \in \mathcal{W}$ can be treated as independent of each other when finding derivatives. Since for Hermitian matrices $\mathcal{D}_{W} g = \operatorname{vec}^T \left(\frac{\partial g}{\partial W} \right)$, it follows from (6.192) that

$$\frac{\partial g}{\partial \boldsymbol{W}} = \left[\frac{\partial g}{\partial \tilde{\boldsymbol{W}}} + \left(\frac{\partial g}{\partial \tilde{\boldsymbol{W}}^*} \right)^T \right]_{\tilde{\boldsymbol{W}}=\boldsymbol{W}}. \tag{6.194}$$

Because $\boldsymbol{W}^* = \boldsymbol{W}^T$, it follows that

$$\frac{\partial g}{\partial \boldsymbol{W}^*} = \frac{\partial g}{\partial \boldsymbol{W}^T} = \left(\frac{\partial g}{\partial \boldsymbol{W}} \right)^T = \left[\frac{\partial g}{\partial \tilde{\boldsymbol{W}}^*} + \left(\frac{\partial g}{\partial \tilde{\boldsymbol{W}}} \right)^T \right]_{\tilde{\boldsymbol{W}}=\boldsymbol{W}}. \tag{6.195}$$

Alternatively, the result in (6.194) can be found by means of (6.78) as follows:

$$\begin{aligned}
\operatorname{vec}\left(\frac{\partial g}{\partial \boldsymbol{W}} \right) &= \boldsymbol{L}_d \frac{\partial h}{\partial \operatorname{vec}_d(\boldsymbol{W})} + \boldsymbol{L}_l \frac{\partial h}{\partial \operatorname{vec}_l(\boldsymbol{W})} + \boldsymbol{L}_u \frac{\partial h}{\partial \operatorname{vec}_u(\boldsymbol{W})} \\
&= \boldsymbol{L}_d (\mathcal{D}_{\boldsymbol{x}} h)^T + \boldsymbol{L}_l (\mathcal{D}_{\boldsymbol{z}} h)^T + \boldsymbol{L}_u (\mathcal{D}_{\boldsymbol{z}^*} h)^T \\
&= \boldsymbol{L}_d \operatorname{vec}_d \left(\frac{\partial g}{\partial \tilde{\boldsymbol{W}}} + \left(\frac{\partial g}{\partial \tilde{\boldsymbol{W}}^*} \right)^T \right) \Bigg|_{\tilde{\boldsymbol{W}}=\boldsymbol{W}} + \boldsymbol{L}_l \operatorname{vec}_l \left(\frac{\partial g}{\partial \tilde{\boldsymbol{W}}} + \left(\frac{\partial g}{\partial \tilde{\boldsymbol{W}}^*} \right)^T \right) \Bigg|_{\tilde{\boldsymbol{W}}=\boldsymbol{W}} \\
&\quad + \boldsymbol{L}_u \operatorname{vec}_u \left(\frac{\partial g}{\partial \tilde{\boldsymbol{W}}} + \left(\frac{\partial g}{\partial \tilde{\boldsymbol{W}}^*} \right)^T \right) \Bigg|_{\tilde{\boldsymbol{W}}=\boldsymbol{W}} = \operatorname{vec} \left(\frac{\partial g}{\partial \tilde{\boldsymbol{W}}} + \left(\frac{\partial g}{\partial \tilde{\boldsymbol{W}}^*} \right)^T \right)_{\tilde{\boldsymbol{W}}=\boldsymbol{W}},
\end{aligned} \tag{6.196}$$

which is equivalent to (6.194).

To find an expression for $\mathcal{D}_{\boldsymbol{W}^*} g$, the method in (6.72) can be used. First, an expression for $\mathcal{D}_{\boldsymbol{W}^*} \boldsymbol{F}^{-1}$ should be found. To achieve this, the expression $\operatorname{vec}\left(\boldsymbol{F}^{-1}\right)$ is studied because all components inside $\boldsymbol{W}^*$ have independent differentials when $\boldsymbol{W}$ is Hermitian. The desired expression can be found as follows:

$$\begin{aligned}
\operatorname{vec}\left(\boldsymbol{F}^{-1}\right) &= \begin{pmatrix} \boldsymbol{x} \\ \boldsymbol{z} \\ \boldsymbol{z}^* \end{pmatrix} = \begin{pmatrix} \boldsymbol{L}_d^T \operatorname{vec}(\boldsymbol{W}) \\ \boldsymbol{L}_l^T \operatorname{vec}(\boldsymbol{W}) \\ \boldsymbol{L}_u^T \operatorname{vec}(\boldsymbol{W}) \end{pmatrix} = \begin{pmatrix} \boldsymbol{L}_d^T \boldsymbol{K}_{N,N} \operatorname{vec}(\boldsymbol{W}^*) \\ \boldsymbol{L}_l^T \boldsymbol{K}_{N,N} \operatorname{vec}(\boldsymbol{W}^*) \\ \boldsymbol{L}_u^T \boldsymbol{K}_{N,N} \operatorname{vec}(\boldsymbol{W}^*) \end{pmatrix} \\
&= \begin{pmatrix} \boldsymbol{L}_d^T \operatorname{vec}(\boldsymbol{W}^*) \\ \boldsymbol{L}_u^T \operatorname{vec}(\boldsymbol{W}^*) \\ \boldsymbol{L}_l^T \operatorname{vec}(\boldsymbol{W}^*) \end{pmatrix} = \begin{pmatrix} \boldsymbol{L}_d^T \\ \boldsymbol{L}_u^T \\ \boldsymbol{L}_l^T \end{pmatrix} \operatorname{vec}(\boldsymbol{W}^*).
\end{aligned} \tag{6.197}$$

Because all elements of $\boldsymbol{W}^*$ have independent differentials, it follows from (6.197) that

$$\mathcal{D}_{\boldsymbol{W}^*} \boldsymbol{F}^{-1} = \begin{pmatrix} \boldsymbol{L}_d^T \\ \boldsymbol{L}_u^T \\ \boldsymbol{L}_l^T \end{pmatrix}. \tag{6.198}$$

When (6.72) and (6.198) are used, it follows that $\mathcal{D}_{W^*} g$ can be expressed as

$$\mathcal{D}_{W^*} g = [\mathcal{D}_x h,\ \mathcal{D}_z h,\ \mathcal{D}_{z^*} h] \mathcal{D}_{W^*} \boldsymbol{F}^{-1} = [\mathcal{D}_x h,\ \mathcal{D}_z h,\ \mathcal{D}_{z^*} h] \begin{pmatrix} \boldsymbol{L}_d^T \\ \boldsymbol{L}_u^T \\ \boldsymbol{L}_l^T \end{pmatrix}$$

$$= \operatorname{vec}_d^T \left(\frac{\partial g}{\partial \tilde{\boldsymbol{W}}} + \left(\frac{\partial g}{\partial \tilde{\boldsymbol{W}}^*} \right)^T \right) \bigg|_{\tilde{\boldsymbol{W}}=\boldsymbol{W}} \boldsymbol{L}_d^T$$

$$+ \operatorname{vec}_l^T \left(\frac{\partial g}{\partial \tilde{\boldsymbol{W}}} + \left(\frac{\partial g}{\partial \tilde{\boldsymbol{W}}^*} \right)^T \right) \bigg|_{\tilde{\boldsymbol{W}}=\boldsymbol{W}} \boldsymbol{L}_u^T$$

$$+ \operatorname{vec}_u^T \left(\frac{\partial g}{\partial \tilde{\boldsymbol{W}}} + \left(\frac{\partial g}{\partial \tilde{\boldsymbol{W}}^*} \right)^T \right) \bigg|_{\tilde{\boldsymbol{W}}=\boldsymbol{W}} \boldsymbol{L}_l^T$$

$$= \operatorname{vec}_d^T \left(\frac{\partial g}{\partial \tilde{\boldsymbol{W}}^*} + \left(\frac{\partial g}{\partial \tilde{\boldsymbol{W}}} \right)^T \right) \bigg|_{\tilde{\boldsymbol{W}}=\boldsymbol{W}} \boldsymbol{L}_d^T$$

$$+ \operatorname{vec}_l^T \left(\frac{\partial g}{\partial \tilde{\boldsymbol{W}}^*} + \left(\frac{\partial g}{\partial \tilde{\boldsymbol{W}}} \right)^T \right) \bigg|_{\tilde{\boldsymbol{W}}=\boldsymbol{W}} \boldsymbol{L}_l^T$$

$$+ \operatorname{vec}_u^T \left(\frac{\partial g}{\partial \tilde{\boldsymbol{W}}^*} + \left(\frac{\partial g}{\partial \tilde{\boldsymbol{W}}} \right)^T \right) \bigg|_{\tilde{\boldsymbol{W}}=\boldsymbol{W}} \boldsymbol{L}_u^T$$

$$= \operatorname{vec}^T \left(\frac{\partial g}{\partial \tilde{\boldsymbol{W}}^*} + \left(\frac{\partial g}{\partial \tilde{\boldsymbol{W}}} \right)^T \right)_{\tilde{\boldsymbol{W}}=\boldsymbol{W}} = \operatorname{vec}^T \left(\left(\frac{\partial g}{\partial \boldsymbol{W}} \right)^T \right), \tag{6.199}$$

which is in agreement with the results found earlier in this example (see (6.195)).

In the next example, the theory developed in the previous example will be used to study the derivatives of a function that is strongly related to the capacity of a Gaussian MIMO system.

Example 6.26 Consider the set of Hermitian matrices, such that $\boldsymbol{W} \in \mathcal{W} = \{\boldsymbol{W} \in \mathbb{C}^{M_t \times M_t} \mid \boldsymbol{W}^H = \boldsymbol{W}\}$. This is the same set of matrices considered in Examples 6.24 and 6.25.

Define the following function $g : \mathbb{C}^{M_t \times M_t} \times \mathbb{C}^{M_t \times M_t} \to \mathbb{C}$, given by:

$$g(\tilde{\boldsymbol{W}}, \tilde{\boldsymbol{W}}^*) = \ln\left(\det\left(\boldsymbol{I}_{M_r} + \boldsymbol{H}\tilde{\boldsymbol{W}}\boldsymbol{H}^H\right)\right), \tag{6.200}$$

where $\tilde{\boldsymbol{W}} \in \mathbb{C}^{M_t \times M_t}$ is an unpatterned complex-valued (not necessarily Hermitian) matrix variable. Using the theory from Chapter 3, it is found that

$$dg = \operatorname{Tr}\left\{ \left(\boldsymbol{I}_{M_r} + \boldsymbol{H}\tilde{\boldsymbol{W}}\boldsymbol{H}^H\right)^{-1} \boldsymbol{H} \left(d\tilde{\boldsymbol{W}}\right) \boldsymbol{H}^H \right\}$$

$$= \operatorname{Tr}\left\{ \boldsymbol{H}^H \left(\boldsymbol{I}_{M_r} + \boldsymbol{H}\tilde{\boldsymbol{W}}\boldsymbol{H}^H\right)^{-1} \boldsymbol{H} d\tilde{\boldsymbol{W}} \right\}. \tag{6.201}$$

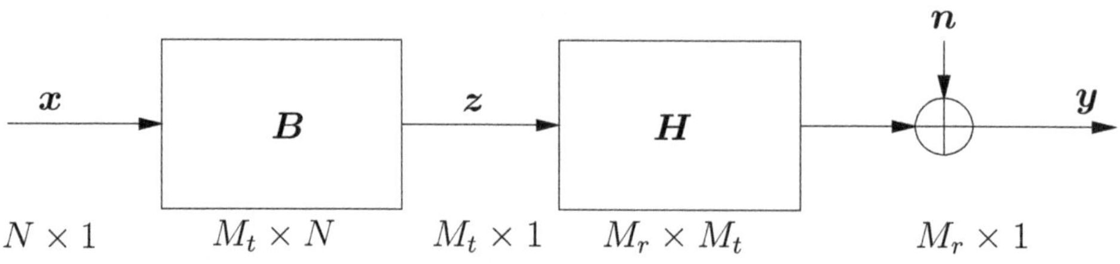

Figure 6.4 Precoded MIMO communication system with M_t transmit and M_r receiver antennas and with correlated additive complex-valued Gaussian noise. The precoder is denoted by $\boldsymbol{B} \in \mathbb{C}^{M_t \times N}$, the original source signal, $\boldsymbol{x} \in \mathbb{C}^{N \times 1}$, the transmitted signal, $\boldsymbol{z} \in \mathbb{C}^{M_t \times 1}$, the Gaussian additive signal-independent channel noise, $\boldsymbol{n} \in \mathbb{C}^{M_r \times 1}$, the received signal, $\boldsymbol{y} \in \mathbb{C}^{M_r \times 1}$, and the MIMO channel transfer matrix, $\boldsymbol{H} \in \mathbb{C}^{M_r \times M_t}$.

From Table 3.2, it follows that

$$\frac{\partial}{\partial \tilde{\boldsymbol{W}}} g\left(\tilde{\boldsymbol{W}}, \tilde{\boldsymbol{W}}^*\right) = \boldsymbol{H}^T \left(\boldsymbol{I}_{M_r} + \boldsymbol{H}^* \tilde{\boldsymbol{W}}^T \boldsymbol{H}^T\right)^{-1} \boldsymbol{H}^*, \tag{6.202}$$

and

$$\frac{\partial}{\partial \tilde{\boldsymbol{W}}^*} g\left(\tilde{\boldsymbol{W}}, \tilde{\boldsymbol{W}}^*\right) = \boldsymbol{0}_{M_t \times M_t}. \tag{6.203}$$

These results are valid *without* enforcing any structure to the matrix $\tilde{\boldsymbol{W}}$. The above results can be rewritten as

$$\mathcal{D}_{\tilde{\boldsymbol{W}}} g = \operatorname{vec}^T\left(\boldsymbol{H}^T \left(\boldsymbol{I}_{M_r} + \boldsymbol{H}^* \tilde{\boldsymbol{W}}^T \boldsymbol{H}^T\right)^{-1} \boldsymbol{H}^*\right), \tag{6.204}$$

$$\mathcal{D}_{\tilde{\boldsymbol{W}}^*} g = \operatorname{vec}^T\left(\frac{\partial g}{\partial \tilde{\boldsymbol{W}}^*}\right) = \boldsymbol{0}_{1 \times M_t^2}. \tag{6.205}$$

Consider now the generalized complex-valued matrix derivatives. Assume that $\boldsymbol{W} \in \mathcal{W}$, and we want to find the generalized derivative with respect to $\boldsymbol{W} \in \mathcal{W}$. From (6.194), (6.202), and (6.203), it follows that

$$\frac{\partial}{\partial \boldsymbol{W}} g(\boldsymbol{W}, \boldsymbol{W}^*) = \boldsymbol{H}^T \left(\boldsymbol{I}_{M_r} + \boldsymbol{H}^* \boldsymbol{W}^T \boldsymbol{H}^T\right)^{-1} \boldsymbol{H}^*. \tag{6.206}$$

Because $\boldsymbol{W}^* = \boldsymbol{W}^T$, it follows that

$$\frac{\partial}{\partial \boldsymbol{W}^*} g(\boldsymbol{W}) = \left(\frac{\partial}{\partial \boldsymbol{W}} g(\boldsymbol{W})\right)^T = \boldsymbol{H}^H \left(\boldsymbol{I}_{M_r} + \boldsymbol{H} \boldsymbol{W} \boldsymbol{H}^H\right)^{-1} \boldsymbol{H}. \tag{6.207}$$

Example 6.27 Consider the precoded MIMO system in Figure 6.4. Assume that the additive complex-valued channel noise $\boldsymbol{n} \in \mathbb{C}^{M_r \times 1}$ is zero-mean, Gaussian, independent of the original input signal vector $\boldsymbol{x} \in \mathbb{C}^{N \times 1}$, and $\boldsymbol{n}$ is correlated with the autocorrelation matrix given by

$$\boldsymbol{\Sigma}_{\boldsymbol{n}} = \mathbb{E}\left[\boldsymbol{n}\boldsymbol{n}^H\right]. \tag{6.208}$$

The $M_r \times M_r$ matrix $\boldsymbol{\Sigma}_{\boldsymbol{n}}$ is Hermitian. The goal of this example is to find the derivative of the mutual information between the input vector $\boldsymbol{x} \in \mathbb{C}^{N\times 1}$ and the output vector $\boldsymbol{y} \in \mathbb{C}^{M_r\times 1}$ when it is considered that $\boldsymbol{\Sigma}_{\boldsymbol{n}}^{-1}$ is Hermitian. Hence, it is here assumed that the inverse of the autocorrelation matrix $\boldsymbol{\Sigma}_{\boldsymbol{n}}$ is a Hermitian matrix, and, for simplicity, it is not taken into consideration that it is a positive definite matrix.

The differential entropy of the Gaussian complex-valued vector $\boldsymbol{n}$ with covariance matrix $\boldsymbol{\Sigma}_{\boldsymbol{n}}$ is given by Telatar (1995, Section 2):

$$H(\boldsymbol{n}) = \ln\left(\det\left(\pi e \boldsymbol{\Sigma}_{\boldsymbol{n}}\right)\right). \tag{6.209}$$

Assume that complex Gaussian signaling is used for $\boldsymbol{x}$. The received vector $\boldsymbol{y} \in \mathbb{C}^{M_r\times 1}$ is complex Gaussian distributed with covariance:

$$\begin{aligned}\boldsymbol{\Sigma}_{\boldsymbol{y}} &= \mathbb{E}\left[\boldsymbol{y}\boldsymbol{y}^H\right] = \mathbb{E}\left[(\boldsymbol{H}\boldsymbol{B}\boldsymbol{x}+\boldsymbol{n})(\boldsymbol{H}\boldsymbol{B}\boldsymbol{x}+\boldsymbol{n})^H\right] \\ &= \mathbb{E}\left[(\boldsymbol{H}\boldsymbol{B}\boldsymbol{x}+\boldsymbol{n})\left(\boldsymbol{x}^H\boldsymbol{B}^H\boldsymbol{H}^H+\boldsymbol{n}^H\right)\right] = \boldsymbol{H}\boldsymbol{B}\boldsymbol{\Sigma}_{\boldsymbol{x}}\boldsymbol{B}^H\boldsymbol{H}^H + \boldsymbol{\Sigma}_{\boldsymbol{n}},\end{aligned} \tag{6.210}$$

where $\boldsymbol{\Sigma}_{\boldsymbol{x}} = \mathbb{E}[\boldsymbol{x}\boldsymbol{x}^H]$ and $\mathbb{E}[\boldsymbol{x}\boldsymbol{n}^H] = \boldsymbol{0}_{N\times M_r}$. The mutual information between $\boldsymbol{x}$ and $\boldsymbol{y}$ is given by Telatar (1995, Section 3):

$$\begin{aligned}I(\boldsymbol{x};\boldsymbol{y}) &= H(\boldsymbol{y}) - H(\boldsymbol{y} \mid \boldsymbol{x}) = H(\boldsymbol{y}) - H(\boldsymbol{n}) = \ln\left(\det\left(\pi e \boldsymbol{\Sigma}_{\boldsymbol{y}}\right)\right) - \ln\left(\det\left(\pi e \boldsymbol{\Sigma}_{\boldsymbol{n}}\right)\right) \\ &= \ln\left(\det\left(\boldsymbol{\Sigma}_{\boldsymbol{y}}\boldsymbol{\Sigma}_{\boldsymbol{n}}^{-1}\right)\right) = \ln\left(\det\left(\boldsymbol{I}_{M_r} + \boldsymbol{H}\boldsymbol{B}\boldsymbol{\Sigma}_{\boldsymbol{x}}\boldsymbol{B}^H\boldsymbol{H}^H\boldsymbol{\Sigma}_{\boldsymbol{n}}^{-1}\right)\right).\end{aligned} \tag{6.211}$$

Let $\tilde{\boldsymbol{W}} \in \mathbb{C}^{M_r\times M_r}$ be a matrix of the same size as $\boldsymbol{\Sigma}_{\boldsymbol{n}}^{-1}$, which represents a matrix with independent matrix components. Define the function $g : \mathbb{C}^{M_r\times M_r} \times \mathbb{C}^{M_r\times M_r} \to \mathbb{C}$ as

$$g(\tilde{\boldsymbol{W}}, \tilde{\boldsymbol{W}}^*) = \ln\left(\det\left(\boldsymbol{I}_{M_r} + \boldsymbol{H}\boldsymbol{B}\boldsymbol{\Sigma}_{\boldsymbol{x}}\boldsymbol{B}^H\boldsymbol{H}^H\tilde{\boldsymbol{W}}\right)\right). \tag{6.212}$$

The differential of f is found as

$$\begin{aligned}dg &= \operatorname{Tr}\left\{\left(\boldsymbol{I}_{M_r} + \boldsymbol{H}\boldsymbol{B}\boldsymbol{\Sigma}_{\boldsymbol{x}}\boldsymbol{B}^H\boldsymbol{H}^H\tilde{\boldsymbol{W}}\right)^{-1}\boldsymbol{H}\boldsymbol{B}\boldsymbol{\Sigma}_{\boldsymbol{x}}\boldsymbol{B}^H\boldsymbol{H}^H d\tilde{\boldsymbol{W}}\right\} \\ &= \operatorname{vec}^T\left(\boldsymbol{H}^*\boldsymbol{B}^*\boldsymbol{\Sigma}_{\boldsymbol{x}}^*\boldsymbol{B}^T\boldsymbol{H}^T\left(\boldsymbol{I}_{M_r} + \tilde{\boldsymbol{W}}^T\boldsymbol{H}^*\boldsymbol{B}^*\boldsymbol{\Sigma}_{\boldsymbol{x}}^*\boldsymbol{B}^T\boldsymbol{H}^T\right)^{-1}\right) d\operatorname{vec}\left(\tilde{\boldsymbol{W}}\right) \\ &= \operatorname{vec}^T\left(\boldsymbol{H}^*\boldsymbol{B}^*\left(\boldsymbol{I}_N + \boldsymbol{\Sigma}_{\boldsymbol{x}}^*\boldsymbol{B}^T\boldsymbol{H}^T\tilde{\boldsymbol{W}}^T\boldsymbol{H}^*\boldsymbol{B}^*\right)^{-1}\boldsymbol{\Sigma}_{\boldsymbol{x}}^*\boldsymbol{B}^T\boldsymbol{H}^T\right) d\operatorname{vec}\left(\tilde{\boldsymbol{W}}\right) \\ &= \operatorname{vec}^T\left(\boldsymbol{H}^*\boldsymbol{B}^*\left(\left(\boldsymbol{\Sigma}_{\boldsymbol{x}}^*\right)^{-1} + \boldsymbol{B}^T\boldsymbol{H}^T\tilde{\boldsymbol{W}}^T\boldsymbol{H}^*\boldsymbol{B}^*\right)^{-1}\boldsymbol{B}^T\boldsymbol{H}^T\right) d\operatorname{vec}\left(\tilde{\boldsymbol{W}}\right),\end{aligned} \tag{6.213}$$

where Lemma 2.5 was used in the third equality. Define the matrix $\boldsymbol{E}(\tilde{\boldsymbol{W}}) \in \mathbb{C}^{N\times N}$ as

$$\boldsymbol{E}(\tilde{\boldsymbol{W}}) \triangleq \left(\boldsymbol{\Sigma}_{\boldsymbol{x}}^{-1} + \boldsymbol{B}^H\boldsymbol{H}^H\tilde{\boldsymbol{W}}\boldsymbol{H}\boldsymbol{B}\right)^{-1}. \tag{6.214}$$

Then it is seen from dg in (6.213) that the derivative of g with respect to the unpatterned matrices $\tilde{\boldsymbol{W}}$ and $\tilde{\boldsymbol{W}}^*$ can be expressed as

$$\mathcal{D}_{\tilde{\boldsymbol{W}}}g = \operatorname{vec}^T\left(\boldsymbol{H}^*\boldsymbol{B}^*\boldsymbol{E}^T(\tilde{\boldsymbol{W}})\boldsymbol{B}^T\boldsymbol{H}^T\right), \tag{6.215}$$

and

$$\mathcal{D}_{\tilde{W}^*} g = \mathbf{0}_{1\times M_r^2}, \tag{6.216}$$

respectively.

Let W be a Hermitian matrix of size $M_r \times M_r$. The parameterization function $F(x, z, z^*)$ is defined in (6.183). From (6.193), it follows that for Hermitian matrices, the following relation can be used to find the generalized matrix derivative in terms of the unpatterned matrix derivative:

$$\mathcal{D}_{W} g = \left[\mathcal{D}_{\tilde{W}} g + \left(\mathcal{D}_{\tilde{W}^*} g\right) K_{M_r,M_r}\right]_{\tilde{W}=W}, \tag{6.217}$$

when the standard basis $\{E_{i,j}\}$ is used to express $\mathcal{D}_W g$. By using this relation, it is found that the generalized derivative of g with respect to the Hermitian matrix W can be expressed as

$$\mathcal{D}_{W} g = \operatorname{vec}^T\left(H^* B^* E^T(W) B^T H^T\right). \tag{6.218}$$

When $\tilde{W} = W$ is used as argument for $g(\tilde{W}, \tilde{W}^*)|_{\tilde{W}=W} = g(W, W^*)$, then the function $g(W, W^*) = I(x; y)$, where $I(x; y)$ is mutual information between x and y, and W is a Hermitian matrix representing the inverse of the autocorrelation matrix of the noise vector n, that is, W represents Σ_n^{-1}. Hence, the derivative of the mutual information $I(x; y)$ with respect to W is given by

$$\mathcal{D}_{W} I = \operatorname{vec}^T\left(H^* B^* E^T(W) B^T H^T\right). \tag{6.219}$$

Because W is Hermitian, $\mathcal{D}_W I = \operatorname{vec}^T\left(\frac{\partial I}{\partial W}\right)$, and it is possible to write

$$\frac{\partial}{\partial W} I = H^* B^* E^T(W) B^T H^T. \tag{6.220}$$

Because I is scalar, it follows that

$$\begin{aligned}\frac{\partial}{\partial (W)^*} I = \frac{\partial}{\partial (W)^T} I = \left(\frac{\partial}{\partial W} I\right)^T &= \left(H^* B^* E^T(W) B^T H^T\right)^T \\ &= H B E(W) B^H H^H. \end{aligned} \tag{6.221}$$

This is in agreement with Palomar and Verdú (2006, Eq. (26)).

From the results in this example, it can be seen that

$$\mathcal{D}_{\tilde{W}^*} g = \mathbf{0}_{1\times M_r^2}, \tag{6.222}$$

$$\mathcal{D}_{W^*} g = \operatorname{vec}^T\left(H B E(W) B^H H^H\right). \tag{6.223}$$

This means that by introducing the Hermitian structure, the derivative $\mathcal{D}_{W^*} g$ is a *nonzero* vector in this example, and the unconstrained derivative $\mathcal{D}_{\tilde{W}^*} g$ is equal to the *zero* vector.

6.5.6 Generalized Matrix Derivative with Respect to Skew-Symmetric Matrices

---

Example 6.28 (Skew-Symmetric) Let the set of $N \times N$ complex-valued skew-symmetric matrices be denoted $\mathcal{S}$ and given by

$$\mathcal{S} = \{\boldsymbol{S} \in \mathbb{C}^{N\times N} \mid \boldsymbol{S}^T = -\boldsymbol{S}\}. \tag{6.224}$$

Skew-symmetric matrices have zero elements along the main diagonal, and the elements strictly below the main diagonal in position (k, l), where $k > l$, are equal to the elements strictly above the main diagonal in position (l, k) with the opposite sign. Notice that no complex conjugation is involved in the definition of skew-symmetric complex-valued matrices; hence, it is enough to parameterize these matrices with only the elements strictly below the main diagonal. Skew-symmetric matrices $\boldsymbol{S} \in \mathcal{S} \subset \mathbb{C}^{N\times N}$ can be parameterized with the parameterization function $\boldsymbol{F} : \mathbb{C}^{\frac{(N-1)N}{2}\times 1} \rightarrow \mathcal{S}$, given by

$$\text{vec}\,(\boldsymbol{S}) = \text{vec}\,(\boldsymbol{F}(\boldsymbol{z})) = \boldsymbol{L}_l\boldsymbol{z} - \boldsymbol{L}_u\boldsymbol{z} = (\boldsymbol{L}_l - \boldsymbol{L}_u)\,\boldsymbol{z}, \tag{6.225}$$

where $\boldsymbol{z} \in \mathbb{C}^{\frac{(N-1)N}{2}\times 1}$ contains all the independent complex-valued variables that are necessary for producing the skew-symmetric matrices $\boldsymbol{S} \in \mathcal{S}$. From (6.225), it is seen that the derivatives of $\boldsymbol{F}$ with respect to $\boldsymbol{z}$ and $\boldsymbol{z}^*$ are given by

$$\mathcal{D}_{\boldsymbol{z}}\boldsymbol{F} = \boldsymbol{L}_l - \boldsymbol{L}_u, \tag{6.226}$$

$$\mathcal{D}_{\boldsymbol{z}^*}\boldsymbol{F} = \boldsymbol{0}_{N^2\times\frac{(N-1)N}{2}}. \tag{6.227}$$

By complex conjugation (6.226) and (6.227), it follows that $\mathcal{D}_{\boldsymbol{z}}\boldsymbol{F}^* = \boldsymbol{0}_{N^2\times\frac{(N-1)N}{2}}$ and $\mathcal{D}_{\boldsymbol{z}^*}\boldsymbol{F}^* = \boldsymbol{L}_l - \boldsymbol{L}_u$. Let the function $g : \mathbb{C}^{N\times N} \times \mathbb{C}^{N\times N} \rightarrow \mathbb{C}$ be denoted $g(\tilde{\boldsymbol{S}}, \tilde{\boldsymbol{S}}^*)$, where $\tilde{\boldsymbol{S}} \in \mathbb{C}^{N\times N}$ is unpatterned, and assume that the two derivatives $\mathcal{D}_{\tilde{\boldsymbol{S}}}g = \text{vec}^T\left(\frac{\partial g}{\partial \tilde{\boldsymbol{S}}}\right)$ and $\mathcal{D}_{\tilde{\boldsymbol{S}}^*}g = \text{vec}^T\left(\frac{\partial g}{\partial \tilde{\boldsymbol{S}}^*}\right)$ are available. Define the composed function $h : \mathbb{C}^{\frac{(N-1)N}{2}\times 1} \rightarrow \mathbb{C}$ as follows:

$$h(\boldsymbol{z}) = g(\tilde{\boldsymbol{S}}, \tilde{\boldsymbol{S}}^*)\Big|_{\tilde{\boldsymbol{S}}=\boldsymbol{S}=\boldsymbol{F}(\boldsymbol{z})} = g(\boldsymbol{S}, \boldsymbol{S}^*) = g(\boldsymbol{F}(\boldsymbol{z}), \boldsymbol{F}^*(\boldsymbol{z})). \tag{6.228}$$

By means of the chain rule, the derivative of h with respect to $\boldsymbol{z}$ is found as follows:

$$\begin{aligned}\mathcal{D}_{\boldsymbol{z}}h &= \mathcal{D}_{\tilde{\boldsymbol{S}}}g\big|_{\tilde{\boldsymbol{S}}=\boldsymbol{S}}\mathcal{D}_{\boldsymbol{z}}\boldsymbol{F} + \mathcal{D}_{\tilde{\boldsymbol{S}}^*}g\big|_{\tilde{\boldsymbol{S}}=\boldsymbol{S}}\mathcal{D}_{\boldsymbol{z}}\boldsymbol{F}^* = \text{vec}^T\left(\frac{\partial g}{\partial \tilde{\boldsymbol{S}}}\right)\Bigg|_{\tilde{\boldsymbol{S}}=\boldsymbol{S}}(\boldsymbol{L}_l - \boldsymbol{L}_u) \\ &= \text{vec}_l^T\left(\frac{\partial g}{\partial \tilde{\boldsymbol{S}}}\right)\Bigg|_{\tilde{\boldsymbol{S}}=\boldsymbol{S}} - \text{vec}_u^T\left(\frac{\partial g}{\partial \tilde{\boldsymbol{S}}}\right)\Bigg|_{\tilde{\boldsymbol{S}}=\boldsymbol{S}}.\end{aligned} \tag{6.229}$$

From (6.225), it follows that $\text{vec}_d\,(\boldsymbol{S}) = \boldsymbol{0}_{N\times 1}$, $\text{vec}_l\,(\boldsymbol{S}) = \boldsymbol{z}$, and $\text{vec}_u\,(\boldsymbol{S}) = -\boldsymbol{z}$. If the result from Exercise 3.7 is used in (6.78), together with Definition 6.7, it is found that

$$\begin{aligned}
\operatorname{vec}\left(\frac{\partial g}{\partial \boldsymbol{S}}\right) &= \boldsymbol{L}_d \frac{\partial h}{\partial \operatorname{vec}_d(\boldsymbol{S})} + \boldsymbol{L}_l \frac{\partial h}{\partial \operatorname{vec}_l(\boldsymbol{S})} + \boldsymbol{L}_u \frac{\partial h}{\partial \operatorname{vec}_u(\boldsymbol{S})} \\
&= \boldsymbol{L}_l (\mathcal{D}_z h)^T + \boldsymbol{L}_u \left(\mathcal{D}_{(-z)} h\right)^T = \boldsymbol{L}_l (\mathcal{D}_z h)^T - \boldsymbol{L}_u (\mathcal{D}_z h)^T = (\boldsymbol{L}_l - \boldsymbol{L}_u)(\mathcal{D}_z h)^T \\
&= (\boldsymbol{L}_l - \boldsymbol{L}_u)\left[\operatorname{vec}_l\left(\frac{\partial g}{\partial \tilde{\boldsymbol{S}}}\right) - \operatorname{vec}_u\left(\frac{\partial g}{\partial \tilde{\boldsymbol{S}}}\right)\right]_{\tilde{\boldsymbol{S}}=\boldsymbol{S}} \\
&= \boldsymbol{L}_d \operatorname{vec}_d\left(\frac{\partial g}{\partial \tilde{\boldsymbol{S}}}\right)\Bigg|_{\tilde{\boldsymbol{S}}=\boldsymbol{S}} + \boldsymbol{L}_l \operatorname{vec}_l\left(\frac{\partial g}{\partial \tilde{\boldsymbol{S}}}\right)\Bigg|_{\tilde{\boldsymbol{S}}=\boldsymbol{S}} + \boldsymbol{L}_u \operatorname{vec}_u\left(\frac{\partial g}{\partial \tilde{\boldsymbol{S}}}\right)\Bigg|_{\tilde{\boldsymbol{S}}=\boldsymbol{S}} \\
&\quad -\boldsymbol{L}_d \operatorname{vec}_d\left(\left(\frac{\partial g}{\partial \tilde{\boldsymbol{S}}}\right)^T\right)\Bigg|_{\tilde{\boldsymbol{S}}=\boldsymbol{S}} -\boldsymbol{L}_l \operatorname{vec}_l\left(\left(\frac{\partial g}{\partial \tilde{\boldsymbol{S}}}\right)^T\right)\Bigg|_{\tilde{\boldsymbol{S}}=\boldsymbol{S}} -\boldsymbol{L}_u \operatorname{vec}_u\left(\left(\frac{\partial g}{\partial \tilde{\boldsymbol{S}}}\right)^T\right)\Bigg|_{\tilde{\boldsymbol{S}}=\boldsymbol{S}} \\
&= \operatorname{vec}\left(\frac{\partial g}{\partial \tilde{\boldsymbol{S}}} - \left(\frac{\partial g}{\partial \tilde{\boldsymbol{S}}}\right)^T\right)\Bigg|_{\tilde{\boldsymbol{S}}=\boldsymbol{S}}. \qquad (6.230)
\end{aligned}$$

This result leads to

$$\frac{\partial g}{\partial \boldsymbol{S}} = \left[\frac{\partial g}{\partial \tilde{\boldsymbol{S}}} - \left(\frac{\partial g}{\partial \tilde{\boldsymbol{S}}}\right)^T\right]_{\tilde{\boldsymbol{S}}=\boldsymbol{S}}. \qquad (6.231)$$

From (6.231), it is observed that $\left(\frac{\partial g}{\partial \boldsymbol{S}}\right)^T = -\frac{\partial g}{\partial \boldsymbol{S}}$; hence, $\frac{\partial g}{\partial \boldsymbol{S}}$ is skew-symmetric, implying that $\frac{\partial g}{\partial \boldsymbol{S}}$ has zeros on its main diagonal.

6.5.7 Generalized Matrix Derivative with Respect to Skew-Hermitian Matrices

---

Example 6.29 (Skew-Hermitian) Let the set of $N \times N$ skew-Hermitian matrices be denoted by $\mathcal{W}$ and given by

$$\mathcal{W} = \{\boldsymbol{W} \in \mathbb{C}^{N\times N} \mid \boldsymbol{W}^H = -\boldsymbol{W}\}. \qquad (6.232)$$

An arbitrary skew-Hermitian matrix $\boldsymbol{W} \in \mathcal{W} \subset \mathbb{C}^{N\times N}$ can be parameterized by an $N \times 1$ real vector $\boldsymbol{x} = \operatorname{vec}_d(\boldsymbol{W})/\jmath$, that will produce the pure imaginary diagonal elements of $\boldsymbol{W}$ and a complex-valued vector $\boldsymbol{z} = \operatorname{vec}_l(\boldsymbol{W}) = -(\operatorname{vec}_u(\boldsymbol{W}))^* \in \mathbb{C}^{\frac{(N-1)N}{2}\times 1}$ that contains the strictly below main diagonal elements in position (k, l), where $k > l$; these are equal to the complex conjugates of the strictly above main diagonal elements in position (l, k) with the opposite sign. One way of generating any skew-Hermitian $N \times N$ matrix is by using the parameterization function $\boldsymbol{F} : \mathbb{R}^{N\times 1} \times \mathbb{C}^{\frac{(N-1)N}{2}\times 1} \times \mathbb{C}^{\frac{(N-1)N}{2}\times 1} \to \mathcal{W} \subset \mathbb{C}^{N\times N}$, given by

$$\operatorname{vec}(\boldsymbol{W}) = \operatorname{vec}\left(\boldsymbol{F}(\boldsymbol{x}, \boldsymbol{z}, \boldsymbol{z}^*)\right) = \jmath \boldsymbol{L}_d \boldsymbol{x} + \boldsymbol{L}_l \boldsymbol{z} - \boldsymbol{L}_u \boldsymbol{z}^*. \tag{6.233}$$

From (6.233), it follows that $\mathcal{D}_{\boldsymbol{x}}\boldsymbol{F} = \jmath \boldsymbol{L}_d$, $\mathcal{D}_{\boldsymbol{z}}\boldsymbol{F} = \boldsymbol{L}_l$, and $\mathcal{D}_{\boldsymbol{z}^*}\boldsymbol{F} = -\boldsymbol{L}_u$. By complex conjugating both sides of (6.233), it follows that $\mathcal{D}_{\boldsymbol{x}}\boldsymbol{F}^* = -\jmath \boldsymbol{L}_d$, $\mathcal{D}_{\boldsymbol{z}}\boldsymbol{F}^* = -\boldsymbol{L}_u$, and $\mathcal{D}_{\boldsymbol{z}^*}\boldsymbol{F}^* = \boldsymbol{L}_l$. The dimension of the tangent space of $\mathcal{W}$ is N^2, and all components of $\boldsymbol{W} \in \mathcal{W}$ can be treated as independent when finding derivatives. If we choose as a basis for $\mathcal{W}$ the N^2 matrices of size $N \times N$ found by reshaping each of the *columns* of $\jmath \boldsymbol{L}_d$, $\boldsymbol{L}_l$, and $-\boldsymbol{L}_u$ by "inverting" the vec-operation, then we can represent $\boldsymbol{W} \in \mathcal{W}$ as $\boldsymbol{W} \triangleq [[\boldsymbol{x}], [\boldsymbol{z}], [\boldsymbol{z}^*]]_{[\jmath \boldsymbol{L}_d, \boldsymbol{L}_l, -\boldsymbol{L}_u]}$. With this representation, the function $\boldsymbol{F}$ is the identity function in a similar manner as in Example 6.24. This means that

$$\mathcal{D}_{\boldsymbol{W}}\boldsymbol{F}^{-1} = \boldsymbol{I}_{N^2}, \tag{6.234}$$

when $[\jmath \boldsymbol{L}_d, \boldsymbol{L}_l, -\boldsymbol{L}_u]$ is used as a basis for $\mathcal{W}$.

Assume that the function $g : \mathbb{C}^{N\times N} \times \mathbb{C}^{N\times N} \to \mathbb{C}$ is given, and that the two derivatives $\mathcal{D}_{\tilde{\boldsymbol{W}}} g = \operatorname{vec}^T\left(\frac{\partial g}{\partial \tilde{\boldsymbol{W}}}\right)$ and $\mathcal{D}_{\tilde{\boldsymbol{W}}^*} g = \operatorname{vec}^T\left(\frac{\partial g}{\partial \tilde{\boldsymbol{W}}^*}\right)$ are available. Define the function $h : \mathbb{R}^{N\times 1} \times \mathbb{C}^{\frac{(N-1)N}{2}\times 1} \times \mathbb{C}^{\frac{(N-1)N}{2}\times 1} \to \mathbb{C}$ as

$$\begin{aligned} h(\boldsymbol{x}, \boldsymbol{z}, \boldsymbol{z}^*) &= g(\tilde{\boldsymbol{W}}, \tilde{\boldsymbol{W}}^*)\Big|_{\tilde{\boldsymbol{W}}=\boldsymbol{W}=\boldsymbol{F}(\boldsymbol{x},\boldsymbol{z},\boldsymbol{z}^*)} = g(\boldsymbol{W}, \boldsymbol{W}^*) \\ &= g(\boldsymbol{F}(\boldsymbol{x}, \boldsymbol{z}, \boldsymbol{z}^*), \boldsymbol{F}^*(\boldsymbol{x}, \boldsymbol{z}, \boldsymbol{z}^*)). \end{aligned} \tag{6.235}$$

By the chain rule, the derivatives of h with respect to $\boldsymbol{x}$, $\boldsymbol{z}$, and $\boldsymbol{z}^*$ can be found as

$$\begin{aligned} \mathcal{D}_{\boldsymbol{x}} h &= \mathcal{D}_{\tilde{\boldsymbol{W}}} g\big|_{\tilde{\boldsymbol{W}}=\boldsymbol{W}} \mathcal{D}_{\boldsymbol{x}}\boldsymbol{F} + \mathcal{D}_{\tilde{\boldsymbol{W}}^*} g\big|_{\tilde{\boldsymbol{W}}=\boldsymbol{W}} \mathcal{D}_{\boldsymbol{x}}\boldsymbol{F}^* \\ &= \operatorname{vec}^T\left(\frac{\partial g}{\partial \tilde{\boldsymbol{W}}}\right)\bigg|_{\tilde{\boldsymbol{W}}=\boldsymbol{W}} \jmath \boldsymbol{L}_d - \operatorname{vec}^T\left(\frac{\partial g}{\partial \tilde{\boldsymbol{W}}^*}\right)\bigg|_{\tilde{\boldsymbol{W}}=\boldsymbol{W}} \jmath \boldsymbol{L}_d \\ &= \jmath \operatorname{vec}_d^T\left(\frac{\partial g}{\partial \tilde{\boldsymbol{W}}}\right)\bigg|_{\tilde{\boldsymbol{W}}=\boldsymbol{W}} - \jmath \operatorname{vec}_d^T\left(\frac{\partial g}{\partial \tilde{\boldsymbol{W}}^*}\right)\bigg|_{\tilde{\boldsymbol{W}}=\boldsymbol{W}}, \end{aligned} \tag{6.236}$$

$$\begin{aligned} \mathcal{D}_{\boldsymbol{z}} h &= \mathcal{D}_{\tilde{\boldsymbol{W}}} g\big|_{\tilde{\boldsymbol{W}}=\boldsymbol{W}} \mathcal{D}_{\boldsymbol{z}}\boldsymbol{F} + \mathcal{D}_{\tilde{\boldsymbol{W}}^*} g\big|_{\tilde{\boldsymbol{W}}=\boldsymbol{W}} \mathcal{D}_{\boldsymbol{z}}\boldsymbol{F}^* \\ &= \operatorname{vec}^T\left(\frac{\partial g}{\partial \tilde{\boldsymbol{W}}}\right)\bigg|_{\tilde{\boldsymbol{W}}=\boldsymbol{W}} \boldsymbol{L}_l - \operatorname{vec}^T\left(\frac{\partial g}{\partial \tilde{\boldsymbol{W}}^*}\right)\bigg|_{\tilde{\boldsymbol{W}}=\boldsymbol{W}} \boldsymbol{L}_u \\ &= \operatorname{vec}_l^T\left(\frac{\partial g}{\partial \tilde{\boldsymbol{W}}}\right)\bigg|_{\tilde{\boldsymbol{W}}=\boldsymbol{W}} - \operatorname{vec}_u^T\left(\frac{\partial g}{\partial \tilde{\boldsymbol{W}}^*}\right)\bigg|_{\tilde{\boldsymbol{W}}=\boldsymbol{W}}, \end{aligned} \tag{6.237}$$

and

$$\begin{aligned}
\mathcal{D}_{z^*} h &= \mathcal{D}_{\tilde{W}} g\big|_{\tilde{W}=W} \mathcal{D}_{z^*} F + \mathcal{D}_{\tilde{W}^*} g\big|_{\tilde{W}=W} \mathcal{D}_{z^*} F^* \\
&= -\operatorname{vec}^T\left(\frac{\partial g}{\partial \tilde{W}}\right)\bigg|_{\tilde{W}=W} L_u + \operatorname{vec}^T\left(\frac{\partial g}{\partial \tilde{W}^*}\right)\bigg|_{\tilde{W}=W} L_l \\
&= -\operatorname{vec}_u^T\left(\frac{\partial g}{\partial \tilde{W}}\right)\bigg|_{\tilde{W}=W} + \operatorname{vec}_l^T\left(\frac{\partial g}{\partial \tilde{W}^*}\right)\bigg|_{\tilde{W}=W}.
\end{aligned} \tag{6.238}$$

Using the results above for finding the generalized complex-valued matrix derivative of g with respect to $W \in \mathcal{W}$ in (6.71) leads to

$$[\mathcal{D}_W g]_{[\jmath L_d, L_l, -L_u]} = [\mathcal{D}_x h, \mathcal{D}_z h, \mathcal{D}_{z^*} h] \mathcal{D}_W F^{-1} = [\mathcal{D}_x h, \mathcal{D}_z h, \mathcal{D}_{z^*} h], \tag{6.239}$$

when $[\jmath L_d, L_l, -L_u]$ is used as a basis for $\mathcal{W}$. By means of Exercise 6.16, it is possible to express the generalized complex-valued matrix derivative $\mathcal{D}_W g$ in terms of the standard basis $\{E_{i,j}\}$ as follows:

$$\begin{aligned}
[\mathcal{D}_W g]_{\{E_{i,j}\}} &= [\mathcal{D}_W g]_{[\jmath L_d, L_l, -L_u]} [\jmath L_d, L_l, -L_u]^{-1} = [\mathcal{D}_W g]_{[\jmath L_d, L_l, -L_u]} \begin{bmatrix} -\jmath L_d^T \\ L_l^T \\ -L_u^T \end{bmatrix} \\
&= [\mathcal{D}_x h, \mathcal{D}_z h, \mathcal{D}_{z^*} h] \begin{bmatrix} -\jmath L_d^T \\ L_l^T \\ -L_u^T \end{bmatrix} \\
&= \left[\jmath \operatorname{vec}_d^T\left(\frac{\partial g}{\partial \tilde{W}}\right) - \jmath \operatorname{vec}_d^T\left(\frac{\partial g}{\partial \tilde{W}^*}\right)\right]_{\tilde{W}=W} (-\jmath L_d^T) \\
&\quad + \left[\operatorname{vec}_l^T\left(\frac{\partial g}{\partial \tilde{W}}\right) - \operatorname{vec}_u^T\left(\frac{\partial g}{\partial \tilde{W}^*}\right)\right]_{\tilde{W}=W} L_l^T \\
&\quad + \left[-\operatorname{vec}_u^T\left(\frac{\partial g}{\partial \tilde{W}}\right) + \operatorname{vec}_l^T\left(\frac{\partial g}{\partial \tilde{W}^*}\right)\right]_{\tilde{W}=W} (-L_u^T) \\
&= \left[\operatorname{vec}_d^T\left(\frac{\partial g}{\partial \tilde{W}}\right)\right]_{\tilde{W}=W} L_d^T - \left[\operatorname{vec}_d^T\left(\left(\frac{\partial g}{\partial \tilde{W}^*}\right)^T\right)\right]_{\tilde{W}=W} L_d^T \\
&\quad + \left[\operatorname{vec}_l^T\left(\frac{\partial g}{\partial \tilde{W}}\right)\right]_{\tilde{W}=W} L_l^T - \left[\operatorname{vec}_l^T\left(\left(\frac{\partial g}{\partial \tilde{W}^*}\right)^T\right)\right]_{\tilde{W}=W} L_l^T \\
&\quad + \left[\operatorname{vec}_u^T\left(\frac{\partial g}{\partial \tilde{W}}\right)\right]_{\tilde{W}=W} L_u^T - \left[\operatorname{vec}_u^T\left(\left(\frac{\partial g}{\partial \tilde{W}^*}\right)^T\right)\right]_{\tilde{W}=W} L_u^T \\
&= \operatorname{vec}^T\left(\frac{\partial g}{\partial \tilde{W}}\right) - \operatorname{vec}^T\left(\frac{\partial g}{\partial \tilde{W}^*}\right).
\end{aligned} \tag{6.240}$$

From (6.233), it is seen that $\text{vec}_d(\boldsymbol{W}) = \jmath \boldsymbol{x}$, $\text{vec}_l(\boldsymbol{W}) = \boldsymbol{z}$, and $\text{vec}_u(\boldsymbol{W}) = -\boldsymbol{z}^*$. By using (6.78) and the result from Exercise 3.7, it is found that

$$
\begin{aligned}
\text{vec}\left(\frac{\partial g}{\partial \boldsymbol{W}}\right) &= \boldsymbol{L}_d \frac{\partial h}{\partial \,\text{vec}_d(\boldsymbol{W})} + \boldsymbol{L}_l \frac{\partial h}{\partial \,\text{vec}_l(\boldsymbol{W})} + \boldsymbol{L}_u \frac{\partial h}{\partial \,\text{vec}_u(\boldsymbol{W})} \\
&= \boldsymbol{L}_d \left(\mathcal{D}_{\jmath \boldsymbol{x}} g\right)^T + \boldsymbol{L}_l \left(\mathcal{D}_{\boldsymbol{z}} g\right)^T + \boldsymbol{L}_u \left(\mathcal{D}_{-\boldsymbol{z}^*} g\right)^T \\
&= \frac{\boldsymbol{L}_d}{\jmath} \left(\mathcal{D}_{\boldsymbol{x}} h\right)^T + \boldsymbol{L}_l \left(\mathcal{D}_{\boldsymbol{z}} h\right)^T - \boldsymbol{L}_u \left(\mathcal{D}_{\boldsymbol{z}^*} h\right)^T \\
&= -\jmath \boldsymbol{L}_d \jmath \left[\text{vec}_d\left(\frac{\partial g}{\partial \tilde{\boldsymbol{W}}}\right) - \text{vec}_d\left(\frac{\partial g}{\partial \tilde{\boldsymbol{W}}^*}\right)\right]_{\tilde{\boldsymbol{W}}=\boldsymbol{W}} \\
&\quad + \boldsymbol{L}_l \left[\text{vec}_l\left(\frac{\partial g}{\partial \tilde{\boldsymbol{W}}}\right) - \text{vec}_u\left(\frac{\partial g}{\partial \tilde{\boldsymbol{W}}^*}\right)\right]_{\tilde{\boldsymbol{W}}=\boldsymbol{W}} \\
&\quad - \boldsymbol{L}_u \left[- \text{vec}_u\left(\frac{\partial g}{\partial \tilde{\boldsymbol{W}}}\right) + \text{vec}_l\left(\frac{\partial g}{\partial \tilde{\boldsymbol{W}}^*}\right)\right]_{\tilde{\boldsymbol{W}}=\boldsymbol{W}} \\
&= \boldsymbol{L}_d \,\text{vec}_d\left(\frac{\partial g}{\partial \tilde{\boldsymbol{W}}}\right)\Bigg|_{\tilde{\boldsymbol{W}}=\boldsymbol{W}} - \boldsymbol{L}_d \,\text{vec}_d\left(\left(\frac{\partial g}{\partial \tilde{\boldsymbol{W}}^*}\right)^T\right)\Bigg|_{\tilde{\boldsymbol{W}}=\boldsymbol{W}} \\
&\quad + \boldsymbol{L}_l \,\text{vec}_l\left(\frac{\partial g}{\partial \tilde{\boldsymbol{W}}}\right)\Bigg|_{\tilde{\boldsymbol{W}}=\boldsymbol{W}} - \boldsymbol{L}_l \,\text{vec}_l\left(\left(\frac{\partial g}{\partial \tilde{\boldsymbol{W}}^*}\right)^T\right)\Bigg|_{\tilde{\boldsymbol{W}}=\boldsymbol{W}} \\
&\quad + \boldsymbol{L}_u \,\text{vec}_u\left(\frac{\partial g}{\partial \tilde{\boldsymbol{W}}}\right)\Bigg|_{\tilde{\boldsymbol{W}}=\boldsymbol{W}} - \boldsymbol{L}_u \,\text{vec}_u\left(\left(\frac{\partial g}{\partial \tilde{\boldsymbol{W}}^*}\right)^T\right)\Bigg|_{\tilde{\boldsymbol{W}}=\boldsymbol{W}} \\
&= \text{vec}\left(\frac{\partial g}{\partial \tilde{\boldsymbol{W}}} - \left(\frac{\partial g}{\partial \tilde{\boldsymbol{W}}^*}\right)^T\right)_{\tilde{\boldsymbol{W}}=\boldsymbol{W}}.
\end{aligned}
\tag{6.241}
$$

From the above expression, it follows that

$$
\frac{\partial g}{\partial \boldsymbol{W}} = \left[\frac{\partial g}{\partial \tilde{\boldsymbol{W}}} - \left(\frac{\partial g}{\partial \tilde{\boldsymbol{W}}^*}\right)^T\right]_{\tilde{\boldsymbol{W}}=\boldsymbol{W}}. \tag{6.242}
$$

It is observed that $\left(\frac{\partial g}{\partial \boldsymbol{W}}\right)^H = -\frac{\partial g}{\partial \boldsymbol{W}}$, that is, $\frac{\partial g}{\partial \boldsymbol{W}}$ is skew-Hermitian. In addition, it is seen that (6.240) and (6.241) are consistent.

As a particular case, consider the function $g : \mathbb{C}^{2\times 2} \times \mathbb{C}^{2\times 2} \rightarrow \mathbb{C}$ as

$$
g(\tilde{\boldsymbol{W}}, \tilde{\boldsymbol{W}}^*) = \det(\tilde{\boldsymbol{W}}) = \det\left(\begin{bmatrix} \tilde{w}_{0,0} & \tilde{w}_{0,1} \\ \tilde{w}_{1,0} & \tilde{w}_{1,1} \end{bmatrix}\right) = \tilde{w}_{0,0}\tilde{w}_{1,1} - \tilde{w}_{0,1}\tilde{w}_{1,0}. \tag{6.243}
$$

In this case, it follows from (3.48) that the derivatives of g with respect to the unpatterned matrices $\tilde{W}$ and $\tilde{W}^*$ are given by

$$\mathcal{D}_{\tilde{W}} g = \text{vec}^T \left(\begin{bmatrix} \tilde{w}_{1,1} & -\tilde{w}_{1,0} \\ -\tilde{w}_{0,1} & \tilde{w}_{0,0} \end{bmatrix} \right), \tag{6.244}$$

$$\mathcal{D}_{\tilde{W}^*} g = \mathbf{0}_{1\times 4}. \tag{6.245}$$

Assume now that $\mathcal{W}$ is the set of 2×2 skew-Hermitian matrices. Then $W \in \mathcal{W}$ can be expressed as

$$W = \begin{pmatrix} w_{0,0} & w_{0,1} \\ w_{1,0} & w_{1,1} \end{pmatrix} = \begin{pmatrix} \jmath x_0 & -z^* \\ z & \jmath x_1 \end{pmatrix}. \tag{6.246}$$

Using (6.242) to find the generalized complex-valued matrix derivative leads to

$$\begin{aligned} \frac{\partial g}{\partial W} &= \left[\frac{\partial g}{\partial \tilde{W}} - \left(\frac{\partial g}{\partial \tilde{W}^*} \right)^T \right]_{\tilde{W}=W} = \left[\begin{pmatrix} \tilde{w}_{1,1} & -\tilde{w}_{1,0} \\ -\tilde{w}_{0,1} & \tilde{w}_{0,0} \end{pmatrix} - \mathbf{0}_{2\times 2} \right]_{\tilde{W}=W} \\ &= \begin{pmatrix} w_{1,1} & -w_{1,0} \\ -w_{0,1} & w_{0,0} \end{pmatrix} = \begin{pmatrix} \jmath x_1 & -z \\ z^* & \jmath x_0 \end{pmatrix}. \end{aligned} \tag{6.247}$$

By direct calculation of $\frac{\partial g}{\partial W}$ using (6.74), it is found that

$$\frac{\partial g}{\partial W} = \begin{bmatrix} \frac{\partial}{\partial w_{0,0}} & \frac{\partial}{\partial w_{0,1}} \\ \frac{\partial}{\partial w_{1,0}} & \frac{\partial}{\partial w_{1,1}} \end{bmatrix} (w_{0,0} w_{1,1} - w_{1,0} w_{0,1}) = \begin{pmatrix} w_{1,1} & -w_{1,0} \\ -w_{0,1} & w_{0,0} \end{pmatrix} = \begin{pmatrix} \jmath x_1 & -z \\ z^* & \jmath x_0 \end{pmatrix}, \tag{6.248}$$

which is in agreement with (6.247).

In Table 6.2, the derivatives of the function $g : \mathbb{C}^{N\times N} \times \mathbb{C}^{N\times N} \rightarrow \mathbb{C}$, denoted by $g(\tilde{W}, \tilde{W}^*)$, are summarized for complex-valued unpatterned, diagonal, symmetric, skew-symmetric, Hermitian, and skew-Hermitian matrices. These results were derived earlier in this chapter, or in Exercise 6.4.

6.5.8 Orthogonal Matrices

Example 6.30 (Orthogonal) Let $Q \in \mathbb{C}^{N\times N}$ be an orthogonal matrix, then Q can be found from

$$Q = \exp(S), \tag{6.249}$$

where $\exp(\cdot)$ is the exponential matrix function stated in Definition 2.5, and $S = -S^T$ is skew-symmetric. In (6.225), a parameterization function for the set of $N \times N$ complex-valued skew-symmetric matrices is given. The matrix Q is orthogonal

Table 6.2 Various results for the generalized derivatives $\frac{\partial g}{\partial W}$ of the function $g : \mathbb{C}^{N\times N} \times \mathbb{C}^{N\times N} \to \mathbb{C}$ denoted by $g(\tilde{W}, \tilde{W}^*)$ when the input matrix variable $\tilde{W}$ belongs to the set $\mathcal{W}$ of unpatterned, diagonal, symmetric, skew-symmetric, Hermitian, and skew-Hermitian complex-valued matrices.

The set W belongs to	$\frac{\partial g}{\partial W}$
Unpatterned $\mathcal{W} = \{W \in \mathbb{C}^{N\times N}\}$	$\frac{\partial g}{\partial \tilde{W}}$
Diagonal $\mathcal{W} = \{W \in \mathbb{C}^{N\times N} \mid W = I_N \odot W\}$	$I_N \odot \left.\frac{\partial g}{\partial \tilde{W}}\right\vert_{\tilde{W}=W}$
Symmetric $\mathcal{W} = \{W \in \mathbb{C}^{N\times N} \mid W = W^T\}$	$\left[\frac{\partial g}{\partial \tilde{W}} + \left(\frac{\partial g}{\partial \tilde{W}}\right)^T - I_N \odot \frac{\partial g}{\partial \tilde{W}}\right]_{\tilde{W}=W}$
Skew-symmetric $\mathcal{W} = \{W \in \mathbb{C}^{N\times N} \mid W = -W^T\}$	$\left[\frac{\partial g}{\partial \tilde{W}} - \left(\frac{\partial g}{\partial \tilde{W}}\right)^T\right]_{\tilde{W}=W}$
Hermitian $\mathcal{W} = \{W \in \mathbb{C}^{N\times N} \mid W = W^H\}$	$\left[\frac{\partial g}{\partial \tilde{W}} + \left(\frac{\partial g}{\partial \tilde{W}^*}\right)^T\right]_{\tilde{W}=W}$
Skew-Hermitian $\mathcal{W} = \{W \in \mathbb{C}^{N\times N} \mid W = -W^H\}$	$\left[\frac{\partial g}{\partial \tilde{W}} - \left(\frac{\partial g}{\partial \tilde{W}^*}\right)^T\right]_{\tilde{W}=W}$

because

$$\begin{aligned} QQ^T &= \exp(S)\exp(S^T) = \exp(S)\exp(-S) = \exp(S - S) \\ &= \exp(0_{N\times N}) = I_N, \end{aligned} \tag{6.250}$$

where it has been shown that $\exp(A)\exp(B) = \exp(B)\exp(A) = \exp(A + B)$ when $AB = BA$ (see Exercise 2.5). However, (6.249) does always return an orthogonal matrix with determinant $+1$ because

$$\det(Q) = \det(\exp(S)) = \exp(\text{Tr}\{S\}) = \exp(0) = 1, \tag{6.251}$$

where Lemma 2.6 was utilized together with the fact that $\text{Tr}\{S\} = 0$ for skew-symmetric matrices. Because there exist infinitely many orthogonal matrices with determinant -1 when $N > 1$, the function in (6.249) does *not* parameterize the whole set of orthogonal matrices.

6.5.9 Unitary Matrices

Example 6.31 (Unitary) Let $\mathcal{W} = \{W \in \mathbb{C}^{N\times N} \mid W^H = -W\}$ be the set of skew-Hermitian $N \times N$ matrices. Any complex-valued unitary $N \times N$ matrix can be parameterized in the following way (Rinehart 1964):

$$U = \exp(W) = \exp\left(F(x, z, z^*)\right), \tag{6.252}$$

where $\exp(\cdot)$ is described in Definition 2.5, and $W \in \mathcal{W} \subset \mathbb{C}^{N\times N}$ is a skew-Hermitian matrix that can be produced by the function $W = F(x, z, z^*)$ given in (6.233). It was shown in Example 6.29 how to find the derivative of the parameterization function $F(x, z, z^*)$ and its complex conjugate with respect to the three vectors x, z, and z^*.

Let $\tilde{W} \in \mathbb{C}^{N\times N}$ be an unpatterned complex-valued $N \times N$ matrix. The derivatives of the two functions $\hat{U} \triangleq \exp(\tilde{W})$ and $\hat{U}^* \triangleq \exp(\tilde{W}^*)$ are[12] now found with respect to the two unpatterned matrices $\tilde{W} \in \mathbb{C}^{N\times N}$ and $\tilde{W}^* \in \mathbb{C}^{N\times N}$. To achieve this, the following results are useful:

$$d\,\mathrm{vec}(\hat{U}) = \sum_{k=0}^{\infty} \frac{1}{(k+1)!} \sum_{i=0}^{k} \left(\left(\tilde{W}^T\right)^{k-i} \otimes \tilde{W}^i \right) d\,\mathrm{vec}(\tilde{W}), \tag{6.253}$$

$$d\,\mathrm{vec}(\hat{U}^*) = \sum_{k=0}^{\infty} \frac{1}{(k+1)!} \sum_{i=0}^{k} \left(\left(\tilde{W}^H\right)^{k-i} \otimes (\tilde{W}^*)^i \right) d\,\mathrm{vec}(\tilde{W}^*), \tag{6.254}$$

following from (4.134) and (4.135) by adjusting the notation to the symbols used here. From (6.253) and (6.254), the following derivatives can be found:

$$\mathcal{D}_{\tilde{W}} \hat{U} = \sum_{k=0}^{\infty} \frac{1}{(k+1)!} \sum_{i=0}^{k} \left(\tilde{W}^T\right)^{k-i} \otimes \tilde{W}^i, \tag{6.255}$$

$$\mathcal{D}_{\tilde{W}^*} \hat{U} = \mathbf{0}_{N^2\times N^2}, \tag{6.256}$$

$$\mathcal{D}_{\tilde{W}} \hat{U}^* = \mathbf{0}_{N^2\times N^2}, \tag{6.257}$$

$$\mathcal{D}_{\tilde{W}^*} \hat{U}^* = \sum_{k=0}^{\infty} \frac{1}{(k+1)!} \sum_{i=0}^{k} \left(\tilde{W}^H\right)^{k-i} \otimes (\tilde{W}^*)^i. \tag{6.258}$$

Consider the *real-valued* function $g : \mathbb{C}^{N\times N} \times \mathbb{C}^{N\times N} \to \mathbb{R}$ denoted by $g(\tilde{U}, \tilde{U}^*)$, where $\tilde{U}$ is unpatterned. Assume that the two derivatives $\mathcal{D}_{\tilde{U}} g = \mathrm{vec}^T\left(\frac{\partial g}{\partial \tilde{U}}\right)$ and $\mathcal{D}_{\tilde{U}^*} g = \mathrm{vec}^T\left(\frac{\partial g}{\partial \tilde{U}^*}\right)$ are available. Define the composed function $h : \mathbb{R}^{N\times 1} \times \mathbb{C}^{\frac{(N-1)N}{2}\times 1} \times \mathbb{C}^{\frac{(N-1)N}{2}\times 1} \to \mathbb{R}$ as

$$\begin{aligned} h(x, z, z^*) &= g\left(\tilde{U}, \tilde{U}^*\right)\Big|_{\tilde{U}=U=\exp(F(x,z,z^*))} = g\left(U, U^*\right) \\ &= g\left(\exp(F(x, z, z^*)), \exp(F^*(x, z, z^*))\right), \end{aligned} \tag{6.259}$$

where $F : \mathbb{R}^{N\times 1} \times \mathbb{C}^{\frac{(N-1)N}{2}\times 1} \times \mathbb{C}^{\frac{(N-1)N}{2}\times 1} \to \mathcal{W} \subset \mathbb{C}^{N\times N}$ is defined in (6.233). The derivatives of h with respect to the independent variables x, z, and z^* can be found

[12] The two symbols $\hat{U}$ and $\hat{U}^*$ are used because the two matrices $\hat{U} = \exp(\tilde{W})$ and $\hat{U}^* = \exp(\tilde{W}^*)$ can be parameterized; hence, they are not unpatterned. The symbol $\hat{U}$ is different from both the unitary matrix U and the unpatterned matrix $\tilde{U} \in \mathbb{C}^{N\times N}$.

by applying the chain rule twice as follows:

$$\mathcal{D}_x h(x, z, z^*) = \mathcal{D}_{\tilde{U}} g(\tilde{U}, \tilde{U}^*)|_{\tilde{U}=U} \mathcal{D}_{\tilde{W}} \hat{U}|_{\tilde{W}=W} \mathcal{D}_x F + \mathcal{D}_{\tilde{U}^*} g(\tilde{U}, \tilde{U}^*)|_{\tilde{U}=U} \mathcal{D}_{\tilde{W}^*} \hat{U}^*|_{\tilde{W}=W} \mathcal{D}_x F^*, \tag{6.260}$$

$$\mathcal{D}_z h(x, z, z^*) = \mathcal{D}_{\tilde{U}} g(\tilde{U}, \tilde{U}^*)|_{\tilde{U}=U} \mathcal{D}_{\tilde{W}} \hat{U}|_{\tilde{W}=W} \mathcal{D}_z F + \mathcal{D}_{\tilde{U}^*} g(\tilde{U}, \tilde{U}^*)|_{\tilde{U}=U} \mathcal{D}_{\tilde{W}^*} \hat{U}^*|_{\tilde{W}=W} \mathcal{D}_z F^*, \tag{6.261}$$

$$\mathcal{D}_{z^*} h(x, z, z^*) = \mathcal{D}_{\tilde{U}} g(\tilde{U}, \tilde{U}^*)|_{\tilde{U}=U} \mathcal{D}_{\tilde{W}} \hat{U}|_{\tilde{W}=W} \mathcal{D}_{z^*} F + \mathcal{D}_{\tilde{U}^*} g(\tilde{U}, \tilde{U}^*)|_{\tilde{U}=U} \mathcal{D}_{\tilde{W}^*} \hat{U}^*|_{\tilde{W}=W} \mathcal{D}_{z^*} F^*, \tag{6.262}$$

where $\mathcal{D}_{\tilde{U}} g(\tilde{U}, \tilde{U}^*)|_{\tilde{U}=U}$ and $\mathcal{D}_{\tilde{U}^*} g(\tilde{U}, \tilde{U}^*)|_{\tilde{U}=U}$ must be found for the function under consideration, and $\mathcal{D}_{\tilde{W}} \hat{U}|_{\tilde{W}=W}$ and $\mathcal{D}_{\tilde{W}^*} \hat{U}^*|_{\tilde{W}=W}$ are found in (6.255) and (6.258), respectively. The derivatives of $F(x, z, z^*)$ and $F^*(x, z, z^*)$ with respect to x, z, and z^* are found in Example 6.29. To use the steepest descent method, the results in (6.260) and (6.262) can be used in (6.40).

More information about unitary matrix optimization can be found in Abrudan, Eriksson, and Koivunen (2008), and Manton (2002) and the references therein.

6.5.10 Positive Semidefinite Matrices

Example 6.32 (Positive Semidefinite) Let $\mathcal{W} = \{S \in \mathbb{C}^{N\times N} \mid S \succeq 0_{N\times N}\}$, where the notation $S \succeq 0_{N\times N}$ means that S is positive semidefinite. If $S \in \mathcal{W} \subset \mathbb{C}^{N\times N}$ is positive semidefinite, then S is Hermitian such that $S^H = S$ and its eigenvalues are non-negative. Define the set $\mathcal{L}$ as

$$\mathcal{L} \triangleq \left\{ L \in \mathbb{C}^{N\times N} \mid \operatorname{vec}(L) = L_d x + L_l z, x \in \{\mathbb{R}^+ \cup \{0\}\}^{N\times 1}, z \in \mathbb{C}^{\frac{(N-1)N}{2}\times 1} \right\}. \tag{6.263}$$

One parameterization of an arbitrary positive semidefinite matrix is the Cholesky decomposition (Barry, Lee, & Messerschmitt 2004, p. 506):

$$S = LL^H, \tag{6.264}$$

where $L \in \mathcal{L} \subset \mathbb{C}^{N\times N}$ is a lower triangular matrix with real non-negative elements on its main diagonal, and independent complex-valued elements below the main diagonal. Therefore, one way to generate L is with the function $F : \{\mathbb{R}^+ \cup \{0\}\}^{N\times 1} \times \mathbb{C}^{\frac{(N-1)N}{2}\times 1} \to \mathcal{L}$ denoted by $F(x, z)$, where $x \in \left\{\mathbb{R}^+ \cup \{0\}\right\}^{N\times 1}$ and $z \in \mathbb{C}^{\frac{(N-1)N}{2}\times 1}$, and $F(x, z)$ is given by

$$\operatorname{vec}(L) = \operatorname{vec}(F(x, z)) = L_d x + L_l z. \tag{6.265}$$

The Cholesky factorization is unique (Bhatia 2007, p. 2) for positive definite matrices, and the number of real dimensions for parameterizing a positive semidefinite matrix is $\dim_{\mathbb{R}}\{\{\mathbb{R}^+ \cup \{0\}\}^{N\times 1}\} + \dim_{\mathbb{R}}\{\mathbb{C}^{\frac{(N-1)N}{2}\times 1}\} = N + 2\frac{(N-1)N}{2} = N^2$.

A positive semidefinite complex-valued $N \times N$ matrix can also be factored as

$$S = U \Delta U^H, \tag{6.266}$$

where $U \in \mathbb{C}^{N \times N}$ is unitary and Δ is diagonal with non-negative elements on the main diagonal. Assuming that the two matrices U and Δ are independent of each other, the number of real variables used to parameterize S as in (6.266) is

$$\dim_{\mathbb{R}}\{\mathbb{R}^{+} \cup \{0\}\}^{N \times 1} + N + 2\frac{(N-1)N}{2} = 2N + N^2 - N = N^2 + N. \tag{6.267}$$

This decomposition does not represent parameterization with a minimum number of variables because too many input variables are used to parameterize the set of positive definite matrices when Δ and U are parameterized independently. It is seen from the Cholesky decomposition that the minimum number of real-valued parameters is N^2, which is strictly less than $N^2 + N$.

In Magnus and Neudecker (1988, pp. 316–317), it is shown how to optimize over the set of symmetric matrices in both an implicit and explicit manner. It is mentioned how to optimize over the set of positive semidefinite matrices; however, they did *not* use the Cholesky decomposition. They stated that a positive semidefinite matrix $W \in \mathbb{C}^{N \times N}$ can be parameterized by the unpatterned matrix $X \in \mathbb{C}^{N \times N}$ as follows:

$$W = X^H X, \tag{6.268}$$

where too many parameters[13] are used compared with the Cholesky decomposition because $\dim_{\mathbb{R}}\{\mathbb{C}^{N \times N}\} = 2N^2$.

6.6 Exercises

6.1 Let $a, b, c \in \mathbb{C}$ be given constants. We want to solve the scalar equation

$$az + bz^* = c, \tag{6.269}$$

where $z \in \mathbb{C}$ is the unknown variable. Show that if $|a|^2 \neq |b|^2$, the solution is given by

$$z = \frac{a^* c - b c^*}{|a|^2 - |b|^2}. \tag{6.270}$$

Show that if $|a|^2 = |b|^2$, then (6.269) might have no solution or infinitely many solutions.

6.2 Show that w stated in (6.126) satisfies

$$J_2 w = w^*. \tag{6.271}$$

[13] If the positive scalar 1 (which is a special case of a positive definite 1×1 matrix) is decomposed by the Cholesky factorization, then it is written as $1 = 1 \cdot 1$, such that the Cholesky factor L (which is denoted l here because it is a scalar) is given by $l = 1$, and this is a unique factorization. However, if the decomposition in (6.268) is used, then $1 = e^{-\jmath\theta} e^{\jmath\theta}$ for any $\theta \in \mathbb{R}$, such that $x = e^{\jmath\theta}$ (where the symbol x is used instead of X because it is a scalar). Therefore, in (6.268), the decomposition is not unique.

6.3 Define the set $\mathcal{W}$ by

$$\mathcal{W} = \left\{ \boldsymbol{w} \in \mathbb{C}^{2N+1} \;\middle|\; \boldsymbol{w} = \begin{bmatrix} \boldsymbol{z} \\ x \\ \boldsymbol{J}_N \boldsymbol{z}^* \end{bmatrix}, \boldsymbol{z} \in \mathbb{C}^{N \times 1},\; x \in \mathbb{R} \right\} \subset \mathbb{C}^{(2N+1)\times 1}. \quad (6.272)$$

It is observed that if $\boldsymbol{w} \in \mathcal{W}$, then $\boldsymbol{J}_{2N+1}\boldsymbol{w}^* = \boldsymbol{w}$. Hence, the set $\mathcal{W}$ can be interpreted as FIR filters that are equal to their own time reverse complex conjugation. These filters are important in signal processing and communications.

Let $g : \mathbb{C}^{(2N+1)\times 1} \times \mathbb{C}^{(2N+1)\times 1} \to \mathbb{R}$ be given by

$$g(\tilde{\boldsymbol{w}}, \tilde{\boldsymbol{w}}^*) = \|\boldsymbol{A}\tilde{\boldsymbol{w}} - \boldsymbol{b}\|^2, \quad (6.273)$$

where $\boldsymbol{A} \in \mathbb{C}^{M\times(2N+1)}$ and $\boldsymbol{b} \in \mathbb{C}^{M\times 1}$, and where it is assumed that $\operatorname{rank}(\boldsymbol{A}) = 2N+1$. Show by solving $\mathcal{D}_{\tilde{\boldsymbol{w}}^*} g = \boldsymbol{0}_{1\times(2N+1)}$, that the optimal solution of the unconstrained optimization problem $\min\limits_{\tilde{\boldsymbol{w}} \in \mathbb{C}^{(2N+1)\times 1}} g(\tilde{\boldsymbol{w}}, \tilde{\boldsymbol{w}}^*)$ is given by

$$\tilde{\boldsymbol{w}} = \left[\boldsymbol{A}^H \boldsymbol{A}\right]^{-1} \boldsymbol{A}^H \boldsymbol{b} = \boldsymbol{A}^+ \boldsymbol{b}. \quad (6.274)$$

A possible parameterization function for $\mathcal{W}$ is defined by $\boldsymbol{f} : \mathbb{R} \times \mathbb{C}^{N\times 1} \times \mathbb{C}^{N\times 1} \to \mathcal{W}$ and is given by

$$\boldsymbol{f}(x, \boldsymbol{z}, \boldsymbol{z}^*) = \begin{bmatrix} \boldsymbol{z} \\ x \\ \boldsymbol{J}_N \boldsymbol{z}^* \end{bmatrix}, \quad (6.275)$$

where $x \in \mathbb{R}$ and $\boldsymbol{z} \in \mathbb{C}^{N\times 1}$.

By using generalized complex-valued vector derivatives, show that the solution of the constrained optimization problem $\min\limits_{\boldsymbol{w} \in \mathcal{W}} g(\boldsymbol{w}, \boldsymbol{w}^*)$ is given by

$$\boldsymbol{w} = \left[\boldsymbol{A}^T \boldsymbol{A}^* \boldsymbol{J}_{2N+1} + \boldsymbol{J}_{2N+1} \boldsymbol{A}^H \boldsymbol{A}\right]^{-1} \left(\boldsymbol{A}^T \boldsymbol{b}^* + \boldsymbol{J}_{2N+1} \boldsymbol{A}^H \boldsymbol{b}\right). \quad (6.276)$$

Show that for the solution in (6.276), $\boldsymbol{J}_{2N+1}\boldsymbol{w}^* = \boldsymbol{w}$ is satisfied.

6.4 Let $\mathcal{W} = \{\boldsymbol{W} \in \mathbb{C}^{N\times N} \mid \boldsymbol{W}^T = \boldsymbol{W}\}$ with the parameterization function given in (6.155). Assume that the function $g : \mathbb{C}^{N\times N} \times \mathbb{C}^{N\times N} \to \mathbb{C}$ is denoted by $g(\tilde{\boldsymbol{W}}, \tilde{\boldsymbol{W}}^*)$, and that the two derivatives $\mathcal{D}_{\tilde{\boldsymbol{W}}} g = \operatorname{vec}^T\left(\frac{\partial g}{\partial \tilde{\boldsymbol{W}}}\right)$ and $\mathcal{D}_{\tilde{\boldsymbol{W}}^*} g = \operatorname{vec}^T\left(\frac{\partial g}{\partial \tilde{\boldsymbol{W}}^*}\right)$ are available, where $\tilde{\boldsymbol{W}} \in \mathbb{C}^{N\times N}$ is unpatterned. Show that the partial derivative of g with respect to the patterned matrix $\boldsymbol{W} \in \mathcal{W}$ can be expressed as

$$\frac{\partial g}{\partial \boldsymbol{W}} = \left[\frac{\partial g}{\partial \tilde{\boldsymbol{W}}} + \left(\frac{\partial g}{\partial \tilde{\boldsymbol{W}}}\right)^T - \boldsymbol{I}_N \odot \frac{\partial g}{\partial \tilde{\boldsymbol{W}}}\right]_{\tilde{\boldsymbol{W}} = \boldsymbol{W}}. \quad (6.277)$$

As an example, set $g(\tilde{\boldsymbol{W}}, \tilde{\boldsymbol{W}}^*) = \det\left(\tilde{\boldsymbol{W}}\right)$. Use (6.277) to show that

$$\frac{\partial g}{\partial \boldsymbol{W}} = 2\boldsymbol{C}(\boldsymbol{W}) - \boldsymbol{I}_N \odot \boldsymbol{C}(\boldsymbol{W}), \quad (6.278)$$

where $\boldsymbol{C}(\boldsymbol{W})$ contains the cofactors of $\boldsymbol{C}$.

For a common example in signal processing, let $g(\tilde{W}, \tilde{W}^*) = \mathrm{Tr}\left\{\tilde{W}^T \tilde{W}\right\} = \|\tilde{W}\|_F^2$, where $\tilde{W} \in \mathbb{C}^{N\times N}$ is unpatterned. Use (6.277) to show that

$$\frac{\partial g}{\partial W} = 4W - 2I_N \odot W. \tag{6.279}$$

6.5 Let $T \in \mathbb{C}^{N\times N}$ be a Toeplitz matrix (Jain 1989, p. 25) that is completely defined by its $2N-1$ elements in the first column and row. Toeplitz matrices are characterized by having the same element along the *diagonals*. The $N \times N$ complex-valued Toeplitz matrix T can be expressed by

$$T = \begin{bmatrix} z_0 & z_{-1} & \cdots & \cdots & z_{-(N-1)} \\ z_1 & z_0 & z_{-1} & \cdots & z_{-(N-2)} \\ z_2 & z_1 & z_0 & \cdots & z_{-(N-3)} \\ \vdots & \vdots & \ddots & \ddots & \vdots \\ z_{N-1} & z_{N-2} & \cdots & z_1 & z_0 \end{bmatrix}, \tag{6.280}$$

where $z_k \in \mathbb{C}$ and $k \in \{0, 1, \ldots, N-1\}$. Let the set of all such $N \times N$ Toeplitz matrices be denoted by $\mathcal{T}$.

One parameterization function for $\mathcal{T}$ is $F : \mathbb{C}^{(2N-1)\times 1} \to \mathcal{T} \subseteq \mathbb{C}^{N\times N}$, given by

$$T = F(z) = \sum_{k=-(N-1)}^{N-1} z_k I_N^{(k)}, \tag{6.281}$$

where $z \in \mathbb{C}^{(2N-1)\times 1}$ contains the $2N-1$ independent complex parameters given by

$$z = [z_{N-1}, z_{N-2}, \ldots, z_1, z_0, z_{-1}, \ldots, z_{-(N-1)}]^T, \tag{6.282}$$

and $I_N^{(k)}$ is defined as the $N \times N$ matrix with zeros everywhere except for $+1$ along the k-th diagonal, where the diagonals are numbered from $N-1$ for the lower diagonal and $-(N-1)$ for the upper diagonal. In this way, the main diagonal is numbered as 0, such that $I_N^{(0)} = I_N$.

Show that the derivative of $T = F(z)$ with respect to z is given by

$$\mathcal{D}_z F = \left[\mathrm{vec}\left(I_N^{(N-1)}\right), \ \mathrm{vec}\left(I_N^{(N-2)}\right), \ \ldots \mathrm{vec}\left(I_N^{(-(N-1))}\right)\right], \tag{6.283}$$

which has size $N^2 \times (2N-1)$.

6.6 Let $T \in \mathbb{C}^{N\times N}$ be a Hermitian Toeplitz matrix that is completely defined by all the N elements in the first column, where the main diagonal contains a real-valued element and the off-diagonal elements are complex-valued. The $N \times N$ complex-valued Hermitian Toeplitz matrix T can be expressed as

$$T = \begin{bmatrix} x & z_1^* & \cdots & \cdots & z_{N-1}^* \\ z_1 & x & z_1^* & \cdots & z_{N-2}^* \\ z_2 & z_1 & x & \cdots & z_{N-3}^* \\ \vdots & \vdots & \ddots & \ddots & \vdots \\ z_{N-1} & z_{N-2} & \cdots & z_1 & x \end{bmatrix}, \tag{6.284}$$

where $x \in \mathbb{R}$ and $z_k \in \mathbb{C}$ for $k \in \{1, 2, \ldots, N-1\}$. Let the set of all such $N \times N$ Hermitian Toeplitz matrices be denoted by $\mathcal{T}$.

A parameterization function for the set of Hermitian Toeplitz matrices $\mathcal{T}$ is $\boldsymbol{F} : \mathbb{R} \times \mathbb{C}^{(N-1)\times 1} \times \mathbb{C}^{(N-1)\times 1} \to \mathcal{T} \subset \mathbb{C}^{N\times N}$, given by

$$\boldsymbol{T} = \boldsymbol{F}(x, \boldsymbol{z}, \boldsymbol{z}^*) = x\boldsymbol{I}_N + \sum_{k=1}^{N-1} z_k \boldsymbol{I}_N^{(k)} + \sum_{k=1}^{N-1} z_k^* \boldsymbol{I}_N^{(-k)}, \tag{6.285}$$

where $\boldsymbol{z} \in \mathbb{C}^{(N-1)\times 1}$ contains the $N-1$ independent complex parameters given by

$$\boldsymbol{z} = [z_1, z_2, \ldots, z_{N-1}]^T, \tag{6.286}$$

and where $\boldsymbol{I}_N^{(k)}$ is defined as in Exercise 6.5. Show that the derivatives of $\boldsymbol{F}(x, \boldsymbol{z}, \boldsymbol{z}^*)$ with respect to x, $\boldsymbol{z}$, and $\boldsymbol{z}^*$ are given by

$$\mathcal{D}_x \boldsymbol{F} = \operatorname{vec}(\boldsymbol{I}_N), \tag{6.287}$$

$$\mathcal{D}_{\boldsymbol{z}} \boldsymbol{F} = \left[\operatorname{vec}\left(\boldsymbol{I}_N^{(1)}\right), \operatorname{vec}\left(\boldsymbol{I}_N^{(2)}\right), \ldots, \operatorname{vec}\left(\boldsymbol{I}_N^{(N-1)}\right)\right], \tag{6.288}$$

$$\mathcal{D}_{\boldsymbol{z}^*} \boldsymbol{F} = \left[\operatorname{vec}\left(\boldsymbol{I}_N^{(-1)}\right), \operatorname{vec}\left(\boldsymbol{I}_N^{(-2)}\right), \ldots, \operatorname{vec}\left(\boldsymbol{I}_N^{(-(N-1))}\right)\right], \tag{6.289}$$

respectively, of sizes $N^2 \times 1$, $N^2 \times (N-1)$, and $N^2 \times (N-1)$.

6.7 Let $\boldsymbol{C} \in \mathbb{C}^{N\times N}$ be a circulant matrix (Gray 2006), that is, row $i+1$ is found by circularly shifting row i one position to the right, where the last element of row i becomes the first element of row $i+1$. The $N \times N$ circulant matrix $\boldsymbol{C}$ can be expressed as

$$\boldsymbol{C} = \begin{bmatrix} z_0 & z_1 & \cdots & \cdots & z_{N-1} \\ z_{N-1} & z_0 & z_1 & \cdots & z_{N-2} \\ z_{N-2} & z_{N-1} & z_0 & \cdots & z_{N-3} \\ \vdots & \vdots & \ddots & \ddots & \vdots \\ z_1 & z_2 & \cdots & z_{N-1} & z_0 \end{bmatrix}, \tag{6.290}$$

where $z_k \in \mathbb{C}$ for all $k \in \{0, 1, \ldots, N-1\}$. Let the set of all such matrices of size $N \times N$ be denoted $\mathcal{C}$. Circulant matrices are used in signal processing and communications, for example, when calculating the discrete Fourier transform and when working on cyclic error correcting codes (Gray 2006).

Let the *primary circular matrix* (Bernstein 2005, p. 213) be denoted by $\boldsymbol{P}_N$; it has size $N \times N$ with zeros everywhere except for ones on the diagonal just above the main diagonal and on the lower left diagonal. As an example for $N = 4$, then $\boldsymbol{P}_N$ is given by

$$\boldsymbol{P}_4 = \begin{bmatrix} 0 & 1 & 0 & 0 \\ 0 & 0 & 1 & 0 \\ 0 & 0 & 0 & 1 \\ 1 & 0 & 0 & 0 \end{bmatrix}. \tag{6.291}$$

Let the transpose of the first row of the circulant matrix $\boldsymbol{C}$ be given by the $N \times 1$ vector $\boldsymbol{z} = [z_0, z_1, \ldots, z_{N-1}]^T$, where $z_k \in \mathbb{C}$, such that $\boldsymbol{z} \in \mathbb{C}^{N\times 1}$. A parameterization function $\boldsymbol{F} : \mathbb{C}^{N\times 1} \to \mathcal{C} \subseteq \mathbb{C}^{N\times N}$ for generating the circulant matrices in $\mathcal{C}$ is given by

$$\boldsymbol{C} = \boldsymbol{F}(\boldsymbol{z}) = \sum_{k=0}^{N-1} z_k \boldsymbol{P}_N^k, \tag{6.292}$$

because the matrix $\boldsymbol{P}_N^k$ contains 0s everywhere except on the k-th diagonal above the main diagonal and the $(N-k)$-th diagonal below the main diagonal. Notice that $\boldsymbol{P}_N^N = \boldsymbol{P}_N^0 = \boldsymbol{I}_N$.

Show that the derivative of $\boldsymbol{C} = \boldsymbol{F}(\boldsymbol{z})$ with respect to the vector $\boldsymbol{z}$ can be expressed as

$$\mathcal{D}_{\boldsymbol{z}}\boldsymbol{F} = \left[\operatorname{vec}(\boldsymbol{I}_N),\ \operatorname{vec}\left(\boldsymbol{P}_N^1\right), \ldots, \operatorname{vec}\left(\boldsymbol{P}_N^{N-1}\right)\right], \tag{6.293}$$

where $\mathcal{D}_{\boldsymbol{z}}\boldsymbol{F}$ has size $N^2 \times N$.

As an application for how to use the results derived in this exercise, consider the problem of finding the closest circulant matrix to an arbitrary matrix $\boldsymbol{C}_0 \in \mathbb{C}^{N\times N}$. This problem can be formulated as

$$\boldsymbol{z} = \underset{\{\boldsymbol{z}\in\mathbb{C}^{N\times 1}\}}{\operatorname{argmin}} \|\boldsymbol{F}(\boldsymbol{z}) - \boldsymbol{C}_0\|_F^2, \tag{6.294}$$

where $\|\boldsymbol{W}\|_F^2 \triangleq \operatorname{Tr}\left\{\boldsymbol{W}\boldsymbol{W}^H\right\}$ denotes the squared Frobenius norm (Bernstein 2005, p. 348) and argmin returns the argument which minimizes the expression stated after argmin. By using generalized complex-valued matrix derivatives, find the necessary conditions for optimality of the problem given in (6.294), and show that the solution of this problem is found as

$$\boldsymbol{z} = \frac{1}{N}\left(\mathcal{D}_{\boldsymbol{z}}\boldsymbol{F}\right)^T \operatorname{vec}(\boldsymbol{C}_0), \tag{6.295}$$

where $\mathcal{D}_{\boldsymbol{z}}\boldsymbol{F}$ is given in (6.293). If the value of $\boldsymbol{z}$ given in (6.295) is used in (6.290), the closest circulant matrix to $\boldsymbol{C}_0$ is found.

6.8 Let $\boldsymbol{H} \in \mathbb{C}^{N\times N}$ be a Hankel matrix (Bernstein 2005, p. 83) that is completely defined by all the $2N-1$ elements in the *first row* and the *last column*. Hankel matrices contain the same elements along the skew-diagonals, that is, $(\boldsymbol{H})_{i,j} = (\boldsymbol{H})_{k,l}$ for all $i + j = k + l$. An $N \times N$ Hankel matrix can be expressed as

$$\boldsymbol{H} = \begin{bmatrix} z_{N-1} & z_{N-2} & z_{N-3} & \cdots & z_0 \\ z_{N-2} & z_{N-3} & z_{N-4} & \iddots & z_{-1} \\ z_{N-3} & z_{N-4} & z_{N-5} & \iddots & z_{-2} \\ \vdots & \iddots & \iddots & \iddots & \vdots \\ z_0 & z_{-1} & z_{-2} & \cdots & z_{-(N-1)} \end{bmatrix}, \tag{6.296}$$

where $z_k \in \mathbb{C}$. Let the set of all $N \times N$ complex-valued Hankel matrices be denoted by $\mathcal{H}$.

A possible parameterization function for producing all Hankel matrices in $\mathcal{H}$ is $\boldsymbol{F}: \mathbb{C}^{(2N-1)\times 1} \rightarrow \mathcal{H} \subseteq \mathbb{C}^{N\times N}$ given by

$$\boldsymbol{H} = \boldsymbol{F}(\boldsymbol{z}) = \sum_{k=-(N-1)}^{N-1} z_k \boldsymbol{J}_N^{(k)}, \tag{6.297}$$

where $\boldsymbol{z} \in \mathbb{C}^{(2N-1)\times 1}$ contains the $2N-1$ independent complex parameters given by

$$\boldsymbol{z} = [z_{N-1}, z_{N-2}, \ldots, z_1, z_0, z_{-1}, \ldots, z_{-(N-1)}]^T, \tag{6.298}$$

and where $\boldsymbol{J}_N^{(k)}$ has size $N \times N$ with zeros everywhere except for $+1$ along the k-th reverse diagonal, where the diagonals are numbered from $N-1$ for the left upper reverse diagonal to $-(N-1)$ for the right lower reverse diagonal. In this way, the reverse identity matrix is numbered 0, such that $\boldsymbol{J}_N^{(0)} = \boldsymbol{J}_N$.

Show that the derivatives of $\boldsymbol{F}(\boldsymbol{z})$ with respect to $\boldsymbol{z}$ can be expressed as

$$\mathcal{D}_{\boldsymbol{z}}\boldsymbol{F} = \left[\operatorname{vec}\left(\boldsymbol{J}_N^{(N-1)}\right), \operatorname{vec}\left(\boldsymbol{J}_N^{(N-2)}\right), \ldots, \operatorname{vec}\left(\boldsymbol{J}_N^{(-(N-1))}\right)\right], \tag{6.299}$$

which has size $N^2 \times (2N-1)$.

6.9 Let $\boldsymbol{V} \in \mathbb{C}^{N\times N}$ be a *Vandermonde matrix* (Horn & Johnson 1985, p. 29). An arbitrary $N \times N$ complex-valued Vandermonde matrix $\boldsymbol{V}$ can be expressed as

$$\boldsymbol{V} = \begin{bmatrix} 1 & z_0 & z_0^2 & \cdots & z_0^{N-1} \\ 1 & z_1 & z_1^2 & \cdots & z_1^{N-1} \\ \vdots & \vdots & \vdots & & \vdots \\ 1 & z_{N-1} & z_{N-1}^2 & \cdots & z_{N-1}^{N-1} \end{bmatrix}, \tag{6.300}$$

where $z_k \in \mathbb{C}$ for all $k \in \{0, 1, \ldots, N-1\}$. Let the set of all such $N \times N$ Vandermonde matrices be denoted by $\mathcal{V}$.

Define $\boldsymbol{z} \in \mathbb{C}^{N\times 1}$ to be the vector given by $\boldsymbol{z} = [z_0, z_1, \ldots, z_{N-1}]^T$. A parameterization function $\boldsymbol{F}: \mathbb{C}^{N\times 1} \rightarrow \mathcal{V} \subset \mathbb{C}^{N\times N}$ for generating the complex-valued Vandermonde matrix $\boldsymbol{V}$ is

$$\boldsymbol{V} = \boldsymbol{F}(\boldsymbol{z}) = \left[\boldsymbol{1}_{N\times 1}, \boldsymbol{z}^{\odot 1}, \boldsymbol{z}^{\odot 2}, \ldots, \boldsymbol{z}^{\odot(N-1)}\right], \tag{6.301}$$

where the special notation $\boldsymbol{A}^{\odot k}$ is defined as $\boldsymbol{A}^{\odot k} \triangleq \boldsymbol{A} \odot \boldsymbol{A} \odot \ldots \odot \boldsymbol{A}$, where $\boldsymbol{A}$ appears k times on the right-hand side. If $\boldsymbol{A} \in \mathbb{C}^{M\times N}$, then $\boldsymbol{A}^{\odot 1} = \boldsymbol{A}$ and $\boldsymbol{A}^{\odot 0} \triangleq \boldsymbol{1}_{M\times N}$, even when $\boldsymbol{A}$ contains zeros.

Show that the derivative of the parameterization function $\boldsymbol{F}$ with respect to $\boldsymbol{z}$ is given by

$$\mathcal{D}_{\boldsymbol{z}}\boldsymbol{F} = \begin{bmatrix} \boldsymbol{0}_{N\times N} \\ \boldsymbol{I}_N \\ 2\operatorname{diag}(\boldsymbol{z}) \\ 3\operatorname{diag}^2(\boldsymbol{z}) \\ \vdots \\ (N-1)\operatorname{diag}^{N-2}(\boldsymbol{z}) \end{bmatrix}, \tag{6.302}$$

where $\text{diag}^k(z) = (\text{diag}(z))^k$ and the operator diag(·) is defined in Definition 2.10. The size of $\mathcal{D}_z F$ is $N^2 \times N$.

6.10 Consider the following capacity function C for Gaussian complex-valued MIMO channels (Telatar 1995):

$$C = \ln\left(\det\left(\boldsymbol{I}_{M_r} + \boldsymbol{H}\boldsymbol{W}\boldsymbol{H}^H\right)\right), \tag{6.303}$$

where $\boldsymbol{H} \in \mathbb{C}^{M_r \times M_t}$ is a channel transfer matrix that is independent of the autocorrelation matrix $\boldsymbol{W} \in \mathbb{C}^{M_t \times M_t}$. The matrix $\boldsymbol{W}$ is a correlation matrix; thus, it is a positive semidefinite matrix. The channel matrix $\boldsymbol{H}$ can be expressed with its singular value decomposition (SVD) as

$$\boldsymbol{H} = \boldsymbol{U}\boldsymbol{\Sigma}\boldsymbol{V}^H, \tag{6.304}$$

where the matrices $\boldsymbol{U} \in \mathbb{C}^{M_r \times M_r}$ and $\boldsymbol{V} \in \mathbb{C}^{M_t \times M_t}$ are unitary matrices while the matrix $\boldsymbol{\Sigma} \in \mathbb{C}^{M_r \times M_t}$ is diagonal and contains the singular values $\sigma_i \geq 0$ in decreasing order on its main diagonal, that is, $(\boldsymbol{\Sigma})_{i,i} = \sigma_i$, and $\sigma_i \geq \sigma_j$ when $i \leq j$.

The capacity function should be minimized over the set of autocorrelation matrices $\boldsymbol{W}$ satisfying

$$\text{Tr}\{\boldsymbol{W}\} = \rho, \tag{6.305}$$

where $\rho > 0$ is a given constant indicating the transmitted power per sent vector. By using Hadamard's inequality in Lemma 2.1, show that the autocorrelation matrix $\boldsymbol{W}$ that is maximizing the capacity is given by

$$\boldsymbol{W} = \boldsymbol{V}\boldsymbol{\Lambda}\boldsymbol{V}^H, \tag{6.306}$$

where $\boldsymbol{\Lambda} \in \mathbb{C}^{M_t \times M_t}$ is a diagonal matrix with non-negative diagonal elements satisfying

$$\text{Tr}\{\boldsymbol{\Lambda}\} = \rho. \tag{6.307}$$

Let $i \in \{0, 1, \ldots, \min\{M_r, M_t\} - 1\}$, and let the i-th diagonal element of $\boldsymbol{\Lambda}$ be denoted by λ_i. Show, by using the positive Lagrange multiplier μ, that the diagonal elements of $\boldsymbol{\Lambda}$ maximizing the capacity can be expressed as

$$\lambda_i = \max\left(0, \frac{1}{\mu} - \frac{1}{\sigma_i^2}\right). \tag{6.308}$$

The Lagrange multiplier $\mu > 0$ should be chosen in such a way that the power constraint in (6.305) is satisfied with equality. The solution in (6.308) is called the *water-filling solution* (Telatar 1995).

6.11 Consider the MIMO communication system illustrated in Figure 6.4. The zero-mean transmitted vector is denoted by $z \in \mathbb{C}^{M_t \times 1}$, and its autocorrelation matrix $\boldsymbol{\Sigma}_z \in \mathbb{C}^{M_t \times M_t}$ is given by

$$\boldsymbol{\Sigma}_z = \mathbb{E}\left[z z^H\right] = \boldsymbol{B}\boldsymbol{\Sigma}_x\boldsymbol{B}^H, \tag{6.309}$$

where $\boldsymbol{B} \in \mathbb{C}^{M_t \times N}$ represents the precoder matrix, and $\boldsymbol{\Sigma}_x \in \mathbb{C}^{N \times N}$ is the autocorrelation matrix of the original signal vector $\boldsymbol{x} \in \mathbb{C}^{N \times 1}$.

Let the set of all Hermitian $M_t \times M_t$ matrices be given by

$$\mathcal{W} = \{\boldsymbol{W} \in \mathbb{C}^{M_t \times M_t} \mid \boldsymbol{W}^H = \boldsymbol{W}\} \subset \mathbb{C}^{M_t \times M_t}. \tag{6.310}$$

The set $\mathcal{W}$ is a manifold of Hermitian $M_t \times M_t$ matrices. Let $\tilde{\boldsymbol{W}} \in \mathbb{C}^{M_t \times M_t}$ be a matrix with complex-valued independent components, such that $\tilde{\boldsymbol{W}}$ is an unpatterned version of $\boldsymbol{W} \in \mathcal{W}$. In this exercise, as a simplification, the Hermitian matrix $\boldsymbol{W}$ is used to represent the autocorrelation matrix $\boldsymbol{\Sigma}_{\boldsymbol{x}}$, even though $\boldsymbol{\Sigma}_{\boldsymbol{x}}$ is positive semidefinite. Hence, we represent $\boldsymbol{\Sigma}_{\boldsymbol{x}}$ with the Hermitian matrix $\boldsymbol{W}$. This is a simplification since the set of Hermitian matrices $\mathcal{W}$ is larger than the set of positive semidefinite matrices.

Let $g : \mathbb{C}^{M_t \times M_t} \times \mathbb{C}^{M_t \times M_t} \to \mathbb{C}$ be given as

$$g(\tilde{\boldsymbol{W}}, \tilde{\boldsymbol{W}}^*) = \ln\left(\det\left(\boldsymbol{I}_{M_r} + \boldsymbol{H}\tilde{\boldsymbol{W}}\boldsymbol{H}^H\boldsymbol{\Sigma}_{\boldsymbol{n}}^{-1}\right)\right). \tag{6.311}$$

This means that $g(\tilde{\boldsymbol{W}}, \tilde{\boldsymbol{W}}^*)$ has a similar shape as the mutual information; in (6.211); however, the autocorrelation matrix $\boldsymbol{\Sigma}_{\boldsymbol{z}} = \boldsymbol{B}\boldsymbol{\Sigma}_{\boldsymbol{x}}\boldsymbol{B}^H$ is replaced by the unpatterned matrix $\tilde{\boldsymbol{W}}$. For an arbitrary unpatterned matrix $\tilde{\boldsymbol{W}}$, the function $g(\tilde{\boldsymbol{W}}, \tilde{\boldsymbol{W}}^*)$ is complex valued in general. Assume that the matrices $\boldsymbol{H}$ and $\boldsymbol{\Sigma}_{\boldsymbol{n}}^{-1}$ are independent of $\tilde{\boldsymbol{W}}$, $\boldsymbol{W}$, and their complex conjugates.

Show that the derivatives of g with respect to $\tilde{\boldsymbol{W}}$ and $\tilde{\boldsymbol{W}}^*$ are given by

$$\mathcal{D}_{\tilde{\boldsymbol{W}}} g = \operatorname{vec}^T\left(\boldsymbol{H}^T\left(\boldsymbol{I}_{M_r} + \boldsymbol{\Sigma}_{\boldsymbol{n}}^{-T}\boldsymbol{H}^*\tilde{\boldsymbol{W}}^T\boldsymbol{H}^T\right)^{-1}\boldsymbol{\Sigma}_{\boldsymbol{n}}^{-T}\boldsymbol{H}^*\right), \tag{6.312}$$

and

$$\mathcal{D}_{\tilde{\boldsymbol{W}}^*} g = \boldsymbol{0}_{1 \times M_t^2}, \tag{6.313}$$

respectively.

Use the results from Example 6.25 to show that when expressed in the standard basis, the generalized matrix derivatives of g with respect to $\boldsymbol{W}$ and $\boldsymbol{W}^*$ are given by

$$\frac{\partial}{\partial \boldsymbol{W}} g = \boldsymbol{H}^T\left(\boldsymbol{I}_{M_r} + \boldsymbol{\Sigma}_{\boldsymbol{n}}^{-T}\boldsymbol{H}^*\boldsymbol{W}^T\boldsymbol{H}^T\right)^{-1}\boldsymbol{\Sigma}_{\boldsymbol{n}}^{-T}\boldsymbol{H}^*, \tag{6.314}$$

$$\frac{\partial}{\partial \boldsymbol{W}^*} g = \boldsymbol{H}^H\boldsymbol{\Sigma}_{\boldsymbol{n}}^{-1}\left(\boldsymbol{I}_{M_r} + \boldsymbol{H}\boldsymbol{W}\boldsymbol{H}^H\boldsymbol{\Sigma}_{\boldsymbol{n}}^{-1}\right)^{-1}\boldsymbol{H}. \tag{6.315}$$

Show that

$$\left[\frac{\partial}{\partial \boldsymbol{W}^*} g\right]_{\boldsymbol{W}=\boldsymbol{B}\boldsymbol{\Sigma}_{\boldsymbol{x}}\boldsymbol{B}^H} \boldsymbol{B}\boldsymbol{\Sigma}_{\boldsymbol{x}} = \boldsymbol{H}^H\boldsymbol{\Sigma}_{\boldsymbol{n}}^{-1}\boldsymbol{H}\boldsymbol{B}\left(\boldsymbol{\Sigma}_{\boldsymbol{x}}^{-1} + \boldsymbol{B}^H\boldsymbol{H}^H\boldsymbol{\Sigma}_{\boldsymbol{n}}^{-1}\boldsymbol{H}\boldsymbol{B}\right)^{-1}. \tag{6.316}$$

Explain why this is in agreement with Palomar and Verdú (2006, Eq. (23)).

6.12 Assume that the function $\boldsymbol{F} : \mathbb{C}^{N \times Q} \times \mathbb{C}^{N \times Q} \to \mathbb{C}^{M \times M}$ is given by

$$\boldsymbol{F}(\boldsymbol{Z}, \boldsymbol{Z}^*) = \boldsymbol{A}\boldsymbol{Z}\boldsymbol{C}\boldsymbol{C}^H\boldsymbol{Z}^H\boldsymbol{A}^H \triangleq \boldsymbol{W}, \tag{6.317}$$

where the matrix $\boldsymbol{W}$ was defined in the last equality, and the two matrices $\boldsymbol{A} \in \mathbb{C}^{M \times N}$ and $\boldsymbol{C} \in \mathbb{C}^{Q \times P}$ are independent of the matrix variables $\boldsymbol{Z}$ and $\boldsymbol{Z}^*$. Show that the derivatives of $\boldsymbol{F}$ and $\boldsymbol{F}^*$ with respect to $\boldsymbol{Z}$ and $\boldsymbol{Z}^*$ are given by

$$\mathcal{D}_{\boldsymbol{Z}}\boldsymbol{F} = \left(\boldsymbol{A}^*\boldsymbol{Z}^*\boldsymbol{C}^*\boldsymbol{C}^T\right) \otimes \boldsymbol{A}, \tag{6.318}$$

$$\mathcal{D}_{\boldsymbol{Z}^*}\boldsymbol{F} = \boldsymbol{K}_{M,M}\left[\left(\boldsymbol{A}\boldsymbol{Z}\boldsymbol{C}\boldsymbol{C}^H\right) \otimes \boldsymbol{A}^*\right], \tag{6.319}$$

$$\mathcal{D}_{\boldsymbol{Z}}\boldsymbol{F}^* = \boldsymbol{K}_{M,M}\left[\left(\boldsymbol{A}^*\boldsymbol{Z}^*\boldsymbol{C}^*\boldsymbol{C}^T\right) \otimes \boldsymbol{A}\right], \tag{6.320}$$

$$\mathcal{D}_{\boldsymbol{Z}^*}\boldsymbol{F}^* = \left(\boldsymbol{A}\boldsymbol{Z}\boldsymbol{C}\boldsymbol{C}^H\right) \otimes \boldsymbol{A}^*. \tag{6.321}$$

Let the function $g : \mathbb{C}^{M\times M} \times \mathbb{C}^{M\times M} \to \mathbb{C}$ be denoted $g(\tilde{\boldsymbol{W}}, \tilde{\boldsymbol{W}}^*)$, where the matrix $\tilde{\boldsymbol{W}} \in \mathbb{C}^{M\times M}$ is an unpatterned version of $\boldsymbol{W}$. Assume that the two derivatives $\mathcal{D}_{\tilde{\boldsymbol{W}}}g = \operatorname{vec}^T\left(\frac{\partial g}{\partial \tilde{\boldsymbol{W}}}\right)$ and $\mathcal{D}_{\tilde{\boldsymbol{W}}^*}g = \operatorname{vec}^T\left(\frac{\partial g}{\partial \tilde{\boldsymbol{W}}^*}\right)$ are available.

Let the composed function $h : \mathbb{C}^{N\times Q} \times \mathbb{C}^{N\times Q} \to \mathbb{C}$ be given by

$$h(\boldsymbol{Z}, \boldsymbol{Z}^*) = g(\tilde{\boldsymbol{W}}, \tilde{\boldsymbol{W}}^*)\Big|_{\tilde{\boldsymbol{W}}=\boldsymbol{W}=\boldsymbol{F}(\boldsymbol{Z},\boldsymbol{Z}^*)} = g(\boldsymbol{F}(\boldsymbol{Z}, \boldsymbol{Z}^*), \boldsymbol{F}^*(\boldsymbol{Z}, \boldsymbol{Z}^*)). \tag{6.322}$$

By using the chain rule, show that the derivatives of h with respect to $\boldsymbol{Z}$ and $\boldsymbol{Z}^*$ are given by

$$\frac{\partial h}{\partial \boldsymbol{Z}} = \boldsymbol{A}^T\left(\frac{\partial g}{\partial \tilde{\boldsymbol{W}}} + \left(\frac{\partial g}{\partial \tilde{\boldsymbol{W}}^*}\right)^T\right)\boldsymbol{A}^*\boldsymbol{Z}^*\boldsymbol{C}^*\boldsymbol{C}^T, \tag{6.323}$$

$$\frac{\partial h}{\partial \boldsymbol{Z}^*} = \boldsymbol{A}^H\left[\left(\frac{\partial g}{\partial \tilde{\boldsymbol{W}}}\right)^T + \frac{\partial g}{\partial \tilde{\boldsymbol{W}}^*}\right]_{\tilde{\boldsymbol{W}}=\boldsymbol{W}}\boldsymbol{A}\boldsymbol{Z}\boldsymbol{C}\boldsymbol{C}^H. \tag{6.324}$$

6.13 Consider the MIMO system shown in Figure 6.4, and let

$$\mathcal{W} = \{\boldsymbol{W} \in \mathbb{C}^{N\times N} \mid \boldsymbol{W}^H = \boldsymbol{W}\}, \tag{6.325}$$

be the manifold of $N \times N$ Hermitian matrices. Let the matrix $\boldsymbol{W} \in \mathcal{W} \subset \mathbb{C}^{N\times N}$ be Hermitian, and let $\boldsymbol{W}$ represent $\boldsymbol{\Sigma}_{\boldsymbol{x}}$. The unpatterned version of $\boldsymbol{W}$ is denoted by $\tilde{\boldsymbol{W}} \in \mathbb{C}^{N\times N}$. Let $g : \mathbb{C}^{N\times N} \times \mathbb{C}^{N\times N} \to \mathbb{C}$ be given by the mutual information function in (6.211), where $\boldsymbol{\Sigma}_{\boldsymbol{x}}$ is replaced by $\tilde{\boldsymbol{W}}$, that is,

$$g(\tilde{\boldsymbol{W}}, \tilde{\boldsymbol{W}}^*) = \ln\left(\det\left(\boldsymbol{I}_{M_r} + \boldsymbol{H}\boldsymbol{B}\tilde{\boldsymbol{W}}\boldsymbol{B}^H\boldsymbol{H}^H\boldsymbol{\Sigma}_{\boldsymbol{n}}^{-1}\right)\right), \tag{6.326}$$

where the three matrices $\boldsymbol{\Sigma}_{\boldsymbol{n}} \in \mathbb{C}^{M_r\times M_r}$ (positive semidefinite), $\boldsymbol{H} \in \mathbb{C}^{M_r\times M_t}$, and $\boldsymbol{B} \in \mathbb{C}^{M_t\times N}$ are independent of $\tilde{\boldsymbol{W}}$ and $\tilde{\boldsymbol{W}}^*$. Notice that the function $g(\tilde{\boldsymbol{W}}, \tilde{\boldsymbol{W}}^*)$ is in general complex valued when the input matrix $\tilde{\boldsymbol{W}}$ is unpatterned. Show that the derivatives of g with respect to $\tilde{\boldsymbol{W}}$ and $\tilde{\boldsymbol{W}}^*$ are given by

$$\mathcal{D}_{\tilde{\boldsymbol{W}}}g = \operatorname{vec}^T\left(\boldsymbol{B}^T\boldsymbol{H}^T\left(\boldsymbol{I}_{M_r} + \boldsymbol{\Sigma}_{\boldsymbol{n}}^{-T}\boldsymbol{H}^*\boldsymbol{B}^*\tilde{\boldsymbol{W}}^T\boldsymbol{B}^T\boldsymbol{H}^T\right)^{-1}\boldsymbol{\Sigma}_{\boldsymbol{n}}^{-T}\boldsymbol{H}^*\boldsymbol{B}^*\right), \tag{6.327}$$

$$\mathcal{D}_{\tilde{\boldsymbol{W}}^*}g = \boldsymbol{0}_{1\times N^2}, \tag{6.328}$$

respectively.

Because $\boldsymbol{W}$ is Hermitian, the results from Example 6.25 can be utilized. Use these results to show that the generalized derivative of g with respect to $\boldsymbol{W}$ and $\boldsymbol{W}^*$ when using the standard basis can be expressed as

$$\frac{\partial g}{\partial \boldsymbol{W}} = \boldsymbol{B}^T \boldsymbol{H}^T \left(\boldsymbol{I}_{M_r} + \boldsymbol{\Sigma}_n^{-T} \boldsymbol{H}^* \boldsymbol{B}^* \boldsymbol{W}^T \boldsymbol{B}^T \boldsymbol{H}^T\right)^{-1} \boldsymbol{\Sigma}_n^{-T} \boldsymbol{H}^* \boldsymbol{B}^*, \tag{6.329}$$

$$\frac{\partial g}{\partial \boldsymbol{W}^*} = \boldsymbol{B}^H \boldsymbol{H}^H \boldsymbol{\Sigma}_n^{-1} \left(\boldsymbol{I}_{M_r} + \boldsymbol{H}\boldsymbol{B}\boldsymbol{W}\boldsymbol{B}^H \boldsymbol{H}^H \boldsymbol{\Sigma}_n^{-1}\right)^{-1} \boldsymbol{H}\boldsymbol{B}. \tag{6.330}$$

Show that

$$\left[\frac{\partial}{\partial \boldsymbol{W}^*} g\right]_{\boldsymbol{W}=\boldsymbol{\Sigma}_x} \boldsymbol{\Sigma}_x = \boldsymbol{B}^H \boldsymbol{H}^H \boldsymbol{\Sigma}_n^{-1} \boldsymbol{H}\boldsymbol{B} \left(\boldsymbol{\Sigma}_x^{-1} + \boldsymbol{B}^H \boldsymbol{H}^H \boldsymbol{\Sigma}_n^{-1} \boldsymbol{H}\boldsymbol{B}\right)^{-1}. \tag{6.331}$$

Explain why (6.331) is in agreement with Palomar and Verdú (2006, Eq. (25)).

6.14 Consider the MIMO system shown in Figure 6.4, and let the Hermitian manifold $\mathcal{W}$ of size $M_r \times M_r$ be given by

$$\mathcal{W} = \{\boldsymbol{W} \in \mathbb{C}^{M_r \times M_r} \mid \boldsymbol{W}^H = \boldsymbol{W}\}. \tag{6.332}$$

Let the autocorrelation matrix of the noise $\boldsymbol{\Sigma}_n \in \mathbb{C}^{M_r \times M_r}$ be represented by the Hermitian matrix $\boldsymbol{W} \in \mathcal{W} \subset \mathbb{C}^{M_r \times M_r}$. The unpatterned version of $\boldsymbol{W}$ is denoted by $\tilde{\boldsymbol{W}} \in \mathbb{C}^{M_r \times M_r}$, where it is assumed that $\tilde{\boldsymbol{W}}$ and $\boldsymbol{W}$ are invertible. Let the function $g : \mathbb{C}^{M_r \times M_r} \times \mathbb{C}^{M_r \times M_r} \to \mathbb{C}$ be defined by replacing $\boldsymbol{\Sigma}_n$ by the unpatterned matrix $\tilde{\boldsymbol{W}}$ in the mutual information function of the MIMO system given in (6.211), that is,

$$g(\tilde{\boldsymbol{W}}, \tilde{\boldsymbol{W}}^*) = \ln\left(\det\left(\boldsymbol{I}_{M_r} + \boldsymbol{H}\boldsymbol{B}\boldsymbol{\Sigma}_x \boldsymbol{B}^H \boldsymbol{H}^H \tilde{\boldsymbol{W}}^{-1}\right)\right). \tag{6.333}$$

Show that the unpatterned derivatives of g with respect to $\tilde{\boldsymbol{W}}$ and $\tilde{\boldsymbol{W}}^*$ can be expressed as

$$\frac{\partial}{\partial \tilde{\boldsymbol{W}}} g = -\tilde{\boldsymbol{W}}^{-T} \boldsymbol{H}^* \boldsymbol{B}^* \boldsymbol{\Sigma}_x^T \boldsymbol{B}^T \boldsymbol{H}^T \left(\boldsymbol{I}_{M_r} + \tilde{\boldsymbol{W}}^{-T} \boldsymbol{H}^* \boldsymbol{B}^* \boldsymbol{\Sigma}_x^T \boldsymbol{B}^T \boldsymbol{H}^T\right)^{-1} \tilde{\boldsymbol{W}}^{-T}, \tag{6.334}$$

$$\frac{\partial}{\partial \tilde{\boldsymbol{W}}^*} g = \boldsymbol{0}_{M_r \times M_r}, \tag{6.335}$$

respectively.

By using the results from Example 6.25, show that the generalized derivative of g with respect to the Hermitian matrices $\boldsymbol{W}$ and $\boldsymbol{W}^*$ can be expressed as

$$\frac{\partial}{\partial \boldsymbol{W}} g = -\boldsymbol{W}^{-T} \boldsymbol{H}^* \boldsymbol{B}^* \boldsymbol{\Sigma}_x^T \boldsymbol{B}^T \boldsymbol{H}^T \left(\boldsymbol{I}_{M_r} + \boldsymbol{W}^{-T} \boldsymbol{H}^* \boldsymbol{B}^* \boldsymbol{\Sigma}_x^T \boldsymbol{B}^T \boldsymbol{H}^T\right)^{-1} \boldsymbol{W}^{-T}, \tag{6.336}$$

$$\frac{\partial}{\partial \boldsymbol{W}^*} g = -\boldsymbol{W}^{-1} \left(\boldsymbol{I}_{M_r} + \boldsymbol{H}\boldsymbol{B}\boldsymbol{\Sigma}_x \boldsymbol{B}^H \boldsymbol{H}^H \boldsymbol{W}^{-1}\right)^{-1} \boldsymbol{H}\boldsymbol{B}\boldsymbol{\Sigma}_x \boldsymbol{B}^H \boldsymbol{H}^H \boldsymbol{W}^{-1}, \tag{6.337}$$

respectively, when the standard basis is used to expand the Hermitian matrices in the manifold $\mathcal{W}$.

Show that

$$\left.\frac{\partial}{\partial \boldsymbol{W}^*} g\right|_{\boldsymbol{W}=\boldsymbol{\Sigma}_n} = -\boldsymbol{\Sigma}_n^{-1} \boldsymbol{H}\boldsymbol{B} \left(\boldsymbol{\Sigma}_x^{-1} + \boldsymbol{B}^H \boldsymbol{H}^H \boldsymbol{\Sigma}_n^{-1} \boldsymbol{H}\boldsymbol{B}\right)^{-1} \boldsymbol{B}^H \boldsymbol{H}^H \boldsymbol{\Sigma}_n^{-1}. \tag{6.338}$$

Explain why this result is in agreement with Palomar and Verdú (2006, Eq. (27)).

6.15 Let the function $g : \mathbb{C}^{N\times N} \times \mathbb{C}^{N\times N} \to \mathbb{R}$ be given by

$$g(\tilde{W}, \tilde{W}^*) = \left\|A\tilde{W} - B\right\|_F^2 = \mathrm{Tr}\left\{(A\tilde{W} - B)\left(\tilde{W}^H A^H - B^H\right)\right\}, \tag{6.339}$$

where $\tilde{W} \in \mathbb{C}^{N\times N}$ is unpatterned, while the two matrices $A \in \mathbb{C}^{M\times N}$ and $B \in \mathbb{C}^{M\times N}$ are independent of $\tilde{W}$ and $\tilde{W}^*$. Assume that $\mathrm{rank}(A) = N$. In this exercise, the function g will be minimized under different constraints on the input matrix variables, and the results derived earlier in this chapter can be used to derive some of these results.

For all cases given below, show that the inverse matrices involved in the expressions exist.

(a) Show that the derivatives of g with respect to $\tilde{W}$ and $\tilde{W}^*$ are given by

$$\mathcal{D}_{\tilde{W}} g = \mathrm{vec}^T\left(A^T\left(A^*\tilde{W}^* - B^*\right)\right), \tag{6.340}$$

$$\mathcal{D}_{\tilde{W}^*} g = \mathrm{vec}^T\left(A^H\left(A\tilde{W} - B\right)\right), \tag{6.341}$$

respectively.

Because $\mathcal{D}_{\tilde{W}} g = \mathrm{vec}^T\left(\frac{\partial g}{\partial \tilde{W}}\right)$ and $\mathcal{D}_{\tilde{W}^*} g = \mathrm{vec}^T\left(\frac{\partial g}{\partial \tilde{W}^*}\right)$, it follows that

$$\frac{\partial g}{\partial \tilde{W}} = A^T\left(A^*\tilde{W}^* - B^*\right), \tag{6.342}$$

$$\frac{\partial g}{\partial \tilde{W}^*} = A^H\left(A\tilde{W} - B\right). \tag{6.343}$$

By using the above derivatives, show that the minimum unconstrained value of $\tilde{W}$ is given by

$$\tilde{W} = \left(A^H A\right)^{-1} A^H B = A^+ B, \tag{6.344}$$

where (2.80) is used in the last equality.

(b) Assume that W is diagonal such that $W \in \mathcal{W} = \{W \in \mathbb{C}^{N\times N} \mid W = I_N \odot W\}$. Show that

$$\frac{\partial}{\partial W} g(W, W^*) = I \odot \left[A^T\left(A^* W^* - B^*\right)\right]. \tag{6.345}$$

By solving $\frac{\partial}{\partial W} g(W, W^*) = \mathbf{0}_{N\times N}$, show that the $N \times N$ diagonal matrix that minimizes g must satisfy

$$\mathrm{vec}_d(W) = \left[I_N \odot \left(A^H A\right)\right]^{-1} \mathrm{vec}_d\left(A^H B\right). \tag{6.346}$$

(c) Assume that W is symmetric, such that $W \in \mathcal{W} = \{W \in \mathbb{C}^{N\times N} \mid W^T = W\}$. Show that

$$\begin{aligned}\frac{\partial}{\partial W} g(W, W^*) = {} & A^T A^* W^* + W^* A^H A - I_N \odot \left[A^T A^* W^*\right] \\ & - A^T B^* - B^H A + I_N \odot \left[A^T B^*\right].\end{aligned} \tag{6.347}$$

By solving $\frac{\partial}{\partial W} g(W, W^*) = \mathbf{0}_{N \times N}$, show that the symmetric W that minimizes g is given by

$$\operatorname{vec}(W) = \left[I_N \otimes \left(A^H A \right) + \left(A^H A \right) \otimes I_N - L_d L_d^T \left\{ I_N \otimes \left(A^H A \right) \right\} \right]^{-1} \times \operatorname{vec}\left(A^H B + B^T A^* - I_N \odot \left(A^H B \right) \right). \tag{6.348}$$

Show that if W is satisfying (6.348), then it is symmetric, that is, $W^T = W$.

(d) Assume that W is skew-symmetric such that $W \in \mathcal{W} = \{W \in \mathbb{C}^{N \times N} \mid W^T = -W\}$. Show that

$$\frac{\partial}{\partial W} g(W, W^*) = A^T A^* W^* + W^* A^H A - A^T B^* + B^H A. \tag{6.349}$$

By solving $\frac{\partial}{\partial W} g(W, W^*) = \mathbf{0}_{N \times N}$, show that the skew-symmetric W that minimizes g is given by

$$\operatorname{vec}(W) = \left[I_N \otimes \left(A^H A \right) + \left(A^H A \right) \otimes I_N \right]^{-1} \operatorname{vec}\left(A^H B - B^T A^* \right). \tag{6.350}$$

Show that if W is satisfying (6.350), then it is skew-symmetric, that is, $W^T = -W$.

(e) Assume that W is Hermitian, such that $W \in \mathcal{W} = \{W \in \mathbb{C}^{N \times N} \mid W^H = W\}$. Show that

$$\frac{\partial}{\partial W} g(W, W^*) = A^T A^* W^T + W^T A^T A^* - A^T B^* - B^T A^*. \tag{6.351}$$

By solving the equation $\frac{\partial}{\partial W} g(W, W^*) = \mathbf{0}_{N \times N}$, show that the Hermitian W that minimizes g is given by

$$\operatorname{vec}(W) = \left[\left(A^T A^* \right) \otimes I_N + I_N \otimes \left(A^H A \right) \right]^{-1} \operatorname{vec}\left(A^H B + B^H A \right). \tag{6.352}$$

Show that if W is satisfying (6.352), then it is Hermitian, that is, $W^H = W$.

(f) Assume that W is skew-Hermitian, such that $W \in \mathcal{W} = \{W \in \mathbb{C}^{N \times N} \mid W^H = -W\}$. Show that

$$\frac{\partial}{\partial W} g(W, W^*) = -A^T A^* W^T - W^T A^T A^* - A^T B^* + B^T A^*. \tag{6.353}$$

By solving $\frac{\partial}{\partial W} g(W, W^*) = \mathbf{0}_{N \times N}$, show that the skew-Hermitian W that minimizes g is given by

$$\operatorname{vec}(W) = \left[\left(A^T A^* \right) \otimes I_N + I_N \otimes \left(A^H A \right) \right]^{-1} \operatorname{vec}\left(A^H B - B^H A \right). \tag{6.354}$$

Show that if W is satisfying (6.354), then it is skew-Hermitian, that is, $W^H = -W$.

6.16 Let $\mathcal{B} = \{b_0, b_1, \cdots, b_{N-1}\}$ be a basis for $\mathbb{C}^{N \times 1}$, such that $b_i \in \mathbb{C}^{N \times 1}$ are linearly independent and they span $\mathbb{C}^{N \times 1}$. Let the matrix $U \in \mathbb{C}^{N \times N}$ be given by

$$U = [b_0, b_1, \ldots, b_{N-1}]. \tag{6.355}$$

We say that the vector $z \in \mathbb{C}^{N \times 1}$ has the coordinates $c_0, c_1, \ldots, c_{N-1}$ with respect to the basis $\mathcal{B}$ if

$$z = \sum_{i=0}^{N-1} c_i \boldsymbol{b}_i = [\boldsymbol{b}_0, \boldsymbol{b}_1, \ldots, \boldsymbol{b}_{N-1}] \begin{bmatrix} c_0 \\ c_1 \\ \vdots \\ c_{N-1} \end{bmatrix} = \boldsymbol{U}\boldsymbol{c}, \tag{6.356}$$

where $\boldsymbol{c} = [c_0, c_1, \ldots, c_{N-1}]^T \in \mathbb{C}^{N \times 1}$. If and only if (6.356) is satisfied, the notation $[z]_{\mathcal{B}} = \boldsymbol{c}$ is used. Show that the coordinates of z with respect to the basis $\mathcal{B}$ are the same as the coordinates of $\boldsymbol{U}^{-1} z$ with respect to the standard basis $\{\boldsymbol{e}_0, \boldsymbol{e}_1, \ldots, \boldsymbol{e}_{N-1}\}$, where $\boldsymbol{e}_i \in \mathbb{Z}_2^{N \times 1}$ is defined in Definition 2.16.

Let $\mathcal{A} = \{\boldsymbol{a}_0, \boldsymbol{a}_1, \ldots, \boldsymbol{a}_{M-1}\}$ be a basis of $\mathbb{C}^{1 \times M}$ where $\boldsymbol{a}_i \in \mathbb{C}^{1 \times M}$. The matrix $\boldsymbol{V} \in \mathbb{C}^{M \times M}$ is given by

$$\boldsymbol{V} = \begin{bmatrix} \boldsymbol{a}_0 \\ \boldsymbol{a}_1 \\ \vdots \\ \boldsymbol{a}_{M-1} \end{bmatrix}. \tag{6.357}$$

Let $\boldsymbol{x} \in \mathbb{C}^{1 \times M}$; then it is said that the vector $\boldsymbol{x}$ has coordinates $d_0, d_1, \ldots, d_{M-1}$ with respect to the basis $\mathcal{A}$ if

$$\boldsymbol{x} = \sum_{i=0}^{M-1} d_i \boldsymbol{a}_i = [d_0, d_1, \ldots, d_{M-1}] \begin{bmatrix} \boldsymbol{a}_0 \\ \boldsymbol{a}_1 \\ \vdots \\ \boldsymbol{a}_{M-1} \end{bmatrix} = \boldsymbol{d}\boldsymbol{V}, \tag{6.358}$$

where $\boldsymbol{d} = [d_0, d_1, \ldots, d_{M-1}] \in \mathbb{C}^{1 \times M}$. If and only if (6.358) holds, then the notation $[\boldsymbol{x}]_{\mathcal{A}} = \boldsymbol{d}$. Show that the coordinates of $\boldsymbol{x}$ with respect to $\mathcal{A}$ are the same as the coordinates of $\boldsymbol{x}\boldsymbol{V}^{-1}$ with respect to the standard basis $\{\boldsymbol{e}_0^T, \boldsymbol{e}_1^T, \ldots, \boldsymbol{e}_{M-1}^T\}$, where $\boldsymbol{e}_i \in \mathbb{Z}_2^{M \times 1}$ is given in Definition 2.16.

7 Applications in Signal Processing and Communications

7.1 Introduction

In this chapter, several examples of how the theory of complex-valued matrix derivatives can be used as an important tool to solve research problems taken from signal processing and communications. The developed theory can be used to solve problems in areas where the unknown matrices are complex-valued matrices. Examples of such areas are signal processing and communications. Often in these areas, the objective function is a real-valued function that depends on a continuous complex-valued matrix and its complex conjugate. In Hjørungnes and Ramstad (1999) and Hjørungnes (2000), matrix derivatives were used to optimize filter banks used for source coding. The book by Vaidyanathan et al. (2010) contains material on how to optimize communication systems by means of complex-valued derivatives. Complex-valued derivatives were applied to find the Cramer-Rao lower bound for complex-valued parameters in van den Bos (1994*b*) and Jagannatham and Rao (2004)

The rest of this chapter is organized as follows: Section 7.2 presents a problem from signal processing on how to find the derivative and the Hessian of a real-valued function that depends on the magnitude of the Fourier transform of the complex-valued argument vector. In Section 7.3, an example from signal processing is studied in which the sums of the squared absolute values of the off-diagonal elements in a covariance matrix are minimized. This problem of minimizing the off-diagonal elements has applications in blind carrier frequency offset (CFO) estimation. A multiple-input multiple-output (MIMO) precoder for coherent detection is designed in Section 7.4 for minimizing the exact symbol error rate (SER) when an orthogonal space-time block code (OSTBC) is used in the transmitter to encode the signal for communication over a correlated Ricean channel. In Section 7.5, a finite impulse response (FIR) MIMO filter system is studied. Necessary conditions for finding the minimum mean square error (MSE) receive filter are developed for a given transmit filter, and vice versa. Finally, exercises related to this chapter are presented in Section 7.6.

7.2 Absolute Value of Fourier Transform Example

The case that was studied in Osherovich, Zibulevsky, and Yavneh (2008) will be considered in this section. In Osherovich, Zibulevsky, and Yavneh (2008), the problem studied is how to reconstruct a signal from the absolute value of the Fourier transform of a signal. This problem has applications in, for example, how to do visualization of nano-structures. In this section, the derivatives and the Hessian of an objective function that depends on the magnitude of the Fourier transform of the original signal will be derived.

The rest of this section is organized as follows: Four special functions and the inverse discrete Fourier transform (DFT) matrix are defined in Subsection 7.2.1. The objective function that should be minimized is defined in Subsection 7.2.2. In Subsection 7.2.3, the first-order differential and the derivatives of the objective function are found. Subsection 7.2.4 contains a calculation of the second-order differential and the Hessian of the objective function.

7.2.1 Special Function and Matrix Definitions

Four special functions and one special matrix are needed in this section, and they are now defined. The four functions are (1) the component-wise absolute value of a vector, (2) the component-wise principal argument of complex vectors, (3) the inverse of the component-wise absolute value of a complex vector, which does not contain any zeros, and (4) the exponential function of a vector. These functions are used to simplify the presentation.

Definition 7.1 *Let the function* $|\cdot| : \mathbb{C}^{N\times 1} \rightarrow \{\mathbb{R}^+ \cup \{0\}\}^{N\times 1}$ *return the* component-wise absolute value *of the vector it is applied to. If* $\mathbf{z} \in \mathbb{C}^{N\times 1}$*, then,*

$$|\mathbf{z}| = \left| \begin{bmatrix} z_0 \\ z_1 \\ \vdots \\ z_{N-1} \end{bmatrix} \right| = \begin{bmatrix} |z_0| \\ |z_1| \\ \vdots \\ |z_{N-1}| \end{bmatrix}, \tag{7.1}$$

where z_i *is the* i*-th component of* $\mathbf{z}$*, where* $i \in \{0, 1, \ldots, N-1\}$.

Definition 7.2 *The function* $\angle(\cdot) : \mathbb{C}^{N\times 1} \rightarrow (-\pi, \pi]^{N\times 1}$ *returns the* component-wise principal argument *of the vector it is applied to. If* $\mathbf{z} \in \mathbb{C}^{N\times 1}$*, then,*

$$\angle\mathbf{z} = \angle \begin{bmatrix} z_0 \\ z_1 \\ \vdots \\ z_{N-1} \end{bmatrix} = \begin{bmatrix} \angle z_0 \\ \angle z_1 \\ \vdots \\ \angle z_{N-1} \end{bmatrix}, \tag{7.2}$$

where the function $\angle\cdot : \mathbb{C} \rightarrow (-\pi, \pi]$ *returns the principal value of the argument (Kreyszig 1988, Section 12.2) of the input.*

Definition 7.3 *The function* $|\cdot|^{-1} : \{\mathbb{C} \setminus \{0\}\}^{N\times 1} \to (\mathbb{R}^{+})^{N\times 1}$ *returns the* component-wise inverse of the absolute values *of the input vector. If* $z \in \{\mathbb{C} \setminus \{0\}\}^{N\times 1}$*, then,*

$$|z|^{-1} = \begin{bmatrix} |z_0|^{-1} \\ |z_1|^{-1} \\ \vdots \\ |z_{N-1}|^{-1} \end{bmatrix}, \tag{7.3}$$

where $z_i \neq 0$ *is the* i*-th component of* z*, where* $i \in \{0, 1, \ldots, N-1\}$*.*

Definition 7.4 *Let* $z \in \mathbb{C}^{N\times 1}$*, then the* exponential function of a vector e^{z} *is defined as the* $N \times 1$ *vector:*

$$e^{z} = \begin{bmatrix} e^{z_0} \\ e^{z_1} \\ \vdots \\ e^{z_{N-1}} \end{bmatrix}, \tag{7.4}$$

where z_i *is the* i*-th component of the vector* z*, where* $i \in \{0, 1, \ldots, N-1\}$*.*

Definitions 7.1, 7.2, 7.3, and 7.4 are presented above for column vectors; however, they are also valid for row vectors.

Using the three functions $|z|$, $\angle z$, and $e^{\jmath \angle z}$ given through Definitions 7.1, 7.2, and 7.4, the vector $z \in \mathbb{C}^{N\times 1}$ can be expressed as

$$z = \exp\left(\jmath \operatorname{diag}\left(\angle z\right)\right) |z| = |z| \odot e^{\jmath \angle z}, \tag{7.5}$$

where $\operatorname{diag}(\cdot)$ is found in Definition 2.10, and where $\exp(\cdot)$ is the exponential matrix function defined in Definition 2.5, such that $\exp\left(\jmath \operatorname{diag}\left(\angle z\right)\right)$ has size $N \times N$. If $z \in (\mathbb{C} \setminus \{0\})^{N\times 1}$, then it follows from (7.5) that

$$e^{\jmath z} = z \odot |z|^{-1}. \tag{7.6}$$

The complex conjugate of z is given by

$$z^{*} = \exp\left(-\jmath \operatorname{diag}\left(\angle z\right)\right) |z| = |z| \odot e^{-\jmath \angle z}. \tag{7.7}$$

If $z \in (\mathbb{C} \setminus \{0\})^{N\times 1}$, then it follows from (7.7) that

$$e^{-\jmath z} = z^{*} \odot |z|^{-1}. \tag{7.8}$$

One frequently used matrix in signal processing is defined next. This is the inverse DFT (Sayed 2003, p. 577).

Definition 7.5 *The* inverse DFT matrix *of size* $N \times N$ *is denoted by* $\boldsymbol{F}_N$*; it is a unitary symmetric matrix with the* (k, l)*-th element given by*

$$(\boldsymbol{F}_N)_{k,l} = \frac{1}{\sqrt{N}} e^{\frac{\jmath k l 2\pi}{N}}, \tag{7.9}$$

where $k, l \in \{0, 1, \ldots, N-1\}$*.*

It is observed that the inverse DFT matrix is symmetric, hence, $\boldsymbol{F}_N^T = \boldsymbol{F}_N$. The DFT matrix of size $N \times N$ is a unitary matrix that is the inverse of the inverse DFT matrix. Therefore, the DFT matrix is given by

$$\boldsymbol{F}_N^{-1} = \boldsymbol{F}_N^H = \boldsymbol{F}_N^*. \tag{7.10}$$

7.2.2 Objective Function Formulation

Let $\boldsymbol{w} \in \mathbb{C}^{N\times 1}$, and let the real-valued function $g : \mathbb{C}^{N\times 1} \times \mathbb{C}^{N\times 1} \to \mathbb{R}$ be given by

$$g(\boldsymbol{w}, \boldsymbol{w}^*) = \| |\boldsymbol{w}| - \boldsymbol{r} \|^2 = \left(|\boldsymbol{w}|^T - \boldsymbol{r}^T \right) (|\boldsymbol{w}| - \boldsymbol{r}) = |\boldsymbol{w}|^T |\boldsymbol{w}| - 2\boldsymbol{r}^T |\boldsymbol{w}| + \|\boldsymbol{r}\|^2, \tag{7.11}$$

where $\boldsymbol{r} \in (\mathbb{R}^+)^{N\times 1}$ is a constant vector that is independent of $\boldsymbol{w}$, and $\boldsymbol{w}^*$, and where $\|\boldsymbol{a}\|$ denotes the Euclidean norm of the vector $\boldsymbol{a} \in \mathbb{C}^{N\times 1}$, i.e., $\|\boldsymbol{a}\|^2 = \boldsymbol{a}^H \boldsymbol{a}$. One goal of this section is to find the derivative of g with respect to $\boldsymbol{w}$ and $\boldsymbol{w}^*$.

The function $h : \mathbb{C}^{N\times 1} \times \mathbb{C}^{N\times 1} \to \mathbb{R}$ is defined as

$$h(\boldsymbol{z}, \boldsymbol{z}^*) = g(\boldsymbol{w}, \boldsymbol{w}^*)\big|_{\boldsymbol{w}=\boldsymbol{F}_N^* \boldsymbol{z}} = g(\boldsymbol{F}_N^* \boldsymbol{z}, \boldsymbol{F}_N \boldsymbol{z}^*) = \left\| |\boldsymbol{F}_N^* \boldsymbol{z}| - \boldsymbol{r} \right\|^2, \tag{7.12}$$

where the vector $\boldsymbol{F}_N^* \boldsymbol{z}$ is the DFT transform of the vector $\boldsymbol{z} \in \mathbb{C}^{N\times 1}$. Another goal of this section is to find the derivative of h with respect to $\boldsymbol{z}$ and $\boldsymbol{z}^*$; this will be achieved by the chain rule presented in Theorem 3.1 by first finding the derivative of g with respect to $\boldsymbol{w}$ and $\boldsymbol{w}^*$. The function h measures the distance between the magnitude of the DFT of the original vector $\boldsymbol{z} \in \mathbb{C}^{N\times 1}$ and the constant vector $\boldsymbol{r} \in (\mathbb{R}^+)^{N\times 1}$.

7.2.3 First-Order Derivatives of the Objective Function

One way to find the derivative of g is through the differential of g, which can be expressed as

$$dg = (d|\boldsymbol{w}|)^T |\boldsymbol{w}| + |\boldsymbol{w}|^T d|\boldsymbol{w}| - 2\boldsymbol{r}^T d|\boldsymbol{w}| = 2\left(|\boldsymbol{w}|^T - \boldsymbol{r}^T \right) d|\boldsymbol{w}|. \tag{7.13}$$

It is seen from (7.13) that an expression of $d|\boldsymbol{w}|$ is needed. From (3.22), (4.13), and (4.14), it follows that the differential of $|w_i|$ is given by

$$d|w_i| = \frac{1}{2} e^{-\jmath \angle w_i} dw_i + \frac{1}{2} e^{\jmath \angle w_i} dw_i^*, \tag{7.14}$$

where $i \in \{0, 1, \ldots, N-1\}$, and where w_i is the i-th component of the vector $\boldsymbol{w}$. Now, $d|\boldsymbol{w}| = [d|w_0|, d|w_1|, \ldots, d|w_{N-1}|]^T$ can be found by

$$d|\boldsymbol{w}| = \begin{bmatrix} d|w_0| \\ d|w_1| \\ \vdots \\ d|w_{N-1}| \end{bmatrix} = \frac{1}{2}\begin{bmatrix} e^{-\jmath\angle w_0} dw_0 \\ e^{-\jmath\angle w_1} dw_1 \\ \vdots \\ e^{-\jmath\angle w_{N-1}} dw_{N-1} \end{bmatrix} + \frac{1}{2}\begin{bmatrix} e^{\jmath\angle w_0} dw_0^* \\ e^{\jmath\angle w_1} dw_1^* \\ \vdots \\ e^{\jmath\angle w_{N-1}} dw_{N-1}^* \end{bmatrix}$$

$$= \frac{1}{2} e^{-\jmath\angle \boldsymbol{w}} \odot d\boldsymbol{w} + \frac{1}{2} e^{\jmath\angle \boldsymbol{w}} \odot d\boldsymbol{w}^*$$

$$= \frac{1}{2} \exp\left(-\jmath \operatorname{diag}\left(\angle \boldsymbol{w}\right)\right) d\boldsymbol{w} + \frac{1}{2} \exp\left(\jmath \operatorname{diag}\left(\angle \boldsymbol{w}\right)\right) d\boldsymbol{w}^*, \tag{7.15}$$

where it has been used that $\exp(\operatorname{diag}(\boldsymbol{a}))\boldsymbol{b} = e^{\boldsymbol{a}} \odot \boldsymbol{b}$ when $\boldsymbol{a}, \boldsymbol{b} \in \mathbb{C}^{N\times 1}$, and that

$$\operatorname{diag}\left(\angle \boldsymbol{w}\right) = \begin{bmatrix} \angle w_0 & 0 & \cdots & 0 \\ 0 & \angle w_1 & \cdots & 0 \\ \vdots & \vdots & \ddots & \vdots \\ 0 & 0 & \cdots & \angle w_{N-1} \end{bmatrix}. \tag{7.16}$$

By inserting the results from (7.15) into (7.13), the first-order differential of g can be expressed as

$$dg = 2\left(|\boldsymbol{w}|^T - \boldsymbol{r}^T\right)\left[\frac{1}{2} \exp\left(-\jmath \operatorname{diag}\left(\angle \boldsymbol{w}\right)\right) d\boldsymbol{w} + \frac{1}{2} \exp\left(\jmath \operatorname{diag}\left(\angle \boldsymbol{w}\right)\right) d\boldsymbol{w}^*\right]$$

$$= \left(|\boldsymbol{w}|^T - \boldsymbol{r}^T\right)\left[\exp\left(-\jmath \operatorname{diag}\left(\angle \boldsymbol{w}\right)\right) d\boldsymbol{w} + \exp\left(\jmath \operatorname{diag}\left(\angle \boldsymbol{w}\right)\right) d\boldsymbol{w}^*\right]. \tag{7.17}$$

From dg, the derivatives of g with respect to $\boldsymbol{w}$ and $\boldsymbol{w}^*$ can be identified as

$$\mathcal{D}_{\boldsymbol{w}} g = \left(|\boldsymbol{w}|^T - \boldsymbol{r}^T\right) \exp\left(-\jmath \operatorname{diag}\left(\angle \boldsymbol{w}\right)\right)$$

$$= |\boldsymbol{w}|^T \exp\left(-\jmath \operatorname{diag}\left(\angle \boldsymbol{w}\right)\right) - \boldsymbol{r}^T \exp\left(-\jmath \operatorname{diag}\left(\angle \boldsymbol{w}\right)\right)$$

$$= \boldsymbol{w}^H - \boldsymbol{r}^T \exp\left(-\jmath \operatorname{diag}\left(\angle \boldsymbol{w}\right)\right) = \boldsymbol{w}^H - \boldsymbol{r}^T \odot e^{-\jmath(\angle \boldsymbol{w})^T}, \tag{7.18}$$

and

$$\mathcal{D}_{\boldsymbol{w}^*} g = \left(|\boldsymbol{w}|^T - \boldsymbol{r}^T\right) \exp\left(\jmath \operatorname{diag}\left(\angle \boldsymbol{w}\right)\right)$$

$$= |\boldsymbol{w}|^T \exp\left(\jmath \operatorname{diag}\left(\angle \boldsymbol{w}\right)\right) - \boldsymbol{r}^T \exp\left(\jmath \operatorname{diag}\left(\angle \boldsymbol{w}\right)\right)$$

$$= \boldsymbol{w}^T - \boldsymbol{r}^T \exp\left(\jmath \operatorname{diag}\left(\angle \boldsymbol{w}\right)\right) = \boldsymbol{w}^T - \boldsymbol{r}^T \odot e^{\jmath(\angle \boldsymbol{w})^T}, \tag{7.19}$$

respectively, where (7.5) and (7.7) have been used.

To use the chain rule to find the derivative of h, let us first define the function that returns the DFT of the input vector $\boldsymbol{f} : \mathbb{C}^{N\times 1} \times \mathbb{C}^{N\times 1} \to \mathbb{C}^{N\times 1}$; $\boldsymbol{f}(\boldsymbol{z}, \boldsymbol{z}^*)$ is given by

$$\boldsymbol{f}(\boldsymbol{z}, \boldsymbol{z}^*) = \boldsymbol{F}_N^* \boldsymbol{z}, \tag{7.20}$$

where $\boldsymbol{F}_N$ is given in Definition 7.5. By calculating the differentials of $\boldsymbol{f}$ and $\boldsymbol{f}^*$, it is found that

$$\mathcal{D}_{z}\boldsymbol{f} = \boldsymbol{F}_N^*, \tag{7.21}$$

$$\mathcal{D}_{z^*}\boldsymbol{f} = \boldsymbol{0}_{N\times N}, \tag{7.22}$$

$$\mathcal{D}_{z}\boldsymbol{f}^* = \boldsymbol{0}_{N\times N}, \tag{7.23}$$

$$\mathcal{D}_{z^*}\boldsymbol{f}^* = \boldsymbol{F}_N. \tag{7.24}$$

The chain rule in Theorem 3.1 is now used to find the derivatives of h in (7.12) with respect to $\boldsymbol{z}$ and $\boldsymbol{z}^*$ as

$$\begin{aligned}\mathcal{D}_{z}h &= \mathcal{D}_{w}g|_{\boldsymbol{w}=\boldsymbol{F}_N^*\boldsymbol{z}}\mathcal{D}_{z}\boldsymbol{f} + \mathcal{D}_{w^*}g|_{\boldsymbol{w}=\boldsymbol{F}_N^*\boldsymbol{z}}\mathcal{D}_{z}\boldsymbol{f}^* \\ &= \left[\boldsymbol{w}^H - \boldsymbol{r}^T\exp(-\jmath\,\mathrm{diag}(\angle\boldsymbol{w}))\right]_{\boldsymbol{w}=\boldsymbol{F}_N^*\boldsymbol{z}}\boldsymbol{F}_N^* \\ &= \boldsymbol{z}^H\boldsymbol{F}_N\boldsymbol{F}_N^* - \boldsymbol{r}^T\exp\left(-\jmath\,\mathrm{diag}\left(\angle\left(\boldsymbol{F}_N^*\boldsymbol{z}\right)\right)\right)\boldsymbol{F}_N^* \\ &= \boldsymbol{z}^H - \boldsymbol{r}^T\exp\left(-\jmath\,\mathrm{diag}\left(\angle\left(\boldsymbol{F}_N^*\boldsymbol{z}\right)\right)\right)\boldsymbol{F}_N^* \\ &= \boldsymbol{z}^H - \left[\boldsymbol{r}^T\odot(\boldsymbol{F}_N\boldsymbol{z}^*)^T\odot\left|\boldsymbol{F}_N\boldsymbol{z}^*\right|^{-T}\right]\boldsymbol{F}_N^*,\end{aligned} \tag{7.25}$$

where $|\boldsymbol{z}|^{-T} \triangleq \left(|\boldsymbol{z}|^{-1}\right)^T = |\boldsymbol{z}^T|^{-1}$ (see Definition 7.3), and

$$\begin{aligned}\mathcal{D}_{z^*}h &= \mathcal{D}_{w}g|_{\boldsymbol{w}=\boldsymbol{F}_N^*\boldsymbol{z}}\mathcal{D}_{z^*}\boldsymbol{f} + \mathcal{D}_{w^*}g|_{\boldsymbol{w}=\boldsymbol{F}_N^*\boldsymbol{z}}\mathcal{D}_{z^*}\boldsymbol{f}^* \\ &= \left[\boldsymbol{w}^T - \boldsymbol{r}^T\exp(\jmath\,\mathrm{diag}(\angle\boldsymbol{w}))\right]_{\boldsymbol{w}=\boldsymbol{F}_N^*\boldsymbol{z}}\boldsymbol{F}_N \\ &= \boldsymbol{z}^T\boldsymbol{F}_N^*\boldsymbol{F}_N - \boldsymbol{r}^T\exp\left(\jmath\,\mathrm{diag}\left(\angle\left(\boldsymbol{F}_N^*\boldsymbol{z}\right)\right)\right)\boldsymbol{F}_N \\ &= \boldsymbol{z}^T - \boldsymbol{r}^T\exp\left(\jmath\,\mathrm{diag}\left(\angle\left(\boldsymbol{F}_N^*\boldsymbol{z}\right)\right)\right)\boldsymbol{F}_N \\ &= \boldsymbol{z}^T - \left[\boldsymbol{r}^T\odot\left(\boldsymbol{F}_N^*\boldsymbol{z}\right)^T\odot\left|\boldsymbol{F}_N^*\boldsymbol{z}\right|^{-T}\right]\boldsymbol{F}_N,\end{aligned} \tag{7.26}$$

where the results from (7.18), (7.19), (7.21), (7.22), (7.23), and (7.24) were used.

7.2.4 Hessians of the Objective Function

The second-order differential can be found by calculating the differential of the first-order differential of g. From (7.17), (7.18), and (7.19), it follows that dg is given by

$$dg = \left[\boldsymbol{w}^H - \boldsymbol{r}^T\exp(-\jmath\,\mathrm{diag}(\angle\boldsymbol{w}))\right]d\boldsymbol{w} + \left[\boldsymbol{w}^T - \boldsymbol{r}^T\exp(\jmath\,\mathrm{diag}(\angle\boldsymbol{w}))\right]d\boldsymbol{w}^*. \tag{7.27}$$

From (7.27), it is seen that to proceed to find d^2g, the following differential is needed:

$$d\exp(\jmath\,\mathrm{diag}(\angle\boldsymbol{w})) = \begin{bmatrix} de^{\jmath\angle w_0} & 0 & \cdots & 0 \\ 0 & de^{\jmath\angle w_1} & \cdots & 0 \\ \vdots & \vdots & \ddots & \vdots \\ 0 & 0 & \cdots & de^{\jmath\angle w_{N-1}} \end{bmatrix}. \tag{7.28}$$

Hence, the differential $de^{\jmath\angle w_i}$ is needed. This expression can be found as

$$\begin{aligned} de^{\jmath\angle w_i} &= \frac{\partial e^{\jmath\angle w_i}}{\partial w_i} dw_i + \frac{\partial e^{\jmath\angle w_i}}{\partial w_i^*} dw_i^* = e^{\jmath\angle w_i} \jmath \left(\frac{-\jmath}{2 w_i} \right) dw_i + e^{\jmath\angle w_i} \jmath \left(\frac{\jmath}{2 w_i^*} \right) dw_i^* \\ &= \frac{1}{2 w_i e^{-\jmath\angle w_i}} dw_i - \frac{1}{2 w_i^* e^{-\jmath\angle w_i}} dw_i^* = \frac{1}{2|w_i|} dw_i - \frac{1}{2|w_i| e^{-2\jmath\angle w_i}} dw_i^*, \end{aligned} \tag{7.29}$$

where (3.22), (4.21), and (4.22) were utilized. By taking the complex conjugate of both sides of (7.29), it is found that

$$de^{-\jmath\angle w_i} = -\frac{1}{2|w_i| e^{2\jmath\angle w_i}} dw_i + \frac{1}{2|w_i|} dw_i^*. \tag{7.30}$$

The second-order differential of g is found by applying the differential operator on both sides of (7.27) to obtain

$$d^2 g = \left[d\boldsymbol{w}^H - \boldsymbol{r}^T d \exp(-\jmath \operatorname{diag}(\angle \boldsymbol{w})) \right] d\boldsymbol{w} + \left[d\boldsymbol{w}^T - \boldsymbol{r}^T d \exp(\jmath \operatorname{diag}(\angle \boldsymbol{w})) \right] d\boldsymbol{w}^*. \tag{7.31}$$

From (7.28) and (7.29), it follows that

$$\begin{aligned} & d \exp(\jmath \operatorname{diag}(\angle \boldsymbol{w})) \\ &= \frac{1}{2} \begin{bmatrix} \frac{dw_0}{|w_0|} & 0 & \cdots & 0 \\ 0 & \frac{dw_1}{|w_1|} & \cdots & 0 \\ \vdots & \vdots & \ddots & \vdots \\ 0 & 0 & \cdots & \frac{dw_{N-1}}{|w_{N-1}|} \end{bmatrix} - \frac{1}{2} \begin{bmatrix} \frac{e^{\jmath 2\angle w_0} dw_0^*}{|w_0|} & 0 & \cdots & 0 \\ 0 & \frac{e^{\jmath 2\angle w_1} dw_1^*}{|w_1|} & \cdots & 0 \\ \vdots & \vdots & \ddots & \vdots \\ 0 & 0 & \cdots & \frac{e^{\jmath 2\angle w_{N-1}} dw_{N-1}^*}{|w_{N-1}|} \end{bmatrix} \\ &= \frac{1}{2} \operatorname{diag}\left(|\boldsymbol{w}|^{-1} \odot d\boldsymbol{w} \right) - \frac{1}{2} \operatorname{diag}\left(e^{2\jmath\angle \boldsymbol{w}} \odot |\boldsymbol{w}|^{-1} \odot d\boldsymbol{w}^* \right), \end{aligned} \tag{7.32}$$

where $\odot$ denotes the Hadamard product defined in Definition 2.7, the special notation $|\boldsymbol{w}|^{-1}$ is defined in Definition 7.3, and $e^{\jmath\angle \boldsymbol{w}} \triangleq \left[e^{\jmath\angle w_0}, e^{\jmath\angle w_1}, \ldots, e^{\jmath\angle w_{N-1}} \right]^T$ follows from Definition 7.4. By complex conjugation of (7.32), it follows that

$$d \exp(-\jmath \operatorname{diag}(\angle \boldsymbol{w})) = -\frac{1}{2} \operatorname{diag}\left(e^{-2\jmath\angle \boldsymbol{w}} \odot |\boldsymbol{w}|^{-1} \odot d\boldsymbol{w} \right) + \frac{1}{2} \operatorname{diag}\left(|\boldsymbol{w}|^{-1} \odot d\boldsymbol{w}^* \right). \tag{7.33}$$

By putting together (7.31), (7.32), and (7.33), it is found that the second-order differential of g can be expressed as

$$d^2g = \left[d\boldsymbol{w}^H - \boldsymbol{r}^T\frac{1}{2}\left[-\operatorname{diag}\left(e^{-2\jmath\angle\boldsymbol{w}}\odot|\boldsymbol{w}|^{-1}\odot d\boldsymbol{w}\right)+\operatorname{diag}\left(|\boldsymbol{w}|^{-1}\odot d\boldsymbol{w}^*\right)\right]\right]d\boldsymbol{w}$$
$$+\left[d\boldsymbol{w}^T - \boldsymbol{r}^T\frac{1}{2}\left[\operatorname{diag}\left(|\boldsymbol{w}|^{-1}\odot d\boldsymbol{w}\right)-\operatorname{diag}\left(e^{2\jmath\angle\boldsymbol{w}}\odot|\boldsymbol{w}|^{-1}\odot d\boldsymbol{w}^*\right)\right]\right]d\boldsymbol{w}^*$$
$$=\left(d\boldsymbol{w}^H\right)\left[2\boldsymbol{I}_N-\operatorname{diag}\left(\boldsymbol{r}\odot|\boldsymbol{w}|^{-1}\right)\right]d\boldsymbol{w}+\left(d\boldsymbol{w}^T\right)\frac{1}{2}\operatorname{diag}\left(\boldsymbol{r}\odot|\boldsymbol{w}|^{-1}\odot e^{-\jmath 2\angle\boldsymbol{w}}\right)d\boldsymbol{w}$$
$$+\left(d\boldsymbol{w}^H\right)\frac{1}{2}\operatorname{diag}\left(\boldsymbol{r}\odot|\boldsymbol{w}|^{-1}\odot e^{\jmath 2\angle\boldsymbol{w}}\right)d\boldsymbol{w}^*, \tag{7.34}$$

where it is used that $\boldsymbol{a}^T\operatorname{diag}(\boldsymbol{b}\odot\boldsymbol{c}) = \boldsymbol{b}^T\operatorname{diag}(\boldsymbol{a}\odot\boldsymbol{c})$ for $\boldsymbol{a},\boldsymbol{b},\boldsymbol{c}\in\mathbb{C}^{N\times 1}$. From the theory developed in Chapter 5, it is possible to identify the Hessians of g from d^2g. This can be done by first noticing that the augmented complex-valued matrix variable $\mathcal{W}$ introduced in Subsection 5.2.2 is now given by

$$\mathcal{W} = \left[\boldsymbol{w}\ \boldsymbol{w}^*\right]\in\mathbb{C}^{N\times 2}. \tag{7.35}$$

To identify the Hessian matrix $\mathcal{H}_{\mathcal{W},\mathcal{W}}g\in\mathbb{C}^{2N\times 2N}$, the second-order differential d^2g should be rearranged into the same form as (5.53). This can be done as follows:

$$d^2g=\left(d\operatorname{vec}^T(\mathcal{W})\right)\begin{bmatrix}\frac{1}{2}\operatorname{diag}\left(\boldsymbol{r}\odot|\boldsymbol{w}|^{-1}\odot e^{-\jmath 2\angle\boldsymbol{w}}\right) & \boldsymbol{I}_N-\frac{1}{2}\operatorname{diag}\left(\boldsymbol{r}\odot|\boldsymbol{w}|^{-1}\right)\\ \boldsymbol{I}_N-\frac{1}{2}\operatorname{diag}\left(\boldsymbol{r}\odot|\boldsymbol{w}|^{-1}\right) & \frac{1}{2}\operatorname{diag}\left(\boldsymbol{r}\odot|\boldsymbol{w}|^{-1}\odot e^{\jmath 2\angle\boldsymbol{w}}\right)\end{bmatrix}$$
$$\times\, d\operatorname{vec}(\mathcal{W}) \triangleq \left(d\operatorname{vec}^T(\mathcal{W})\right)\boldsymbol{A}\,d\operatorname{vec}(\mathcal{W}), \tag{7.36}$$

where the middle matrix $\boldsymbol{A}\in\mathbb{C}^{2N\times 2N}$ was defined. It is observed that the matrix $\boldsymbol{A}$ is *symmetric* (i.e., $\boldsymbol{A}^T = \boldsymbol{A}$). The Hessian matrix $\mathcal{H}_{\mathcal{W},\mathcal{W}}g$ can now be identified from (5.55) as $\boldsymbol{A}$, hence,

$$\mathcal{H}_{\mathcal{W},\mathcal{W}}g = \boldsymbol{A} = \begin{bmatrix}\frac{1}{2}\operatorname{diag}\left(\boldsymbol{r}\odot|\boldsymbol{w}|^{-1}\odot e^{-\jmath 2\angle\boldsymbol{w}}\right) & \boldsymbol{I}_N-\frac{1}{2}\operatorname{diag}\left(\boldsymbol{r}\odot|\boldsymbol{w}|^{-1}\right)\\ \boldsymbol{I}_N-\frac{1}{2}\operatorname{diag}\left(\boldsymbol{r}\odot|\boldsymbol{w}|^{-1}\right) & \frac{1}{2}\operatorname{diag}\left(\boldsymbol{r}\odot|\boldsymbol{w}|^{-1}\odot e^{\jmath 2\angle\boldsymbol{w}}\right)\end{bmatrix}. \tag{7.37}$$

It remains to identify the complex Hessian matrix of the function h in (7.12). This complex Hessian matrix is denoted by $\mathcal{H}_{\mathcal{Z},\mathcal{Z}}h$, where the augmented complex-valued matrix variable is given as

$$\mathcal{Z} = [\boldsymbol{z},\ \boldsymbol{z}^*]\in\mathbb{C}^{N\times 2}. \tag{7.38}$$

The complex Hessian matrix $\mathcal{H}_{\mathcal{Z},\mathcal{Z}}h$ can be identified by the chain rule for complex Hessian matrices in Theorem 5.1. The function $\boldsymbol{F}:\mathbb{C}^{N\times 2}\to\mathbb{C}^{N\times 2}$ is first identified as

$$\boldsymbol{F}(\mathcal{Z}) = \left[\boldsymbol{F}_N^*\boldsymbol{z},\ \boldsymbol{F}_N\boldsymbol{z}^*\right] = \left[\left[\boldsymbol{F}_N^*\ \boldsymbol{0}_{N\times N}\right]\operatorname{vec}(\mathcal{Z}),\left[\boldsymbol{0}_{N\times N}\ \boldsymbol{F}_N\right]\operatorname{vec}(\mathcal{Z})\right]. \tag{7.39}$$

The derivative of $\boldsymbol{F}$ with respect to $\mathcal{Z}$ can be identified from

$$d\operatorname{vec}(\boldsymbol{F}) = \begin{bmatrix}\boldsymbol{F}_N^* & \boldsymbol{0}_{N\times N}\\ \boldsymbol{0}_{N\times N} & \boldsymbol{F}_N\end{bmatrix}d\operatorname{vec}(\mathcal{Z}). \tag{7.40}$$

From this expression of $d\,\text{vec}(\boldsymbol{F})$, it is seen that $d^2\boldsymbol{F} = \boldsymbol{0}_{2N\times 2N}$, and that

$$\mathcal{D}_{\mathcal{Z}}\boldsymbol{F} = \begin{bmatrix} \boldsymbol{F}_N^* & \boldsymbol{0}_{N\times N} \\ \boldsymbol{0}_{N\times N} & \boldsymbol{F}_N \end{bmatrix}. \tag{7.41}$$

The scalars R and S in Theorem 5.1 are identified as $R = S = 1$; hence, it follows from (5.91) that $\mathcal{H}_{\mathcal{Z},\mathcal{Z}}h$ is given by

$$\begin{aligned}\mathcal{H}_{\mathcal{Z},\mathcal{Z}}h &= \left[1\otimes(\mathcal{D}_{\mathcal{Z}}\boldsymbol{F})^T\right]\left[\mathcal{H}_{\mathcal{W},\mathcal{W}}g\right]\mathcal{D}_{\mathcal{Z}}\boldsymbol{F} = \begin{bmatrix} \boldsymbol{F}_N^* & \boldsymbol{0}_{N\times N} \\ \boldsymbol{0}_{N\times N} & \boldsymbol{F}_N \end{bmatrix} \boldsymbol{A} \begin{bmatrix} \boldsymbol{F}_N^* & \boldsymbol{0}_{N\times N} \\ \boldsymbol{0}_{N\times N} & \boldsymbol{F}_N \end{bmatrix} \\ &= \begin{bmatrix} \frac{1}{2}\boldsymbol{F}_N^*\,\text{diag}\left(\boldsymbol{r}\odot|\boldsymbol{w}|^{-1}\odot e^{-\jmath 2\angle \boldsymbol{w}}\right)\boldsymbol{F}_N^* & \boldsymbol{I}_N - \frac{1}{2}\boldsymbol{F}_N^*\,\text{diag}\left(\boldsymbol{r}\odot|\boldsymbol{w}|^{-1}\right)\boldsymbol{F}_N \\ \boldsymbol{I}_N - \frac{1}{2}\boldsymbol{F}_N\,\text{diag}\left(\boldsymbol{r}\odot|\boldsymbol{w}|^{-1}\right)\boldsymbol{F}_N^* & \frac{1}{2}\boldsymbol{F}_N\,\text{diag}\left(\boldsymbol{r}\odot|\boldsymbol{w}|^{-1}\odot e^{\jmath 2\angle \boldsymbol{w}}\right)\boldsymbol{F}_N \end{bmatrix}.\end{aligned} \tag{7.42}$$

It is observed that the final form of $\mathcal{H}_{\mathcal{Z},\mathcal{Z}}h$ in (7.42) is symmetric.

7.3 Minimization of Off-Diagonal Covariance Matrix Elements

This application example is related to the problem studied in Roman & Koivunen (2004) and Roman, Visuri, and Koivunen (2006), where blind CFO estimation in orthogonal frequency-division multiplexing (OFDM) is studied. Let the $N\times N$ covariance matrix $\boldsymbol{\Phi}$ be given by[1]

$$\boldsymbol{\Phi} = \boldsymbol{F}_N^H \boldsymbol{C}^H(\mu)\boldsymbol{R}\boldsymbol{C}(\mu)\boldsymbol{F}_N, \tag{7.43}$$

where $\boldsymbol{F}_N$ denotes the symmetric unitary $N\times N$ inverse DFT matrix (see Definition 7.5). The matrix $\boldsymbol{R}$ is a given $N\times N$ positive definite autocorrelation matrix, the diagonal $N\times N$ matrix $\boldsymbol{C}:\mathbb{R}\to\mathbb{C}^{N\times N}$ is dependent on the real variable μ, and $\boldsymbol{C}(\mu)$ is given by

$$\boldsymbol{C}(\mu) = \begin{bmatrix} 1 & 0 & \cdots & 0 \\ 0 & e^{\frac{\jmath 2\pi\mu}{N}} & \cdots & 0 \\ \vdots & \vdots & \ddots & \vdots \\ 0 & 0 & \cdots & e^{\frac{\jmath 2\pi\mu(N-1)}{N}} \end{bmatrix}, \tag{7.44}$$

where $\mu\in\mathbb{R}$. It can be shown that $\boldsymbol{\Phi}^H = \boldsymbol{\Phi}$ and $\boldsymbol{C}^H(\mu) = \boldsymbol{C}^*(\mu) = \boldsymbol{C}(-\mu)$. The covariance matrix $\boldsymbol{\Phi}$ in (7.43) is a simplified version of Roman and Koivunen (2004, Eq. (10)). If the matrix $\boldsymbol{\Phi}$ is diagonal, it means that the frequency offset is perfectly compensated. Hence, the objective is to choose μ such that the matrix $\boldsymbol{\Phi}$ becomes as close to a diagonal matrix as possible. One possible real scalar objective function $f(\mu)$ that could be minimized to make the matrix $\boldsymbol{\Phi}$ become as diagonal as possible is the sum of the squared magnitude of the *off-diagonal elements* of the matrix $\boldsymbol{\Phi}$. The term $\text{Tr}\left\{\boldsymbol{\Phi}\boldsymbol{\Phi}^H\right\}$ is the squared Frobenius norm of $\boldsymbol{\Phi}$, hence, it is the sum of the absolute squared value

[1] This covariance matrix corresponds to Roman and Koivunen (2004, Eq. (11)) when the cyclic prefix is set to 0.

of all elements of $\boldsymbol{\Phi}$. Furthermore, the term $\operatorname{Tr}\left\{\boldsymbol{\Phi} \odot \boldsymbol{\Phi}^*\right\}$ is the sum of all the squared absolute values of the *diagonal* elements of $\boldsymbol{\Phi}$. Hence, the objective function $f(\mu)$ can be expressed as

$$f(\mu) = \sum_{k=0}^{N-1} \sum_{\substack{l=0 \\ l \neq k}}^{N-1} \left|(\boldsymbol{\Phi})_{k,l}\right|^2 = \operatorname{Tr}\left\{\boldsymbol{\Phi}\boldsymbol{\Phi}^H\right\} - \operatorname{Tr}\left\{\boldsymbol{\Phi} \odot \boldsymbol{\Phi}^*\right\} = \operatorname{Tr}\left\{\boldsymbol{\Phi}^2\right\} - \operatorname{Tr}\left\{\boldsymbol{\Phi} \odot \boldsymbol{\Phi}\right\}, \tag{7.45}$$

where it has been used that $\boldsymbol{\Phi}$ is Hermitian and, therefore, has real diagonal elements.

The goal of this section is to find an expression of the derivative of $f(\mu)$ with respect to $\mu \in \mathbb{R}$. This will be accomplished by the developed theory in this book; in particular, the chain rule will be used.

Define the function $\boldsymbol{\Phi} : \mathbb{C}^{N \times N} \times \mathbb{C}^{N \times N} \to \mathbb{C}^{N \times N}$ given by

$$\boldsymbol{\Phi}(\tilde{\boldsymbol{C}}, \tilde{\boldsymbol{C}}^*) = \boldsymbol{F}_N^* \tilde{\boldsymbol{C}}^* \boldsymbol{R} \tilde{\boldsymbol{C}} \boldsymbol{F}_N, \tag{7.46}$$

where the matrix $\tilde{\boldsymbol{C}} \in \mathbb{C}^{N \times N}$ is an unpatterned version[2] of the matrix $\boldsymbol{C}$ given in (7.44). Define another function $g : \mathbb{C}^{N \times N} \to \mathbb{R}$ by

$$g(\tilde{\boldsymbol{\Phi}}) = \operatorname{Tr}\left\{\tilde{\boldsymbol{\Phi}}^2\right\} - \operatorname{Tr}\left\{\tilde{\boldsymbol{\Phi}} \odot \tilde{\boldsymbol{\Phi}}\right\}, \tag{7.47}$$

where $\tilde{\boldsymbol{\Phi}} \in \mathbb{C}^{N \times N}$ is a matrix with independent components. The total objective function $f(\mu)$ can be expressed as

$$f(\mu) = g(\tilde{\boldsymbol{\Phi}})\Big|_{\tilde{\boldsymbol{\Phi}} = \boldsymbol{\Phi}(\tilde{\boldsymbol{C}}, \tilde{\boldsymbol{C}}^*)}\Big|_{\tilde{\boldsymbol{C}} = \boldsymbol{C}(\mu)} = g(\tilde{\boldsymbol{\Phi}})\Big|_{\tilde{\boldsymbol{\Phi}} = \boldsymbol{\Phi}(\boldsymbol{C}(\mu), \boldsymbol{C}^*(\mu))} = g(\boldsymbol{\Phi}(\boldsymbol{C}(\mu), \boldsymbol{C}^*(\mu))). \tag{7.48}$$

Applying the chain rule in Theorem 3.1 leads to

$$\mathcal{D}_\mu f = \left(\mathcal{D}_{\tilde{\boldsymbol{\Phi}}} g\right)\Big|_{\tilde{\boldsymbol{\Phi}} = \boldsymbol{\Phi}(\boldsymbol{C}(\mu), \boldsymbol{C}^*(\mu))} \Big\{ \left(\mathcal{D}_{\tilde{\boldsymbol{C}}} \boldsymbol{\Phi}\right)\Big|_{\tilde{\boldsymbol{C}} = \boldsymbol{C}(\mu)} \mathcal{D}_\mu \boldsymbol{C} + \left(\mathcal{D}_{\tilde{\boldsymbol{C}}^*} \boldsymbol{\Phi}\right)\Big|_{\tilde{\boldsymbol{C}} = \boldsymbol{C}(\mu)} \mathcal{D}_\mu \boldsymbol{C}^* \Big\}. \tag{7.49}$$

All the derivatives in (7.49) are found in the rest of this section. The derivatives $\mathcal{D}_\mu \boldsymbol{C}$ and $\mathcal{D}_\mu \boldsymbol{C}^*$ are obtained by component-wise derivation, and they can be expressed as

$$\mathcal{D}_\mu \boldsymbol{C}(\mu) = \operatorname{vec}\left(\begin{bmatrix} 0 & 0 & \cdots & 0 \\ 0 & \frac{\jmath 2\pi}{N} e^{\frac{\jmath 2\pi\mu}{N}} & \cdots & 0 \\ \vdots & \vdots & \ddots & \vdots \\ 0 & 0 & \cdots & \frac{\jmath 2\pi(N-1)}{N} e^{\frac{\jmath 2\pi\mu(N-1)}{N}} \end{bmatrix} \right), \tag{7.50}$$

$$\mathcal{D}_\mu \boldsymbol{C}^*(\mu) = -\mathcal{D}_\mu \boldsymbol{C}(-\mu) = \left(\mathcal{D}_\mu \boldsymbol{C}(\mu)\right)^*. \tag{7.51}$$

The differential of the function $\boldsymbol{\Phi}$ is calculated as

$$d\boldsymbol{\Phi} = \boldsymbol{F}_N^* \tilde{\boldsymbol{C}}^* \boldsymbol{R} \left(d\tilde{\boldsymbol{C}}\right) \boldsymbol{F}_N + \boldsymbol{F}_N^* \left(d\tilde{\boldsymbol{C}}^*\right) \boldsymbol{R} \tilde{\boldsymbol{C}} \boldsymbol{F}_N. \tag{7.52}$$

[2] The same convention as used in Chapter 6 is used here to show that a matrix contains independent components. Hence, the symbol $\tilde{\boldsymbol{C}}$ is used for an unpatterned version of the matrix $\boldsymbol{C}$, which is a diagonal matrix (see (7.44)).

Applying the vec(·) operator to this equation leads to

$$d\,\text{vec}(\boldsymbol{\Phi}) = \left[\boldsymbol{F}_N \otimes \left(\boldsymbol{F}_N^* \tilde{\boldsymbol{C}}^* \boldsymbol{R}\right)\right] d\,\text{vec}\left(\tilde{\boldsymbol{C}}\right) + \left[\left(\boldsymbol{F}_N \tilde{\boldsymbol{C}}^T \boldsymbol{R}^T\right) \otimes \boldsymbol{F}_N^*\right] d\,\text{vec}\left(\tilde{\boldsymbol{C}}^*\right). \tag{7.53}$$

From this equation, the derivatives $\mathcal{D}_{\tilde{\boldsymbol{C}}}\boldsymbol{\Phi}$ and $\mathcal{D}_{\tilde{\boldsymbol{C}}^*}\boldsymbol{\Phi}$ are identified as follows:

$$\mathcal{D}_{\tilde{\boldsymbol{C}}}\boldsymbol{\Phi} = \boldsymbol{F}_N \otimes \left(\boldsymbol{F}_N^* \tilde{\boldsymbol{C}}^* \boldsymbol{R}\right), \tag{7.54}$$

$$\mathcal{D}_{\tilde{\boldsymbol{C}}^*}\boldsymbol{\Phi} = \left(\boldsymbol{F}_N \tilde{\boldsymbol{C}}^T \boldsymbol{R}^T\right) \otimes \boldsymbol{F}_N^*. \tag{7.55}$$

The derivative that remains to be found in (7.49) is $\mathcal{D}_{\tilde{\boldsymbol{\Phi}}}g$, and $\mathcal{D}_{\tilde{\boldsymbol{\Phi}}}g$ can be found through the differential of g in the following way:

$$\begin{aligned} dg &= 2\,\text{Tr}\left\{\tilde{\boldsymbol{\Phi}} d\tilde{\boldsymbol{\Phi}}\right\} - 2\,\text{Tr}\left\{\tilde{\boldsymbol{\Phi}} \odot d\tilde{\boldsymbol{\Phi}}\right\} \\ &= 2\,\text{vec}^T\left(\tilde{\boldsymbol{\Phi}}^T\right) d\,\text{vec}\left(\tilde{\boldsymbol{\Phi}}\right) - 2\,\text{vec}^T\left(\text{diag}(\text{vec}_d(\tilde{\boldsymbol{\Phi}}))\right) d\,\text{vec}\left(\tilde{\boldsymbol{\Phi}}\right) \\ &= 2\,\text{vec}^T\left(\tilde{\boldsymbol{\Phi}}^T - \text{diag}(\text{vec}_d(\tilde{\boldsymbol{\Phi}}))\right) d\,\text{vec}\left(\tilde{\boldsymbol{\Phi}}\right) \\ &= 2\,\text{vec}^T\left(\tilde{\boldsymbol{\Phi}}^T - \boldsymbol{I}_N \odot \tilde{\boldsymbol{\Phi}}\right) d\,\text{vec}\left(\tilde{\boldsymbol{\Phi}}\right), \end{aligned} \tag{7.56}$$

where the identities from Exercises 7.1 and 7.2 were utilized in the second and last equalities, respectively. Hence,

$$\mathcal{D}_{\tilde{\boldsymbol{\Phi}}}g = 2\,\text{vec}^T\left(\tilde{\boldsymbol{\Phi}}^T - \boldsymbol{I}_N \odot \tilde{\boldsymbol{\Phi}}\right), \tag{7.57}$$

and now the expression $\mathcal{D}_{\mu}f$ can be found by (7.49) because all the derivatives on the right-hand side of the equation have been found.

7.4 MIMO Precoder Design for Coherent Detection

This section follows the presentation given in Hjørungnes and Gesbert (2007*c*) and shows in greater detail how the theory of complex-valued matrix derivatives can be used to derive a fixed point method for precoder optimization. For an arbitrary given orthogonal space-time block code (OSTBC), exact symbol error rate (SER) expressions will be derived for a precoded MIMO system for communication over a correlated Ricean channel. The receiver employs maximum likelihood decoding (MLD) and has knowledge of the exact MIMO channel coefficients; the transmitter knows only the channel statistics, that is, the Ricean factor, the line-of-sight (LOS) component, and the autocorrelation matrix of the fading component of the channel. An iterative method is derived for finding the exact minimum SER precoder for M-PSK, M-PAM, and M-QAM signaling based on complex-valued matrix derivatives.

The rest of this section is organized as follows: In Subsection 7.4.1, the block model of the MIMO system, which constitutes a transmitter containing an OSTBC and a precoder, the MIMO channel, and the MLD in the receiver, is presented. A model for the

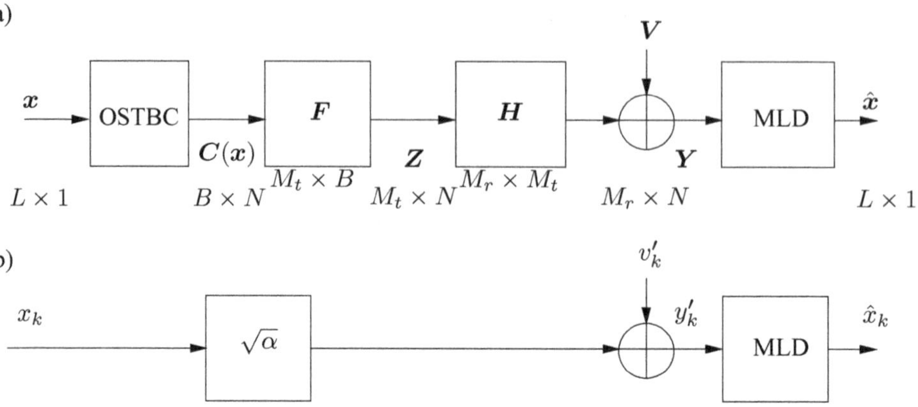

Figure 7.1 (a) Block model of the linearly precoded OSTBC MIMO system. (b) The equivalent system is given by L SISO system of this type. Adapted from Hjørungnes and Gesbert (2007*c*), © 2007 IEEE.

correlated Ricean MIMO channel is presented in Subsection 7.4.2. The studied MIMO system is equivalent to a single-input single-output (SISO) system, which is presented in Subsection 7.4.3. Exact SER expressions are derived in Subsection 7.4.4. The problem of finding the minimum SER precoder under a power constraint is formulated and solved in Subsection 7.4.5.

7.4.1 Precoded OSTBC System Model

Figure 7.1 (a) shows the block MIMO system model with M_t transmit and M_r receive antennas. One block of L symbols $x_0, x_1, \ldots, x_{L-1}$ is transmitted by means of an OSTBC matrix $\boldsymbol{C}(\boldsymbol{x})$ of size $B \times N$, where B and N are the space and time dimensions of the given OSTBC, respectively, and $\boldsymbol{x} = [x_0, x_1, \ldots, x_{L-1}]^T$. It is assumed that the OSTBC is given. Let $x_i \in \mathcal{A}$, where $\mathcal{A}$ is a signal constellation set such as M-PAM, M-QAM, or M-PSK. If bits are used as inputs to the system, $L \log_2 |\mathcal{A}|$ bits are used to produce the vector $\boldsymbol{x}$, where $|\cdot|$ denotes cardinality. Assume that $\mathbb{E}\left[|x_i|^2\right] = \sigma_x^2$ for all $i \in \{0, 1, \ldots, L-1\}$. Since the OSTBC $\boldsymbol{C}(\boldsymbol{x})$ is orthogonal, the following holds:

$$\boldsymbol{C}(\boldsymbol{x})\boldsymbol{C}^H(\boldsymbol{x}) = a \sum_{i=0}^{L-1} |x_i|^2 \boldsymbol{I}_B, \tag{7.58}$$

where the constant a is OSTBC dependent. For example, $a = 1$ if $\boldsymbol{C}(\boldsymbol{x}) = \mathcal{G}_2^T$, $\boldsymbol{C}(\boldsymbol{x}) = \mathcal{H}_3^T$, or $\boldsymbol{C}(\boldsymbol{x}) = \mathcal{H}_4^T$ in Tarokh, Jafarkhani, and Calderbank (1999, pp. 452–453), and $a = 2$ if $\boldsymbol{C}(\boldsymbol{x}) = (\mathcal{G}_c^3)^T$ or $\boldsymbol{C}(\boldsymbol{x}) = (\mathcal{G}_c^4)^T$ in Tarokh et al. (1999, p. 1464). However, the presented theory holds for any given OSTBC. The spatial rate of the code is L/N.

Before each code word $\boldsymbol{C}(\boldsymbol{x})$ is launched into the MIMO channel $\boldsymbol{H}$, it is precoded with a memoryless complex-valued matrix $\boldsymbol{F}$ of size $M_t \times B$, such that the $M_r \times N$

receive signal matrix $\boldsymbol{Y}$ becomes

$$\boldsymbol{Y} = \boldsymbol{HFC}(\boldsymbol{x}) + \boldsymbol{V}, \tag{7.59}$$

where the additive noise is contained in the block matrix $\boldsymbol{V}$ of size $M_r \times N$, where all the components are complex Gaussian circularly symmetric and distributed with independent components having variance N_0; $\boldsymbol{H}$ is the channel transfer MIMO matrix. The receiver is assumed to know the channel matrix $\boldsymbol{H}$ and the precoder matrix $\boldsymbol{F}$ exactly, and it performs MLD of block $\boldsymbol{Y}$ of size $M_r \times N$.

7.4.2 Correlated Ricean MIMO Channel Model

In this section, it is assumed that a quasi-static non–frequency selective correlated Ricean fading channel model (Paulraj et al. 2003) is used. Let $\boldsymbol{R}$ be the general $M_t M_r \times M_t M_r$ positive definite autocorrelation matrix for the fading part of the channel coefficients, and let $\sqrt{\frac{K}{1+K}}\bar{\boldsymbol{H}}$ be the mean value of the channel coefficients. The mean value represents the LOS component of the MIMO channel. The factor $K \geq 0$ is called the Ricean factor (Paulraj et al. 2003). A channel realization of the correlated channel is found from

$$\begin{aligned} \mathrm{vec}(\boldsymbol{H}) &= \sqrt{\frac{K}{1+K}}\,\mathrm{vec}\left(\bar{\boldsymbol{H}}\right) + \sqrt{\frac{1}{1+K}}\,\mathrm{vec}\left(\boldsymbol{H}_{\text{Fading}}\right) \\ &= \sqrt{\frac{K}{1+K}}\,\mathrm{vec}\left(\bar{\boldsymbol{H}}\right) + \sqrt{\frac{1}{1+K}}\,\boldsymbol{R}^{1/2}\,\mathrm{vec}(\boldsymbol{H}_w), \end{aligned} \tag{7.60}$$

where $\boldsymbol{R}^{1/2}$ is the unique positive definite matrix square root (Horn & Johnson 1991) of the assumed invertible matrix $\boldsymbol{R}$, where $\boldsymbol{R} = \mathbb{E}\left[\mathrm{vec}\left(\boldsymbol{H}_{\text{Fading}}\right)\mathrm{vec}^H\left(\boldsymbol{H}_{\text{Fading}}\right)\right]$ is the autocorrelation matrix of the $M_r \times M_t$ fading component $\boldsymbol{H}_{\text{Fading}}$ of the channel, and $\boldsymbol{H}_w$ of size $M_r \times M_t$ is complex Gaussian circularly symmetric distributed with independent components having zero mean and unit variance. The notation $\mathrm{vec}(\boldsymbol{H}_w) \sim \mathcal{CN}\left(\boldsymbol{0}_{M_t M_r \times 1}, \boldsymbol{I}_{M_t M_r}\right)$ is used to show that the distribution of the vector $\mathrm{vec}(\boldsymbol{H}_w)$ is circularly symmetric complex Gaussian with mean value $\boldsymbol{0}_{M_t M_r \times 1}$ given in the first argument in $\mathcal{CN}(\cdot,\cdot)$, and its autocovariance matrix $\boldsymbol{I}_{M_t M_r}$ in the second argument in $\mathcal{CN}(\cdot,\cdot)$. When using this notation, $\mathrm{vec}(\boldsymbol{H}) \sim \mathcal{CN}\left(\sqrt{\frac{K}{1+K}}\,\mathrm{vec}\left(\bar{\boldsymbol{H}}\right), \frac{1}{1+K}\boldsymbol{R}\right)$.

7.4.3 Equivalent Single-Input Single-Output Model

Define the positive semidefinite matrix $\boldsymbol{\Phi}$ of size $M_t M_r \times M_t M_r$ as

$$\boldsymbol{\Phi} = \boldsymbol{R}^{1/2}\left[\left(\boldsymbol{F}^*\boldsymbol{F}^T\right) \otimes \boldsymbol{I}_{M_r}\right]\boldsymbol{R}^{1/2}. \tag{7.61}$$

This matrix plays an important role in the presented theory in finding the instantaneous effective channel gain and the exact average SER. Let the eigenvalue decomposition of this Hermitian positive semidefinite matrix $\boldsymbol{\Phi}$ be given by

$$\boldsymbol{\Phi} = \boldsymbol{U}\boldsymbol{\Lambda}\boldsymbol{U}^H, \tag{7.62}$$

where $\boldsymbol{U} \in \mathbb{C}^{M_t M_r \times M_t M_r}$ is unitary and $\boldsymbol{\Lambda} \in \mathbb{R}^{M_t M_r \times M_t M_r}$ is a diagonal matrix containing the non-negative eigenvalues λ_i of $\boldsymbol{\Phi}$ on its main diagonal.

It is assumed that $\boldsymbol{R}$ is invertible. Define the real non-negative scalar α by

$$\begin{aligned}\alpha &\triangleq \|\boldsymbol{HF}\|_F^2 = \mathrm{Tr}\left\{\boldsymbol{I}_{M_r}\boldsymbol{HFF}^H\boldsymbol{H}^H\right\} = \mathrm{vec}^H(\boldsymbol{H})\left[\left(\boldsymbol{F}^*\boldsymbol{F}^T\right)\otimes\boldsymbol{I}_{M_r}\right]\mathrm{vec}(\boldsymbol{H})\\ &= \left[\sqrt{\frac{1}{1+K}}\,\mathrm{vec}^H(\boldsymbol{H}_w)\,\boldsymbol{R}^{1/2}+\sqrt{\frac{K}{1+K}}\,\mathrm{vec}^H\left(\bar{\boldsymbol{H}}\right)\right]\\ &\quad\times\left[\left(\boldsymbol{F}^*\boldsymbol{F}^T\right)\otimes\boldsymbol{I}_{M_r}\right]\left[\sqrt{\frac{1}{1+K}}\boldsymbol{R}^{1/2}\mathrm{vec}(\boldsymbol{H}_w)+\sqrt{\frac{K}{1+K}}\mathrm{vec}\left(\bar{\boldsymbol{H}}\right)\right]\\ &= \frac{1}{1+K}\left[\mathrm{vec}^H(\boldsymbol{H}_w)+\sqrt{K}\,\mathrm{vec}^H\left(\bar{\boldsymbol{H}}\right)\boldsymbol{R}^{-1/2}\right]\boldsymbol{\Phi}\\ &\quad\times\left[\mathrm{vec}(\boldsymbol{H}_w)+\sqrt{K}\boldsymbol{R}^{-1/2}\,\mathrm{vec}\left(\bar{\boldsymbol{H}}\right)\right],\end{aligned} \tag{7.63}$$

where (2.116) was used in the third equality. The scalar α can be rewritten by means of the eigen-decomposition of $\boldsymbol{\Phi}$ as

$$\alpha = \sum_{i=0}^{M_t M_r - 1}\frac{\lambda_i}{1+K}\left|\left(\mathrm{vec}\left(\boldsymbol{H}'_w\right)+\sqrt{K}\boldsymbol{U}^H\boldsymbol{R}^{-1/2}\,\mathrm{vec}\left(\bar{\boldsymbol{H}}\right)\right)_i\right|^2, \tag{7.64}$$

where $\mathrm{vec}\left(\boldsymbol{H}'_w\right) \sim \mathcal{CN}\left(\boldsymbol{0}_{M_t M_r\times 1}, \boldsymbol{I}_{M_t M_r}\right)$ has the same distribution as $\mathrm{vec}(\boldsymbol{H}_w)$.

By generalizing the approach given in Shin and Lee (2002) and Li, Luo, Yue, and Yin (2001) to include a *full* complex-valued precoder $\boldsymbol{F}$ of size $M_t \times B$ and having the channel correlation matrix $1/(1+K)\boldsymbol{R}$ and mean $\sqrt{K/(1+K)}\bar{\boldsymbol{H}}$, the MIMO system can be shown to be equivalent to a system having the following input-output relationship:

$$y'_k = \sqrt{\alpha}x_k + v'_k, \tag{7.65}$$

for $k \in \{0, 1, \ldots, L-1\}$, and where $v'_k \sim \mathcal{CN}(0, N_0/a)$ is complex circularly symmetric distributed. This signal is fed into a memory-less MLD that is designed from the signal constellation of the source symbol $\mathcal{A}$. The equivalent SISO model given in (7.65) is shown in Figure 7.1 (b). The equivalent SISO model is valid for any realization of $\boldsymbol{H}$.

7.4.4 Exact SER Expressions for Precoded OSTBC

By considering the SISO system in Figure 7.1 (b), it is seen that the instantaneous received signal-to-noise ratio (SNR) γ per source symbol is given by $\gamma \triangleq \frac{a\sigma_x^2\alpha}{N_0} = \delta\alpha$, where $\delta \triangleq \frac{a\sigma_x^2}{N_0}$. To simplify the expressions, the following three signal constellation–dependent constants are defined:

$$g_{\mathrm{PSK}} = \sin^2\frac{\pi}{M},\quad g_{\mathrm{PAM}} = \frac{3}{M^2-1},\quad g_{\mathrm{QAM}} = \frac{3}{2(M-1)}. \tag{7.66}$$

The symbol error probability $\mathrm{SER}_\gamma \triangleq \Pr\{\mathrm{Error}|\gamma\}$ for a given γ for M-PSK, M-PAM, and M-QAM signaling is given, respectively, by Simon and Alouini (2005, Eqs. (8.23),

(8.5), (8.12)).

$$\mathrm{SER}_\gamma = \frac{1}{\pi}\int_0^{\frac{(M-1)\pi}{M}} e^{-\frac{g_{\mathrm{PSK}}\gamma}{\sin^2\theta}}\,d\theta, \tag{7.67}$$

$$\mathrm{SER}_\gamma = \frac{2}{\pi}\frac{M-1}{M}\int_0^{\frac{\pi}{2}} e^{-\frac{g_{\mathrm{PAM}}\gamma}{\sin^2\theta}}\,d\theta, \tag{7.68}$$

$$\mathrm{SER}_\gamma = \frac{4}{\pi}\left(1-\frac{1}{\sqrt{M}}\right)\left[\frac{1}{\sqrt{M}}\int_0^{\frac{\pi}{4}} e^{-\frac{g_{\mathrm{QAM}}\gamma}{\sin^2\theta}}\,d\theta + \int_{\frac{\pi}{4}}^{\frac{\pi}{2}} e^{-\frac{g_{\mathrm{QAM}}\gamma}{\sin^2\theta}}\,d\theta\right]. \tag{7.69}$$

The moment generating function of the probability density function $p_\gamma(\gamma)$ is defined as $\phi_\gamma(s) = \int_0^\infty p_\gamma(\gamma)e^{s\gamma}d\gamma$. Because all L source symbols go through the same SISO system in Figure 7.1 (b), the average SER of the MIMO system can be found as

$$\mathrm{SER} \triangleq \Pr\{\mathrm{Error}\} = \int_0^\infty \mathrm{SER}_\gamma\, p_\gamma(\gamma)d\gamma. \tag{7.70}$$

This integral can be rewritten in terms of the moment generating function of γ. Since $\mathrm{vec}\left(\boldsymbol{H}'_w\right) + \sqrt{K}\boldsymbol{U}^H\boldsymbol{R}^{-1/2}\,\mathrm{vec}\left(\bar{\boldsymbol{H}}\right) \sim \mathcal{CN}\left(\sqrt{K}\boldsymbol{U}^H\boldsymbol{R}^{-1/2}\,\mathrm{vec}\left(\bar{\boldsymbol{H}}\right), \boldsymbol{I}_{M_tM_r}\right)$, it follows by straightforward manipulations from Turin (1960, Eq. (4a)), that the moment generating function of α can be written as

$$\phi_\alpha(s) = \frac{e^{-K\,\mathrm{vec}^H\left(\bar{\boldsymbol{H}}\right)\boldsymbol{R}^{-1/2}\left[\boldsymbol{I}_{M_tM_r} - \left[\boldsymbol{I}_{M_tM_r} - \frac{s}{1+K}\boldsymbol{\Phi}\right]^{-1}\right]\boldsymbol{R}^{-1/2}\,\mathrm{vec}\left(\bar{\boldsymbol{H}}\right)}}{\det\left(\boldsymbol{I}_{M_tM_r} - \frac{s}{1+K}\boldsymbol{\Phi}\right)}. \tag{7.71}$$

Because $\gamma = \delta\alpha$, the moment generating function of γ is given by

$$\phi_\gamma(s) = \phi_\alpha(\delta s). \tag{7.72}$$

By using (7.70) and the definition of the moment generating function together with (7.72), it is possible to express the exact SER for all signal constellations in terms of the eigenvalues λ_i and eigenvectors $\boldsymbol{u}_i$ of the matrix $\boldsymbol{\Phi}$:

$$\mathrm{SER} = \frac{1}{\pi}\int_0^{\frac{M-1}{M}\pi} \phi_\gamma\left(-\frac{g_{\mathrm{PSK}}}{\sin^2\theta}\right)d\theta, \tag{7.73}$$

$$\mathrm{SER} = \frac{2}{\pi}\frac{M-1}{M}\int_0^{\frac{\pi}{2}} \phi_\gamma\left(-\frac{g_{\mathrm{PAM}}}{\sin^2\theta}\right)d\theta, \tag{7.74}$$

$$\mathrm{SER} = \frac{4}{\pi}\left(1-\frac{1}{\sqrt{M}}\right)\left[\frac{1}{\sqrt{M}}\int_0^{\frac{\pi}{4}} \phi_\gamma\left(-\frac{g_{\mathrm{QAM}}}{\sin^2\theta}\right)d\theta + \int_{\frac{\pi}{4}}^{\frac{\pi}{2}} \phi_\gamma\left(-\frac{g_{\mathrm{QAM}}}{\sin^2\theta}\right)d\theta\right], \tag{7.75}$$

for PSK, PAM, and QAM signaling, respectively.

Define the positive definite matrix $\boldsymbol{A}$ of size $M_tM_r \times M_tM_r$ as

$$\boldsymbol{A} = \boldsymbol{I}_{M_tM_r} + \frac{\delta g}{(1+K)\sin^2(\theta)}\boldsymbol{\Phi}, \tag{7.76}$$

where g takes one of the forms in (7.66). The symbols $\boldsymbol{A}^{(\mathrm{PSK})}$, $\boldsymbol{A}^{(\mathrm{PAM})}$, and $\boldsymbol{A}^{(\mathrm{QAM})}$ are used for the PSK, PAM, and QAM constellations, respectively.

To present the SER expressions compactly, define the following real non-negative scalar function, which is dependent on the LOS component $\bar{\boldsymbol{H}}$, the Ricean factor K, and the correlation of the channel $\boldsymbol{R}$, as

$$f(\boldsymbol{X}) = \frac{e^{K \operatorname{vec}^H(\bar{\boldsymbol{H}})\boldsymbol{R}^{-1/2}\boldsymbol{X}^{-1}\boldsymbol{R}^{-1/2}\operatorname{vec}(\bar{\boldsymbol{H}})}}{|\det(\boldsymbol{X})|}, \tag{7.77}$$

where the argument matrix $\boldsymbol{X} \in \mathbb{C}^{M_t M_r \times M_t M_r}$ is nonsingular and Hermitian.

By inserting (7.72) into (7.73), (7.74), and (7.75) and utilizing the function defined in (7.77), the following exact SER expressions are found

$$\text{SER} = \frac{f(-\boldsymbol{I}_{M_t M_r})}{\pi} \int_0^{\frac{M-1}{M}\pi} f(\boldsymbol{A}^{(\text{PSK})}) d\theta, \tag{7.78}$$

$$\text{SER} = \frac{2 f(-\boldsymbol{I}_{M_t M_r})}{\pi} \frac{M-1}{M} \int_0^{\frac{\pi}{2}} f(\boldsymbol{A}^{(\text{PAM})}) d\theta, \tag{7.79}$$

$$\text{SER} = \frac{4 f(-\boldsymbol{I}_{M_t M_r})}{\pi} \left(1 - \frac{1}{\sqrt{M}}\right) \left[\frac{1}{\sqrt{M}} \int_0^{\frac{\pi}{4}} f(\boldsymbol{A}^{(\text{QAM})}) d\theta + \int_{\frac{\pi}{4}}^{\frac{\pi}{2}} f(\boldsymbol{A}^{(\text{QAM})}) d\theta\right], \tag{7.80}$$

for PSK, PAM, and QAM signaling, respectively.

7.4.5 Precoder Optimization Problem Statement and Optimization Algorithm

This subsection contains two parts. The first part formulates the precoder optimization problem. The second part shows how the problem can be solved by a fixed point iteration, and this iteration is derived using complex-valued matrix derivatives.

7.4.5.1 Optimal Precoder Problem Formulation

When an OSTBC is used, (7.58) holds, and the average power constraint on the transmitted block $\boldsymbol{Z} \triangleq \boldsymbol{F}\boldsymbol{C}(\boldsymbol{x})$ is given by $\operatorname{Tr}\{\boldsymbol{Z}\boldsymbol{Z}^H\} = P$; this is equivalent to

$$a L \sigma_x^2 \operatorname{Tr}\left\{\boldsymbol{F}\boldsymbol{F}^H\right\} = P, \tag{7.81}$$

where P is the average power used by the transmitted block $\boldsymbol{Z}$. The goal is to find the precoder matrix $\boldsymbol{F}$, such that the exact SER is minimized under the power constraint. Note that the same precoder is used over all realizations of the fading channel, as it is assumed that only channel statistics are fed back to the transmitter. In general, the optimal precoder is dependent on N_0 and, therefore, also on the SNR. The optimal precoder is given by the following optimization problem:

Problem 7.1

$$\min_{\{\boldsymbol{F} \in \mathbb{C}^{M_t \times B}\}} \quad \text{SER},$$

$$\text{subject to} \tag{7.82}$$

$$L a \sigma_x^2 \operatorname{Tr}\left\{\boldsymbol{F}\boldsymbol{F}^H\right\} = P.$$

7.4.5.2 Precoder Optimization Algorithm

The constrained minimization in Problem 7.1 can be converted into an unconstrained optimization problem by introducing a Lagrange multiplier $\mu' > 0$. This is done by defining the following Lagrangian function:

$$\mathcal{L}(\boldsymbol{F}, \boldsymbol{F}^*) = \text{SER} + \mu' \operatorname{Tr}\left\{\boldsymbol{F}\boldsymbol{F}^H\right\}. \tag{7.83}$$

Define the $M_t^2 \times M_t^2 M_r^2$ matrix,

$$\boldsymbol{\Pi} \triangleq \left[\boldsymbol{I}_{M_t^2} \otimes \operatorname{vec}^T\left(\boldsymbol{I}_{M_r}\right)\right]\left[\boldsymbol{I}_{M_t} \otimes \boldsymbol{K}_{M_t,M_r} \otimes \boldsymbol{I}_{M_r}\right]. \tag{7.84}$$

To present the results compactly, define the $BM_t \times 1$ vector $\boldsymbol{q}(\boldsymbol{F}, \theta, g, \mu)$ as follows:

$$\begin{aligned}\boldsymbol{q}(\boldsymbol{F}, \theta, g, \mu) = \mu &\left[\boldsymbol{F}^T \otimes \boldsymbol{I}_{M_t}\right] \boldsymbol{\Pi} \left[\boldsymbol{R}^{1/2} \otimes \left(\boldsymbol{R}^{1/2}\right)^T\right] \\ &\times \operatorname{vec}^*\left(\boldsymbol{A}^{-1} + K\boldsymbol{A}^{-1}\boldsymbol{R}^{-1/2}\operatorname{vec}\left(\bar{\boldsymbol{H}}\right)\operatorname{vec}^H\left(\bar{\boldsymbol{H}}\right)\boldsymbol{R}^{-1/2}\boldsymbol{A}^{-1}\right) \\ &\times \frac{e^{K\operatorname{vec}^H(\bar{\boldsymbol{H}})\boldsymbol{R}^{-1/2}\boldsymbol{A}^{-1}\boldsymbol{R}^{-1/2}\operatorname{vec}(\bar{\boldsymbol{H}})}}{\sin^2(\theta)\det(\boldsymbol{A})}.\end{aligned} \tag{7.85}$$

Theorem 7.1 *The precoder that is optimal in* (7.82) *must satisfy*

$$\operatorname{vec}(\boldsymbol{F}) = \int_0^{\frac{M-1}{M}\pi} \boldsymbol{q}(\boldsymbol{F}, \theta, g_{\text{PSK}}, \mu)d\theta, \tag{7.86}$$

$$\operatorname{vec}(\boldsymbol{F}) = \int_0^{\frac{\pi}{2}} \boldsymbol{q}(\boldsymbol{F}, \theta, g_{\text{PAM}}, \mu)d\theta, \tag{7.87}$$

$$\operatorname{vec}(\boldsymbol{F}) = \frac{1}{\sqrt{M}}\int_0^{\frac{\pi}{4}} \boldsymbol{q}(\boldsymbol{F}, \theta, g_{\text{QAM}}, \mu)d\theta + \int_{\frac{\pi}{4}}^{\frac{\pi}{2}} \boldsymbol{q}(\boldsymbol{F}, \theta, g_{\text{QAM}}, \mu)d\theta. \tag{7.88}$$

for the M-PSK, M-PAM, and M-QAM constellations, respectively. The scalar μ is positive and is chosen such that the power constraint in (7.81) *is satisfied.*

Proof The necessary conditions for the optimality of (7.82) are found by setting the derivative of the Lagrangian function $\mathcal{L}(\boldsymbol{F}, \boldsymbol{F}^*)$ in (7.83) with respect to $\operatorname{vec}(\boldsymbol{F}^*)$ equal to the zero vector of size $M_t B \times 1$.

The simplest part of the Lagrangian function $\mathcal{L}(\boldsymbol{F}, \boldsymbol{F}^*)$ to differentiate is the second term $\mu' \operatorname{Tr}\{\boldsymbol{F}\boldsymbol{F}^H\}$, which has derivative wrt. $\boldsymbol{F}^*$ given by

$$\begin{aligned}\mathcal{D}_{\boldsymbol{F}^*}\left[\mu' \operatorname{Tr}\{\boldsymbol{F}\boldsymbol{F}^H\}\right] &= \mu'\mathcal{D}_{\boldsymbol{F}^*}\left[\operatorname{vec}^H(\boldsymbol{F})\operatorname{vec}(\boldsymbol{F})\right] = \mu'\mathcal{D}_{\boldsymbol{F}^*}\left[\operatorname{vec}^T(\boldsymbol{F})\operatorname{vec}(\boldsymbol{F}^*)\right] \\ &= \mu' \operatorname{vec}^T(\boldsymbol{F}).\end{aligned} \tag{7.89}$$

It is observed from the exact expressions of the SER in (7.78), (7.79), and (7.80) that all these expressions have similar forms. Hence, it is enough to consider only the M-PSK case; the M-PAM and M-QAM cases follow in a similar manner.

When finding the derivative of the SER for M-PSK with respect to $\boldsymbol{F}^*$, it is first seen that the factor in front of the integral expression in (7.78), that is, $\frac{f(-\boldsymbol{I}_{M_t M_r})}{\pi}$, is a non-negative scalar, which is independent of the precoder matrix $\boldsymbol{F}$ and its complex conjugated $\boldsymbol{F}^*$.

From Table 3.1, we know that $de^z = e^z dz$ and $d\det(\boldsymbol{Z}) = \det(\boldsymbol{Z})\,\mathrm{Tr}\{\boldsymbol{Z}^{-1}d\boldsymbol{Z}\}$, and these will now be used. The differential of the integral of the SER in (7.78) can be written as

$$
\begin{aligned}
d\int_0^{\frac{M-1}{M}\pi} f(\boldsymbol{A}^{(\mathrm{PSK})})d\theta &= d\int_0^{\frac{M-1}{M}\pi} \frac{e^{K\,\mathrm{vec}^H(\bar{\boldsymbol{H})}\boldsymbol{R}^{-1/2}\left[\boldsymbol{A}^{(\mathrm{PSK})}\right]^{-1}\boldsymbol{R}^{-1/2}\,\mathrm{vec}(\bar{\boldsymbol{H}})}}{\det\left(\boldsymbol{A}^{(\mathrm{PSK})}\right)}d\theta \\
&= \int_0^{\frac{M-1}{M}\pi}\left[\frac{d\left\{e^{K\,\mathrm{vec}^H(\bar{\boldsymbol{H}})\boldsymbol{R}^{-1/2}\left[\boldsymbol{A}^{(\mathrm{PSK})}\right]^{-1}\boldsymbol{R}^{-1/2}\,\mathrm{vec}(\bar{\boldsymbol{H}})}\right\}}{\det\left(\boldsymbol{A}^{(\mathrm{PSK})}\right)}\right. \\
&\quad \left. + e^{K\,\mathrm{vec}^H(\bar{\boldsymbol{H}})\boldsymbol{R}^{-1/2}\left[\boldsymbol{A}^{(\mathrm{PSK})}\right]^{-1}\boldsymbol{R}^{-1/2}\,\mathrm{vec}(\bar{\boldsymbol{H}})} d\left\{\frac{1}{\det\left(\boldsymbol{A}^{(\mathrm{PSK})}\right)}\right\}\right]d\theta \\
&= \int_0^{\frac{M-1}{M}\pi}\left[\frac{e^h K\,\mathrm{vec}^H\left(\bar{\boldsymbol{H}}\right)\boldsymbol{R}^{-1/2}\left(d\left[\boldsymbol{A}^{(\mathrm{PSK})}\right]^{-1}\right)\boldsymbol{R}^{-1/2}\,\mathrm{vec}\left(\bar{\boldsymbol{H}}\right)}{\det\left(\boldsymbol{A}^{(\mathrm{PSK})}\right)}\right. \\
&\quad \left. - \frac{e^h}{\det^2\left(\boldsymbol{A}^{(\mathrm{PSK})}\right)}d\det\left(\boldsymbol{A}^{(\mathrm{PSK})}\right)\right]d\theta \\
&= \int_0^{\frac{M-1}{M}\pi}\frac{e^h}{\det\left(\boldsymbol{A}^{(\mathrm{PSK})}\right)}\left[K\,\mathrm{vec}^H\left(\bar{\boldsymbol{H}}\right)\boldsymbol{R}^{-1/2}\left(d\left[\boldsymbol{A}^{(\mathrm{PSK})}\right]^{-1}\right)\boldsymbol{R}^{-1/2}\,\mathrm{vec}\left(\bar{\boldsymbol{H}}\right)\right. \\
&\quad \left. - \mathrm{Tr}\left\{\left(\boldsymbol{A}^{(\mathrm{PSK})}\right)^{-1}d\boldsymbol{A}^{(\mathrm{PSK})}\right\}\right]d\theta,
\end{aligned} \tag{7.90}
$$

where the exponent of the exponential function h is introduced to simplify the expressions, and h is defined as

$$
h \triangleq K\,\mathrm{vec}^H\left(\bar{\boldsymbol{H}}\right)\boldsymbol{R}^{-1/2}\left[\boldsymbol{A}^{(\mathrm{PSK})}\right]^{-1}\boldsymbol{R}^{-1/2}\,\mathrm{vec}\left(\bar{\boldsymbol{H}}\right). \tag{7.91}
$$

To proceed, it is seen that an expression of $d\boldsymbol{A}^{(\mathrm{PSK})}$ is needed, where $\boldsymbol{A}^{(\mathrm{PSK})}$ can be expressed as

$$
\boldsymbol{A}^{(\mathrm{PSK})} = \boldsymbol{I}_{M_t M_r} + \frac{\delta g_{\mathrm{PSK}}}{(1+K)\sin^2(\theta)}\boldsymbol{R}^{1/2}\left[\left(\boldsymbol{F}^*\boldsymbol{F}^T\right)\otimes\boldsymbol{I}_{M_r}\right]\boldsymbol{R}^{1/2}. \tag{7.92}
$$

The differential $d\,\mathrm{vec}\left(\boldsymbol{A}^{(\mathrm{PSK})}\right)$ is derived in Exercise 7.4 and is stated in (7.142).

Using the fact that $d\boldsymbol{A}^{-1} = -\boldsymbol{A}^{-1}(d\boldsymbol{A})\boldsymbol{A}^{-1}$ and (2.205), it is seen that (7.90) can be rewritten as

$$
\begin{aligned}
&\int_0^{\frac{M-1}{M}\pi}\frac{-e^h}{\det(\boldsymbol{A})}\left[\mathrm{Tr}\left\{K\boldsymbol{R}^{-1/2}\,\mathrm{vec}\left(\bar{\boldsymbol{H}}\right)\mathrm{vec}^H\left(\bar{\boldsymbol{H}}\right)\boldsymbol{R}^{-1/2}\boldsymbol{A}^{-1}(d\boldsymbol{A})\boldsymbol{A}^{-1}\right\}\right. \\
&\quad \left. + \mathrm{Tr}\left\{\boldsymbol{A}^{-1}d\boldsymbol{A}\right\}\right]d\theta = \int_0^{\frac{M-1}{M}\pi}\frac{-e^h}{\det(\boldsymbol{A})}\mathrm{vec}^H\left(K\boldsymbol{A}^{-1}\boldsymbol{R}^{-1/2}\,\mathrm{vec}\left(\bar{\boldsymbol{H}}\right)\right. \\
&\quad \left.\mathrm{vec}^H\left(\bar{\boldsymbol{H}}\right)\boldsymbol{R}^{-1/2}\boldsymbol{A}^{-1} + \boldsymbol{A}^{-1}\right)d\,\mathrm{vec}(\boldsymbol{A})\,d\theta,
\end{aligned} \tag{7.93}
$$

where the dependency on PSK has been dropped for simplicity, and it has been used that $\boldsymbol{A}^H = \boldsymbol{A}$. After putting together the results derived above and using the results from

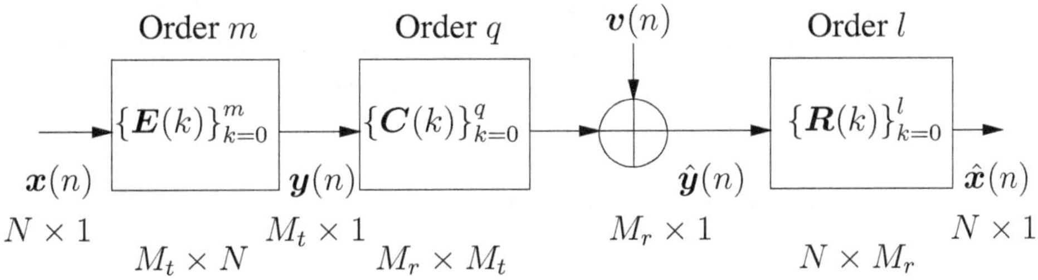

Figure 7.2 FIR MIMO block system model.

Exercise 7.4, it is seen that

$$\begin{aligned}\mathcal{D}_{\boldsymbol{F}^*}\left[\int_0^{\frac{M-1}{M}\pi} \frac{e^h}{\det(\boldsymbol{A})} d\theta\right] &= \int_0^{\frac{M-1}{M}\pi} \frac{-e^h}{\det(\boldsymbol{A})} \frac{\delta g}{(1+K)\sin^2\theta} \\ &\quad \operatorname{vec}^H\left(K\boldsymbol{A}^{-1}\boldsymbol{R}^{-1/2}\operatorname{vec}\left(\bar{\boldsymbol{H}}\right)\operatorname{vec}^H\left(\bar{\boldsymbol{H}}\right)\boldsymbol{R}^{-1/2}\boldsymbol{A}^{-1} + \boldsymbol{A}^{-1}\right) d\theta \\ &\quad \times \left[\left(\boldsymbol{R}^{1/2}\right)^T \otimes \boldsymbol{R}^{1/2}\right] \boldsymbol{\Pi}^T \left[\boldsymbol{F} \otimes \boldsymbol{I}_{M_t}\right]. \end{aligned} \tag{7.94}$$

By using the results in (7.89) and (7.94) in the equation

$$\mathcal{D}_{\boldsymbol{F}^*}\mathcal{L} = \boldsymbol{0}_{1\times M_t B}, \tag{7.95}$$

it is seen that the fixed point equation in (7.86) follows.

The fixed point equations for PAM and QAM in (7.87) and (7.88), respectively, can be derived in a similar manner. ■

Precoder Optimization Algorithm: Equations (7.86), (7.87), and (7.88) can be used in a fixed point iteration (Naylor & Sell 1982) to find the precoder that solves Problem 7.1. This is done by inserting an initial precoder value in the right-hand side of the equations of Theorem 7.1, that is, (7.86), (7.87), and (7.88), and by evaluating the corresponding integrals to obtain an improved value of the precoder. This process is repeated until the one-step change in $\boldsymbol{F}$ is less than some preset threshold. Notice that the positive constants μ' and μ are different. When we used this algorithm, convergence was always observed.

7.5 Minimum MSE FIR MIMO Transmit and Receive Filters

The application example presented in this section is based on a simplified version of the system studied in Hjørungnes, de Campos, and Diniz (2005). We will consider the FIR MIMO system in Figure 7.2. The transmit and receive filters are minimized with respect to the MSE between the output signal and a delayed version of the original input signal subject to an average transmitted power constraint.

The rest of this section is organized as follows: In Subsection 7.5.1, the FIR MIMO system model is introduced. Subsection 7.5.2 contains special notation that is useful for presenting compact expressions when working with FIR MIMO filters. The problem of finding the FIR MIMO transmit and receive filters is formulated in Subsection 7.5.3. In Subsection 7.5.4, it is shown how to find the equation for the minimum MSE FIR MIMO receive filter when the FIR MIMO transmit filter is fixed. For a fixed FIR MIMO receive filter, the minimum MSE FIR MIMO transmit filter is derived in Subsection 7.5.5 under a constraint on the average transmitted power.

7.5.1 FIR MIMO System Model

Consider the FIR MIMO system model shown in Figure 7.2. As explained in detail in Scaglione, Giannakis, and Barbarossa (1999), the system in Figure 7.2 includes time division multiple access (TDMA), OFDM, code division multiple access (CDMA), and several other structures as special cases. In this section, the symbol n is used as a time index, and n is an integer, that is, $n \in \mathbb{Z} = \{\ldots, -2, -1, 0, 1, 2, \ldots\}$.

The sizes of all time-series and FIR MIMO filters in the system are shown below the corresponding mathematical symbols within Figure 7.2. Two input time-series are included in the system in Figure 7.2, and these are the original time-series $\boldsymbol{x}(n)$ of size $N \times 1$ and the additive channel noise time-series $\boldsymbol{v}(n)$ of size $M_r \times 1$. The channel input time-series $\boldsymbol{y}(n)$ and the channel output time-series $\hat{\boldsymbol{y}}(n)$ have sizes $M_t \times 1$ and $M_r \times 1$, respectively. The output of the system is the time-series $\hat{\boldsymbol{x}}(n)$ of size $N \times 1$. There are no constraints on the values of N, M_t, and M_r, except that they need to be positive integers. It is assumed that all vector time-series in Figure 7.2 are jointly wide sense stationary.

FIR MIMO filters are used to model the transfer functions of the transmitter, the channel, and the receiver. The three boxes shown from left to right in Figure 7.2 are the $M_t \times N$ transmit FIR MIMO $\boldsymbol{E}$ with coefficients $\{\boldsymbol{E}(k)\}_{k=0}^{m}$, the $M_r \times M_t$ channel FIR MIMO filter $\boldsymbol{C}$ with coefficients $\{\boldsymbol{C}(k)\}_{k=0}^{q}$, and the $N \times M_r$ receive FIR MIMO filter $\boldsymbol{R}$ with coefficients $\{\boldsymbol{R}(k)\}_{k=0}^{l}$. The orders of the transmitter, channel, and receiver are m, q, and l, respectively, and they are assumed to be known non-negative integers. The sizes of the filter coefficient matrices $\boldsymbol{E}(k)$, $\boldsymbol{C}(k)$, and $\boldsymbol{R}(k)$ are $M_t \times N$, $M_r \times M_t$, and $N \times M_r$, respectively. The channel matrix coefficients $\boldsymbol{C}(k)$ are assumed to be known both at the transmitter and at the receiver.

7.5.2 FIR MIMO Filter Expansions

Four expansion operators for FIR MIMO filters and one operator for vector time-series are useful for a compact mathematical description of linear FIR MIMO systems.

Let $\{\boldsymbol{A}(i)\}_{i=0}^{\eta}$ be the filter coefficients of an FIR MIMO filter of order η and size $M_0 \times M_1$. The z-transform (Vaidyanathan 1993) of this FIR MIMO filter is given by $\sum_{i=0}^{\eta} \boldsymbol{A}(i) z^{-i}$. The matrix $\boldsymbol{A}(i)$ is the i-th coefficient of the FIR MIMO filter denoted by $\boldsymbol{A}$, and it has size $M_0 \times M_1$.

Definition 7.6 *The* row-expanded *matrix $\boldsymbol{A}_{-}$ of the FIR MIMO filter $\boldsymbol{A}$ with filter coefficients $\{\boldsymbol{A}(i)\}_{i=0}^{\eta}$, where $\boldsymbol{A}(i) \in \mathbb{C}^{M_0 \times M_1}$, is defined as the $M_0 \times (\eta+1)M_1$ matrix:*

$$\boldsymbol{A}_{-} = [\boldsymbol{A}(0)\ \boldsymbol{A}(1)\ \cdots\ \boldsymbol{A}(\eta)]. \tag{7.96}$$

Definition 7.7 *The* column-expanded *matrix $\boldsymbol{A}_{|}$ of the FIR MIMO filter $\boldsymbol{A}$ with filter coefficients $\{\boldsymbol{A}(i)\}_{i=0}^{\eta}$, where $\boldsymbol{A}(i) \in \mathbb{C}^{M_0 \times M_1}$, is defined as the $(\eta+1)M_0 \times M_1$ matrix given by*

$$\boldsymbol{A}_{|} = \begin{bmatrix} \boldsymbol{A}(\eta) \\ \boldsymbol{A}(\eta-1) \\ \vdots \\ \boldsymbol{A}(1) \\ \boldsymbol{A}(0) \end{bmatrix}. \tag{7.97}$$

Definition 7.8 *Let q be a non-negative integer. The* row-diagonal-expanded *matrix $\boldsymbol{A}_{\neg}^{(q)}$ of order q of the FIR MIMO filter $\boldsymbol{A}$ with filter coefficients $\{\boldsymbol{A}(i)\}_{i=0}^{\eta}$, where $\boldsymbol{A}(i) \in \mathbb{C}^{M_0 \times M_1}$, is defined as the $(q+1)M_0 \times (\eta+q+1)M_1$ matrix given by*

$$\boldsymbol{A}_{\neg}^{(q)} = \begin{bmatrix} \boldsymbol{A}(0) & \boldsymbol{A}(1) & \boldsymbol{A}(2) & \cdots & \boldsymbol{A}(\eta) & \boldsymbol{0}_{M_0 \times M_1} & \cdots & \boldsymbol{0}_{M_0 \times M_1} \\ \boldsymbol{0}_{M_0 \times M_1} & \boldsymbol{A}(0) & \boldsymbol{A}(1) & \cdots & \boldsymbol{A}(\eta-1) & \boldsymbol{A}(\eta) & \cdots & \boldsymbol{0}_{M_0 \times M_1} \\ \vdots & & \ddots & \ddots & & \ddots & \ddots & \vdots \\ \boldsymbol{0}_{M_0 \times M_1} & \boldsymbol{0}_{M_0 \times M_1} & \cdots & \boldsymbol{A}(0) & \boldsymbol{A}(1) & \cdots & \boldsymbol{A}(\eta-1) & \boldsymbol{A}(\eta) \end{bmatrix}. \tag{7.98}$$

Definition 7.9 *Let q be a non-negative integer. The* column-diagonal-expanded *matrix $\boldsymbol{A}_{\llcorner}^{(q)}$ of order q of the FIR MIMO filter $\boldsymbol{A}$ with filter coefficients $\{\boldsymbol{A}(i)\}_{i=0}^{\eta}$, where $\boldsymbol{A}(i) \in \mathbb{C}^{M_0 \times M_1}$, is defined as the $(\eta+q+1)M_0 \times (q+1)M_1$ matrix given by*

$$\boldsymbol{A}_{\llcorner}^{(q)} = \begin{bmatrix} \boldsymbol{A}(\eta) & \boldsymbol{0}_{M_0 \times M_1} & \cdots & \boldsymbol{0}_{M_0 \times M_1} \\ \boldsymbol{A}(\eta-1) & \boldsymbol{A}(\eta) & & \boldsymbol{0}_{M_0 \times M_1} \\ \boldsymbol{A}(\eta-2) & \boldsymbol{A}(\eta-1) & \ddots & \vdots \\ \vdots & \vdots & \ddots & \boldsymbol{A}(\eta) \\ \boldsymbol{A}(0) & \boldsymbol{A}(1) & & \boldsymbol{A}(\eta-1) \\ \boldsymbol{0}_{M_0 \times M_1} & \boldsymbol{A}(0) & \ddots & \vdots \\ \vdots & & \ddots & \boldsymbol{A}(1) \\ \boldsymbol{0}_{M_0 \times M_1} & \boldsymbol{0}_{M_0 \times M_1} & \cdots & \boldsymbol{A}(0) \end{bmatrix}. \tag{7.99}$$

Definition 7.10 *Let ν be a non-negative integer, and let $\boldsymbol{x}(n)$ be a time-series of size $N \times 1$. The column-expansion of a vector time-series $\boldsymbol{x}(n)$ of order ν is denoted by $\boldsymbol{x}(n)_{|}^{(\nu)}$*

and it has size $(\nu+1)N \times 1$. *It is defined as*

$$\boldsymbol{x}(n)_{\shortmid}^{(\nu)} = \begin{bmatrix} \boldsymbol{x}(n) \\ \boldsymbol{x}(n-1) \\ \vdots \\ \boldsymbol{x}(n-\nu) \end{bmatrix}. \tag{7.100}$$

The column-expansion operator for vector time-series has a certain order. In each case, the correct size of the column-expansion of the vector time-series can be deduced from the notation. The size of the column-expansion of a FIR MIMO filter is given by the filter order and the size of the FIR MIMO filter.

Remark *Notice that the block vectorization in Definition 2.13 and the column-expansion in* (7.97) *are different but related. The main difference is that in the block vectorization, the indexes of the block matrices are* increasing *when going from the top to the bottom (see* (2.46)*). However, in the column-expansion of FIR MIMO filters, the indexes are* decreasing *when going from the top to the bottom of the output block matrix (see* (7.97)*).*

Next, the connection between the column-expansion of FIR MIMO filters in Definition 7.7 and the block vectorization operator in Definition 2.13 will be shown mathematically for square *FIR MIMO filters. The reason for considering square FIR MIMO filters is that the block vectorization operator in Definition 2.13 is defined only when the submatrices are square. Let* $\{\boldsymbol{C}(i)\}_{i=0}^{M-1}$, *where* $\boldsymbol{C}(i) \in \mathbb{C}^{N\times N}$ *is square. The row-expansion of this FIR MIMO filter is given by*

$$\boldsymbol{C}_{-} = [\boldsymbol{C}(0)\ \boldsymbol{C}(1)\ \cdots\ \boldsymbol{C}(M-1)], \tag{7.101}$$

and $\boldsymbol{C}_{-} \in \mathbb{C}^{N\times NM}$. *Let the* $MN \times MN$ *matrix* $\boldsymbol{J}$ *be given by*

$$\boldsymbol{J} \triangleq \begin{bmatrix} \boldsymbol{0}_{N\times N} & \cdots & \boldsymbol{0}_{N\times N} & \boldsymbol{I}_N \\ \boldsymbol{0}_{N\times N} & \cdots & \boldsymbol{I}_N & \boldsymbol{0}_{N\times N} \\ \vdots & \vdots & \vdots & \vdots \\ \boldsymbol{I}_N & \cdots & \boldsymbol{0}_{N\times N} & \boldsymbol{0}_{N\times N} \end{bmatrix}. \tag{7.102}$$

The block vectorization of the row-expansion of $\boldsymbol{C}$ *can be expressed as*

$$\operatorname{vecb}(\boldsymbol{C}_{-}) = \begin{bmatrix} \boldsymbol{C}(0) \\ \boldsymbol{C}(1) \\ \vdots \\ \boldsymbol{C}(M-1) \end{bmatrix}, \tag{7.103}$$

and $\operatorname{vecb}(\boldsymbol{C}_{-}) \in \mathbb{C}^{MN\times N}$. *The column-expansion of the FIR MIMO filter* $\boldsymbol{C}$ *is given by*

$$\boldsymbol{C}_{\shortmid} = \begin{bmatrix} \boldsymbol{C}(M-1) \\ \vdots \\ \boldsymbol{C}(1) \\ \boldsymbol{C}(0) \end{bmatrix}, \tag{7.104}$$

and $\boldsymbol{C}_{\vert} \in \mathbb{C}^{MN \times N}$. *By multiplying out* $\boldsymbol{J}\,\mathrm{vecb}\,(\boldsymbol{C}_{-})$, *it is seen that the connection between the block vectorization operator and the column-expansion is given through the following relation:*

$$\boldsymbol{J}\,\mathrm{vecb}\,(\boldsymbol{C}_{-}) = \boldsymbol{C}_{\vert}, \tag{7.105}$$

which is equivalent to $\boldsymbol{J}\boldsymbol{C}_{\vert} = \mathrm{vecb}\,(\boldsymbol{C}_{-})$. *These relations are valid for any square FIR MIMO filter.*

Let the vector time-series $\boldsymbol{x}(n)$ of size $N \times 1$ be the input to the causal FIR MIMO filter $\boldsymbol{E}$. The FIR MIMO coefficients $\{\boldsymbol{E}(k)\}_{k=0}^{m}$ of $\boldsymbol{E}$ have size $M_t \times N$. Denote the $M_t \times 1$ output vector time-series from the filter $\boldsymbol{E}$ as $\boldsymbol{y}(n)$ (see Figure 7.2). Assuming that the FIR MIMO filter $\{\boldsymbol{E}(k)\}_{k=0}^{m}$ is linear time-invariant (LTI), then convolution can be used to find $\boldsymbol{y}(n)$ in the following way:

$$\begin{aligned}\boldsymbol{y}(n) &= \sum_{k=0}^{m} \boldsymbol{E}(k)\boldsymbol{x}(n-k) = [\boldsymbol{E}(0)\ \boldsymbol{E}(1)\ \cdots\ \boldsymbol{E}(m)] \begin{bmatrix} \boldsymbol{x}(n) \\ \boldsymbol{x}(n-1) \\ \vdots \\ \boldsymbol{x}(n-m) \end{bmatrix} \\ &= \boldsymbol{E}_{-}\boldsymbol{x}(n)_{\vert}^{(m)}, \end{aligned} \tag{7.106}$$

where the notations in (7.96) and (7.100) have been used, and the size of the column-expanded vector $\boldsymbol{x}(n)_{\vert}^{(m)}$ is $(m+1)N \times 1$. In Exercise 7.5, it is shown that

$$\boldsymbol{y}(n)_{\vert}^{(l)} = \boldsymbol{E}_{\urcorner}^{(l)}\boldsymbol{x}(n)_{\vert}^{(m+l)}, \tag{7.107}$$

where l is a non-negative integer. Let the FIR MIMO filter $\boldsymbol{C}$ have size $M_r \times M_t$ and filter coefficients given by $\{\boldsymbol{C}(k)\}_{k=0}^{q}$. The FIR MIMO filter $\boldsymbol{B}$ is given by the convolution between the filters $\boldsymbol{C}$ and $\boldsymbol{E}$, and it has size $M_r \times N$ and order $m+q$. The row- and column-expansions of the filter $\boldsymbol{B}$ have sizes $M_r \times (m+q+1)N$ and $(m+q+1)M_r \times N$, respectively, and they are given by

$$\boldsymbol{B}_{-} = \boldsymbol{C}_{-}\boldsymbol{E}_{\urcorner}^{(q)}, \tag{7.108}$$

$$\boldsymbol{B}_{\vert} = \boldsymbol{C}_{\llcorner}^{(m)}\boldsymbol{E}_{\vert}. \tag{7.109}$$

The relations in (7.108) and (7.109) are proven in Exercise 7.6. Furthermore, it can be shown that

$$\boldsymbol{B}_{\urcorner}^{(l)} = \boldsymbol{C}_{\urcorner}^{(l)}\boldsymbol{E}_{\urcorner}^{(q+l)}, \tag{7.110}$$

$$\boldsymbol{B}_{\llcorner}^{(l)} = \boldsymbol{C}_{\llcorner}^{(m+l)}\boldsymbol{E}_{\llcorner}^{(l)}. \tag{7.111}$$

The two results in (7.110) and (7.111) are shown in Exercise 7.7.

7.5.3 FIR MIMO Transmit and Receive Filter Problems

The system should be designed such that the MSE between a delayed version of the input and the output of the system is minimized with respect to the FIR MIMO transmit and receive filters subject to the average transmit power. It is assumed that all input vector time-series, that is, $\boldsymbol{x}(n)$ and $\boldsymbol{v}(n)$, have zero-mean, and all second-order statistics of the vector time-series are assumed to be known. The vector time-series $\boldsymbol{x}(n)$ and $\boldsymbol{v}(n)$ are assumed to be uncorrelated.

The autocorrelation matrix of size $(\nu+1)N \times (\nu+1)N$ of the $(\nu+1)N \times 1$ vector $\boldsymbol{x}(n)_{|}^{(\nu)}$ is defined as

$$\boldsymbol{\Phi}_{\boldsymbol{x}}^{(\nu,N)} = \mathbb{E}\left[\boldsymbol{x}(n)_{|}^{(\nu)} \left(\boldsymbol{x}(n)_{|}^{(\nu)}\right)^{H}\right]. \tag{7.112}$$

The autocorrelation matrix of $\boldsymbol{v}(n)_{|}^{(\nu)}$ is defined in a similar way. Let the $(m+1)N \times (m+1)N$ matrix $\boldsymbol{\Psi}_{\boldsymbol{x}}^{(m,N)}(i)$ be defined as follows:

$$\boldsymbol{\Psi}_{\boldsymbol{x}}^{(m,N)}(i) = E\left[\left(\boldsymbol{x}(n)_{|}^{(m)}\right)^{*} \left(\boldsymbol{x}(n+i)_{|}^{(m)}\right)^{T}\right], \tag{7.113}$$

where $i \in \{-q-l, -q-l+1, \ldots, q+l\}$. From (7.112) and (7.113), it is seen that the following relationship is valid:

$$\boldsymbol{\Psi}_{\boldsymbol{x}}^{(m,N)}(0) = \left(\boldsymbol{\Phi}_{\boldsymbol{x}}^{(m,N)}\right)^{*} = \left(\boldsymbol{\Phi}_{\boldsymbol{x}}^{(m,N)}\right)^{T}. \tag{7.114}$$

The desired receiver output signal $\boldsymbol{d}(n) \in \mathbb{C}^{N\times 1}$ is often chosen as the vector time-series given by

$$\boldsymbol{d}(n) = \boldsymbol{x}(n-\delta), \tag{7.115}$$

where the integer δ denotes the non-negative *vector delay* through the overall communication system; it should be chosen carefully depending on the channel $\boldsymbol{C}$ and the orders of the transmit and receive filters, that is, m and l. The cross-covariance matrix $\boldsymbol{\phi}_{\boldsymbol{x},\boldsymbol{d}}^{(\nu,N)}$ of size $(\nu+1)N \times N$ is defined as

$$\boldsymbol{\phi}_{\boldsymbol{x},\boldsymbol{d}}^{(\nu,N)} = \mathbb{E}\left[\boldsymbol{x}(n)_{|}^{(\nu)} \boldsymbol{d}^{H}(n)\right]. \tag{7.116}$$

The block MSE, denoted by $\mathcal{E}$, is defined as

$$\mathcal{E} = \mathbb{E}\left[\|\hat{\boldsymbol{x}}(n) - \boldsymbol{d}(n)\|^{2}\right] = \operatorname{Tr}\left\{\left(\hat{\boldsymbol{x}}(n) - \boldsymbol{d}(n)\right)\left(\hat{\boldsymbol{x}}^{H}(n) - \boldsymbol{d}^{H}(n)\right)\right\}. \tag{7.117}$$

By rewriting the convolution sum with the notations and relations introduced in Subsection 7.5.2, it is possible to express the output vector $\hat{\boldsymbol{x}}(n)$ of the receive filter as follows:

$$\hat{\boldsymbol{x}}(n) = \boldsymbol{R}_{-} \boldsymbol{C}_{\urcorner}^{(l)} \boldsymbol{E}_{\urcorner}^{(q+l)} \boldsymbol{x}(n)_{|}^{(m+q+l)} + \boldsymbol{R}_{-} \boldsymbol{v}(n)_{|}^{(l)}. \tag{7.118}$$

In Exercise 7.8, it is shown that the MSE $\mathcal{E}$ in (7.117) can be expressed as

$$\begin{aligned}
\mathcal{E} = \operatorname{Tr}\Big\{ & \boldsymbol{\Phi}_{\boldsymbol{d}}^{(0,N)} + \boldsymbol{R}_{-} \boldsymbol{\Phi}_{\boldsymbol{v}}^{(l,M_r)} \boldsymbol{R}_{-}^{H} - \boldsymbol{R}_{-} \boldsymbol{C}_{\urcorner}^{(l)} \boldsymbol{E}_{\urcorner}^{(q+l)} \boldsymbol{\phi}_{\boldsymbol{x},\boldsymbol{d}}^{(m+q+l,N)} \\
& - \left(\boldsymbol{\phi}_{\boldsymbol{x},\boldsymbol{d}}^{(m+q+l,N)}\right)^{H} \left(\boldsymbol{E}_{\urcorner}^{(q+l)}\right)^{H} \left(\boldsymbol{C}_{\urcorner}^{(l)}\right)^{H} \boldsymbol{R}_{-}^{H} \\
& + \boldsymbol{R}_{-} \boldsymbol{C}_{\urcorner}^{(l)} \boldsymbol{E}_{\urcorner}^{(q+l)} \boldsymbol{\Phi}_{\boldsymbol{x}}^{(m+q+l,N)} \left(\boldsymbol{E}_{\urcorner}^{(q+l)}\right)^{H} \left(\boldsymbol{C}_{\urcorner}^{(l)}\right)^{H} \boldsymbol{R}_{-}^{H} \Big\}.
\end{aligned} \tag{7.119}$$

The receiver (equalizer) design problem can be formulated as follows:

Problem 7.2 (FIR MIMO Receive Filter)

$$\min_{\{\{R(k)\}_{k=0}^{l}\}} \mathcal{E}. \tag{7.120}$$

The average power constraint for the channel input time-series $y(n)$ can be expressed as

$$\mathbb{E}\left[\|y(n)\|^2\right] = \text{Tr}\left\{E_{-}\Phi_{x}^{(m,N)}E_{-}^{H}\right\} = P, \tag{7.121}$$

where (7.106) and (7.112) have been used.

The transmitter design problem is the following:

Problem 7.3 (FIR MIMO Transmit Filter)

$$\min_{\{\{E(k)\}_{k=0}^{m}\}} \mathcal{E},$$

subject to (7.122)

$$\text{Tr}\left\{E_{-}\Phi_{x}^{(m,N)}E_{-}^{H}\right\} = P.$$

The constrained optimization in Problem 7.3 can be converted into an unconstrained optimization problem by using a Lagrange multiplier. The unconstrained Lagrangian function $\mathcal{L}$ can be expressed as

$$\mathcal{L}(E_{-}, E_{-}^{*}) = \mathcal{E} + \mu \,\text{Tr}\left\{E_{-}\Phi_{x}^{(m,N)}E_{-}^{H}\right\}, \tag{7.123}$$

where μ is the positive Lagrange multiplier. Necessary conditions for optimality are found through complex-valued matrix derivatives of the positive Lagrangian function $\mathcal{L}$ with respect to the conjugate of the complex unknown parameters.

7.5.4 FIR MIMO Receive Filter Optimization

In the optimization for the FIR MIMO receiver, the following three relations are needed:

$$\frac{\partial}{\partial R_{-}^{*}} \text{Tr}\left\{R_{-}C_{\urcorner}^{(l)}E_{\urcorner}^{(q+l)}\phi_{x,d}^{(m+q+l,N)}\right\} = \mathbf{0}_{N\times(l+1)M_r}, \tag{7.124}$$

which follows from the fact that R_{-} and R_{-}^{*} should be treated independently when finding complex-valued matrix derivatives,

$$\frac{\partial}{\partial R_{-}^{*}} \text{Tr}\left\{R_{-}\Phi_{v}^{(l,M_r)}R_{-}^{H}\right\} = R_{-}\Phi_{v}^{(l,M_r)}, \tag{7.125}$$

which follows from Table 4.3, and

$$\frac{\partial}{\partial R_{-}^{*}} \text{Tr}\left\{\left(\phi_{x,d}^{(m+q+l,N)}\right)^{H}\left(E_{\urcorner}^{(q+l)}\right)^{H}\left(C_{\urcorner}^{(l)}\right)^{H}R_{-}^{H}\right\}$$
$$= \left(\phi_{x,d}^{(m+q+l,N)}\right)^{H}\left(E_{\urcorner}^{(q+l)}\right)^{H}\left(C_{\urcorner}^{(l)}\right)^{H}, \tag{7.126}$$

which also follows from Table 4.3.

The derivative of the MSE $\mathcal{E}$ with respect to $\boldsymbol{R}_{-}^{*}$ can be found by using the results in (7.124), (7.125), and (7.126), and $\frac{\partial}{\partial \boldsymbol{R}_{-}^{*}}\mathcal{E}$ is given by

$$\begin{aligned}\frac{\partial}{\partial \boldsymbol{R}_{-}^{*}}\mathcal{E} &= \boldsymbol{R}_{-}\boldsymbol{\Phi}_{\boldsymbol{v}}^{(l,M_r)} - \left(\boldsymbol{\phi}_{\boldsymbol{x},\boldsymbol{d}}^{(m+q+l,N)}\right)^{H}\left(\boldsymbol{E}_{\urcorner}^{(q+l)}\right)^{H}\left(\boldsymbol{C}_{\urcorner}^{(l)}\right)^{H} \\ &\quad + \boldsymbol{R}_{-}\boldsymbol{C}_{\urcorner}^{(l)}\boldsymbol{E}_{\urcorner}^{(q+l)}\boldsymbol{\Phi}_{\boldsymbol{x}}^{(m+q+l,N)}\left(\boldsymbol{E}_{\urcorner}^{(q+l)}\right)^{H}\left(\boldsymbol{C}_{\urcorner}^{(l)}\right)^{H}. \end{aligned} \tag{7.127}$$

By solving the equation $\frac{\partial}{\partial \boldsymbol{R}_{-}^{*}}\mathcal{E} = \boldsymbol{0}_{N\times(l+1)M_r}$, it is seen that the minimum MSE FIR MIMO receiver is given by

$$\begin{aligned}\boldsymbol{R}_{-} &= \left(\boldsymbol{\phi}_{\boldsymbol{x},\boldsymbol{d}}^{(m+q+l,N)}\right)^{H}\left(\boldsymbol{E}_{\urcorner}^{(q+l)}\right)^{H}\left(\boldsymbol{C}_{\urcorner}^{(l)}\right)^{H} \\ &\quad \times\left[\boldsymbol{C}_{\urcorner}^{(l)}\boldsymbol{E}_{\urcorner}^{(q+l)}\boldsymbol{\Phi}_{\boldsymbol{x}}^{(m+q+l,N)}\left(\boldsymbol{E}_{\urcorner}^{(q+l)}\right)^{H}\left(\boldsymbol{C}_{\urcorner}^{(l)}\right)^{H} + \boldsymbol{\Phi}_{\boldsymbol{v}}^{(l,M_r)}\right]^{-1}. \end{aligned} \tag{7.128}$$

7.5.5 FIR MIMO Transmit Filter Optimization

The *reshape operator*, denoted by $\mathcal{T}^{(k)}$, which is needed in this subsection is introduced next.[3] The operator $\mathcal{T}^{(k)} : \mathbb{C}^{N\times(m+k+1)N} \rightarrow \mathbb{C}^{(k+1)N\times(m+1)N}$ produces a $(k+1)N \times (m+1)N$ block Toeplitz matrix from an $N \times (m+k+1)N$ matrix. Let $\boldsymbol{W}_{-}$ be an $N \times (m+k+1)N$ matrix, where the i-th $N \times N$ block is given by $\boldsymbol{W}(i) \in \mathbb{C}^{N\times N}$, where $i \in \{0, 1, \ldots, m+k\}$. Then, the operator $\mathcal{T}^{(k)}$ acting on the matrix $\boldsymbol{W}_{-}$ yields

$$\mathcal{T}^{(k)}\left\{\boldsymbol{W}_{-}\right\} = \begin{bmatrix} \boldsymbol{W}(k) & \boldsymbol{W}(k+1) & \cdots & \boldsymbol{W}(m+k) \\ \vdots & \vdots & \ddots & \vdots \\ \boldsymbol{W}(1) & \boldsymbol{W}(2) & \cdots & \boldsymbol{W}(m+1) \\ \boldsymbol{W}(0) & \boldsymbol{W}(1) & \cdots & \boldsymbol{W}(m) \end{bmatrix}, \tag{7.129}$$

where k is a non-negative integer.

All terms of the unconstrained objective function $\mathcal{L}$ given in (7.123) that depend on the transmit filter can be rewritten by means of the vec$(\cdot)$ operator. The block MSE $\mathcal{E}$ can be written in terms of $\text{vec}\left(\boldsymbol{E}_{-}\right)$ by means of the following three relations:

$$\begin{aligned}&\text{Tr}\left\{\boldsymbol{R}_{-}\boldsymbol{C}_{\urcorner}^{(l)}\boldsymbol{E}_{\urcorner}^{(q+l)}\boldsymbol{\phi}_{\boldsymbol{x},\boldsymbol{d}}^{(m+q+l,N)}\right\} \\ &\quad = \text{vec}^{T}\left(\boldsymbol{C}_{|}^{T}\left(\boldsymbol{R}_{\llcorner}^{(q)}\right)^{T}\mathcal{T}^{(q+l)}\left\{\left(\boldsymbol{\phi}_{\boldsymbol{x},\boldsymbol{d}}^{(m+q+l,N)}\right)^{T}\right\}\right)\text{vec}\left(\boldsymbol{E}_{-}\right), \end{aligned} \tag{7.130}$$

$$\begin{aligned}&\text{Tr}\left\{\left(\boldsymbol{\phi}_{\boldsymbol{x},\boldsymbol{d}}^{(m+q+l,N)}\right)^{H}\left(\boldsymbol{E}_{\urcorner}^{(q+l)}\right)^{H}\left(\boldsymbol{C}_{\urcorner}^{(l)}\right)^{H}\boldsymbol{R}_{-}^{H}\right\} \\ &\quad = \text{vec}^{H}\left(\boldsymbol{E}_{-}\right)\text{vec}\left(\boldsymbol{C}_{|}^{H}\left(\boldsymbol{R}_{\llcorner}^{(q)}\right)^{H}\mathcal{T}^{(q+l)}\left\{\left(\boldsymbol{\phi}_{\boldsymbol{x},\boldsymbol{d}}^{(m+q+l,N)}\right)^{H}\right\}\right), \end{aligned} \tag{7.131}$$

[3] This is an example of a reshape$(\cdot)$ operator introduced in Proposition 3.9 where the output has multiple copies of certain input components.

and

$$\operatorname{Tr}\left\{\boldsymbol{R}_{-}\boldsymbol{C}_{\neg}^{(l)}\boldsymbol{E}_{\neg}^{(q+l)}\boldsymbol{\Phi}_{\boldsymbol{x}}^{(m+q+l,N)}\left(\boldsymbol{E}_{\neg}^{(q+l)}\right)^{H}\left(\boldsymbol{C}_{\neg}^{(l)}\right)^{H}\boldsymbol{R}_{-}^{H}\right\}$$

$$= \operatorname{vec}^{H}\left(\boldsymbol{E}_{-}\right)\sum_{i_0=0}^{q}\sum_{i_1=0}^{l}\sum_{i_2=0}^{l}\sum_{i_3=0}^{q}\boldsymbol{\Psi}_{\boldsymbol{x}}^{(m,N)}(i_0+i_1-i_2-i_3)\otimes$$

$$\left[\boldsymbol{C}^{H}(i_0)\boldsymbol{R}^{H}(i_1)\boldsymbol{R}(i_2)\boldsymbol{C}(i_3)\right]\operatorname{vec}\left(\boldsymbol{E}_{-}\right), \tag{7.132}$$

where the operator $\mathcal{T}^{(q+l)}$ is defined in (7.129). The three relations in (7.130), (7.131), and (7.132) are shown in Exercises 7.9, 7.10, and 7.11, respectively.

To find the derivative of the power constraint with respect to $\boldsymbol{E}_{-}^{*}$, the following equation is useful:

$$\operatorname{Tr}\left\{\boldsymbol{E}_{-}\boldsymbol{\Phi}_{\boldsymbol{x}}^{(m,N)}\boldsymbol{E}_{-}^{H}\right\} = \operatorname{Tr}\left\{\boldsymbol{\Phi}_{\boldsymbol{x}}^{(m,N)}\boldsymbol{E}_{-}^{H}\boldsymbol{E}_{-}\right\} = \operatorname{vec}^{H}\left(\boldsymbol{E}_{-}\boldsymbol{\Phi}_{\boldsymbol{x}}^{(m,N)}\right)\operatorname{vec}\left(\boldsymbol{E}_{-}\right)$$

$$= \operatorname{vec}^{H}\left(\boldsymbol{I}_{M_t}\boldsymbol{E}_{-}\boldsymbol{\Phi}_{\boldsymbol{x}}^{(m,N)}\right)\operatorname{vec}\left(\boldsymbol{E}_{-}\right)=\left[\left\{\left(\boldsymbol{\Phi}_{\boldsymbol{x}}^{(m,N)}\right)^{T}\otimes\boldsymbol{I}_{M_t}\right\}\operatorname{vec}\left(\boldsymbol{E}_{-}\right)\right]^{H}\operatorname{vec}\left(\boldsymbol{E}_{-}\right)$$

$$= \operatorname{vec}^{H}\left(\boldsymbol{E}_{-}\right)\left[\left(\boldsymbol{\Phi}_{\boldsymbol{x}}^{(m,N)}\right)^{*}\otimes\boldsymbol{I}_{M_t}\right]\operatorname{vec}\left(\boldsymbol{E}_{-}\right)$$

$$= \operatorname{vec}^{T}\left(\boldsymbol{E}_{-}\right)\left[\boldsymbol{\Phi}_{\boldsymbol{x}}^{(m,N)}\otimes\boldsymbol{I}_{M_t}\right]\operatorname{vec}\left(\boldsymbol{E}_{-}^{*}\right). \tag{7.133}$$

By using the above relations, taking the derivative of the Lagrangian function $\mathcal{L}$ in (7.123) with respect to $\boldsymbol{E}_{-}^{*}$, and setting the result equal to the zero vector, one obtains the necessary conditions for the optimal FIR MIMO transmit filter. The derivative of the Lagrangian function $\mathcal{L}$ with respect to $\boldsymbol{E}_{-}^{*}$ is given by

$$\mathcal{D}_{\boldsymbol{E}_{-}^{*}}\mathcal{L} = \operatorname{vec}^{T}\left(\boldsymbol{E}_{-}\right)\sum_{i_0=0}^{q}\sum_{i_1=0}^{l}\sum_{i_2=0}^{l}\sum_{i_3=0}^{q}\left(\boldsymbol{\Psi}_{\boldsymbol{x}}^{(m,N)}(i_0+i_1-i_2-i_3)\right)^{T}$$

$$\otimes\left[\boldsymbol{C}^{H}(i_0)\boldsymbol{R}^{H}(i_1)\boldsymbol{R}(i_2)\boldsymbol{C}(i_3)\right]^{T} + \mu\operatorname{vec}^{T}\left(\boldsymbol{E}_{-}\right)\left[\boldsymbol{\Phi}_{\boldsymbol{x}}^{(m,N)}\otimes\boldsymbol{I}_{M_t}\right]$$

$$-\operatorname{vec}^{T}\left(\boldsymbol{C}_{\mathsf{I}}^{H}\left(\boldsymbol{R}_{\llcorner}^{(q)}\right)^{H}\mathcal{T}^{(q+l)}\left\{\left(\boldsymbol{\Phi}_{\boldsymbol{x},\boldsymbol{d}}^{(m+q+l,N)}\right)^{H}\right\}\right). \tag{7.134}$$

For a given FIR MIMO receive filter $\boldsymbol{R}$, the necessary condition for optimality of the optimal transmitter is found by solving $\mathcal{D}_{\boldsymbol{E}_{-}^{*}}\mathcal{L} = \boldsymbol{0}_{1\times(m+1)NM_t}$, which is equivalent to

$$\boldsymbol{A}\cdot\operatorname{vec}(\boldsymbol{E}_{-}) = \boldsymbol{b}, \tag{7.135}$$

where matrix $\boldsymbol{A}$ is an $(m+1)M_tN \times (m+1)M_tN$ matrix given by

$$\boldsymbol{A} = \sum_{i_0=0}^{q}\sum_{i_1=0}^{l}\sum_{i_2=0}^{l}\sum_{i_3=0}^{q}\boldsymbol{\Psi}_{\boldsymbol{x}}^{(m,N)}(i_0+i_1-i_2-i_3)\otimes\left(\boldsymbol{C}^{H}(i_0)\boldsymbol{R}^{H}(i_1)\boldsymbol{R}(i_2)\boldsymbol{C}(i_3)\right)$$

$$+\boldsymbol{\Psi}_{\boldsymbol{x}}^{(m,N)}(0)\otimes\mu\boldsymbol{I}_{M_t}, \tag{7.136}$$

and the vector $\boldsymbol{b}$ of size $(m+1)M_t N \times 1$ is given by

$$\boldsymbol{b} = \operatorname{vec}\left(\boldsymbol{C}_{\mathsf{I}}^H \left(\boldsymbol{R}_{\mathsf{L}}^{(q)}\right)^H \mathcal{T}^{(q+l)} \left\{ \left(\boldsymbol{\Phi}_{\boldsymbol{x},\boldsymbol{d}}^{(m+q+l,N)}\right)^H \right\}\right). \tag{7.137}$$

7.6 Exercises

7.1 Show that

$$\operatorname{Tr}\{\boldsymbol{A} \odot \boldsymbol{B}\} = \operatorname{vec}^T\left(\operatorname{diag}\left(\operatorname{vec}_d\left(\boldsymbol{A}\right)\right)\right)\operatorname{vec}\left(\boldsymbol{B}\right), \tag{7.138}$$

where $\boldsymbol{A}, \boldsymbol{B} \in \mathbb{C}^{N\times N}$.

7.2 Let $\boldsymbol{A} \in \mathbb{C}^{N\times N}$. Show that

$$\operatorname{diag}\left(\operatorname{vec}_d\left(\boldsymbol{A}\right)\right) = \boldsymbol{I}_N \odot \boldsymbol{A}. \tag{7.139}$$

7.3 Using the result in (4.123), show that

$$\operatorname{vec}\left(\left[\left(d\boldsymbol{F}^*\right)\boldsymbol{F}^T\right] \otimes \boldsymbol{I}_{M_r}\right) = \boldsymbol{\Pi}^T \left[\boldsymbol{F} \otimes \boldsymbol{I}_{M_t}\right] d\operatorname{vec}\left(\boldsymbol{F}^*\right), \tag{7.140}$$

and

$$\operatorname{vec}\left(\left[\boldsymbol{F}^*\left(d\boldsymbol{F}^T\right)\right] \otimes \boldsymbol{I}_{M_r}\right) = \boldsymbol{\Pi}^T \left[\boldsymbol{I}_{M_t} \otimes \boldsymbol{F}^*\right] \boldsymbol{K}_{M_t,B} d\operatorname{vec}\left(\boldsymbol{F}\right), \tag{7.141}$$

where the matrix $\boldsymbol{\Pi}$ is defined in (7.84).

7.4 Let $\boldsymbol{A}$ be defined in (7.92), where the indication of PSK is now dropped for simplicity. Using the results from Exercise 7.3, show that

$$\begin{aligned} d\operatorname{vec}\left(\boldsymbol{A}\right) = {} & \frac{\delta g}{(1+K)\sin^2\theta}\left[\left(\boldsymbol{R}^{1/2}\right)^T \otimes \boldsymbol{R}^{1/2}\right] \boldsymbol{\Pi}^T \left[\boldsymbol{I}_{M_t} \otimes \boldsymbol{F}^*\right] \boldsymbol{K}_{M_t,B} d\operatorname{vec}\left(\boldsymbol{F}\right) \\ & + \frac{\delta g}{(1+K)\sin^2\theta}\left[\left(\boldsymbol{R}^{1/2}\right)^T \otimes \boldsymbol{R}^{1/2}\right] \boldsymbol{\Pi}^T \left[\boldsymbol{F} \otimes \boldsymbol{I}_{M_t}\right] d\operatorname{vec}\left(\boldsymbol{F}^*\right), \end{aligned} \tag{7.142}$$

where the matrix $\boldsymbol{\Pi}$ is defined in (7.84).

7.5 Let $\boldsymbol{y}(n)$ of $M_t \times 1$ be the output of the LTI FIR MIMO filter $\{\boldsymbol{E}(k)\}_{k=0}^m$ of size $M_t \times N$ and order m when the vector time-series $\boldsymbol{x}(n)$ of size $N \times 1$ is the input signal. If l is a non-negative integer, show that the column-expanded vector of order l of the output of the filter is given by (7.107).

7.6 Let the FIR MIMO filter coefficients $\{\boldsymbol{E}(k)\}_{k=0}^m$ of $\boldsymbol{E}$ have size $M_t \times N$, and the FIR MIMO filter $\boldsymbol{C}$ have size $M_r \times M_t$ with filter coefficients $\{\boldsymbol{C}(k)\}_{k=0}^q$. If the FIR MIMO filter $\boldsymbol{B}$ is the convolution between the filters $\boldsymbol{C}$ and $\boldsymbol{E}$, then the filter coefficients $\{\boldsymbol{B}(k)\}_{k=0}^{m+q}$ of $\boldsymbol{B}$ have size $M_r \times N$. Show that the row- and column-expansions of $\boldsymbol{B}$ are given by (7.108) and (7.109), respectively.

7.7 Let the three FIR MIMO filters with matrix coefficients $\{\boldsymbol{E}(k)\}_{k=0}^m$, $\{\boldsymbol{C}(k)\}_{k=0}^q$, and $\{\boldsymbol{B}(k)\}_{k=0}^{m+q}$ be defined as in Exercise 7.6. If l is a non-negative integer, then show that (7.110) and (7.111) hold.

7.8 Show by inserting the result from (7.118) into (7.117) such that the block MSE of the system in Figure 7.2 is given by (7.119).

7.9 Show that (7.130) is valid.

7.10 Show that (7.131) holds.

7.11 Show that (7.132) is valid.

References

Abadir, K. M. and Magnus, J. R. (2005), *Matrix Algebra*, Cambridge University Press, New York, USA.

Abatzoglou, T. J., Mendel, J. M., and Harada, G. A. (1991), "The constrained total least squares technique and its applications to harmonic superresolution," *IEEE Trans. Signal Process.*, vol. 39, no. 5, pp. 1070–1087, May.

Abrudan, T., Eriksson, J., and Koivunen, V. (2008), "Steepest descent algorithms for optimization under unitary matrix constraint," *IEEE Trans. Signal Process.*, vol. 56, no. 3, pp. 1134–1147, March.

Absil, P.-A., Mahony, R., and Sepulchre, R. (2008), *Optimization Algorithms on Matrix Manifolds*, Princeton University Press, Princeton, NJ, USA.

Alexander, S. (1984), "A derivation of the complex fast Kalman algorithm," *IEEE Trans. Acoust., Speech, Signal Process.*, vol. 32, no. 6, pp. 1230–1232, December.

Barry, J. R., Lee, E. A., and Messerschmitt, D. G. (2004), *Digital Communication*, 3rd ed., Kluwer Academic Publishers, Dordrecht, The Netherlands.

Bernstein, D. S. (2005), *Matrix Mathematics: Theory, Facts, and Formulas with Application to Linear System Theory*, Princeton University Press, Princeton, NJ, USA.

Bhatia, R. (2007), *Positive Definite Matrices*, Princeton Series in Applied Mathematics, Princeton University Press, Princeton, NJ, USA.

Boyd, S. and Vandenberghe, L. (2004), *Convex Optimization*, Cambridge University Press, Cambridge, UK.

Brandwood, D. H. (1983), "A complex gradient operator and its application in adaptive array theory," *IEEE Proc., Parts F and H*, vol. 130, no. 1, pp. 11–16, February.

Brewer, J. W. (1978), "Kronecker products and matrix calculus in system theory," *IEEE Trans. Circuits, Syst.*, vol. CAS-25, no. 9, pp. 772–781, September.

Brookes, M. (2009, July 25), "The matrix reference manual," [Online]. Available: http://www.ee.ic.ac.uk/hp/staff/dmb/matrix/intro.html.

de Campos, M. L. R., Werner, S., and Apolinário Jr., J. A. (2004), Constrained adaptive filters. In S. Chandran, ed., *Adaptive Antenna Arrays: Trends and Applications*, Springer Verlag, Berlin, Germany, pp. 46–62.

Diniz, P. S. R. (2008), *Adaptive Filtering: Algorithms and Practical Implementations*, 3rd ed., Springer Verlag, Boston, MA, USA.

Dwyer, P. S. (1967), "Some applications of matrix derivatives in multivariate analysis," *J. Am. Stat. Ass.*, vol. 62, no. 318, pp. 607–625, June.

Dwyer, P. S. and Macphail, M. S. (1948), "Symbolic matrix derivatives," *Annals of Mathematical Statistics*, vol. 19, no. 4, pp. 517–534.

Edwards, C. H. and Penney, D. E. (1986), *Calculus and Analytic Geometry*, 2nd ed., Prentice-Hall, Inc., Englewood Cliffs, NJ, USA.

Eriksson, J., Ollila, E., and Koivunen, V. (2009), Statistics for complex random variables revisited. In "*Proc. IEEE Int. Conf. Acoust., Speech, Signal Proc.*," Taipei, Taiwan, pp. 3565–3568, April.

Feiten, A., Hanly, S., and Mathar, R. (2007), Derivatives of mutual information in Gaussian vector channels with applications. In "*Proc. Int. Symp. on Information Theory*," Nice, France, pp. 2296–2300, June.

Fischer, R. (2002), *Precoding and Signal Shaping for Digital Transmission*, Wiley-Interscience, New York, NY, USA.

Fong, C. K. (2006), "Course Notes for MATH 3002, Winter 2006: §3 Complex Differentials and the $\bar{\partial}$-operator," [Online]. Available: http://mathstat.carleton.ca/~ckfong/S31.pdf.

Fränken, D. (1997), "Complex digital networks: A sensitivity analysis based on the Wirtinger calculus," *IEEE Trans. Circuits, Syst. I: Fundamental Theory and Applications*, vol. 44, no. 9, pp. 839–843, September.

Fritzsche, K. and Grauert, H. (2002), *From Holomorphic Functions to Complex Manifolds*, Springer-Verlag, New York, NY, USA.

Gantmacher, F. R. (1959a), *The Theory of Matrices*, vol. 1, AMS Chelsea Publishing, New York, NY, USA.

Gantmacher, F. R. (1959b), *The Theory of Matrices*, vol. 2, AMS Chelsea Publishing, New York, NY, USA.

Golub, G. H. and van Loan, C. F. (1989), *Matrix Computations*, 2nd ed., The Johns Hopkins University Press, Baltimore, MD, USA.

González-Vázquez, F. J. (1988), "The differentiation of functions of conjugate complex variables: Application to power network analysis," *IEEE Trans. Educ.*, vol. 31, no. 4, pp. 286–291, November.

Graham, A. (1981), *Kronecker Products and Matrix Calculus with Applications*, Ellis Horwood Limited, England.

Gray, R. M. (2006), "Toeplitz and Circulant Matrices: A review," *Foundations and Trends in Communications and Information Theory*, vol. 2, no. 3, Now Publishers, Boston, MA, USA.

Guillemin, V. and Pollack, A. (1974), *Differential Topology*, Prentice-Hall, Inc., Englewood Cliffs, NJ, USA.

Han, Z. and Liu, K. J. R. (2008), *Resource Allocation for Wireless Networks: Basics, Techniques, and Applications*, Cambridge University Press, Cambridge, UK.

Hanna, A. I. and Mandic, D. P. (2003), "A fully adaptive normalized nonlinear gradient descent algorithm for complex-valued nonlinear adaptive filters," *IEEE Trans. Signal Proces.*, vol. 51, no. 10, pp. 2540–2549, October.

Harville, D. A. (1997), *Matrix Algebra from a Statistician's Perspective*, Springer-Verlag, New York, NY, corrected second printing, 1999.

Hayes, M. H. (1996), *Statistical Digital Signal Processing and Modeling*, John Wiley & Sons, Inc., New York, NY, USA.

Haykin, S. (2002), *Adaptive Filter Theory*, 4th ed., Prentice Hall, Englewood Cliffs, NJ, USA.

Hjørungnes, A. (2000), Optimal Bit and Power Constrained Filter Banks, Ph.D. dissertation, Norwegian University of Science and Technology (NTNU), Trondheim, Norway.
Available: http://www.unik.no/~ arehj/publications/thesis.pdf.

Hjørungnes, A. (2005), Minimum symbol error rate transmitter and receiver FIR MIMO filters for multilevel PSK signaling. In "*Proc. Int. Symp. on Wireless Communication Systems*," Siena, Italy, September 2005, IEEE, pp. 27–31.

Hjørungnes, A., de Campos, M. L. R., and Diniz, P. S. R. (2005), "Jointly optimized transmitter and receiver FIR MIMO filters in the presence of near-end crosstalk," *IEEE Trans. Signal Proces.*, vol. 53, no. 1, pp. 346–359, January.

Hjørungnes, A. and Gesbert, D. (2007a), "Complex-valued matrix differentiation: Techniques and key results," *IEEE Trans. Signal Proces.*, vol. 55, no. 6, pp. 2740–2746, June.

Hjørungnes, A. and Gesbert, D. (2007b), Hessians of scalar functions which depend on complex-valued matrices. In "*Proc. Int. Symp. on Signal Proc. and Its Applications*," Sharjah, United Arab Emirates, February.

Hjørungnes, A. and Gesbert, D. (2007c), "Precoded orthogonal space-time block codes over correlated Ricean MIMO channels," *IEEE Trans. Signal Proces.*, vol. 55, no. 2, pp. 779–783, February.

Hjørungnes, A. and Gesbert, D. (2007d), "Precoding of orthogonal space-time block codes in arbitrarily correlated MIMO channels: Iterative and closed-form solutions," *IEEE Trans. Wirel. Commun.*, vol. 6, no. 3, pp. 1072–1082, March.

Hjørungnes, A. and Palomar, D. P. (2008a), Finding patterned complex-valued matrix derivatives by using manifolds. In "*Proc. Int. Symp. on Applied Sciences in Biomedical and Communication Technologies*," Aalborg, Denmark, October. Invited paper.

Hjørungnes, A. and Palomar, D. P. (2008b), Patterned complex-valued matrix derivatives. In "*Proc. IEEE Int. Workshop on Sensor Array and Multi-Channel Signal Processing*," Darmstadt, Germany, pp. 293–297, July.

Hjørungnes, A. and Ramstad, T. A. (1999), Algorithm for jointly optimized analysis and synthesis FIR filter banks. In "*Proc. of the 6th IEEE Int. Conf. Electronics, Circuits and Systems*," vol. 1, Paphos, Cyprus, pp. 369–372, September.

Horn, R. A. and Johnson, C. R. (1985), *Matrix Analysis*, Cambridge University Press, Cambridge, UK. Reprinted 1999.

Horn, R. A. and Johnson, C. R. (1991), *Topics in Matrix Analysis*, Cambridge University Press, Cambridge, UK. Reprinted 1999.

Huang, Y. and Benesty, J. (2003), "A class of frequency-domain adaptive approaches to blind multichannel identification," *IEEE Trans. Signal Proces.*, vol. 51, no. 1, pp. 11–24, January.

Jaffer, A. G. and Jones, W. E. (1995), "Weighted least-squares design and characterization of complex FIR filters," *IEEE Trans. Signal Proces.*, vol. 43, no. 10, pp. 2398–2401, October.

Jagannatham, A. K. and Rao, B. D. (2004), "Cramer-Rao lower bound for constrained complex parameters," *IEEE Signal Proces. Lett.*, vol. 11, no. 11, pp. 875–878, November.

Jain, A. K. (1989), *Fundamentals of Digital Image Processing*, Prentice-Hall, Englewood Cliffs, NJ, USA.

Jonhson, D. H. and Dudgeon, D. A. (1993), *Array Signal Processing: Concepts and Techniques*, Prentice-Hall, Inc., Englewood Cliffs, NJ, USA.

Kailath, T., Sayed, A. H., and Hassibi, B. (2000), *Linear Estimation*, Prentice-Hall, Upper Saddle River, NJ, USA.

Kreutz-Delgado, K. (2008), "Real vector derivatives, gradients, and nonlinear least-squares," [Online]. Available: http://dsp.ucsd.edu/~kreutz/PEI05.html.

Kreutz-Delgado, K. (2009, June 25), "The complex gradient operator and the $\mathbb{CR}$-calculus," [Online]. Available: http://arxiv.org/PS_cache/arxiv/pdf/0906/0906.4835v1.pdf. Course Lecture Supplement No. ECE275A, Dept. of Electrical and Computer Engineering, UC San Diego, CA, USA.

Kreyszig, E. (1988), *Advanced Engineering Mathematics*, 6th ed., John Wiley & Sons, Inc., New York, NY, USA.

Li, X., Luo, T., Yue, G., and Yin, C. (2001), "A squaring method to simplify the decoding of orthogonal space-time block codes," *IEEE Trans. Commun.*, vol. 49, no. 10, pp. 1700–1703, October.

Luenberger, D. G. (1973), *Introduction to Linear and Nonlinear Programming*, Addison–Wesley, Reading, MA, USA.

Lütkepohl, H. (1996), *Handbook of Matrices*, John Wiley & Sons, Inc., New York, NY, USA.

Magnus, J. R. (1988), *Linear Structures*, Charles Griffin & Company Limited, London, UK.

Magnus, J. R. and Neudecker, H. (1988), *Matrix Differential Calculus with Application in Statistics and Econometrics*, John Wiley & Sons, Inc., Essex, UK.

Mandic, D. P. and Goh, V. S. L. (2009), *Complex Valued Nonlinear Adaptive Filters: Noncircularity, Widely Linear and Neural Models*, Adaptive and Learning Systems for Signal Processing, Communications and Control Series, Wiley, Noida, India.

Manton, J. H. (2002), "Optimization algorithms exploiting unitary constraints," *IEEE Trans. Signal Proces.*, vol. 50, no. 3, pp. 635–650, March.

Minka, T. P. (2000, December 28), "Old and new matrix algebra useful for statistics," [Online]. Available: http://research.microsoft.com/~minka/papers/matrix/.

Moon, T. K. and Stirling, W. C. (2000), *Mathematical Methods and Algorithms for Signal Processing*, Prentice Hall, Inc., Englewood Cliffs, NJ, USA.

Munkres, J. R. (2000), *Topology*, 2nd ed., Prentice-Hall, Inc., Upper Saddle River, NJ, USA.

Naylor, A. W. and Sell, G. R. (1982), *Linear Operator Theory in Engineering and Science*, Springer-Verlag, New York, NY, USA.

Nel, D. G. (1980), "On matrix differentiation in statistics," *South African Statistical J.*, vol. 14, pp. 137–193.

Osherovich, E., Zibulevsky, M., and Yavneh, I. (2008), Signal reconstruction from the modulus of its Fourier transform, Technical report, Technion.

Palomar, D. P. and Eldar, Y. C., eds. (2010), *Convex Optimization in Signal Processing and Communications*, Cambridge University Press, Cambridge, UK.

Palomar, D. P. and Verdú, S. (2006), "Gradient of mutual information in linear vector Gaussian channels," *IEEE Trans. Inform. Theory*, vol. 52, no. 1, pp. 141–154, January.

Palomar, D. P. and Verdú, S. (2007), "Representation of mutual information via input estimates," *IEEE Trans. Inform. Theory*, vol. 53, no. 2, pp. 453–470, February.

Paulraj, A., Nabar, R., and Gore, D. (2003), *Introduction to Space-Time Wireless Communications*, Cambridge University Press, Cambridge, UK.

Payaró, M. and Palomar, D. P. (2009), "Hessian and concavity of mutual information, entropy, and entropy power in linear vector Gaussian channels," *IEEE Trans. Inform. Theory*, vol. 55, no. 8, pp. 3613–3628, August.

Petersen, K. B. and Pedersen, M. S. (2008), "The matrix cookbook," [Online]. Available: http://matrixcookbook.com/.

Remmert, R. (1991), *Theory of Complex Functions*, Springer-Verlag, Herrisonburg, VA, USA. Translated by Robert B. Burckel.

Rinehart, R. F. (1964), "The exponential representation of unitary matrices," *Mathematics Magazine*, vol. 37, no. 2, pp. 111–112, March.

Roman, T. and Koivunen, V. (2004), Blind CFO estimation in OFDM systems using diagonality criterion. In "*Proc. IEEE Int. Conf. Acoust., Speech, Signal Proc.*," vol. IV, Montreal, Canada, pp. 369–372, May.

Roman, T., Visuri, S., and Koivunen, V. (2006), "Blind frequency synchronization in OFDM via diagonality criterion," *IEEE Trans. Signal Proces.*, vol. 54, no. 8, pp. 3125–3135, August.

Sayed, A. H. (2003), *Fundamentals of Adaptive Filtering*, John Wiley & Sons, Inc., Hoboken, NJ, USA.

Sayed, A. H. (2008), *Adaptive Filters*, John Wiley & Sons, Inc., Hoboken, NJ, USA.

Scaglione, A., Giannakis, G. B., and Barbarossa, S. (1999), "Redundant filterbank precoders and equalizers, Part I: Unification and optimal designs," *IEEE Trans. Signal Proces.*, vol. 47, no. 7, pp. 1988–2006, July.

Schreier, P. and Scharf, L. (2010), *Statistical Signal Processing of Complex-Valued Data: The Theory of Improper and Noncircular Signals*, Cambridge University Press, Cambridge, UK.

Shin, H. and Lee, J. H. (2002), Exact symbol error probability of orthogonal space-time block codes. In "*Proc. IEEE GLOBECOM*," vol. 2, pp. 1197–1201, November.

Simon, M. K. and Alouini, M.-S. (2005), *Digital Communication over Fading Channels*, 2nd ed., John Wiley & Sons, Inc., Hoboken, NJ, USA.

Spivak, M. (2005), *A Comprehensive Introduction to Differential Geometry*, vol. 1, 3rd ed., Publish or Perish, Inc., Houston, TX, USA.

Strang, G. (1988), *Linear Algebra and Its Applications*, 3rd ed., Harcourt Brave Jovanovich, Inc., San Diego, CA, USA.

Tarokh, V., Jafarkhani, H., and Calderbank, A. R. (1999), "Space-time block coding for wireless communications: Performance results," *IEEE J. Sel. Area Comm.*, vol. 17, no. 3, pp. 451–460, March.

Telatar, İ. E. (1995), "Capacity of multi-antenna Gaussian channels," *AT&T-Bell Laboratories Internal Technical Memo*, June.

Therrien, C. W. (1992), *Discrete Random Signals and Statistical Signal Processing*, Prentice–Hall Inc., Englewood Cliffs, NJ, USA.

Tracy, D. S. and Jinadasa, K. G. (1988), "Patterned matrix derivatives," *Can. J. Stat.*, vol. 16, no. 4, pp. 411–418.

Trees, H. L. V. (2002), *Optimum Array Processing: Part IV of Detection Estimation and Modulation Theory*, Wiley Interscience, New York, NY, USA.

Tsipouridou, D. and Liavas, A. P. (2008), "On the sensitivity of the transmit MIMO Wiener filter with respect to channel and noise second-order statistics uncertainties," *IEEE Trans. Signal Proces.*, vol. 56, no. 2, pp. 832–838, February.

Turin, G. L. (1960), "The characteristic function of Hermitian quadradic forms in complex normal variables," *Biometrica*, vol. 47, pp. 199–201, June.

Vaidyanathan, P. P. (1993), *Multirate Systems and Filter Banks*, Prentice Hall, Englewood Cliffs, NJ, USA.

Vaidyanathan, P. P., Phoong, S.-M., and Lin, Y.-P. (2010), *Signal Processing and Optimization for Transceiver Systems*, Cambridge University Press, Cambridge, UK.

van den Bos, A. (1994a), "Complex gradient and Hessian," *Proc. IEE Vision, Image and Signal Process.*, vol. 141, no. 6, pp. 380–383, December.

van den Bos, A. (1994b), "A Cramér Rao lower bound for complex parameters," *IEEE Trans. Signal Proces.*, vol. 42, no. 10, pp. 2859, October.

van den Bos, A. (1995a), "Estimation of complex Fourier coefficients," *Proc. IEE Control Theory Appl.*, vol. 142, no. 3, pp. 253–256, May.

van den Bos, A. (1995b), "The multivariate complex normal distribution – A generalization," *IEEE Trans. Inform. Theory*, vol. 41, no. 2, pp. 537–539, March.

van den Bos, A. (1998), "The real-complex normal distribution," *IEEE Trans. Inform. Theory*, vol. 44, no. 4, pp. 1670–1672, July.

Wells, Jr., R. O. (2008), *Differential Analysis on Complex Manifolds*, 3rd ed., Springer-Verlag, New York, NY, USA.

Wiens, D. P. (1985), "On some pattern-reduction matrices which appear in statistics," *Linear Algebra and Its Applications*, vol. 68, pp. 233–258.

Wirtinger, W. (1927), "Zur formalen Theorie der Funktionen von mehr komplexen Veränderlichen," *Mathematische Annalen*, vol. 97, pp. 357–375.

Yan, G. and Fan, H. (2000), "A Newton-like algorithm for complex variables with applications in blind equalization," *IEEE Trans. Signal Proces.*, vol. 48, no. 2, pp. 553–556, February.

Young, N. (1990), *An Introduction to Hilbert Space*, Press Syndicate of the University of Cambridge, The Pitt Building, Trumpington Street, Cambridge, UK. Reprinted edition.

Index

Note: The letter *n* after a page number indicates a footnote.

www.ingramcontent.com/pod-product-compliance
Lightning Source LLC
LaVergne TN
LVHW062208200926
848484LV00006B/1351

* 9 7 8 0 5 2 1 1 9 2 6 4 4 *